Cancer Immunology

Cancer Immunology

Editors

Xianda Zhao
Subbaya Subramanian

MDPI • Basel • Beijing • Wuhan • Barcelona • Belgrade • Manchester • Tokyo • Cluj • Tianjin

Editors
Xianda Zhao
Surgery
University of Minnesota
Minneapolis
United States

Subbaya Subramanian
Surgery
University of Minnesota
Minneapolis
United States

Editorial Office
MDPI
St. Alban-Anlage 66
4052 Basel, Switzerland

This is a reprint of articles from the Special Issue published online in the open access journal *Cancers* (ISSN 2072-6694) (available at: www.mdpi.com/journal/cancers/special_issues/Cancer_II).

For citation purposes, cite each article independently as indicated on the article page online and as indicated below:

LastName, A.A.; LastName, B.B.; LastName, C.C. Article Title. *Journal Name* **Year**, *Volume Number*, Page Range.

ISBN 978-3-0365-2573-0 (Hbk)
ISBN 978-3-0365-2572-3 (PDF)

© 2021 by the authors. Articles in this book are Open Access and distributed under the Creative Commons Attribution (CC BY) license, which allows users to download, copy and build upon published articles, as long as the author and publisher are properly credited, which ensures maximum dissemination and a wider impact of our publications.

The book as a whole is distributed by MDPI under the terms and conditions of the Creative Commons license CC BY-NC-ND.

Contents

About the Editors vii

Xianda Zhao and Subbaya Subramanian
Cancer Immunology and Immunotherapies: Mechanisms That Affect Antitumor Immune Response and Treatment Resistance
Reprinted from: *Cancers* **2021**, *13*, 5655, doi:10.3390/cancers13225655 1

Yu Yuan, Abdalla Adam, Chen Zhao and Honglei Chen
Recent Advancements in the Mechanisms Underlying Resistance to PD-1/PD-L1 Blockade Immunotherapy
Reprinted from: *Cancers* **2021**, *13*, 663, doi:10.3390/cancers13040663 5

Navid Sobhani, Dana Rae Tardiel-Cyril, Aram Davtyan, Daniele Generali, Raheleh Roudi and Yong Li
CTLA-4 in Regulatory T Cells for Cancer Immunotherapy
Reprinted from: *Cancers* **2021**, *13*, 1440, doi:10.3390/cancers13061440 23

Ali Mehdi and Shafaat A. Rabbani
Role of Methylation in Pro- and Anti-Cancer Immunity
Reprinted from: *Cancers* **2021**, *13*, 545, doi:10.3390/cancers13030545 41

Shivani Krishnamurthy, David Gilot, Seong Beom Ahn, Vincent Lam, Joo-Shik Shin, Gilles Jackie Guillemin and Benjamin Heng
Involvement of Kynurenine Pathway in Hepatocellular Carcinoma
Reprinted from: *Cancers* **2021**, *13*, 5180, doi:10.3390/cancers13205180 67

Hongmei Zheng, Sumit Siddharth, Sheetal Parida, Xinhong Wu and Dipali Sharma
Tumor Microenvironment: Key Players in Triple Negative Breast Cancer Immunomodulation
Reprinted from: *Cancers* **2021**, *13*, 3357, doi:10.3390/cancers13133357 85

Mariarosaria Marseglia, Adriana Amaro, Nicola Solari, Rosaria Gangemi, Elena Croce, Enrica Teresa Tanda, Francesco Spagnolo, Gilberto Filaci, Ulrich Pfeffer and Michela Croce
How to Make Immunotherapy an Effective Therapeutic Choice for Uveal Melanoma
Reprinted from: *Cancers* **2021**, *13*, 2043, doi:10.3390/cancers13092043 107

Dechen Wangmo, Prem K. Premsrirut, Ce Yuan, William S. Morris, Xianda Zhao and Subbaya Subramanian
ACKR4 in Tumor Cells Regulates Dendritic Cell Migration to Tumor-Draining Lymph Nodes and T-Cell Priming
Reprinted from: *Cancers* **2021**, *13*, 5021, doi:10.3390/cancers13195021 129

Frank Liang, Azar Rezapour, Louis Szeponik, Samuel Alsén, Yvonne Wettergren, Elinor Bexe Lindskog, Marianne Quiding-Järbrink and Ulf Yrlid
Antigen Presenting Cells from Tumor and Colon of Colorectal Cancer Patients Are Distinct in Activation and Functional Status, but Comparably Responsive to Activated T Cells
Reprinted from: *Cancers* **2021**, *13*, 5247, doi:10.3390/cancers13205247 143

Aishwarya Gokuldass, Arianna Draghi, Krisztian Papp, Troels Holz Borch, Morten Nielsen, Marie Christine Wulff Westergaard, Rikke Andersen, Aimilia Schina, Kalijn Fredrike Bol, Christopher Aled Chamberlain, Mario Presti, Özcan Met, Katja Harbst, Martin Lauss, Samuele Soraggi, Istvan Csabai, Zoltán Szállási, Göran Jönsson, Inge Marie Svane and Marco Donia
Qualitative Analysis of Tumor-Infiltrating Lymphocytes across Human Tumor Types Reveals a Higher Proportion of Bystander CD8$^+$ T Cells in Non-Melanoma Cancers Compared to Melanoma
Reprinted from: *Cancers* **2020**, *12*, 3344, doi:10.3390/cancers12113344 159

Kawaljit Kaur, Tahmineh Safaie, Meng-Wei Ko, Yuhao Wang and Anahid Jewett
ADCC against MICA/B Is Mediated against Differentiated Oral and Pancreatic and Not Stem-Like/Poorly Differentiated Tumors by the NK Cells; Loss in Cancer Patients due to Down-Modulation of CD16 Receptor
Reprinted from: *Cancers* **2021**, *13*, 239, doi:10.3390/cancers13020239 175

Cong He, Ying Zhou, Zhenlong Li, Muhammad Asad Farooq, Iqra Ajmal, Hongmei Zhang, Li Zhang, Lei Tao, Jie Yao, Bing Du, Mingyao Liu and Wenzheng Jiang
Co-Expression of IL-7 Improves NKG2D-Based CAR T Cell Therapy on Prostate Cancer by Enhancing the Expansion and Inhibiting the Apoptosis and Exhaustion
Reprinted from: *Cancers* **2020**, *12*, 1969, doi:10.3390/cancers12071969 195

Lei Tao, Muhammad Asad Farooq, Yaoxin Gao, Li Zhang, Congyi Niu, Iqra Ajmal, Ying Zhou, Cong He, Guixia Zhao, Jie Yao, Mingyao Liu and Wenzheng Jiang
CD19-CAR-T Cells Bearing a KIR/PD-1-Based Inhibitory CAR Eradicate CD19$^+$HLA-C1$^-$ Malignant B Cells While Sparing CD19$^+$HLA-C1$^+$ Healthy B Cells
Reprinted from: *Cancers* **2020**, *12*, 2612, doi:10.3390/cancers12092612 213

Fei-Ting Hsu, Yu-Chang Liu, Chang-Liang Tsai, Po-Fu Yueh, Chih-Hsien Chang and Keng-Li Lan
Preclinical Evaluation of Recombinant Human IL15 Protein Fused with Albumin Binding Domain on Anti-PD-L1 Immunotherapy Efficiency and Anti-Tumor Immunity in Colon Cancer and Melanoma
Reprinted from: *Cancers* **2021**, *13*, 1789, doi:10.3390/cancers13081789 231

Lucas A. Horn, Kristen Fousek, Duane H. Hamilton, James W. Hodge, John A. Zebala, Dean Y. Maeda, Jeffrey Schlom and Claudia Palena
Vaccine Increases the Diversity and Activation of Intratumoral T Cells in the Context of Combination Immunotherapy
Reprinted from: *Cancers* **2021**, *13*, 968, doi:10.3390/cancers13050968 257

Xuefei Bai, Wenhui Liu, Shijie Jin, Wenbin Zhao, Yingchun Xu, Zhan Zhou, Shuqing Chen and Liqiang Pan
Facile Generation of Potent Bispecific Fab via Sortase A and Click Chemistry for Cancer Immunotherapy
Reprinted from: *Cancers* **2021**, *13*, 4540, doi:10.3390/cancers13184540 275

Lina Benajiba, Jérôme Lambert, Roberta La Selva, Delphine Cochereau, Barouyr Baroudjian, Jennifer Roux, Jérôme Le Goff, Cécile Pages, Maxime Battistella, Julie Delyon and Céleste Lebbé
Systemic Treatment Initiation in Classical and Endemic Kaposi's Sarcoma: Risk Factors and Global Multi-State Modelling in a Monocentric Cohort Study
Reprinted from: *Cancers* **2021**, *13*, 2519, doi:10.3390/cancers13112519 289

Eline S. Zwart, Esen Yüksel, Anne Pannekoek, Ralph de Vries, Reina E. Mebius and Geert Kazemier
De Novo Carcinoma after Solid Organ Transplantation to Give Insight into Carcinogenesis in General—A Systematic Review and Meta-Analysis
Reprinted from: *Cancers* **2021**, *13*, 1122, doi:10.3390/cancers13051122 . . . . **303**

About the Editors

Xianda Zhao

Xianda Zhao is an M.D. Ph.D. and researcher in the fields of gastrointestinal cancer pathogenesis and novel treatment development at the University of Minnesota. Dr. Zhao has been studying gastric cancer, pancreatic cancer, and colorectal cancer for over ten years. He is a pioneer of advanced gastrointestinal cancer models. His current research focuses on tumor-extracellular vesicles mediated immune responses and extracellular vesicles-based treatment for colorectal cancers. Dr. Zhao has published over 30 full-length original research articles and review articles in top-ranking journals. He has also authored one book chapter focusing on cancer immunotherapy resistance. Dr. Zhao serves as an ad hoc reviewer for multiple journals.

Subbaya Subramanian

Dr. Subbaya Subramanian is currently an Associate Professor in the Department of Surgery at the University of Minnesota, United States. As a faculty at the University of Minnesota, Dr. Subramanian has established an internationally recognized cancer research program focused on deciphering the molecular mechanisms of antitumor immune evasion in cancer. His current research focuses on understanding the interactions between cancer and the host immune system and how cancer cells manipulate the anti-tumor immune response. Dr. Subramanian has authored over 100 peer-reviewed publications and has written numerous book chapters and review articles. He has delivered over 80 national and international invited lectures/seminars and received the American Cancer Society Research Scholar award in 2013. He serves as the section Editor-in-Chief of the journal *Vaccines* and the Associate section Editor-in-Chief of *Cancers*. Dr. Subramanian also serves on the Editorial Board of several journals.

Editorial

Cancer Immunology and Immunotherapies: Mechanisms That Affect Antitumor Immune Response and Treatment Resistance

Xianda Zhao [1,*] and Subbaya Subramanian [1,2,3,*]

1. Department of Surgery, University of Minnesota Medical School, Minneapolis, MN 55455, USA
2. Masonic Cancer Center, University of Minnesota, Minneapolis, MN 55455, USA
3. Center for Immunology, University of Minnesota, Minneapolis, MN 55455, USA
* Correspondence: zhaox714@umn.edu (X.Z.); subree@umn.edu (S.S.)

Citation: Zhao, X.; Subramanian, S. Cancer Immunology and Immunotherapies: Mechanisms That Affect Antitumor Immune Response and Treatment Resistance. *Cancers* **2021**, *13*, 5655. https://doi.org/10.3390/cancers13225655

Received: 9 November 2021
Accepted: 10 November 2021
Published: 12 November 2021

Publisher's Note: MDPI stays neutral with regard to jurisdictional claims in published maps and institutional affiliations.

Copyright: © 2021 by the authors. Licensee MDPI, Basel, Switzerland. This article is an open access article distributed under the terms and conditions of the Creative Commons Attribution (CC BY) license (https://creativecommons.org/licenses/by/4.0/).

The past decade has seen immunotherapy rise to the forefront of cancer treatment. This Special Issue of *Cancers* aims to elaborate on the latest developments, cutting-edge technologies, and prospects in cancer immunology and immunotherapy. Seventeen exceptional studies, including original contributions and review articles, written by international scientists and physicians, primarily concerning the fields of tumor biology, cancer immunology, therapeutics, and drug development comprise the main body of this special issue.

Over the last few years, an increasing understanding has emerged on molecular mechanisms that regulate the anti-tumor immune response and an exponential increase in the use of novel cancer immunotherapies in various cancer types. The field of Cancer immunology and Immunotherapies presents promising therapeutic opportunities for developing novel cancer treatments and improving patient survival outcomes. Chemotherapy is still used as a primary method for treatment, and the standard of care for many cancer types is relatively unselective and presents with the rapid development of treatment resistance. In contrast, cancer immunotherapies stimulate the antitumor immune response via the activation of lymphocytes that can recognize neoantigens, resulting in durable treatment response.

A successful antitumor immune response involves interactions between various cell types that coordinately function to prevent tumor cell proliferation or to effectively eradicate tumor cells. A coordinated functioning of the lymphoid and myeloid lineage cells is critical for killing tumor cells, and is performed by enhancing the activity of cytotoxic cells. Myeloid lineage cells, such as dendritic cells, provide tumor antigens to T cells and secrete cytokines that regulate the activation and function of cytotoxic cells. Despite the demonstrated successes of cancer immunotherapy, most patients do not respond, and the development of resistance has occurred in patients who initially respond to immunotherapies. Recent studies have uncovered novel immune escape mechanisms that affect immune cell infiltration, poor antigen presentation, and tumor intrinsic silencing of the immune response via cytokines and the release of immune suppressive exosomes [1]. Additional mechanisms of antitumor immune escape and immunotherapy resistance are continuously being discovered [2–4].

Based on these factors, significant attention has been directed towards the recent advances in cancer immunology [5–10]. In the past decades, the discovery of Programmed cell death protein 1 (PD-1) and the Cytotoxic T-lymphocyte–associated antigen 4 (CTLA-4) has helped to develop immune checkpoint blockade therapies. The articles by Yuan et al. [5] and Sobhani et al. [6] provide an overview and include recent findings on PD-1 and CTLA-4. The review article by Mehdi et al. [7] focuses on the role of methylation in manipulating cancer immunity. In addition to these general cancer immunology topics, reviews by Krishnamurthy et al. [8], Zheng et al. [9], and Marseglia et al. [10] summarize immune regulation in specific cancer types, such as hepatocellular carcinoma, triple-negative breast cancer, and uveal melanoma.

The second series of articles mainly presents original work deciphering the novel regulatory mechanisms of cancer immunity. For the first time, our group (Wangmo et al. [11])

reported that Atypical Chemokine Receptor 4 (ACKR4) determines the migration of dendritic cells from tumor tissue to the tumor-draining lymph nodes. The loss of ACKR4 expression in tumor cells can affect the migration of dendritic cells and their retention in the tumor microenvironment, impairing T-cell priming in tumor-draining lymph nodes. This finding uncovers a novel mechanism that regulates dendritic cells' migration from the tumor tissue, a critical factor in antigen presentation and in antitumor immune responses. Liang et al. [12] further contribute to the body of research regarding antigen-presenting cells. The authors performed an in-depth analysis of antigen-presenting cells in the human colorectal cancer microenvironment. Interestingly, they observed that antigen-presenting cells within distinct intratumoral and colonic milieus showed different functional statuses but were similarly responsive to induced T-cell activation. The third article in this section focuses on the bystander T-cells in cancers. In a hybrid study of bioinformatics and laboratory analyses, Gokuldass et al. [13] revealed a higher proportion of bystander $CD8^+$ T cells in non-melanoma cancers than in melanoma cancers. This observation helps to establish a new theory to explain the different immune strengths of various tumors. In the context of innate immunity, Kaur et al. [14] reported on the function of CD16 receptors in both direct cytotoxicity and antibody-dependent cell cytotoxicity, making the use of these receptors as a cancer treatment seem promising.

The overarching objective of studying tumor immunity is to develop the next-generation cancer immunotherapies. In the third series of articles, several novel cancer immunotherapy strategies are proposed. Two original research articles from Jiang's group [15,16] provide modified CAR T Cell therapies to treat malignant B-cell neoplasms and prostate cancer. Their modified CAR T cells are better directed to kill malignant B-cells, while sparing the $CD19^+HLA-C1^+$ healthy B Cells. The next study by Hsu et al. [17] developed a recombinant fusion IL15 protein composed of human IL15 (hIL15) and albumin-binding domain (hIL15-ABD) which has been successfully tested with anti-PD-L1 on CT26 murine colon cancer and B16-F10 murine melanoma models. Horn et al. [18] also reported on the use of IL15 as an agonist adjuvant for other cancer immunotherapies. Utilizing colon and mammary carcinoma models, the study showed that a recombinant adenovirus-based vaccine, targeting tumor-associated antigens with an IL-15 superagonist adjuvant is effective when combined with other immunotherapy regimens. This study also validated the idea that providing tumor-associated antigens as a vaccine helps to overcome immune checkpoint blockade resistance. Another feature in this issue is that we include a report on a new method called the 'chemo-enzymatic conjugation approach' (Bai et al. [19]) to generate bispecific antibodies (BiFab). Using this method, the authors produced $BiFab^{Her2/CD3}$ and $BiFab^{CD20/CD3}$ to conjugate both the target and effector cells (T-cells). These BiFabs demonstrated a strong considerable effect for inducing T-cell activation and killing target cancer cells upon conjugation. The $BiFab^{CD20/CD3}$ also showed anti-tumor activity in vivo.

The findings of Benajiba et al. [20] and Zwart et al. [21] highlight clinical observations relevant to cancer and immunology. Disseminated Kaposi's sarcoma is usually treated by interferons, which is a type of immunotherapy. Benajiba et al. [20] performed a retrospective cohort study to evaluate global disease evolution and to identify the risk factors for systemic treatment initiation, including the use of interferons. They found that 41.2% of classic/endemic Kaposi's sarcoma patients require systemic treatment. They also reported that the mean treatment-free time during the first five years following interferon is similar to that of chemotherapy. Lastly, Zwart et al. [21] contribute through a meta-analysis on immunosuppressive therapy after solid organ transplantation and on the development of cancers. Interestingly, the meta-analysis indicated that patients receiving cyclosporine A and Azathioprine after a solid organ transplant are at a higher risk than patients receiving other immunosuppressive drugs of developing certain types of cancers.

In conclusion, the original research articles and reviews included in this special issue ensure that the key aspects of the next generation of cancer immunology and immunotherapy have been covered. We hope that the novel findings in these articles will inform the readers and provide useful references for developing next-generation cancer immunotherapies.

Funding: This research received no external funding.

Conflicts of Interest: The authors declare no conflict of interest.

References

1. Zhao, X.; Yuan, C.; Wangmo, D.; Subramanian, S. Tumor-Secreted Extracellular Vesicles Regulate T-Cell Costimulation and Can Be Manipulated To Induce Tumor-Specific T-Cell Responses. *Gastroenterology* **2021**, *161*, 560–574.e11. [CrossRef]
2. Zhao, X.; Wangmo, D.; Robertson, M.; Subramanian, S. Acquired Resistance to Immune Checkpoint Blockade Therapies. *Cancers* **2020**, *12*, 1161. [CrossRef]
3. Zhao, X.; Subramanian, S. Oncogenic pathways that affect antitumor immune response and immune checkpoint blockade therapy. *Pharmacol. Ther.* **2018**, *181*, 76–84. [CrossRef]
4. Zhao, X.; Subramanian, S. Intrinsic Resistance of Solid Tumors to Immune Checkpoint Blockade Therapy. *Cancer Res.* **2017**, *77*, 817–822. [CrossRef]
5. Yuan, Y.; Adam, A.; Zhao, C.; Chen, H. Recent Advancements in the Mechanisms Underlying Resistance to PD-1/PD-L1 Blockade Immunotherapy. *Cancers* **2021**, *13*, 663. [CrossRef]
6. Sobhani, N.; Tardiel-Cyril, D.; Davtyan, A.; Generali, D.; Roudi, R.; Li, Y. CTLA-4 in Regulatory T Cells for Cancer Immunotherapy. *Cancers* **2021**, *13*, 1440. [CrossRef]
7. Mehdi, A.; Rabbani, S. Role of Methylation in Pro- and Anti-Cancer Immunity. *Cancers* **2021**, *13*, 545. [CrossRef]
8. Krishnamurthy, S.; Gilot, D.; Ahn, S.B.; Lam, V.; Shin, J.-S.; Guillemin, G.J.; Heng, B. Involvement of Kynurenine Pathway in Hepatocellular Carcinoma. *Cancers* **2021**, *13*, 5180. [CrossRef]
9. Zheng, H.; Siddharth, S.; Parida, S.; Wu, X.; Sharma, D. Tumor Microenvironment: Key Players in Triple Negative Breast Cancer Immunomodulation. *Cancers* **2021**, *13*, 3357. [CrossRef]
10. Marseglia, M.; Amaro, A.; Solari, N.; Gangemi, R.; Croce, E.; Tanda, E.; Spagnolo, F.; Filaci, G.; Pfeffer, U.; Croce, M. How to Make Immunotherapy an Effective Therapeutic Choice for Uveal Melanoma. *Cancers* **2021**, *13*, 2043. [CrossRef]
11. Wangmo, D.; Premsrirut, P.K.; Yuan, C.; Morris, W.S.; Zhao, X.; Subramanian, S. ACKR4 in Tumor Cells Regulates Dendritic Cell Migration to Tumor-Draining Lymph Nodes and T-Cell Priming. *Cancers* **2021**, *13*, 5021. [CrossRef]
12. Liang, F.; Rezapour, A.; Szeponik, L.; Alsén, S.; Wettergren, Y.; Lindskog, E.B.; Quiding-Järbrink, M.; Yrlid, U. Antigen Presenting Cells from Tumor and Colon of Colorectal Cancer Patients Are Distinct in Activation and Functional Status, but Comparably Responsive to Activated T Cells. *Cancers* **2021**, *13*, 5247. [CrossRef]
13. Gokuldass, A.; Draghi, A.; Papp, K.; Borch, T.H.; Nielsen, M.; Westergaard, M.C.W.; Andersen, R.; Schina, A.; Bol, K.F.; Chamberlain, C.A.; et al. Qualitative Analysis of Tumor-Infiltrating Lymphocytes across Human Tumor Types Reveals a Higher Proportion of Bystander CD8(+) T Cells in Non-Melanoma Cancers Compared to Melanoma. *Cancers* **2020**, *12*, 3344. [CrossRef]
14. Kaur, K.; Safaie, T.; Ko, M.-W.; Wang, Y.; Jewett, A. ADCC against MICA/B Is Mediated against Differentiated Oral and Pancreatic and Not Stem-Like/Poorly Differentiated Tumors by the NK Cells; Loss in Cancer Patients due to Down-Modulation of CD16 Receptor. *Cancers* **2021**, *13*, 239. [CrossRef]
15. He, C.; Zhou, Y.; Li, Z.; Farooq, M.A.; Ajmal, I.; Zhang, H.; Zhang, L.; Tao, L.; Yao, J.; Du, B.; et al. Co-Expression of IL-7 Improves NKG2D-Based CAR T Cell Therapy on Prostate Cancer by Enhancing the Expansion and Inhibiting the Apoptosis and Exhaustion. *Cancers* **2020**, *12*, 1969. [CrossRef]
16. Tao, L.; Farooq, M.A.; Gao, Y.; Zhang, L.; Niu, C.; Ajmal, I.; Zhou, Y.; He, C.; Zhao, G.; Yao, J.; et al. CD19-CAR-T Cells Bearing a KIR/PD-1-Based Inhibitory CAR Eradicate CD19(+)HLA-C1(-) Malignant B Cells While Sparing CD19(+)HLA-C1(+) Healthy B Cells. *Cancers* **2020**, *12*, 2612. [CrossRef]
17. Hsu, F.-T.; Liu, Y.-C.; Tsai, C.-L.; Yueh, P.-F.; Chang, C.-H.; Lan, K.-L. Preclinical Evaluation of Recombinant Human IL15 Protein Fused with Albumin Binding Domain on Anti-PD-L1 Immunotherapy Efficiency and Anti-Tumor Immunity in Colon Cancer and Melanoma. *Cancers* **2021**, *13*, 1789. [CrossRef]
18. Horn, L.; Fousek, K.; Hamilton, D.; Hodge, J.; Zebala, J.; Maeda, D.; Schlom, J.; Palena, C. Vaccine Increases the Diversity and Activation of Intratumoral T Cells in the Context of Combination Immunotherapy. *Cancers* **2021**, *13*, 968. [CrossRef]
19. Bai, X.; Liu, W.; Jin, S.; Zhao, W.; Xu, Y.; Zhou, Z.; Chen, S.; Pan, L. Facile Generation of Potent Bispecific Fab via Sortase A and Click Chemistry for Cancer Immunotherapy. *Cancers* **2021**, *13*, 4540. [CrossRef]
20. Benajiba, L.; Lambert, J.; La Selva, R.; Cochereau, D.; Baroudjian, B.; Roux, J.; Le Goff, J.; Pages, C.; Battistella, M.; Delyon, J.; et al. Systemic Treatment Initiation in Classical and Endemic Kaposi's Sarcoma: Risk Factors and Global Multi-State Modelling in a Monocentric Cohort Study. *Cancers* **2021**, *13*, 2519. [CrossRef]
21. Zwart, E.; Yüksel, E.; Pannekoek, A.; de Vries, R.; Mebius, R.; Kazemier, G. De Novo Carcinoma after Solid Organ Transplantation to Give Insight into Carcinogenesis in General-A Systematic Review and Meta-Analysis. *Cancers* **2021**, *13*, 1122. [CrossRef]

Review

Recent Advancements in the Mechanisms Underlying Resistance to PD-1/PD-L1 Blockade Immunotherapy

Yu Yuan [1], Abdalla Adam [1], Chen Zhao [2] and Honglei Chen [1,*]

1. Department of Pathology and Hubei Province Key Laboratory of Allergy and Immune-Related Diseases, School of Basic Medical Sciences, Wuhan University, Wuhan 430071, China; kelly_yuanyu@whu.edu.cn (Y.Y.); adamabdalla341@gmail.com (A.A.)
2. Department of Oncology, Renmin Hospital of Wuhan University, Wuhan 430060, China; chen_zhao@whu.edu.cn
* Correspondence: hl-chen@whu.edu.cn; Tel.: +86-27-67811732

Simple Summary: Immune checkpoint blockade targeting PD-1/PD-L1 has a promising therapeutic efficacy in different tumors, but a significant percentage of patients cannot benefit from this therapy due to primary and acquired resistance during treatment. This review summarizes the recent findings of PD-L1 role in resistance to therapies through the PD-1/PD-L1 pathway and other correlating signaling pathways. A special focus will be given to the key mechanisms underlying resistance to the PD-1/PD-L1 blockade in cancer immunotherapy. Furthermore, we also discuss the promising combination of therapeutic strategies for patients resistant to the PD-1/PD-L1 blockade in order to enhance the efficacy of immune checkpoint inhibitors.

Abstract: Release of immunoreactive negative regulatory factors such as immune checkpoint limits antitumor responses. PD-L1 as a significant immunosuppressive factor has been involved in resistance to therapies such as chemotherapy and target therapy in various cancers. Via interacting with PD-1, PD-L1 can regulate other factors or lead to immune evasion of cancer cells. Besides, immune checkpoint blockade targeting PD-1/PD-L1 has promising therapeutic efficacy in the different tumors, but a significant percentage of patients cannot benefit from this therapy due to primary and acquired resistance during treatment. In this review, we described the utility of PD-L1 expression levels for predicting poor prognosis in some tumors and present evidence for a role of PD-L1 in resistance to therapies through PD-1/PD-L1 pathway and other correlating signaling pathways. Afterwards, we elaborate the key mechanisms underlying resistance to PD-1/PD-L1 blockade in cancer immunotherapy. Furthermore, promising combination of therapeutic strategies for patients resistant to PD-1/PD-L1 blockade therapy or other therapies associated with PD-L1 expression was also summarized.

Keywords: PD-L1; resistance; immune checkpoints; immunotherapy

1. Introduction

T-cell activation and proliferation induced by antigens is regulated by expression of both co-stimulatory and co-inhibitory receptors and their ligands [1]. Inhibitory pathways in the immune system can prevent autoimmunity through maintaining self-tolerance and regulating immunity [2]. While in tumors inhibitory pathways known as "checkpoints" can evade immune surveillance. Programmed cell death -1(PD-1) interacting with its corresponding ligand PD-L1 leads to immune suppression via preventing the T-cell activation in the tumor [3]. PD-1 is expressed on activated CD8$^+$ T-cells as well as B cells and natural killer cells, and inhibits T-cell receptor (TCR) signaling and CD28 co-stimulation under chronic antigen exposure. As ligands of PD-1, PD-L2 is primarily expressed on antigen-presenting cells (APC) while PD-L1 is expressed on various types of cells including tumor cells and immune cells. Evidence of PD-L1 expression increase and spontaneous immune resistance is proved in several types of human cancers [4]. Besides, predictive and prognostic value of PD-L1 immunohistochemical expression has been reported in certain cancers. Moreover, PD-L1 as an inhibitory factor is also involved in other signaling pathways underlying mechanisms in resistance to tyrosine kinase inhibitors (TKIs).

Immunotherapy identified as the most promising approach in cancer treatment compared with chemotherapy and targeted therapy, immune checkpoint inhibitors have reported higher rates of response, remission, and better overall survival rates in a variety of tumors [5]. Immunotherapy has received the US Food and Drug Administration (FDA) approval for 57 indications in 17 solid tumors in less than 10 years, while over 80% are PD-1/PD-L1-targeted antibodies. Beneficial function of the PD-1/PD-L1 axis blockade is confirmed in treating many different types of cancers such as non-small cell lung cancer (NSCLC), melanoma and bladder carcinoma [6,7]. So far, six immune checkpoint inhibitors targeting PD-1/PD-L1 have been approved by the FDA for the first and second line of patients with non-small cell lung cancer including monoclonal antibodies (mAb) pembrolizumab, nivolumab and cemiplimab targeting PD-1 and mAb atezolizumab, avelumab and durvalumab targeting PD-L1. However, limited efficacy has been reported in PD-1/PD-L1 blockade therapy which rarely exceeds 40% in most cancer types and a large number of patients show partial responsiveness [8,9]. Even if there is a consistent rate of initial responses, the majority of patients develop therapeutic resistance and disease progression [10,11]. Focusing on PD-L1, we described all these concepts in this review including its predictive and prognostic value, immune resistance induced by PD-L1 and key mechanisms underlying resistance to PD-1/PD-L1 blockade therapy.

2. The Expression of PD-L1 Levels Predicting Resistance and Poor Prognosis

PD-L1 expression is increased in many types of human cancers and is regarded as a predictive and prognostic marker in cancer tissues. Prior data have demonstrated that PD-L1 expression is upregulated in cisplatin-resistant lung cancer cells compared with parent cells [12–14]. Resistance to epigenetic therapy is associated with enhanced PD-L1 expression in myeloid malignancies [15]. For example, 7 myelodysplastic syndrome and 6 acute myeloid leukemia patients received treatments with either azacytidine (Aza) or combined Aza and the histone deacetylase inhibitor LBH-589 to investigate the PD-L1 expression levels. Non-responders showed a more than two-fold increase in PD-L1 expression after treatment commenced, and except for two patients, none of the responders demonstrated increased expression of PD-L1.

PD-L1 expression is correlated with poor prognosis in different cancers. In chemotherapy and radiotherapy-treated patients with head and neck squamous cell carcinoma (HNSCC), high PD-L1 mRNA (>125 FPKM) from The Cancer Genome Atlas database had significantly reduced the 5-year survival rate [16]. Other data regarded PD-L1 as a potential biomarker for radiation therapy failure of HNSCC [17]. Following radiotherapy, a panel of radiation-resistant human papilloma virus (HPV)-negative HNSCC cell lines exhibited increased expression of PD-L1, three cohorts of HPV-negative HNSCC tumors with high expression of PD-L1 had much higher failure rates compared to the PD-L1-low expression group. Similar results have been reported in metastatic melanoma patients (MMP) [18]. Forty six and thirty four BRAFi-treated MMP harboring mutant BRAFV600 received vemurafenib and dabrafenib respectively. Patients with PD-L1 expression and an absence of tumor-infiltrating immune cells (TIMC) are related to shorter progression-free-survival compared to those with TIMC and absence of PD-L1. This study also identified PD-L1 overexpression and loss of TIMC as independent prognostic factors for melanoma-specific survival.

Interestingly, an experiment involving 18 patients with epidermal growth factor receptor (EGFR)-mutant NSCLC investigated the change of PD-L1 expression following gefitinib. A proportion of 38.9% (7/18) of NSCLC patients had a significant increase in the median H-score (marked as group A) of PD-L1, while the rest (61.1%) did not vary (group B). Besides, MET positivity by immunohistochemistry in biopsies is significantly correlated with group A. The results described a marked increase in expression of PD-L1 in tumor cells of a subset of patients after gefitinib treatment. Though EGFR-mutated NSCLC is prone to express less PD-L1 than wild type [19]. Similar results in several studies indicated that PD-L1 expression as a biomarker predicts resistance and poor prognosis after gefitinib treatment, rebiopsy should be considered [20]. Nevertheless, combination therapy with Durvalumab targeting PD-L1 and gefitinib has been proved to be more toxic and does not demonstrate a significant augmentation in progression-free survival (PFS) [21]. As a crucial factor predicting resistance and poor prognosis, PD-L1 has absolutely specific mechanisms for leading to resistance.

3. PD-L1-Induced Resistance

PD-L1 as an inhibitor in the immune system that induces immune resistance through interacting with its ligand PD-1. Besides, it is also involved in other signaling pathways generating resistance to TKIs.

3.1. PD-L1-Mediated Immune Resistance

In certain cancers, efficacy of antitumor treatment has always been found to be limited, due to the activation of immune checkpoints such as PD-1 and PD-L1. Once recognizing the tumor antigen, T-cells produce an anti-tumor immune response, which eventually leads to PD-1 lymphocyte expression and interferon release. To evade this immune attack, PD-L1 expression is adaptively upregulated by cancer cells and other inflammatory cells in the tumor microenvironment (TME) [22]. IFN-γ is secreted by tumor-infiltrating lymphocytes (TILs) and induces PD-L1 expression in the TME, thus T-cell cytotoxic function is impaired through the interaction of PD-L1 and PD-1. A similar pattern has been observed in other cancers including gastric cancer [23]. Fractionated radiation therapy can lead to increased tumor cell expression of PD-L1 in response to CD8$^+$ T-cell production of IFN-γ [24]. In HPV-HNSCC, which is highly infiltrated by lymphocytes, IFN-γ-induced PD-L1 on tumor cells and CD68$^+$ tumor-associated macrophages (TAMs) and highly expressed PD-L1 by CTLs, are found located at the same site [1].

In prior studies, PD-L1 expression is also upregulated followed by drug treatment and mediates an immune resistance. For example, in glioblastoma a compensatory recruitment of tumor-infiltrating myeloid cells elicited by antitumor immune response induced by dendritic cell (DC) vaccination contributed to the majority of PD-L1 expression [25]. Placenta-specific protein 8 (PLAC8) as an oncogene promoting cancer growth and progression is abnormally upregulated in gallbladder carcinoma. Overexpression of PLAC8 conferred resistance to gemcitabine and liplatin (OXA), mainstays of chemotherapy by upregulating PD-L1 expression [26]. 5-Fluorouracil selectively depletes myeloid-derived suppressor cells (MDSCs) and OXA triggers an immunogenic form of tumor cell death. A combined chemotherapy Folfox, 5-Fluorouracil plus OXA, has routinely been regarded as a first line of treatment for advanced colorectal cancer. However, Folfox up-regulates high expression of PD-1 on activated CD8$^+$ TILs, and induces CD8$^+$ T-cells to secret IFN-γ which upregulates PD-L1 expression on tumor cells [27]. CD40 stimulation on APC directly activates CTLs without the help of CD4$^+$ T-cells. Agonistic anti-CD40 antibodies induce antitumor responses and upregulation of PD-L1 on tumor-infiltrating monocytes and macrophages, which are extremely dependent on T-cells and IFN-γ [28]. When co-cultured with human PBMC, trastuzumab, the anti-human epidermal growth factor receptor-2 (HER2) antibody, is shown to upregulate PD-L1 in HER2-overexpressing breast cancer cells via mediating stimulation of IFN-γ secretion on immune cells [29]. Inhibitors of mTOR approved by the Food and Drug Administration to treat advanced metastatic renal cancers and enhance nuclear translocation of transcription factor EB, was bound to PD-L1 promoter and thereby led to increased PD-L1 expression [30].

3.2. Signaling Pathways and Factors Involved in PD-L1-Induced Resistance

Despite immune resistance, PD-L1 has generated resistance to TKIs in certain cancers. Possible mechanisms by which PD-L1 induced acquired resistance through upregulating Yes-associated protein1 (YAP1), [31] Bcl-2-associated athanogene-1 (BAG-1), [32] and DNA methyltransferase 1 (DNMT1), [33] and generated primary resistance by inducing epithelial-to-mesenchymal transition (EMT) have been reported [34] (Figure 1).

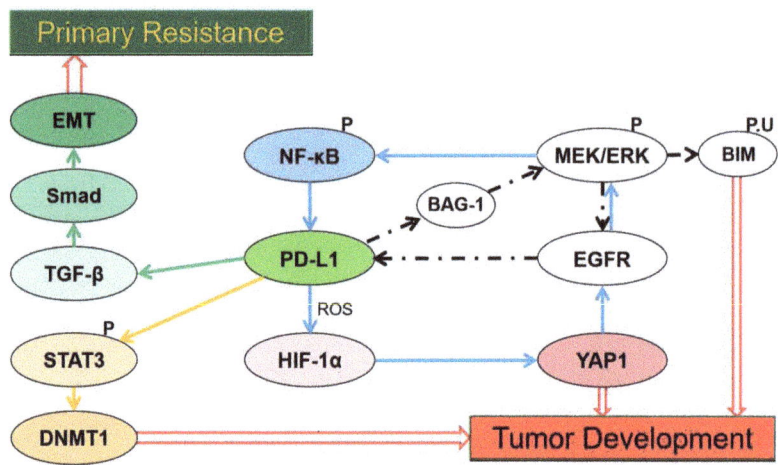

Figure 1. Signaling pathways and factors involved in programmed cell death ligand-1(PD-L1)-induced resistance. (1) PD-L1 expression induced by epidermal growth factor receptor (EGFR) mutation activation via extracellular single-regulated kinase (ERK) signaling, indirectly promotes expression of Bcl-2-associated athanogene-1 (BAG-1), the EGFR/ERK/PD-L1/BAG-1 feedback loop reaches the reactivation of ERK signaling which promotes Bcl-2-interacting mediator of cell death (BIM) phosphorylation to help cells escape from apoptosis. (2) PD-L1-induced hypoxia-inducible factor-1α (HIF-1α) expression is stimulated by reactive oxygen species (ROS), hypoxia increases YAP-1 expression which confers resistance via a YAP1/EGFR/ERK/NF-κB loop. (3) PD-L1 regulates DNA methyltransferase 1(DNMT1) via Signal transducer and activator of transcription 3 (STAT3) signaling and thus induces DNMT1-dependent DNA hypomethylation which promotes cancer development. (4) Activation of transforming growth factor-beta (TGF-β)/Smad pathway induced by PD-L1 is crucial in epithelial-to-mesenchymal transition (EMT) expression which leads to resistance to TKIs.

Activation of MEK/extracellular single-regulated kinase (ERK) signaling furthers phosphorylation and ubiquitination of the Bcl-2-interacting mediator of cell death (BIM), a BH-3-only protein, thereby preventing cells from apoptosis [35]. Resistance to TKI in NSCLC generally occurs through reactivating ERK signaling [36]. EGFR mutation activation induces expression of PD-L1 in NSCLC cells via ERK-signaling [37]. Once triggered by ERK signaling, phosphorylated C/EBPβ induced by PD-L1 can enhance binding to the BAG-1 promoter, thus promoting BAG-1 expression. The PD-L1/BAG-1 axis confers TKI resistance through persistent activated ERK signaling via the EGFR/ERK/PD-L1/BAG-1 feedback loop [32]. Thus combining treatment with TKIs and anti-PD-L1 therapy may provide a promising strategy for tumors with a high expression of PD-L1 and BAG-1, though this has not been researched yet.

YAP1 is another factor known to confer EGFR-TKI resistance in lung cancer cells [38]. Distinct experiments utilizing reactive oxygen species (ROS) scavengers and inducers demonstrated a concomitant change of expression of PD-L1 and hypoxia-inducible factor-1α (HIF-1α), YAP1 [31]. While prior reports described that PD-L1-induced HIF-1α is stimulated by the generation of ROS [39,40], hypoxia promotes formation of YAP1 and HIF-1α complex via regulating SIAH2 ubiquitin E3 ligase and increases YAP1 gene expression [41,42]. TKI resistance may be conferred by PD-L1/ROS/HIF-1α/YAP1 axis and a YAP1/EGFR/ERK/NF-κB loop [31]. Markedly high expression of YAP and PD-L1 are observed in EGFR-TKI-resistant cells in another study, and they demonstrate a positively related change in expression when given a knockdown of YAP [43]. Thereby, giving an anti-PD-L1 or anti-YAP1 may overcome the EGFR-TKI resistance.

The PD-L1/DNMT1 axis is also a critical mechanism leading to acquired resistance [33]. DNMT1, as a member of the DNA methyltransferase family, maintains the DNA methylation pattern [44]. Signal transducer and activator of transcription 3 (STAT3), a well-characterized transcription factor that binds to DNMT1 promoter and positively regulates transcription of DNMT1 [45], since phosphorylated STAT3 induces transcriptional activation via binding with specific DNA elements. PD-L1 regulates DNMT1 through the STAT3-signaling pathway and induces DNMT1-dependent DNA hypomethylation to promote development of cancers [46], thereby resulting in acquired resistance [33]. Currently, a combination therapy with oxaliplatin and decitabine inhibiting DNA demethylation was proved to have a synergistic effect in enhancing anti-PD-L1 therapeutic efficacy in colorectal cancer [47].

The transforming growth factor-beta (TGF-β)/Smad signaling pathway plays a role in PD-L1-induced primary resistance to EGFR-TKIs [34]. EMT can decrease efficacy of drug treatment in NSCLC [48,49]. PD-L1 upregulates phosphorylation of Smad3, which significantly participates in the transcriptional regulation mediated by TGF-β1 [50], and the TGF-β/Smad-signaling pathway has been reported to be crucial in EMT progression [51]. The mechanism of primary resistance to EGFR-TKIs in EGFR-mutant NSCLC may confer through the PD-L1/TGF-β/Smad/EMT axis [34]. In addition, in Kirsten rat sarcoma viral oncogene homolog (KRAS)-mutant NSCLC, KRAS G12 mutation is reported to promote PD-L1 expression via a TGF-β/EMT-signaling pathway [52]. Apparently, PD-L1 expression plays a key role in poor prognosis and resistance after treatment in several types of cancers, thereby adding an anti-PD-1 or anti-PD-L1 therapy may improve the efficacy and become a promising therapeutic strategy.

4. Key Mechanisms Underlying Resistance to PD-1/PD-L1 Blockade

PD-1/PD-L1 blockade therapy has been approved as a significantly helpful treatment in certain cancers, a problem of its limited efficacy has occurred and the targeting solution is urgently discussed and provided. Focusing on PD-L1, we described key mechanisms underlying resistance to PD-1/PD-L1. Surprisingly, abnormally upregulated PD-L1 expression and a lack of PD-L1 can both lead to inefficacy of PD-1/PD-L1 inhibitors (Figure 2).

4.1. Aberrant PD-L1 Expression

PD-L1 is generally regulated by tumor cells in two ways: the first is innate immune resistance in which constitutive oncogenic signaling is correlated with PD-L1 expression, the second is an adaptive immune resistance through which IFN-γ produced by TILs induces PD-L1 expression.

K-ras mutation as a common oncogenic driver in the lung adenocarcinoma (LUAD) and upregulates PD-L1 through p-ERK instead of p-AKT signaling [53]. Different subgroups of KRAS-mutant LUAD are dependent on STK11/LKB1 or TP53 mutations, and alterations of the former has been confirmed as a major factor that leads to primary resistance to PD-1 blockade [54]. Besides, EGFR-mutant or ALK-rearranged patients had a PD-L1 tumor proportion score of ≥50% and turned out not to respond to PD-1/PD-L1 inhibitors [55].

The transcription factor Yin Yang 1 (YY1); a major regulator reported participating in various pathways, is involved in cell growth, survival and metastasis. YY1 upregulates PD-L1 expression on tumor cells via signaling pathways, including p53, STAT3, NF-κB and PI3K/AKT/mTOR [56]. PD-L1v242 and PD-L1v229, two secreted PD-L1 C-terminal splicing variants, could capture the aPD-L1 antibody and function as a "decoy" to prevent antibodies from binding to PD-L1 [57].

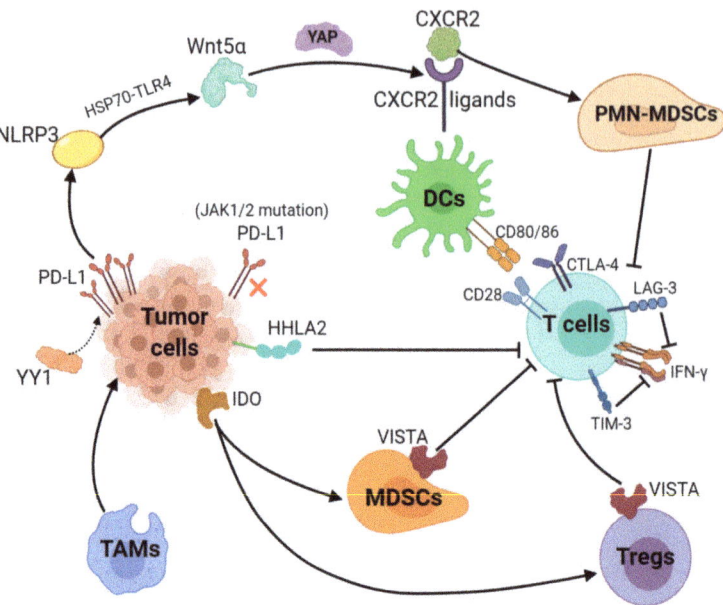

Figure 2. Key mechanisms underlying resistance to PD-L1 (1). The transcription factor Yin Yang 1 (YY1)-induced upregulation of PD-L1 expression triggers NOD-, LRR- and pyrin domain-containing 3 (NLRP3) inflammasome to promote tumor Wnt5α expression via HSP70-TLR4 signaling, and non-canonical WNT ligands activate the YAP pathway to induce chemokine (C-X-C motif) receptor 2 (CXCR2) ligands, while granulocytic subset of myeloid-derived suppressor cells (PMN-MDSCs) relied on CXCR2 to suppress T-cell function. (2) Loss-of-function mutations in JAK1/2 leads to the paucity of PD-L1 expression. (3) Tumor-suppressing microenvironment. Tumor-associated macrophages (TAMs) promote tumor progression, while Indole 2,3-dioxygenase (IDO) generated by tumors enhances Tregs and MDSCs activity, which suppress immunity. (4) Activation of alternative immune checkpoints. T-cell immunoglobulin mucin 3 (TIM-3) and Lymphocyte activation gene-3 (LAG-3) produced by T-cells impair generation of IFN-γ, which activates T-cells. CTLA-4 demonstrates a higher affinity and avidity in conjunction with CD80 and CD86 than CD28 to antagonize costimulation. VISTA is found to be related to MDSC mainly derived CD33 expression and HHLA2 decreases T-cell proliferation.

Besides, a tumor-intrinsic signaling pathway involved with NLRP3 inflammasome in response to upregulated expression of PD-L1 was found to drive adaptive resistance to anti-PD-1 antibody immunotherapy [58]. NLRP3 inflammasome triggered by PD-L1 induces tumor Wnt5α expression via HSP70-TLR4 signaling, while non-canonical WNT ligands promote production of CXCR2 ligands through the activated YAP pathway [59,60]. CXCR2-relied migration and recruitment of a granulocytic subset of MDSCs (PMN-MDSCs) play a role in suppressing CD8$^+$ T-cell infiltration and function, therefore leading to adaptive resistance [61,62].

Previous study showed that tumors can be divided into four categories according to positive/negative tumor PD-L1 expression and presence/absence of TILs. For instance, patients with PD-L1 positive and TILs indicate adaptive immune resistance and those with PD-L1 negative and without TILs show immune ignorance [63]. Among these four types, type I with PD-L1 positive and TILs is the most likely to respond to PD-1/PD-L1 blockade therapy, whilst other types may show unresponsiveness to this monotherapy [64].

4.2. Paucity of PD-L1 Expression

The interaction between PD-L1 and its receptor PD-1 leads to immune escape and inhibits T-cell function and blockade of PD-L1 and PD-1 enhances the antitumor immunity in several cancers. However, the expression of PD-L1 or PD-1 is a prerequisite for the therapeutic efficacy. Evidence of the relation of rare PD-L1 expression and poorer responses to PD-1 blockade has been proved in prostate cancer [65]. DNA hypomethylating agent upregulate PD-L1 gene expression [66]. Anti-PD-1 therapy curbs the expression of PD-L1 through either eliminating the tumor cells that overexpress PD-L1 and possess a hypomethylated PD-L1 promoter or switching off the PD-L1 expression through epigenetic modulation, therefore leading to resistance [67]. Loss-of-function mutations in JAK1/2 can lead to primary resistance to anti-PD-1 therapy due to the inability to respond to IFN-γ for a lack of PD-L1 expressions [68]. Despite the effect of aberrant PD-L1 expression, an abnormal process from antigen expression to T-cell activation can result in resistance to PD-1/PD-L1 inhibitors. Moreover, a recent study demonstrated that PD-L1 expression is enhanced via nicotinamide adenine dinucleotide (NAD^+) metabolism, in which nicotinamide phosphoribosyltransferase (NAMPT) functions as the rate-limiting enzyme [69]. NAMPT increases PD-L1 expression induced by IFN-γ and leads to immune escape in tumors with the help of $CD8^+$ T-cells. Thus NAD^+ metabolism is a promising strategy for resistance to anti-PD-L1 therapy [69].

4.3. Aberrant Antigen Expression, Presentation and Recognition

Tumors with a higher tumor mutation burden (TMB) are likely to have more neoantigens, which can be recognized by the immune system as "non-self" in response to checkpoint inhibition. In Naiyer's study, the result of the treatment of PD-1 targeting antibody pembrolizumab in NSCLC described that a higher burden of nonsynonymous tumors is correlated with a better response and PFS [70]. Besides, strong immunogenicity and extensive expression of immune checkpoint ligands make the microsatellite instability subtype more susceptible to immunotherapeutic methods, for example, with anti-PD-L1 and anti-cytotoxic T lymphocyte-associated antigen-4 (CTLA-4) antibodies [71]. Tumors with defective mismatch repair possess more DNA mutations and show an improved responsiveness to anti-PD-1 therapy [72]. In short, a low mutational burden, microsatellite stability and efficient DNA repair mechanisms are involved in innate resistance to immune-checkpoint blockade therapy. Moreover, evolution of neoantigen loss can produce an acquired resistance [73]. A study also demonstrates that deficiency of heterogeneity in HLA genes is observed in cancer development, a high level of HLA loss results in acquired resistance during immunotherapy [74].

Resistance to immune checkpoint blockades also involves impaired DC maturation, which is an essential process in T-cell activation, through it is displayed in various co-stimulatory factors expression including MHC class I/II, CD80, CD86 and CD40 [75]. IL37b decreases CD80 and CD86 expression through the ERK/S6K/NF-κB axis and suppresses DC maturation [76]. A transcription factor STAT3 that facilitates tumor growth and metastasis leads to the induction of other immunosuppressive factors that possess a suppressive function on DC maturation, including IL10, Tregs and TGF-β [77–80].

Despite inducing PD-L1, IFNs have been reported to (re-)activate T-cells to control the tumor development via advancing DC cross-priming [81–83]. It is well-known that CTLs recognize MHC class I-presented peptide antigens on the surface of tumor cells. Heterozygous mutations, deletions or deficiency in β-2-microglobulin (β2M); a crucial factor in MHC class I antigen presentation, generally reduces antigen recognition by antitumor $CD8^+$ T-cells and mutation of β2M gene leads to resistance to anti-PD-1 therapy [84,85]. IFN-γ can induce tumor cells to express MHC class I molecules, significantly promoting CTL differentiation and enhancing apoptosis. Mutations or loss of IFN-γ pathway-related proteins on tumor cells (such as STATs, IFN-γ receptor chain JAK1 and JAK2) can cause escape from immune recognition and resistance to immune checkpoint inhibitions [68,86].

4.4. Aberrant Immunity of T-Cells

Despite normal antigen expression, presentation, recognition and successfully activated T-cells, resistance to the PD-1/PD-L1 blockade inhibitors may occur owing to the T-cell itself. The aberrant immunity of T-cells include insufficient T lymphocytes infiltration, dysfunction of T-cell and exhausted T-cells.

4.4.1. Insufficient T Lymphocytes Infiltration

Despite the expression of PD-L1, a lack of T lymphocyte infiltration can cause unresponsiveness to anti-PD-L1 therapy. A crucial prerequisite for the therapeutic efficacy is the existing and tumor-infiltrated anti-tumor CTLs [87]. LIGHT, a member of the tumor necrosis factor superfamily, may activate lymphotoxin β-receptor signaling, resulting in the generation of chemokines that recruit a huge number of T-cells [88].

The PI3K-AKT-mTOR pathway, a crucial oncogenic signaling pathway, is involved in a multitude of cellular processes including cell survival, proliferation, and differentiation. PTEN, a lipid phosphatase, inhibits the PI3K signaling activity which activates the pathway. Loss of PTEN has been reported to reduce $CD8^+$ T-cells infiltrating into tumors and lead to resistance to PD-1 blockade therapy. A selective PI3Kβ inhibitor treatment enhanced the efficacy of anti-PD-1 antibodies [89]. The MAPK pathway also plays a major role in cell proliferation, inhibits T-cell recruitment and functions by inducing VEGF and IL-8 [90]. An inhibited MAPK pathway promotes $CD8^+$ T-cell activation and infiltration in melanoma [91,92]. Furthermore, studies showed that the combination of PD-1 blockade, BRAFi and MEKi enhances tumor immune infiltration and improves treatment outcomes [93].

A crucial oncogenic signaling pathway Wnt/β-catenin has been highly related to immune escape [94,95]. An activated Wnt/β-catenin pathway is correlated to loss of T-cell gene expression in metastatic melanoma [96]. Another study reported that the activation of the Wnt/β-catenin pathway in tumors brings about a non-inflammatory environment via numerous mechanisms. For instance, it acts on $CD103^+$ DCs of the Batf3 lineage and induces the transcription inhibitor ATF3 (activating transcription factor 3) expression to decrease production of Chemokine (C-C motif) ligand 4 (CCL4), thereby reducing initiated and infiltrated CTLs. Moreover, the Treg survival rate is enhanced by β-catenin [97].

Recently immune tumors are divided into three phenotypes: immune-desert, excluded and inflamed. Among these, the first and second phenotypes, which are non-inflamed tumors, show a low density of CTLs in the tumors and poor prognosis in immune checkpoint blockade therapy [98].

4.4.2. Dysfunction of T-Cells

Accumulation of extracellular adenosine is exploited by tumors to escape immunosurveillance through the activation of purinergic receptors [99]. CD38 expression expressed on Tregs and MDSCs is infiltrated in the tumor microenvironment and stimulated adenosine production via the CD38–CD203a-CD73 axis, and therefore inhibits CTL function [100,101].

4.4.3. Exhausted T-Cells

In vitro studies have reported that the PD-1 signal intensity determines the severity of T-cell exhaustion, which in turn affects the efficacy of anti-PD-1 treatment. In Nigow's animal model, high expression of PD-1 and extremely unresponsive T-cells showed relevance with resistance of anti-PD-1 therapy [102]. PD-1 treatment helps patients with low or moderate PD-1 expression to re-invigorate exhausted $CD8^+$ T cells and exert their immune effects. However, the cellular, transcriptional, and epigenetic changes following the PD-1 pathway blockade suggested limited storage potential after TEX re-invigoration, which means re-exhaustion following PD-L1 blockade [103].

5. Tumor-Suppressing Microenvironment

Apart from abnormal T-cells and PD-L1 expression, there are some other types of cells and cytokines that benefit tumor development inside the tumor microenvironment, they form the tumor-suppressing microenvironment to play a key role in resistance to the PD-1/PD-L1 blockade.

5.1. Tregs

Tregs are involved in maintaining self-tolerance, and inhibit autoimmunity through secreting cytokines, including TGF-β1. The ratio of CD8$^+$ Teff cells/Tregs is strongly associated with the prognosis of immunotherapy [104,105]. The administration of low-dose TLR-7 agonist resiquimod could transform Treg accumulation-caused resistance to the PD-L1 blockade [106]. Combination of radiation therapy and dual immune checkpoint blockade restores antitumor immunity of consumed Tregs [107]. Currently, anti-CD25 therapy is believed to take effect through Treg depletion when combined with PD-1 blockade therapy [108].

5.2. MDSCs

MDSCs suppress immunity mainly through preventing T-cell activation and function, Arg1 and ROS are the common molecules used. Besides, they downregulate macrophage production of the type I cytokine IL-12 to polarize macrophages toward a tumor-promoting phenotype [109,110], suppress tumor cell lysis mediated by NK cells and induce and recruit Tregs [111–114]. In the presence of MDSCs, the levels of PD-1 expression show a decrease, while PD-L1 expression shows an increase [115]. MDSC-targeted therapy, which decreases MDSC frequency and transforms its function, is studied to overcome the resistance to immune checkpoint inhibitors, thus combining MDSC-targeted therapy and immune checkpoint blockades is considered a promising strategy for the future [116].

5.3. TAMs

Protumor macrophages are differentiated through interaction with tumor cells and turn to polarize into M2-like TAM, which play a significant role in immunosuppression, invasion and metastasis. For the sake of overcoming the latent resistance of macrophages, CSF-1R blockade reduces the frequency of TAMs, therefore increasing production of interferon and tumor regression [117], and synergizing with immune checkpoint blockades [118].

5.4. IDO

Indole 2,3-dioxygenase is generated by tumors and immune cells to enhance Tregs and MDSCs production and activity. IDO, an enzyme catalyzing the degradation of tryptophan along the kynurenine pathway, is induced in response to inflammatory stimuli and its activity is known to have an inhibition of effector on T-cell immunity [119]. A report conducted on B16 melanoma demonstrated that following PD-1 blockade treatment, a subset of mice with IDO knockout had an obviously slower tumor development and better overall survival rates compared with wild type [120]. Thus, a combination therapy of IDO inhibitors and PD-1/PD-L1 antibodies may demonstrate a better efficacy than single agent [121].

5.5. VEGFA

TMB with hypoxia and hyper-angiogenesis is obviously crucial for tumor growth and progression, and vascular endothelial growth factor A (VEGFA) plays a significant role in it. High expression of VEGFA is reported to impair infiltration of effective anti-tumor T-cells, thus leads to innate resistance in PD-1/PD-L1 blockade [122]. Unfortunately, combining treatment with inhibiting the VEGFA and PD-1/PD-L1 blockade demonstrates more toxic and harbors more adverse effects than monotherapy.

5.6. Immunosuppressive Cytokines

TGF-β inhibits the expansion and function of many components of the immune system, either by stimulating or inhibiting their differentiation and function, therefore it maintains immune homeostasis and tolerance. Specific chemokines are capable of recruiting cells into tumors. CXCL9, CXCL10, CXCL11, CCL3, CCL4 and other chemokines and their receptors are recruited to cause antitumor response via recruiting CTL and NK cells while CCL2 CCL22, CCL5, CCL7 and CXCL8 recruit immunosuppressive cells to suppress the immune response. Research reveals that epigenetic silencing of CXCL9 and CXCL10 can suppresses T-cell homing [123].

6. Activation of Alternative Immune Checkpoints

As one of the most prospective approaches in cancer treatment, immunotherapy has reached notable achievements, especially with the PD-L1 blockade. However, the efficacy of PD-L1 inhibitor therapy has been found to be limited due to activation of other immune checkpoints including TIM-3 and VISTA. So far, some studies have reported that the combination therapy targeting distinct types of immune checkpoints has been proved effective in several cancers.

6.1. TIM-3

T-cell immunoglobulin mucin 3 (TIM-3) has been identified as a critical regulator of CTL exhaustion with co-expression of PD-1 [124]. TILs with co-expression of TIM-3 and PD-1 do not produce IL-2 and IFN-γ, and they are prone to exhaust. In response to radiotherapy and PD-L1 inhibition, TIM-3 is upregulated and subsequently caused acquired resistance in HNSCC [107]. Combination therapy targeting TIM-3 and PD-1 signaling pathways simultaneously is proved to be effective against cancer [124].

6.2. HHLA2

HHLA2, a member of the B7 family, can predict poor overall survival in several cancers, including human clear cell renal cell carcinoma and colorectal carcinoma [125]. HHLA2 can suppress T-cell activation and proliferation in the presence of TCR and CD28 signaling [126], and can do this more robustly than PD-L1 [127].

6.3. VISTA

V-domain Ig suppressor of T-cell activation (VISTA) expression induced by IL-10 and IFN-γ is observed to be higher in immature DCs, MDSCs and Tregs compared with peripheral tissues [128,129]. The synergistic effect of the combining VISTA and PD-L1 monoclonal therapy in colon cancer can be taken as an example, reduction of tumor growth and better OS are observed compared with monotherapy [130].

6.4. LAG-3

Lymphocyte activation gene-3 (LAG-3) is responsible for maintaining immune homeostasis through repressing activation of T-cells and cytokines secretion [131]. Interaction between LAG-3 and Galectin-3, a soluble lectin regulating antigen-specific T-cell activation, expands the immunomodulatory effect of LAG-3 on tumor-infiltrating CD8$^+$ T-cells in the TME [132]. Sinusoidal endothelial cell lectin binds to LAG-3 to reduce IFN-γ expression produced by activated T-cells [133]. An amazing synergistic effect in suppressing immune responses is found in LAG-3 with PD-1 under distinct conditions [134].

6.5. CTLA-4

CD28 interacting with the CD80 dimer and the CD86 monomer mediates T-cell co-stimulation along with TCR signals, while CTLA-4 demonstrates a higher affinity and avidity in conjunction with the two ligands than with CD28, which in turn antagonizes CD28-mediated co-stimulation [135]. A combination of PD-1-targeted mAb nivolumab and CTLA-4-targeted mAb ipilimumab has been approved as the first-line treatment for renal clear cell cancer patients with moderate or poor prognosis [136].

6.6. Siglec-15

As a member of the sialic acid-binding immunoglobulin-like lectin (Siglec) gene family, Siglec-15 is found to impair anti-tumor immunity through suppressing T-cell functions. Siglec-15 is expressed only on some myeloid cells normally, while it is upregulated on TAMs and tumor cells [137]. Interestingly, an antagonistic relationship between Siglec-15 and PD-L1 has been reported, mainly due to regulation of IFN-γ [138]. M-CSF induces expression of Siglec-15 on macrophages and IFN-γ, identified as a crucial factor promoting PD-L1 expression, inversely decreases it [137].

6.7. TIGHT

T-cell immunoglobulin and ITIM domain (TIGIT), expressed mainly on Tregs, is a co-inhibitory checkpoint receptor which has a significantly higher affinity in binding to CD155 than the co-stimulatory receptor CD226 [139]. TIGIT/CD155 signaling causes T-cell exhaustion to impair anti-tumor immunity in several types of cancer, including melanoma and HNSCC [140,141]. Furthermore, the phenomenon that TIGIT expression often accompanies PD-1 has been observed in both normal tissues and tumors [142].

6.8. BTLA

B and T lymphocyte attenuator (BTLA), expressed mostly on B-cells, is upregulated on $CD19^+$ high B-cells through AKT and STAT3 pathways once triggered by IL-6 and IL-10 [143]. BTLA is regarded as one of the factors leading to resistance to anti-PD-1 therapy, though they do not suppress T-cell signaling through an identical mechanism related with src-homology-2 domain-containing phosphatase (SHP)1 and SHP2 [143,144].

7. Current Combination Therapies with PD-1/PD-L1 Inhibitors

With regard to clinical the limitations of anti-PD-1/PD-L1 monotherapy, it exists more and more in combination therapies based on mechanisms underlying resistance to the PD-1/PD-L1 blockade. Among all of them, chemotherapy, VEGF/VEGFR-targeted therapy and anti-CTLA-4 rank in the top three. Other treatments that are considered to combine with PD-1/PD-L1 blockade include radiotherapy, vaccines, cytokine therapy and chemokine inhibition. Radiotherapy is identified as alter differentiation and function of T-cells and promote the expression of PD-L1, which means adding radiotherapy may enhance the effects of anti-PD-L1 treatment [145]. A triple therapy with anti-PD-1 antagonist antibody, anti-CD137 agonist antibody and vaccine therapy has been reported to significantly enhance T-cell activation in pancreatic ductal adenocarcinoma in a preclinical study [146]. Recently, another immune checkpoint inhibitor tiragolumab targeting TIGIT has been granted breakthrough therapy designation by the FDA and combining anti-PD-L1 and anti-TIGIT has been reported as highly effective in clinic with metastatic NSCLC patients [147]. Combining TNF-α-loaded liposomes and anti-PD-1/PD-L1 further enhances the anti-tumor immunity [148]. Even utilizing newly emerged neoantigens may improve the therapeutic efficacy of immune checkpoint blockade treatment [148].

8. Conclusions

As an inhibitor in the immune system, PD-L1 plays multiple roles in tumors. PD-L1 has been confirmed as a prospective and prognostic biomarker in certain cancers, while rebiopsy should be considered when PD-L1 expression is increased due to treatment (such as gefitinib treatment). Immune resistance induced by PD-L1 following various therapies inspired a combination therapy of PD-L1 blockade and these therapies. To date, immunotherapy, especially PD-1/PD-L1 blockade, which is at forefront of clinical therapy, has benefited many patients. However, primary and acquired resistance to this blockade therapy still exists and limits its efficacy. So far, key mechanisms suggest complement approaches for patients who cannot respond well to PD-1/PD-L1 antibodies. For example, modulating the immunosuppressive tumor microenvironment, such as depletion of Tregs, IDO, or MDSCs, interfering suppressive cytokines and inhibiting alternative immune checkpoints, may enhance the therapeutic efficacy of the PD-1/PD-L1 blockade. Other mechanisms underlying resistance to this blockade therapy and individual treatments for more patients requires further investigation.

Author Contributions: Conceptualization: Y.Y., C.Z., H.C.; Writing—Original: Draft Y.Y., A.A.; Writing—Review and Editing: A.A., C.Z., H.C.; Supervision: H.C. All authors have read and agreed to the published version of the manuscript.

Funding: This research was partly funded by the National Natural Science Foundation of China [No. 81872443].

Institutional Review Board Statement: Not applicable.

Informed Consent Statement: Not applicable.

Conflicts of Interest: The authors declare no conflict of interest. The funders had no role in the design of the study; in the collection, analyses, or interpretation of data; in the writing of the manuscript, or in the decision to publish the results.

References

1. Lyford-Pike, S.; Peng, S.; Young, G.D.; Taube, J.M.; Westra, W.H.; Akpeng, B.; Bruno, T.C.; Richmon, J.D.; Wang, H.; Bishop, J.A.; et al. Evidence for a Role of the PD-1:PD-L1 Pathway in Immune Resistance of HPV-Associated Head and Neck Squamous Cell Carcinoma. *Cancer Res.* **2013**, *73*, 1733–1741. [CrossRef]
2. Pardoll, D.M. The blockade of immune checkpoints in cancer immunotherapy. *Nat. Rev. Cancer* **2012**, *12*, 252–264. [CrossRef]
3. Chen, D.S.; Irving, B.A.; Hodi, F.S. Molecular Pathways: Next-Generation Immunotherapy—Inhibiting Programmed Death-Ligand 1 and Programmed Death-1. *Clin. Cancer Res.* **2012**, *18*, 6580–6587. [CrossRef]
4. Dong, H.; Strome, S.E.; Salomao, D.R.; Tamura, H.; Hirano, F.; Flies, D.B.; Roche, P.C.; Lu, J.; Zhu, G.; Tamada, K.; et al. Tumor-associated B7-H1 promotes T-cell apoptosis: A potential mechanism of immune evasion. *Nat. Med.* **2002**, *8*, 793–800. [CrossRef]
5. Robert, C.; Thomas, L.; Bondarenko, I.; O'Day, S.; Weber, J.; Garbe, C.; Lebbe, C.; Baurain, J.-F.; Testori, A.; Grob, J.-J.; et al. Ipilimumab plus Dacarbazine for Previously Untreated Metastatic Melanoma. *N. Engl. J. Med.* **2011**, *364*, 2517–2526. [CrossRef] [PubMed]
6. Wellenstein, M.D.; de Visser, K.E. Cancer-Cell-Intrinsic Mechanisms Shaping the Tumor Immune Landscape. *Immunity* **2018**, *48*, 399–416. [CrossRef] [PubMed]
7. Postow, M.A.; Callahan, M.K.; Wolchok, J.D. Immune Checkpoint Blockade in Cancer Therapy. *J. Clin. Oncol.* **2015**, *33*, 1974–1982. [CrossRef] [PubMed]
8. O'Donnell, J.S.; Smyth, M.J.; Teng, M.W.L. Acquired resistance to anti-PD1 therapy: Checkmate to checkpoint blockade? *Genome Med.* **2016**, *8*, 111. [CrossRef]
9. Jiang, X.; Zhou, J.; Giobbie-Hurder, A.; Wargo, J.; Hodi, F.S. The Activation of MAPK in Melanoma Cells Resistant to BRAF Inhibition Promotes PD-L1 Expression That Is Reversible by MEK and PI3K Inhibition. *Clin. Cancer Res.* **2013**, *19*, 598–609. [CrossRef]
10. Jenkins, R.W.; A Barbie, D.; Flaherty, K.T. Mechanisms of resistance to immune checkpoint inhibitors. *Br. J. Cancer* **2018**, *118*, 9–16. [CrossRef]
11. Syn, N.L.; Teng, M.W.L.; Mok, T.S.K.; A Soo, R. De-novo and acquired resistance to immune checkpoint targeting. *Lancet Oncol.* **2017**, *18*, e731–e741. [CrossRef]
12. Yan, F.; Pang, J.; Peng, Y.; Molina, J.R.; Yang, P.; Liu, S. Elevated Cellular PD1/PD-L1 Expression Confers Acquired Resistance to Cisplatin in Small Cell Lung Cancer Cells. *PLoS ONE* **2016**, *11*, e0162925. [CrossRef]
13. M, W.; H, K.; Yy, L.; C, W.; Djm, N.; Lg, F.; Mt, K.; Savaraj, N. Relationship of Metabolic Alterations and PD-L1 Expression in Cisplatin Resistant Lung Cancer. *Cell Dev. Biol.* **2017**, *6*, 183. [CrossRef] [PubMed]
14. Zhang, P.; Ma, Y.; Lv, C.; Huang, M.; Li, M.; Dong, B.; Liu, X.; An, G.; Zhang, W.; Zhang, J.; et al. Upregulation of programmed cell death ligand 1 promotes resistance response in non-small-cell lung cancer patients treated with neo-adjuvant chemotherapy. *Cancer Sci.* **2016**, *107*, 1563–1571. [CrossRef] [PubMed]
15. Liu, H.; Hu, Y.; Rimoldi, R.; Von Hagt, C.; Khong, E.W.; Lee, N.; Flemming, S.; Spencer, A.; Dear, A.E. Epigenetic treatment-mediated modulation of PD-L1 predicts potential therapy resistance over response markers in myeloid malignancies: A molecular mechanism involving effectors of PD-L1 reverse signaling. *Oncol. Lett.* **2018**, *17*, 2543–2550. [CrossRef]
16. Zhang, P.; Liu, J.; Li, W.; Li, S.; Han, X. Lactoferricin B reverses cisplatin resistance in head and neck squamous cell carcinoma cells through targeting PD-L1. *Cancer Med.* **2018**, *7*, 3178–3187. [CrossRef]

17. Skinner, H.D.; Giri, U.; Yang, L.P.; Kumar, M.; Liu, Y.; Story, M.D.; Pickering, C.R.; Byers, L.A.; Williams, M.D.; Wang, J.; et al. Integrative Analysis Identifies a Novel AXL–PI3 Kinase–PD-L1 Signaling Axis Associated with Radiation Resistance in Head and Neck Cancer. *Clin. Cancer Res.* **2017**, *23*, 2713–2722. [CrossRef] [PubMed]
18. Massi, D.; Brusa, D.; Merelli, B.; Falcone, C.; Xue, G.; Carobbio, A.; Nassini, R.; Baroni, G.; Tamborini, E.; Cattaneo, L.; et al. The status of PD-L1 and tumor-infiltrating immune cells predict resistance and poor prognosis in BRAFi-treated melanoma patients harboring mutant BRAFV600. *Ann. Oncol.* **2015**, *26*, 1980–1987. [CrossRef]
19. Miyawaki, E.; Murakami, H.; Mori, K.; Mamesaya, N.; Kawamura, T.; Kobayashi, H.; Omori, S.; Wakuda, K.; Ono, A.; Kenmotsu, H.; et al. PD-L1 expression and response to pembrolizumab in patients with EGFR-mutant non-small cell lung cancer. *Jpn. J. Clin. Oncol.* **2020**, *50*, 617–622. [CrossRef]
20. Han, J.J.; Kim, D.-W.; Koh, J.; Keam, B.; Kim, T.M.; Jeon, Y.K.; Lee, S.-H.; Chung, D.H.; Heo, D.S. Change in PD-L1 Expression After Acquiring Resistance to Gefitinib in EGFR-Mutant Non–Small-Cell Lung Cancer. *Clin. Lung Cancer* **2016**, *17*, 263–270.e2. [CrossRef]
21. Creelan, B.C.; Yeh, T.C.; Kim, S.-W.; Nogami, N.; Kim, D.-W.; Chow, L.Q.M.; Kanda, S.; Taylor, R.; Tang, W.; Tang, M.; et al. A Phase 1 study of gefitinib combined with durvalumab in EGFR TKI-naive patients with EGFR mutation-positive locally advanced/metastatic non-small-cell lung cancer. *Br. J. Cancer* **2020**, *124*, 1–8. [CrossRef]
22. Berry, S.; Taube, J.M. Innate vs. Adaptive: PD-L1-mediated immune resistance by melanoma. *OncoImmunology* **2015**, *4*, e1029704. [CrossRef] [PubMed]
23. Mimura, K.; Teh, J.L.; Okayama, H.; Shiraishi, K.; Kua, L.-F.; Koh, V.; Smoot, D.T.; Ashktorab, H.; Oike, T.; Suzuki, Y.; et al. PD-L1 expression is mainly regulated by interferon gamma associated with JAK-STAT pathway in gastric cancer. *Cancer Sci.* **2017**, *109*, 43–53. [CrossRef] [PubMed]
24. Dovedi, S.J.; Adlard, A.L.; Lipowska-Bhalla, G.; McKenna, C.; Jones, S.; Cheadle, E.J.; Stratford, I.J.; Poon, E.; Morrow, M.; Stewart, R.; et al. Acquired Resistance to Fractionated Radiotherapy Can Be Overcome by Concurrent PD-L1 Blockade. *Cancer Res.* **2014**, *74*, 5458–5468. [CrossRef]
25. Antonios, J.P.; Soto, H.; Everson, R.G.; Moughon, D.; Orpilla, J.R.; Shin, N.P.; Sedighim, S.; Treger, J.; Odesa, S.; Tucker, A.; et al. Immunosuppressive tumor-infiltrating myeloid cells mediate adaptive immune resistance via a PD-1/PD-L1 mechanism in glioblastoma. *Neuro-Oncology* **2017**, *19*, 796–807. [CrossRef]
26. Gong, K.; Gong, Z.-J.; Lu, P.-X.; Ni, X.-L.; Shen, S.; Liu, H.; Wang, J.-W.; Zhang, D.-X.; Liu, H.-B.; Suo, T. PLAC8 overexpression correlates with PD-L1 upregulation and acquired resistance to chemotherapies in gallbladder carcinoma. *Biochem. Biophys. Res. Commun.* **2019**, *516*, 983–990. [CrossRef]
27. Dosset, M.; Vargas, T.R.; Lagrange, A.; Boidot, R.; Végran, F.; Roussey, A.; Chalmin, F.; Dondaine, L.; Paul, C.; Marie-Joseph, E.L.; et al. PD-1/PD-L1 pathway: An adaptive immune resistance mechanism to immunogenic chemotherapy in colorectal cancer. *OncoImmunology* **2018**, *7*, e1433981. [CrossRef]
28. Zippelius, A.; Schreiner, J.; Herzig, P.; Müller, P. Induced PD-L1 Expression Mediates Acquired Resistance to Agonistic Anti-CD40 Treatment. *Cancer Immunol. Res.* **2015**, *3*, 236–244. [CrossRef]
29. Chaganty, B.K.R.; Qiu, S.; Gest, A.; Lu, Y.; Ivan, C.; Calin, G.A.; Weiner, L.M.; Fan, Z. Trastuzumab upregulates PD-L1 as a potential mechanism of trastuzumab resistance through engagement of immune effector cells and stimulation of IFNgamma secretion. *Cancer Lett.* **2018**, *430*, 47–56. [CrossRef] [PubMed]
30. Zhang, C.; Duan, Y.; Xia, M.; Dong, Y.; Chen, Y.; Zheng, L.; Chai, S.; Zhang, Q.; Wei, Z.; Liu, N.; et al. TFEB Mediates Immune Evasion and Resistance to mTOR Inhibition of Renal Cell Carcinoma via Induction of PD-L1. *Clin. Cancer Res.* **2019**, *25*, 6827–6838. [CrossRef] [PubMed]
31. Tung, J.-N.; Lin, P.-L.; Wang, Y.-C.; Wu, D.-W.; Chen, C.-Y.; Lee, H. PD-L1 confers resistance to EGFR mutation-independent tyrosine kinase inhibitors in non-small cell lung cancer via upregulation of YAP1 expression. *Oncotarget* **2017**, *9*, 4637–4646. [CrossRef]
32. Lin, P.-L.; Wu, T.-C.; Wu, D.-W.; Wang, L.; Chen, C.-Y.; Lee, H. An increase in BAG-1 by PD-L1 confers resistance to tyrosine kinase inhibitor in non–small cell lung cancer via persistent activation of ERK signalling. *Eur. J. Cancer* **2017**, *85*, 95–105. [CrossRef] [PubMed]
33. Liu, J.; Liu, Y.; Meng, L.; Liu, K.; Ji, B. Targeting the PD-L1/DNMT1 axis in acquired resistance to sorafenib in human hepatocellular carcinoma. *Oncol. Rep.* **2017**, *38*, 899–907. [CrossRef] [PubMed]
34. Zhang, Y.; Zeng, Y.; Liu, T.; Du, W.; Zhu, J.; Liu, Z.; Huang, J.A. The canonical TGF-β/Smad signalling pathway is involved in PD-L1-induced primary resistance to EGFR-TKIs in EGFR-mutant non-small-cell lung cancer. *Respir. Res.* **2019**, *20*, 164. [CrossRef] [PubMed]
35. Ley, R.; Balmanno, K.; Hadfield, K.; Weston, C.; Cook, S.J. Activation of the ERK1/2 signaling pathway promotes phosphorylation and proteasome-dependent degradation of the BH3-only protein, Bim. *J. Biol. Chem.* **2003**, *278*, 18811–18816. [CrossRef]
36. Ercan, D.; Xu, C.; Yanagita, M.; Monast, C.S.; Pratilas, C.A.; Montero, J.; Butaney, M.; Shimamura, T.; Sholl, L.; Ivanova, E.V.; et al. Reactivation of ERK Signaling Causes Resistance to EGFR Kinase Inhibitors. *Cancer Discov.* **2012**, *2*, 934–947. [CrossRef]
37. Chen, N.; Fang, W.; Zhan, J.; Hong, S.; Tang, Y.; Kang, S.; Zhang, Y.; He, X.; Zhou, T.; Qin, T.; et al. Upregulation of PD-L1 by EGFR Activation Mediates the Immune Escape in EGFR-Driven NSCLC: Implication for Optional Immune Targeted Therapy for NSCLC Patients with EGFR Mutation. *J. Thorac. Oncol.* **2015**, *10*, 910–923. [CrossRef]

38. Cheng, H.; Zhang, Z.; Rodriguez-Barrueco, R.; Borczuk, A.; Liu, H.; Yu, J.; Silva, J.M.; Cheng, S.K.; Perez-Soler, R.; Halmos, B. Functional genomics screen identifies YAP1 as a key determinant to enhance treatment sensitivity in lung cancer cells. *Oncotarget* **2015**, *7*, 28976–28988. [CrossRef]
39. Barsoum, I.B.; Smallwood, C.A.; Siemens, D.R.; Graham, C.H. A Mechanism of Hypoxia-Mediated Escape from Adaptive Immunity in Cancer Cells. *Cancer Res.* **2014**, *74*, 665–674. [CrossRef]
40. Chandel, N.S.; McClintock, D.S.; Feliciano, C.E.; Wood, T.M.; Melendez, J.A.; Rodriguez, A.M.; Schumacker, P.T. Reactive oxygen species generated at mitochondrial complex III stabilize hypoxia-inducible factor-1alpha during hypoxia: A mechanism of O2 sensing. *J. Biol. Chem.* **2000**, *275*, 25130–25138. [CrossRef]
41. Ma, B.; Chen, Y.; Chen, L.; Cheng, H.; Mu, C.; Li, J.; Gao, R.; Zhou, C.; Cao, L.; Liu, J.; et al. Hypoxia regulates Hippo signalling through the SIAH2 ubiquitin E3 ligase. *Nat. Cell Biol.* **2014**, *17*, 95–103. [CrossRef]
42. Li, Q.; Tsuneki, M.; Krauthammer, M.; Couture, R.; Schwartz, M.L.; Madri, J.A. Modulation of Sox10, HIF-1α, Survivin, and YAP by Minocycline in the Treatment of Neurodevelopmental Handicaps following Hypoxic Insult. *Am. J. Pathol.* **2015**, *185*, 2364–2378. [CrossRef] [PubMed]
43. Lee, B.S.; Park, D.I.; Lee, D.H.; Lee, J.E.; Yeo, M.-K.; Park, Y.H.; Lim, D.S.; Choi, W.; Yoo, G.; Kim, H.-B.; et al. Hippo effector YAP directly regulates the expression of PD-L1 transcripts in EGFR-TKI-resistant lung adenocarcinoma. *Biochem. Biophys. Res. Commun.* **2017**, *491*, 493–499. [CrossRef] [PubMed]
44. Jin, B.; Robertson, K.D. DNA Methyltransferases, DNA Damage Repair, and Cancer. *In Vivo Immunol.* **2013**, *754*, 3–29. [CrossRef]
45. Thomas, N.S.B. The STAT3-DNMT1 connection. *JAK-STAT* **2012**, *1*, 257–260. [CrossRef]
46. Baylin, S.B. DNA methylation and gene silencing in cancer. *Nat. Clin. Pract. Oncol.* **2005**, *2*, S4–S11. [CrossRef]
47. Huang, K.C.-Y.; Chiang, S.-F.; Chen, W.T.-L.; Chen, T.-W.; Hu, C.-H.; Yang, P.-C.; Ke, T.-W.; Chao, K.S.C. Decitabine Augments Chemotherapy-Induced PD-L1 Upregulation for PD-L1 Blockade in Colorectal Cancer. *Cancers* **2020**, *12*, 462. [CrossRef]
48. Yauch, R.L.; Januario, T.; Eberhard, D.A.; Cavet, G.; Zhu, W.; Fu, L.; Pham, T.Q.; Soriano, R.; Stinson, J.; Seshagiri, S.; et al. Epithelial versus Mesenchymal Phenotype Determines In vitro Sensitivity and Predicts Clinical Activity of Erlotinib in Lung Cancer Patients. *Clin. Cancer Res.* **2005**, *11*, 8686–8698. [CrossRef] [PubMed]
49. Thomson, S.; Buck, E.; Petti, F.; Griffin, G.; Brown, E.; Ramnarine, N.; Iwata, K.K.; Gibson, N.; Haley, J.D. Epithelial to Mesenchymal Transition Is a Determinant of Sensitivity of Non–Small-Cell Lung Carcinoma Cell Lines and Xenografts to Epidermal Growth Factor Receptor Inhibition. *Cancer Res.* **2005**, *65*, 9455–9462. [CrossRef] [PubMed]
50. Tang, P.M.-K.; Zhou, S.; Meng, X.-M.; Wang, Q.-M.; Li, C.-J.; Lian, G.-Y.; Huang, X.-R.; Tang, Y.-J.; Guan, X.Y.; Yan, B.P.-Y.; et al. Smad3 promotes cancer progression by inhibiting E4BP4-mediated NK cell development. *Nat. Commun.* **2017**, *8*, 14677. [CrossRef]
51. Muraoka-Cook, R.S.; Dumont, N.; Arteaga, C.L. Dual role of transforming growth factor β in mammary tumorigenesis and metastatic progression. *Clin. Cancer Res.* **2005**, *11*, 937s–943s.
52. Pan, L.; Ma, Y.; Li, Z.; Hu, J.; Xu, Z. KRAS G12V mutation upregulates PD-L1 expression via TGF-β/EMT signaling pathway in human non-small-cell lung cancer. *Cell Biol. Int.* **2020**. [CrossRef]
53. Chen, N.; Fang, W.; Lin, Z.; Peng, P.; Wang, J.; Zhan, J.; Hong, S.; Huang, J.; Liu, L.; Sheng, J.; et al. KRAS mutation-induced upregulation of PD-L1 mediates immune escape in human lung adenocarcinoma. *Cancer Immunol. Immunother.* **2017**, *66*, 1175–1187. [CrossRef] [PubMed]
54. Skoulidis, F.; Goldberg, M.E.; Greenawalt, D.M.; Hellmann, M.D.; Awad, M.M.; Gainor, J.F.; Schrock, A.B.; Hartmaier, R.J.; Trabucco, S.E.; Gay, L.; et al. STK11/LKB1 Mutations and PD-1 Inhibitor Resistance in KRAS-Mutant Lung Adenocarcinoma. *Cancer Discov.* **2018**, *8*, 822–835. [CrossRef] [PubMed]
55. Yoneshima, Y.; Ijichi, K.; Anai, S.; Ota, K.; Otsubo, K.; Iwama, E.; Tanaka, K.; Oda, Y.; Nakanishi, Y.; Okamoto, I. PD-L1 expression in lung adenocarcinoma harboring EGFR mutations or ALK rearrangements. *Lung Cancer* **2018**, *118*, 36–40. [CrossRef]
56. Hays, E.; Bonavida, B. YY1 regulates cancer cell immune resistance by modulating PD-L1 expression. *Drug Resist. Updat.* **2019**, *43*, 10–28. [CrossRef]
57. Gong, B.; Kiyotani, K.; Sakata, S.; Nagano, S.; Kumehara, S.; Baba, S.; Besse, B.; Yanagitani, N.; Friboulet, L.; Nishio, M.; et al. Secreted PD-L1 variants mediate resistance to PD-L1 blockade therapy in non–small cell lung cancer. *J. Exp. Med.* **2019**, *216*, 982–1000. [CrossRef]
58. Qu, J.; Tao, X.-Y.; Teng, P.; Zhang, Y.; Guo, C.-L.; Chun-Yi, J.; Qian, Y.; Jiang, C.-Y.; Liu, W.-T. Blocking ATP-sensitive potassium channel alleviates morphine tolerance by inhibiting HSP70-TLR4-NLRP3-mediated neuroinflammation. *J. Neuroinflamm.* **2017**, *14*, 1–17. [CrossRef]
59. Park, H.W.; Kim, Y.C.; Yu, B.; Moroishi, T.; Mo, J.S.; Plouffe, S.W.; Meng, Z.; Lin, K.C.; Yu, F.X.; Alexander, C.M.; et al. Alternative Wnt Signaling Activates YAP/TAZ. *Cell* **2015**, *162*, 780–794. [CrossRef]
60. Wang, G.; Lu, X.; Dey, P.; Deng, P.; Wu, C.C.; Jiang, S.; Fang, Z.; Zhao, K.; Konaparthi, R.; Hua, S.; et al. Targeting YAP-Dependent MDSC Infiltration Impairs Tumor Progression. *Cancer Discov.* **2016**, *6*, 80–95. [CrossRef]
61. Highfill, S.L.; Cui, Y.; Giles, A.J.; Smith, J.P.; Zhang, H.; Morse, E.; Kaplan, R.N.; Mackall, C.L. Disruption of CXCR2-Mediated MDSC Tumor Trafficking Enhances Anti-PD1 Efficacy. *Sci. Transl. Med.* **2014**, *6*, 237ra267. [CrossRef]
62. Li, J.; Byrne, K.T.; Yan, F.; Yamazoe, T.; Chen, Z.; Baslan, T.; Richman, L.P.; Lin, J.H.; Sun, Y.H.; Rech, A.J.; et al. Tumor Cell-Intrinsic Factors Underlie Heterogeneity of Immune Cell Infiltration and Response to Immunotherapy. *Immunity* **2018**, *49*, 178–193.e7. [CrossRef] [PubMed]

63. Teng, M.W.L.; Ngiow, S.F.; Ribas, A.; Smyth, M.J. Classifying Cancers Based on T-cell Infiltration and PD-L1. *Cancer Res.* **2015**, *75*, 2139–2145. [CrossRef] [PubMed]
64. Dong, Z.-Y.; Wu, S.-P.; Liao, R.-Q.; Huang, S.-M.; Wu, Y.-L. Potential biomarker for checkpoint blockade immunotherapy and treatment strategy. *Tumor Biol.* **2016**, *37*, 4251–4261. [CrossRef] [PubMed]
65. Zou, W.; Chen, L. Inhibitory B7-family molecules in the tumour microenvironment. *Nat. Rev. Immunol.* **2008**, *8*, 467–477. [CrossRef] [PubMed]
66. Wrangle, J.; Wang, W.; Koch, A.; Easwaran, H.; Mohammad, H.P.; Pan, X.; Vendetti, F.; VanCriekinge, W.; Demeyer, T.; Du, Z.; et al. Alterations of immune response of non-small cell lung cancer with Azacytidine. *Oncotarget* **2013**, *4*, 2067–2079. [CrossRef]
67. Zhang, Y.; Xiang, C.; Wang, Y.; Duan, Y.; Liu, C.; Zhang, Y. PD-L1 promoter methylation mediates the resistance response to anti-PD-1 therapy in NSCLC patients with EGFR-TKI resistance. *Oncotarget* **2017**, *8*, 101535–101544. [CrossRef]
68. Shin, D.S.; Zaretsky, J.M.; Escuin-Ordinas, H.; Garcia-Diaz, A.; Hu-Lieskovan, S.; Kalbasi, A.; Grasso, C.S.; Hugo, W.; Sandoval, S.; Torrejon, D.Y.; et al. Primary Resistance to PD-1 Blockade Mediated by JAK1/2 Mutations. *Cancer Discov.* **2017**, *7*, 188–201. [CrossRef]
69. Lv, H.; Lv, G.; Chen, C.; Zong, Q.; Jiang, G.; Ye, D.; Cui, X.; He, Y.; Xiang, W.; Han, Q.; et al. NAD+ Metabolism Maintains Inducible PD-L1 Expression to Drive Tumor Immune Evasion. *Cell Metab.* **2021**, *33*, 110–127.e5. [CrossRef] [PubMed]
70. Rizvi, N.A.; Hellmann, M.D.; Snyder, A.; Kvistborg, P.; Makarov, V.; Havel, J.J.; Lee, W.; Yuan, J.; Wong, P.; Ho, T.S.; et al. Faculty Opinions recommendation of Cancer immunology. Mutational landscape determines sensitivity to PD-1 blockade in non-small cell lung cancer. *Fac. Opin.* **2017**, *348*, 124–128. [CrossRef]
71. Ratti, M.; Lampis, A.; Hahne, J.C.; Passalacqua, R.; Valeri, N. Microsatellite instability in gastric cancer: Molecular bases, clinical perspectives, and new treatment approaches. *Cell. Mol. Life Sci.* **2018**, *75*, 4151–4162. [CrossRef] [PubMed]
72. Le, D.T.; Durham, J.N.; Smith, K.N.; Wang, H.; Bartlett, B.R.; Aulakh, L.K.; Lu, S.; Kemberling, H.; Wilt, C.; Luber, B.S.; et al. Mismatch repair deficiency predicts response of solid tumors to PD-1 blockade. *Science* **2017**, *357*, 409–413. [CrossRef] [PubMed]
73. Anagnostou, V.; Smith, K.N.; Forde, P.M.; Niknafs, N.; Bhattacharya, R.; White, J.; Zhang, T.; Adleff, V.; Phallen, J.; Wali, N.; et al. Evolution of Neoantigen Landscape during Immune Checkpoint Blockade in Non–Small Cell Lung Cancer. *Cancer Discov.* **2017**, *7*, 264–276. [CrossRef]
74. Yu, S.; Zhao, Z.; Chen, L.; Gu, T.; Yu, H.; Tang, H.; Wang, Q.; Wu, Y. HLA loss of heterozygosity-mediated discordant responses to immune checkpoint blockade in squamous cell lung cancer with renal metastasis. *Immunotherapy* **2021**, *13*, 195–200. [CrossRef] [PubMed]
75. Li, X.; Shao, C.; Shi, Y.; Han, W. Lessons learned from the blockade of immune checkpoints in cancer immunotherapy. *J. Hematol. Oncol.* **2018**, *11*, 1–26. [CrossRef] [PubMed]
76. Wu, W.; Wang, W.; Wang, Y.; Li, W.; Yu, G.; Li, Z.; Fang, C.; Shen, Y.; Sun, Z.; Han, L.; et al. IL-37b suppresses T cell priming by modulating dendritic cell maturation and cytokine production via dampening ERK/NF-kappaB/S6K signalings. *Acta Biochim. Biophys. Sin.* **2015**, *47*, 597–603. [CrossRef] [PubMed]
77. Emeagi, P.U.; Maenhout, S.; Dang, N.; Heirman, C.; Thielemans, K.; Breckpot, K. Downregulation of Stat3 in melanoma: Reprogramming the immune microenvironment as an anticancer therapeutic strategy. *Gene Ther.* **2013**, *20*, 1085–1092. [CrossRef]
78. Chattopadhyay, G.; Shevach, E.M. Antigen-specific induced T regulatory cells impair dendritic cell function via an IL-10/MARCH1-dependent mechanism. *J. Immunol.* **2013**, *191*, 5875–5884. [CrossRef]
79. Hargadon, K.M.; Bishop, J.D.; Brandt, J.P.; Hand, Z.C.; Ararso, Y.T.; A Forrest, O. Melanoma-derived factors alter the maturation and activation of differentiated tissue-resident dendritic cells. *Immunol. Cell Biol.* **2015**, *94*, 24–38. [CrossRef]
80. Lindenberg, J.J.; Van De Ven, R.; Lougheed, S.M.; Zomer, A.; Santegoets, S.J.; Griffioen, A.W.; Hooijberg, E.; Eertwegh, A.J.V.D.; Thijssen, V.L.; Scheper, R.J.; et al. Functional characterization of a STAT3-dependent dendritic cell-derived CD14+cell population arising upon IL-10-driven maturation. *OncoImmunology* **2013**, *2*, e23837. [CrossRef]
81. Deng, L.; Liang, H.; Xu, M.; Yang, X.; Burnette, B.; Arina, A.; Li, X.-D.; Mauceri, H.; Beckett, M.; Darga, T.; et al. STING-Dependent Cytosolic DNA Sensing Promotes Radiation-Induced Type I Interferon-Dependent Antitumor Immunity in Immunogenic Tumors. *Immunity* **2014**, *41*, 843–852. [CrossRef]
82. Ren, Z.; Guo, J.; Liao, J.; Luan, Y.; Liu, Z.; Sun, Z.; Liu, X.; Liang, Y.; Peng, H.; Fu, Y.-X. CTLA-4 Limits Anti-CD20–Mediated Tumor Regression. *Clin. Cancer Res.* **2017**, *23*, 193–203. [CrossRef]
83. Sistigu, A.; Yamazaki, T.; Vacchelli, E.; Chaba, K.; Enot, D.P.; Adam, J.; Vitale, I.; Goubar, A.; Baracco, E.E.; Remédios, C.; et al. Cancer cell–autonomous contribution of type I interferon signaling to the efficacy of chemotherapy. *Nat. Med.* **2014**, *20*, 1301–1309. [CrossRef]
84. Sucker, A.; Zhao, F.; Real, B.; Heeke, C.; Bielefeld, N.; Maβen, S.; Horn, S.; Moll, I.; Maltaner, R.; Horn, P.A.; et al. Genetic Evolution of T-cell Resistance in the Course of Melanoma Progression. *Clin. Cancer Res.* **2014**, *20*, 6593–6604. [CrossRef]
85. Sade-Feldman, M.; Jiao, Y.J.; Chen, J.H.; Rooney, M.S.; Barzily-Rokni, M.; Eliane, J.-P.; Bjorgaard, S.L.; Hammond, M.R.; Vitzthum, H.; Blackmon, S.M.; et al. Resistance to checkpoint blockade therapy through inactivation of antigen presentation. *Nat. Commun.* **2017**, *8*, 1–11. [CrossRef] [PubMed]
86. Sucker, A.; Zhao, F.; Pieper, N.; Heeke, C.; Maltaner, R.; Stadtler, N.; Real, B.; Bielefeld, N.; Howe, S.; Weide, B.; et al. Acquired IFNgamma resistance impairs anti-tumor immunity and gives rise to T-cell-resistant melanoma lesions. *Nat. Commun.* **2017**, *8*, 15440. [CrossRef] [PubMed]

87. Tumeh, P.C.; Harview, C.L.; Yearley, J.H.; Shintaku, I.P.; Taylor, E.J.M.; Robert, L.; Chmielowski, B.; Spasic, M.; Henry, G.; Ciobanu, V.; et al. PD-1 blockade induces responses by inhibiting adaptive immune resistance. *Nature* **2014**, *515*, 568–571. [CrossRef]
88. Tang, H.; Wang, Y.; Chlewicki, L.K.; Zhang, Y.; Guo, J.; Liang, W.; Wang, J.; Wang, X.; Fu, Y.-X. Facilitating T Cell Infiltration in Tumor Microenvironment Overcomes Resistance to PD-L1 Blockade. *Cancer Cell* **2016**, *30*, 500. [CrossRef] [PubMed]
89. Peng, W.; Chen, J.Q.; Liu, C.; Malu, S.; Creasy, C.; Tetzlaff, M.T.; Xu, C.; McKenzie, J.A.; Zhang, C.; Liang, X.; et al. Loss of PTEN Promotes Resistance to T Cell–Mediated Immunotherapy. *Cancer Discov.* **2016**, *6*, 202–216. [CrossRef]
90. Liu, C.; Peng, W.; Xu, C.; Lou, Y.; Zhang, M.; Wargo, J.A.; Chen, J.Q.; Li, H.S.; Watowich, S.S.; Yang, Y.; et al. BRAF Inhibition Increases Tumor Infiltration by T cells and Enhances the Antitumor Activity of Adoptive Immunotherapy in Mice. *Clin. Cancer Res.* **2013**, *19*, 393–403. [CrossRef]
91. Donia, M.; Fagone, P.; Nicoletti, F.; Andersen, R.S.; Hogdall, E.; Straten, P.T.; Andersen, M.H.; Svane, I.M. BRAF inhibition improves tumor recognition by the immune system: Potential implications for combinatorial therapies against melanoma involving adoptive T-cell transfer. *Oncoimmunology* **2012**, *1*, 1476–1483. [CrossRef]
92. Hugo, W.; Shi, H.; Sun, L.; Piva, M.; Song, C.; Kong, X.; Moriceau, G.; Hong, A.; Dahlman, K.B.; Johnson, D.B.; et al. Non-genomic and Immune Evolution of Melanoma Acquiring MAPKi Resistance. *Cell* **2015**, *162*, 1271–1285. [CrossRef] [PubMed]
93. Deken, M.A.; Gadiot, J.; Jordanova, E.S.; Lacroix, R.; Van Gool, M.; Kroon, P.; Pineda, C.; Foppen, M.H.G.; Scolyer, R.; Song, J.-Y.; et al. Targeting the MAPK and PI3K pathways in combination with PD1 blockade in melanoma. *OncoImmunology* **2016**, *5*, e1238557. [CrossRef] [PubMed]
94. Fu, C.; Liang, X.; Cui, W.; Ober-Blobaum, J.L.; Vazzana, J.; Shrikant, P.A.; Lee, K.P.; Clausen, B.E.; Mellman, I.; Jiang, A. beta-Catenin in dendritic cells exerts opposite functions in cross-priming and maintenance of CD8+ T cells through regulation of IL-10. *Proc. Natl. Acad. Sci. USA* **2015**, *112*, 2823–2828. [CrossRef] [PubMed]
95. Spranger, S.; Gajewski, T.F. A new paradigm for tumor immune escape: Beta-Catenin-Driven immune exclusion. *J. Immunother. Cancer* **2015**, *3*, 43. [CrossRef]
96. Spranger, S.; Bao, R.; Gajewski, T.F. Melanoma-Intrinsic beta-catenin signalling prevents anti-tumour immunity. *Nature* **2015**, *523*, 231–235. [CrossRef]
97. Pai, S.G.; Carneiro, B.A.; Mota, J.M.; Costa, R.; Leite, C.A.; Barroso-Sousa, R.; Kaplan, J.B.; Chae, Y.K.; Giles, F.J. Wnt/beta-catenin pathway: Modulating anticancer immune response. *J. Hematol. Oncol.* **2017**, *10*, 1–12. [CrossRef]
98. Chen, D.S.; Mellman, I. Elements of cancer immunity and the cancer-immune set point. *Nature* **2017**, *541*, 321–330. [CrossRef]
99. Allard, B.; A Beavis, P.; Darcy, P.K.; Stagg, J. Immunosuppressive activities of adenosine in cancer. *Curr. Opin. Pharmacol.* **2016**, *29*, 7–16. [CrossRef]
100. Mittal, D.; Vijayan, D.; Smyth, M.J. Overcoming Acquired PD-1/PD-L1 Resistance with CD38 Blockade. *Cancer Discov.* **2018**, *8*, 1066–1068. [CrossRef]
101. Vijayan, D.; Young, A.; Teng, M.W.; Smyth, M.J. Targeting immunosuppressive adenosine in cancer. *Nat. Rev. Cancer* **2017**, *17*, 709–724. [CrossRef] [PubMed]
102. Ngiow, S.F.; Young, A.; Jacquelot, N.; Yamazaki, T.; Enot, D.; Zitvogel, L.; Smyth, M.J. A threshold level of intratumor CD8 + T-cell PD1 expression dictates therapeutic response to anti-PD1. *Cancer Res.* **2015**, *75*, 3800–3811. [CrossRef] [PubMed]
103. Pauken, K.E.; Sammons, M.A.; Odorizzi, P.M.; Manne, S.; Godec, J.; Khan, O.; Drake, A.M.; Chen, Z.; Sen, D.R.; Kurachi, M.; et al. Epigenetic stability of exhausted T cells limits durability of reinvigoration by PD-1 blockade. *Science* **2016**, *354*, 1160–1165. [CrossRef] [PubMed]
104. Kondo, Y.; Ohno, T.; Nishii, N.; Harada, K.; Yagita, H.; Azuma, M. Differential contribution of three immune checkpoint (VISTA, CTLA-4, PD-1) pathways to antitumor responses against squamous cell carcinoma. *Oral Oncol.* **2016**, *57*, 54–60. [CrossRef]
105. Victor, C.T.-S.; Rech, A.J.; Maity, A.; Rengan, R.; Pauken, K.E.; Stelekati, E.; Benci, J.L.; Xu, B.; Dada, H.; Odorizzi, P.M.; et al. Radiation and dual checkpoint blockade activate non-redundant immune mechanisms in cancer. *Nature* **2015**, *520*, 373–377. [CrossRef] [PubMed]
106. Nishii, N.; Tachinami, H.; Kondo, Y.; Xia, Y.; Kashima, Y.; Ohno, T.; Nagai, S.; Li, L.; Lau, W.; Harada, H.; et al. Systemic administration of a TLR7 agonist attenuates regulatory T cells by dendritic cell modification and overcomes resistance to PD-L1 blockade therapy. *Oncotarget* **2018**, *9*, 13301–13312. [CrossRef]
107. Oweida, A.; Hararah, M.K.; Phan, A.V.; Binder, D.C.; Bhatia, S.; Lennon, S.; Bukkapatnam, S.; Van Court, B.; Uyanga, N.; Darragh, L.; et al. Resistance to Radiotherapy and PD-L1 Blockade Is Mediated by TIM-3 Upregulation and Regulatory T-Cell Infiltration. *Clin. Cancer Res.* **2018**, *24*, 5368–5380. [CrossRef]
108. Kugel, C.H., 3rd; Douglass, S.M.; Webster, M.R.; Kaur, A.; Liu, Q.; Yin, X.; Weiss, S.A.; Darvishian, F.; Al-Rohil, R.N.; Ndoye, A.; et al. Age correlates with response to Anti-PD1, reflecting age-related differences in intratumoral effector and regulatory T-Cell populations. *Clin. Cancer Res.* **2018**, *24*, 5347–5356. [CrossRef]
109. Sinha, P.; Clements, V.K.; Bunt, S.K.; Albelda, S.M.; Ostrand-Rosenberg, S. Cross-Talk between Myeloid-Derived Suppressor Cells and Macrophages Subverts Tumor Immunity toward a Type 2 Response. *J. Immunol.* **2007**, *179*, 977–983. [CrossRef]
110. Ostrand-Rosenberg, S. Myeloid-derived suppressor cells: More mechanisms for inhibiting antitumor immunity. *Cancer Immunol. Immunother.* **2010**, *59*, 1593–1600. [CrossRef]

111. Elkabets, M.; Ribeiro, V.S.; Dinarello, C.A.; Ostrand-Rosenberg, S.; Di Santo, J.P.; Apte, R.N.; Vosshenrich, C.A. IL-1beta regulates a novel myeloid-derived suppressor cell subset that impairs NK cell development and function. *Eur. J. Immunol.* **2010**, *40*, 3347–3357. [CrossRef] [PubMed]
112. Schlecker, E.; Stojanovic, A.; Eisen, C.; Quack, C.; Falk, C.S.; Umansky, V.; Cerwenka, A.; Xu, M.; Hadinoto, V.; Appanna, R.; et al. Tumor-Infiltrating Monocytic Myeloid-Derived Suppressor Cells Mediate CCR5-Dependent Recruitment of Regulatory T Cells Favoring Tumor Growth. *J. Immunol.* **2012**, *189*, 5602–5611. [CrossRef]
113. Huang, B.; Pan, P.-Y.; Li, Q.; Sato, A.I.; Levy, D.E.; Bromberg, J.; Divino, C.M.; Chen, S.-H. Gr-1+CD115+ Immature Myeloid Suppressor Cells Mediate the Development of Tumor-Induced T Regulatory Cells and T-Cell Anergy in Tumor-Bearing Host. *Cancer Res.* **2006**, *66*, 1123–1131. [CrossRef] [PubMed]
114. Serafini, P.; Mgebroff, S.; Noonan, K.; Borrello, I. Myeloid-Derived Suppressor Cells Promote Cross-Tolerance in B-Cell Lymphoma by Expanding Regulatory T Cells. *Cancer Res.* **2008**, *68*, 5439–5449. [CrossRef] [PubMed]
115. Shen, M.; Wang, J.; Yu, W.; Zhang, C.; Liu, M.; Wang, K.; Yang, L.; Wei, F.; Wang, S.E.; Sun, Q.; et al. A novel MDSC-induced PD-1−PD-L1+ B-cell subset in breast tumor microenvironment possesses immuno-suppressive properties. *OncoImmunology* **2018**, *7*, e1413520. [CrossRef]
116. Hou, A.; Hou, K.; Huang, Q.; Lei, Y.; Chen, W. Targeting Myeloid-Derived Suppressor Cell, a Promising Strategy to Overcome Resistance to Immune Checkpoint Inhibitors. *Front. Immunol.* **2020**, *11*, 783. [CrossRef]
117. Dammeijer, F.; Lievense, L.A.; Kaijen-Lambers, M.E.; Van Nimwegen, M.; Bezemer, K.; Hegmans, J.P.; Van Hall, T.; Hendriks, R.W.; Aerts, J.G. Depletion of Tumor-Associated Macrophages with a CSF-1R Kinase Inhibitor Enhances Antitumor Immunity and Survival Induced by DC Immunotherapy. *Cancer Immunol. Res.* **2017**, *5*, 535–546. [CrossRef]
118. Ceci, C.; Atzori, M.G.; Lacal, P.M.; Graziani, G. Targeting Tumor-Associated Macrophages to Increase the Efficacy of Immune Checkpoint Inhibitors: A Glimpse into Novel Therapeutic Approaches for Metastatic Melanoma. *Cancers* **2020**, *12*, 3401. [CrossRef]
119. Puccetti, P.; Grohmann, U. IDO and regulatory T cells: A role for reverse signalling and non-canonical NF-kappaB activation. *Nat. Rev. Immunol.* **2007**, *7*, 817–823. [CrossRef]
120. Holmgaard, R.B.; Zamarin, D.; Munn, D.H.; Wolchok, J.D.; Allison, J.P. Indoleamine 2,3-dioxygenase is a critical resistance mechanism in antitumor T cell immunotherapy targeting CTLA-4. *J. Exp. Med.* **2013**, *210*, 1389–1402. [CrossRef]
121. Khair, D.O.; Bax, H.J.; Mele, S.; Crescioli, S.; Pellizzari, G.; Khiabany, A.; Nakamura, M.; Harris, R.J.; French, E.; Hoffmann, R.M.; et al. Combining Immune Checkpoint Inhibitors: Established and Emerging Targets and Strategies to Improve Outcomes in Melanoma. *Front. Immunol.* **2019**, *10*, 453. [CrossRef]
122. Wang, Q.; Gao, J.; Di, W.; Wu, X. Anti-angiogenesis therapy overcomes the innate resistance to PD-1/PD-L1 blockade in VEGFA-overexpressed mouse tumor models. *Cancer Immunol. Immunother.* **2020**, *69*, 1781–1799. [CrossRef] [PubMed]
123. Peng, D.; Kryczek, I.; Nagarsheth, N.; Zhao, L.; Wei, S.; Wang, W.; Sun, Y.; Zhao, E.; Vatan, L.; Szeliga, W.; et al. Epigenetic silencing of TH1-type chemokines shapes tumour immunity and immunotherapy. *Nat. Cell Biol.* **2015**, *527*, 249–253. [CrossRef] [PubMed]
124. Sakuishi, K.; Apetoh, L.; Sullivan, J.M.; Blazar, B.R.; Kuchroo, V.K.; Anderson, A.C. Targeting Tim-3 and PD-1 pathways to reverse T cell exhaustion and restore anti-tumor immunity. *J. Exp. Med.* **2010**, *207*, 2187–2194. [CrossRef] [PubMed]
125. Zhou, Q.-H.; Li, K.-W.; Chen, X.; He, H.-X.; Peng, S.-M.; Peng, S.-R.; Wang, Q.; Li, Z.-A.; Tao, Y.-R.; Cai, W.-L.; et al. HHLA2 and PD-L1 co-expression predicts poor prognosis in patients with clear cell renal cell carcinoma. *J. Immunother. Cancer* **2019**, *8*, e000157. [CrossRef]
126. Rieder, S.A.; Wang, J.; White, N.; Qadri, A.; Menard, C.; Stephens, G.; Karnell, J.L.; Rudd, C.E.; Kolbeck, R. B7-H7 (HHLA2) inhibits T-cell activation and proliferation in the presence of TCR and CD28 signaling. *Cell. Mol. Immunol.* **2020**. [CrossRef] [PubMed]
127. Cheng, H.; Borczuk, A.; Janakiram, M.; Ren, X.; Lin, J.; Assal, A.; Halmos, B.; Pérez-Soler, R.; Zang, X. Wide Expression and Significance of Alternative Immune Checkpoint Molecules, B7x and HHLA2, in PD-L1–Negative Human Lung Cancers. *Clin. Cancer Res.* **2018**, *24*, 1954–1964. [CrossRef]
128. Le Mercier, I.; Chen, W.; Lines, J.L.; Day, M.; Li, J.; Sergent, P.; Noelle, R.J.; Wang, L. VISTA Regulates the Development of Protective Antitumor Immunity. *Cancer Res.* **2014**, *74*, 1933–1944. [CrossRef]
129. Bharaj, P.; Chahar, H.S.; Alozie, O.K.; Rodarte, L.; Bansal, A.; Goepfert, P.A.; Dwivedi, A.; Manjunath, N.; Shankar, P. Characterization of Programmed Death-1 Homologue-1 (PD-1H) Expression and Function in Normal and HIV Infected Individuals. *PLoS ONE* **2014**, *9*, e109103. [CrossRef] [PubMed]
130. Liu, J.; Yuan, Y.; Chen, W.; Putra, J.; Suriawinata, A.A.; Schenk, A.D.; Miller, H.E.; Guleria, I.; Barth, R.J.; Huang, Y.H.; et al. Immune-checkpoint proteins VISTA and PD-1 nonredundantly regulate murine T-cell responses. *Proc. Natl. Acad. Sci. USA* **2015**, *112*, 6682–6687. [CrossRef] [PubMed]
131. Andrews, L.P.; Marciscano, A.E.; Drake, C.G.; Vignali, D.A.A. LAG3 (CD223) as a cancer immunotherapy target. *Immunol. Rev.* **2017**, *276*, 80–96. [CrossRef]
132. Dumic, J.; Dabelic, S.; Flögel, M. Galectin-3: An open-ended story. *Biochim. Biophys. Acta (BBA)—Gen. Subj.* **2006**, *1760*, 616–635. [CrossRef]
133. Xu, F.; Liu, J.; Liu, D.; Liu, B.; Wang, M.; Hu, Z.; Du, X.; Tang, L.; He, F. LSECtin Expressed on Melanoma Cells Promotes Tumor Progression by Inhibiting Antitumor T-cell Responses. *Cancer Res.* **2014**, *74*, 3418–3428. [CrossRef] [PubMed]

134. Woo, S.-R.; Turnis, M.E.; Goldberg, M.V.; Bankoti, J.; Selby, M.; Nirschl, C.J.; Bettini, M.L.; Gravano, D.M.; Vogel, P.; Liu, C.L.; et al. Immune Inhibitory Molecules LAG-3 and PD-1 Synergistically Regulate T-cell Function to Promote Tumoral Immune Escape. *Cancer Res.* **2012**, *72*, 917–927. [CrossRef] [PubMed]
135. Rowshanravan, B.; Halliday, N.; Sansom, D.M. CTLA-4: A moving target in immunotherapy. *Blood* **2018**, *131*, 58–67. [CrossRef]
136. D'Aniello, C.; Berretta, M.; Cavaliere, C.; Rossetti, S.; Facchini, B.A.; Iovane, G.; Mollo, G.; Capasso, M.; Della Pepa, C.; Pesce, L.; et al. Biomarkers of Prognosis and Efficacy of Anti-angiogenic Therapy in Metastatic Clear Cell Renal Cancer. *Front. Oncol.* **2019**, *9*, 1400. [CrossRef] [PubMed]
137. Wang, J.; Sun, J.; Liu, L.N.; Flies, D.B.; Nie, X.; Toki, M.; Zhang, J.; Song, C.; Zarr, M.; Zhou, X.; et al. Siglec-15 as an immune suppressor and potential target for normalization cancer immunotherapy. *Nat. Med.* **2019**, *25*, 656–666. [CrossRef]
138. Kim, T.K.; Herbst, R.S.; Chen, L. Defining and Understanding Adaptive Resistance in Cancer Immunotherapy. *Trends Immunol.* **2018**, *39*, 624–631. [CrossRef]
139. Lucca, L.E.; Axisa, P.-P.; Singer, E.R.; Nolan, N.M.; Dominguez-Villar, M.; Hafler, D.A. TIGIT signaling restores suppressor function of Th1 Tregs. *JCI Insight* **2019**, *4*, 124427. [CrossRef]
140. Wu, L.; Mao, L.; Liu, J.-F.; Chen, L.; Yu, G.-T.; Yang, L.-L.; Wu, H.; Bu, L.-L.; Kulkarni, A.B.; Zhang, W.-F.; et al. Blockade of TIGIT/CD155 Signaling Reverses T-cell Exhaustion and Enhances Antitumor Capability in Head and Neck Squamous Cell Carcinoma. *Cancer Immunol. Res.* **2019**, *7*, 1700–1713. [CrossRef]
141. Chauvin, J.-M.; Ka, M.; Pagliano, O.; Menna, C.; Ding, Q.; DeBlasio, R.; Sanders, C.; Hou, J.; Li, X.-Y.; Ferrone, S.; et al. IL15 Stimulation with TIGIT Blockade Reverses CD155-mediated NK-Cell Dysfunction in Melanoma. *Clin. Cancer Res.* **2020**, *26*, 5520–5533. [CrossRef] [PubMed]
142. Blessin, N.C.; Simon, R.; Kluth, M.; Fischer, K.; Hube-Magg, C.; Li, W.; Makrypidi-Fraune, G.; Wellge, B.; Mandelkow, T.; Debatin, N.F.; et al. Patterns of TIGIT Expression in Lymphatic Tissue, Inflammation, and Cancer. *Dis. Markers* **2019**, *2019*, 1–13. [CrossRef]
143. Chen, Y.-L.; Lin, H.-W.; Chien, C.-L.; Lai, Y.-L.; Sun, W.-Z.; Chen, C.-A.; Cheng, W.-F. BTLA blockade enhances Cancer therapy by inhibiting IL-6/IL-10-induced CD19high B lymphocytes. *J. Immunother. Cancer* **2019**, *7*, 313. [CrossRef]
144. Xu, X.; Hou, B.; Fulzele, A.; Masubuchi, T.; Zhao, Y.; Wu, Z.; Hu, Y.; Jiang, Y.; Ma, Y.; Wang, H.; et al. PD-1 and BTLA regulate T cell signaling differentially and only partially through SHP1 and SHP2. *J. Cell Biol.* **2020**, *219*, 201905085. [CrossRef] [PubMed]
145. Chen, C.; Liu, Y.; Cui, B. Effect of radiotherapy on T cell and PD-1 / PD-L1 blocking therapy in tumor microenvironment. *Hum. Vaccines Immunother.* **2021**, 1–13. [CrossRef]
146. Muth, S.T.; Saung, M.T.; Blair, A.B.; Henderson, M.G.; Thomas, D.L.; Zheng, L. CD137 agonist-based combination immunotherapy enhances activated, effector memory T cells and prolongs survival in pancreatic adenocarcinoma. *Cancer Lett.* **2021**, *499*, 99–108. [CrossRef]
147. Tiragolumab Impresses in Multiple Trials. *Cancer Discov.* **2020**, *10*, 1086–1087. [CrossRef]
148. Xia, G.-Q.; Lei, T.-R.; Yu, T.-B.; Zhou, P.-H. Nanocarrier-based activation of necroptotic cell death potentiates cancer immunotherapy. *Nanoscale* **2021**, *13*, 1220–1230. [CrossRef] [PubMed]

Review
CTLA-4 in Regulatory T Cells for Cancer Immunotherapy

Navid Sobhani [1,*], Dana Rae Tardiel-Cyril [1], Aram Davtyan [2], Daniele Generali [3], Raheleh Roudi [4] and Yong Li [1,*]

1 Department of Medicine, Section of Epidemiology and Population Sciences, Baylor College of Medicine, Houston, TX 77030, USA; dcyril@bcm.edu
2 Atomwise, 717 Market St, San Francisco, CA 94103, USA; aram@atomwise.com
3 Department of Medical, Surgery and Health Sciences, University of Trieste, 34147 Trieste, Italy; dgenerali@units.it
4 Department of Medicine, University of Minnesota Medical School, Minneapolis, MN 55455, USA; Roudi002@umn.edu
* Correspondence: Navid.Sobhani@bcm.edu (N.S.); Yong.Li@bcm.edu (Y.L.)

Simple Summary: In the fight against cancer, immunotherapies have given great hope after encouraging results in clinical investigations showing complete remission in some patients with melanoma. In fact, directing the immune system against cancer has been a very innovative strategy fostered during the past three decades. Despite this fact, the disease is serious, the mortality is still very high, and only a minority of patients are responsive to immunotherapies. Therefore, there is a need for a better understanding of the molecular mechanisms of resistance to immune checkpoint inhibitors such as antibodies against cytotoxic T-lymphocyte-associated protein 4 (CTLA-4). In this article, we discuss the molecular mechanism of CTLA-4 in T regulatory cell inhibition, while highlighting the knowledge gap.

Abstract: Immune checkpoint inhibitors (ICIs) have obtained durable responses in many cancers, making it possible to foresee their potential in improving the health of cancer patients. However, immunotherapies are currently limited to a minority of patients and there is a need to develop a better understanding of the basic molecular mechanisms and functions of pivotal immune regulatory molecules. Immune checkpoint cytotoxic T-lymphocyte-associated protein 4 (CTLA-4) and regulatory T (T_{reg}) cells play pivotal roles in hindering the anticancer immunity. T_{reg} cells suppress antigen-presenting cells (APCs) by depleting immune stimulating cytokines, producing immunosuppressive cytokines and constitutively expressing CTLA-4. CTLA-4 molecules bind to CD80 and CD86 with a higher affinity than CD28 and act as competitive inhibitors of CD28 in APCs. The purpose of this review is to summarize state-of-the-art understanding of the molecular mechanisms underlining CTLA-4 immune regulation and the correlation of the ICI response with CTLA-4 expression in T_{reg} cells from preclinical and clinical studies for possibly improving CTLA-4-based immunotherapies, while highlighting the knowledge gap.

Keywords: CTLA-4; T_{reg} cells: immune checkpoint inhibitors; CD28; antigen-presenting cells

1. Introduction

Globally, cancer remains the leading cause of mortality and morbidity, with nearly 9 million deaths every year [1]. Early diagnosis and advances in cancer treatment have improved the survival of cancer patients, but there were more than 1.7 million new cases of cancer in the United States in 2019 [1]. A considerable percentage of these patients manifested drug resistance, metastasis, and recurrence [2].

A promising paradigm in the dilemma and challenge of cancer therapy is immunotherapy, and the T cell population has generated considerable enthusiasm among scientists due to its ability to kill malignant tumor cells directly [3].

There are two major types of T cell: Conventional adaptive T cells (including helper CD4+ T cells [Th1, Th2, Th17, Th9, and Tfh], cytotoxic CD8+ T cells, memory T cells, and regulatory CD4+ T cells [T_{reg}]) and innate-like T cells (including natural killer T cells, mucosal associated invariant T cells, and gamma delta T cells ($\gamma\delta$ T cells)) [4]. The CD4+ T cells subset can target malignant tumor cells using different approaches, either by directly killing tumor cells or indirectly modulating tumor microenvironments (TME) [5,6]. These cells can increase the response of cytotoxic T cells (CTL) and quality of B cells [7]. The major killers of tumor cells are cytotoxic CD8+ T cells [8].

Innate-like T cells, representing one of the major groups of T cells, can be grouped into natural killer T cells (NKT cells), mucosal associated invariant cells (MAIT), and gamma delta T cells ($\gamma\delta$ T cells) [9–11]. During development, innate-like T cells, called innate lymphoid cells (ILCs)-Natural Killer (NK) cells, acquire an effector function, whereas conventional T cells remain in a naive state [12]. The first group, NKT cells, express T-cell receptors (TCRs) and cell surface markers of NK cell lineages [13]. They are involved in the recognition of glycolipid antigens and present them to antigen-presenting cells (APCs) in the context of major histocompatibility complex (MHC) class I-associated protein CD1d [14]. T cells with $\gamma\delta$ expression, representing the first layer of defense, constitute nearly 2% of the T cell population in peripheral blood and secondary lymphoid organs, while they are mainly found in the epithelia of the skin, gut, lung, and other organs [15,16]. Another group of innate-like T cells, called MAIT cells, constitute approximately 5% of all T cells and have considerable similarities to NKT cells [17,18].

T_{reg} cells are one of the most fascinating immunosuppressive subsets of CD4+ (CD25+) T cells, mainly represented by master transcription factor 3 (FOXP3), and they account for nearly 5% of the total CD4+ T cell population under normal conditions [19]. T_{reg} cells increase dramatically in response to the early stages of malignant tumor initiation and growth [20]. In the tumor microenvironment, T_{reg} cells can suppress the immune system activity of cytotoxic T lymphocytes (CTLs) [21]. A panel of immune-modulatory receptors expressed on the T_{reg} cell population includes cytotoxic T lymphocyte antigen 4 (CTLA-4), the vascular endothelial growth factor receptor (VEGFR), and programmed cell death protein 1 (PD1) [22]. CTLA-4 is expressed on activated T and T_{reg} cells [23,24] https://paperpile.com/c/d61gxv/defR (accessed on 5 February 2021). Atkins et al. showed that an immune checkpoint blockade of CTLA-4 improved the survival rate of renal cell carcinoma, melanoma, non-small cell lung cancer (NSCLC), and head and neck squamous cell cancer [25]. This protein was the second receptor of the T-cell costimulatory ligand CD80/86 and, therefore, an immune checkpoint whose function is critical for downmodulating the immune response. In contrast to the first receptor (CD28), which is antigen-dependent, CTLA-4 is antigen-independent [26]. In 2011, ipilimumab was the first immunotherapy drug targeting CTLA-4 to receive FDA approval to treat late-stage melanoma [27]. This approval came after encouraging results of a large randomized phase III clinical trial improving patients' survival compared to standard therapy. Since then, several immunotherapies targeting the PD-1/PD-L1 axis have received FDA approval to treat multiple types of cancer [27].

This review will describe the mechanisms of CTLA-4 immune checkpoint inhibition, the role of T_{reg} cells in tumorigenesis, and how anti-CTLA-4 antibodies can provoke an alteration in the expression of CTLA-4 on T_{reg} cells while exerting anti-cancer therapeutic activity.

2. Mechanism of CTLA-4 Immune System Inhibition

A better understanding of the biological mechanisms and functions of negative and positive co-stimulatory molecules has been shown to be essential for improving current and potentially new CTLA-4 or Programmed Cell Death 1 (PD-1) inhibitors for anti-cancer immunotherapies.

Once bound to B7-1 (CD80) or B7-2 (CD86), CTLA-4 switches-off antigen-presenting cells [28]. CTLA-4 was immediately increased after T-cell receptor (TCR) engagement, reaching its highest level of expression as a homodimer at 2–3 days after the activation of conventional CD4+ and CD8+ T cells [29,30]. CTLA-4 competes with costimulatory molecule CD28 for the CD80/86 ligands CD80 and CD86, for which it has a higher affinity and avidity [31,32]. It is necessary to inhibit interactions with both CD80 and CD86 with antibodies to optimally block the CD28-dependent proliferation of T cells in an allogenic mixed lymphocyte reaction stimulated with B lymphoblastoid cell lines. Since both CD80 and CD86 exert a positive costimulatory signal through CD28, the role played by CTLA-4 in the competitive inhibition of CD28 is important for attenuating T-cell activation, thereby fine-tuning the immune response [33]. Rapid binding kinetics with a very fast dissociation rate constant (k_{off}) of both CTLA-4 and CD28 to CD80 has been observed ($k_{off} \geq 1.6$ and ≥ 0.43 s^{-1}) [34], which permits their instant competition. The function of T cells can be suppressed by T_{reg} cells through multiple mechanisms [35]. T_{reg} cells constitutively express CTLA-4 on their suppressive functions. CTLA-4-expressing T cells (T_{reg} or activated conventional T cells) have been shown to lower levels of CD80/86 costimulatory molecules available on APCs by CTLA-4-dependent sequestration via trans-endocytosis [36]. This event can negatively regulate the proliferation of non-T_{reg} T cells, as well as the production of cytokines.

RAG2-deficient mice reconstituted with CTLA-4-deficient bone marrow developed lethal inflammation of multiple organs and died around 10 weeks after reconstitution, whereas control mice (reconstituted with normal bone marrow) were healthy. Intriguingly, the mouse chimeras reconstituted with a mixture of normal and CTLA-4-deficient bone marrow remained healthy, without developing any disease [37]. The authors concluded that the disease observed in CTLA-4$^{-/-}$ mice is not due to a T cell autonomous defect and that CTLA-4 triggering on normal T cells produces factors inhibiting the disease induced by CTLA-4-deficient T cells. It has been shown that mice selectively deficient in CTLA-4 in T_{reg} cells (Foxp3+) develop systemic lymphoproliferation and fatal T cell-mediated autoimmune disease, indicating that T_{reg} cells critically require CTLA-4 to suppress immune responses and maintain immunological self-tolerance [38,39].

Additionally, after T-cell activation by TCR, CTLA-4 within intracellular compartments is immediately transported to the immunologic synapse [40]. The stronger the TCR signaling, the more CTLA-4 transported to the immunological synapse [40]. After reaching the synapse, CTLA-4 becomes stable through its binding to the CD80 and CD86 ligands, leading to its accumulation and effective out-competition against CD28 [28]. Differences in both the affinity and avidity in ligand-binding cause selective CD28 or CTLA-4 recruitment to the immunological synapse. The major ligand leading to CTLA-4 localization in the synapse is CD80, while for CD28, it is CD86 [28]. In this way, CTLA-4 attenuates the positive co-stimulation of CD28, thereby limiting the downstream signaling of CD28, which is primarily achieved through PI3K and AKT [41,42]. This mechanism allows a fine-tuning of TCR signaling and therefore T-cell activity. The negative co-stimulation of CTLA-4 is intrinsically linked to CD80/86 and CD28 positive co-stimulations. CTLA-4 mainly regulates T cells at priming sites (e.g., gut or lymphoid organs such as spleen and lymph nodes). Since CTLA-4 plays a crucial function in the activation of T cells, its negative co-stimulation plays a critical role in tolerance. As a matter of fact, the biallelic genetic *Ctla-4* deletion in mice leads to their death at 3–4 weeks of age because of pronounced lymphoproliferation with multi-organ lymphocytic infiltration and tissue destruction, particularly with pancreatitis and myocarditis [43–45]. Mice lethality can therefore be prevented by normal T cell factors. Several groups foster the idea that extrinsic cell suppressive functions of CTLA-4 are mainly mediated through T_{reg} cells [38,46]. Others support the idea that CTLA-4's ability to inhibit T cells is T_{reg} cell-independent [47,48]. An argument for the first line of thought is that a particular loss of CTLA-4 in T_{reg} cells was enough to induce abnormal T-cell activation and autoimmunity [38,49]. In fact, Wing et al. showed that the loss of CTLA-4 in T_{reg} cells was capable of hyper producing immunoglobulin E, systemic lymphoproliferation, fatal T cell-mediated autoimmune disease, and powerful tumor immunity [38]. After losing the CTLA-4-expressing subpopulation, the T_{reg} cells were not capable of exerting their T cell suppressive functions; in particular, they were not able to down-modulate the dendritic cell expressions of CD80 and CD86 [38]. It must be noted that the lack of CTLA-4 in T_{reg} cells also leads to an aberrant expression and expansion of T_{conv} cells, which can cause the latter cells to infiltrate and fatally damage nonlymphoid tissues and cells [49]. Therefore, CTLA-4 in T_{reg} cells is also needed to prevent the accumulation of T cells that may harm vital organs.

As a hypothetical molecular biology explanation, it is possible that T_{reg} cells with CTLA-4 may limit the availability of CD80/86 ligands for the positive co-stimulation of CD28 in effector T cells. Through such a mechanism, the CTLA-4 would indirectly and cell-extrinsically dampen T-cell activation. It is also known that CTLA-4 on effector T cells can trans-compete for CD80/86 ligands [50]. Another mechanism by which CTLA-4 can lower the total availability of CD80/86 ligands is through APC-mediated trans-endocytosis of CD80/86 ligands [36]. The last two mechanisms explain how CTLA-4 could prevent anti-cancer immune reactions without the need for T_{reg} cells. Overall, it is noteworthy that these mechanisms are not yet fully understood and each contribution remains elusive in the context of cancer immunity and drug design.

Furthermore, unexpectedly, the depletion of CTLA-4 from a T_{reg} cell population of adult mice conferred resistance to autoimmune encephalomyelitis (EAE) and did not enhance anti-tumor immunity [51]. This was accompanied by an expansion of functional CTLA-4-deficient T_{reg} cells expressing immunosuppressive molecules (IL-10, LAG-3, and PD-1) capable of protecting them from EAE, demonstrating that CTLA-4, in addition to previously described mechanisms of action, has a T_{reg}-intrinsic effect in limiting T_{reg} expansion.

Additionally, since CTLA-4 expression has been correlated with the TCR signal strength, high T_{reg} cell and CTLA-4 expressions are concomitant [52,53]. The inhibition efficacy of any cell by CTLA-4 depends on the affinity between the major histocompatibility complex (pMHC) ligand and its TCR. The higher the affinity of TCRs, the more those cells can be inhibited through CTLA-4 [54,55]. Additionally, the induction of CTLA-4 also restricts CD4+ T-helper clonal expansion. Ultimately, through such a mechanism of action of CTLA-4, the TCR signal is fine-tuned in response to specific immunological threats.

Furthermore, a number of structures of the extracellular domain of human CTLA-4 are available in Protein Data Bank (PDB), including apo structures and various complexes. The very first structure of CTLA-4 was determined using solution NMR spectroscopy (PDB ID: 1AH1), revealing an Ig-like V (variable)-type domain, where two beta-sheets of the V-fold are connected by two disulfide bonds (21 to 94 and 48 to 68) [56]. Another apo structure of CTLA-4 was later published in the physiological dimeric state (PDB ID: 3OSK) [57]. CTLA-4 binds its native ligands CD80 and CD86 at the A 'GFCC' face, which contains a number of charged residues that are highly conserved between CTLA-4 and CD28 (and across species). A key role in these interactions is also played by the $_{99}MYPPPY_{104}$ loop connecting F and G strands [56]. The structures of CTLA-4 with CD80 and CD86 (PDB IDs: 1I8L and 1I85) manifested a mostly convex binding surface at CTLA-4, free of any notable cavities that could have been targeted with traditional small-molecule campaigns [58,59]. It is also interesting to note that while the CD80-bound conformation of CTLA-4 is very similar to the apo form, CD86 binding requires some structural rearrangement, most significantly, in the FG loop [57–59]. Finally, several structures of CTLA-4 bound to monoclonal antibodies have also recently been reported (PDB IDs: 5GGV, 5TRU, 5XJ3, and 6RP8) [60–62]. These structures reveal that ipilimumab and tremelimumab directly compete with CD80 and CD86 at their binding surface, sterically displacing and preventing their interactions with CTLA-4. Moreover, subtle differences in the CTLA-4 structure, such as a slightly larger distance between G and F stands, and extended interactions of antibodies with non-conserved residues on the opposite side of the FG loop, enable selectivity between CTLA-4 and CD28 [61]. Interestingly, the amino acid sequence of the intracellular tail of CTLA-4 is conserved in 100% of all mammalian species, meaning that its intracellular domain must have an important role in the inhibition of T-cell activation [63,64]. In fact, the inhibitory functions of CTLA-4, by competing with CD28 for CD80 and CD86 or through its transmission of negative signals, can be accomplished because of its intracellular domain, but such a downstream mechanism of CTLA-4 signal transduction deserves further investigations [64,65]. Based on the primary amino acid sequence of the CTLA-4 cytoplasmic region, there are two potential binding sites for Src homology domain 2 (SH2) and an SH3 potential binding motif [66]. CTLA-4 was found to be capable of becoming associated with SH2-containing tyrosine phosphatase-2 (SHP-2) through the SH2 domain of SHP-2. Such an association resulted in phosphatase activity against Ras regulatory protein p52SHC [67]. Therefore, CTLA-4 might be able to start a signal transduction cascade leading to the dephosphorylation of TCR-associated kinases or substrates.

While the antitumor activity and clinical benefits of antibodies such as ipilimumab that block CTLA-4 interactions with ligands have been demonstrated [61], it is always desirable to have bioavailable and cheaper options in the form of small molecules or peptides. In cases of traditionally undruggable targets, such as CTLA-4, where no suitable small-molecule binding pockets can be immediately identified at the ligand-binding interface, peptide drugs can present a viable alternative. Like antibodies, peptides can achieve a high affinity and specificity by capturing a larger interaction area with the target. At the same time, they are easier to synthesize and have greater tissue penetration due to their smaller size compared to the antibodies. Moreover, peptides have recommended themselves in a variety of therapeutic areas, including cancer [68,69]. In addition, targets similar to CTLA-4 can be amenable to less-standard small molecule campaigns. One such approach is allosteric modulation. In this case, a small molecule bound to a distant site can activate or inhibit the protein function or its interactions with other molecules as a result of structural changes that it induces at a distance [70]. However, for CTLA-4, such sites still have to be determined through either experimental or computational techniques [71,72].

3. Regulatory T Cells and Anticancer Immunity

3.1. First Insights into T_{reg} Cells

T_{reg} cells are a population of CD4 T cells constitutively expressing CTLA-4. They are crucial for both immune-oncology and autoimmunity, as we will describe in this review. The focus of this article is on the CTLA-4-positive population of T_{reg} cells in cancer. After T_{reg} cells were discovered for the first time in the CD4+ CD25+ T cell subpopulation in 1995 [73], mutations of *FOXP3* recapitulated the impaired formation or improper function of T_{reg} cells, causing an immune dysregulation syndrome in mice, termed polyendocrinopathy enteropathy X-linked syndrome, which ultimately leads to multiple autoimmune disorders [74]. Corroborating the importance of T_{reg} cells for a functional immune response, mice carrying spontaneous alterations of *Foxp3*—that ultimately lacked T_{reg} cells—died due to systemic autoimmunity [75,76]. As expected, the external expression of FOXP3 bestowed naïve CD4+ T cells (T_{conv}, without T_{reg} cells) with the same immune-suppressive capacity typical of T_{reg} cells. Therefore, FOXP3 is a master transcription factor that regulates T_{reg} cell phenotypes and their function as immunosuppressants. The role of T_{reg} cells in cancer is mainly observed at inflammatory sites, where they migrate and inactivate different types of effector T cells, such as CD4$^+$ T helper (T_H) cells and CD8$^+$ cytotoxic T cells (CTLs) [77–80]. As a consequence, intervening in this activity of T_{reg} cells could induce the immune system in the fight against cancer.

3.2. Inhibitory Effects of Treg Cells on APC

T_{reg} cells represent a crucial component of the immune system, being essential for controlling self-tolerance, and thereby play essential roles in various medical conditions. T_{reg} cells have a crucial role in the suppression of the immune response in cancer [73,75,81–85]. T_{reg} cells inhibit APC by three main mechanisms: (1) Depleting immune-stimulating cytokines [86–89]; (2) producing immunosuppressive cytokines (like TGF-β, IL-10, and IL-35); and (3) constitutively expressing CTLA-4. T_{reg} cells express Interleukin 2 (IL2) receptors that bind to IL2, thereby limiting the amount of this cytokine available for T_{conv}

cells [90,91]. As a consequence, the constitutive expression of CTLA-4 blocks the priming and activation of T_{conv} cells to APCs [38,92].

Figure 1 summarizes the role of CTLA-4 in T_{reg} cells modulating T_{conv} activation.

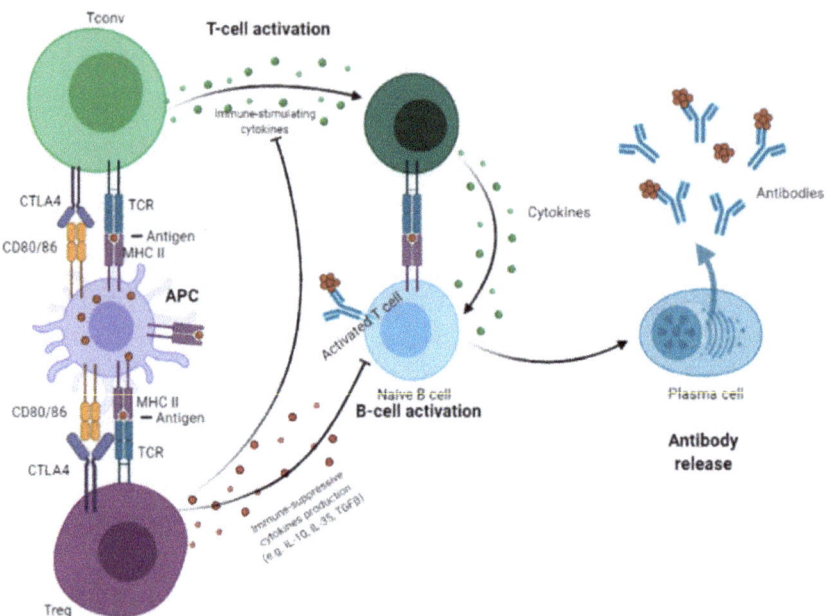

Figure 1. Regulatory T (T_{reg}) cells inhibit antigen-presenting cells (APC) by three main mechanisms: (1) Depleting immune-stimulating cytokines; (2) producing immunosuppressive cytokines (e.g., TGF-β, IL-10, and IL-35); and (3) constitutively expressing CTLA-4, which blocks the priming and activation of naïve CD4+ T (T_{conv}) cells to APCs.

T_{reg} cells block normal protective immune-surveillance and inhibit the antitumor immune response in cancer patients. Thereby, if T_{conv} cells are like tumor suppressors, T_{reg} cells could be considered as oncogenes because they are suppressing antitumor immunity [81,82,93,94], although the definitions of oncogenes and tumor suppressors refer to genes in tumors that, when expressed, cause or prevent cancer, respectively [95]. Likewise, CTLA-4 and PD-1 immune checkpoints, since they block the immune system's recognition of cancer cells, could also be comparable as tumor suppressors.

3.3. Conflicting Roles of T_{reg} Cells in Malignant Tumors

The role of T_{reg} cells in immunoncology was discovered by two Japanese groups in 1999 [93,94]. The two groups independently reported that anti-CD25 antibodies, capable of depleting CD4+CD25+ T_{reg} cells, led to higher tumor rejection and retarded tumor growth in normal and T cell reconstituted nude mice [93,94]. CD25 is the α chain of the interleukin-2 receptor. Onizuka et al. showed that a single dose (less than 0.125 mg) of anti-CD25 was capable of causing the regression of multiple tumors derived from four different inbred mouse strains (five leukemias, myeloma, and two sarcomas) [93]. Similarly, Shimizu et al. showed that the elimination of CD25-expressing T cells caused a powerful immune response in syngeneic tumors in mice, leading to tumor regression within 1 month, thereby allowing the host to survive > 80 days [94]. Among CD4+ T cells, the percentage of T_{reg} cells is higher in the blood of cancer patients compared to that of healthy individuals [83,96,97]. Expectedly, the relatively higher T_{reg} cell levels in the tumor microenvironment correlated with a poor prognosis in various cancer types, such as melanoma and non-small cell lung, ovarian, and gastric cancers [82,83]. The T_{reg} cell population is not large in the periphery blood of cancer patients compared with the TME, implying that T cells' interaction with tumor cells is important [97]. On the contrary, in certain tumors, such as colorectal cancer (CRC), a high level of FOXP3+ T cells is correlated with a better prognosis [98]. This is because the accumulation of FOXP3+ occurs together with inflammatory cytokines, possibly implying that T_{reg} cells play a role in repressing tumor inflammation. It was brought to light that two populations of FOXP3 (+) CD4 (+) T cells had distinct roles in controlling the prognosis of CRCs, contributing in opposing ways. FOXP3 (hi) T_{reg} cells are correlated with worse survival, whereas FOXP3 (lo) non-T_{reg} T cells are correlated with better survival. This is possibly because the FOXP3+ (lo) non-T_{reg} T cell population leads to an inflammatory TME against the tumor. In fact, it was observed that FOXP3+ non-T_{reg} T cells in CRCs are correlated with high levels of tumor necrosis factor (TNF), IL2, and TGFβ [96]. Depleting FOXP3 (hi) T_{reg} cells from tumor tissues could be deployed to increase the antitumor immunity to treat CRC or other cancers, whereas other strategies enhancing the levels of FOXP3(lo) non-T_{reg} T cells could also be used to suppress or prevent tumorigenesis [96].

There are conflicting reports regarding the prognostic value of tumor-infiltrating T_{reg} cells. Shang et al. demonstrated that FOXP3+ T_{reg} cells are correlated with shorter overall survival in breast, hepatocellular, gastric, melanoma, renal, and cervical cancers, and longer overall survival in head and neck, colorectal, and esophageal cancers, whilst they display no correlation for pancreatic and ovarian cancers [99].

In conclusion, T_{reg} cells inhibit anti-cancer immunity, blocking the immune surveillance of tumors, which ultimately leads to cancer spreading [81–83,93,94]. Immunosuppressive T_{reg} cells, producing cytokines, are observed in both human chronic inflammatory disease and cancers, where they promote tumorigenesis through a mechanism similar to that of chronic inflammation [48,100,101]. The depletion of T_{reg} cells in mice is capable of promoting lymphocyte recruitment and as a consequence, a decrease in the tumor growth rate and the presence of high endothelial venules, indicating destruction of the tumor tissues [102,103].

3.4. T_{reg} Cells and the Tumor Microenvironment

The TME is mainly comprised of a subpopulation of T_{reg} cells called bona fide T_{reg} cells that enhance the expression of immunosuppressant molecules such as CTLA-4 and T-cell immunoreceptors with Ig and ITIM domains (also called TIGIT), whose expression is very low in naïve T_{reg} cells [83,96,104]. A transcriptome analysis of 15 human lung cancer samples and 14 colorectal cancer samples demonstrated that tumor-infiltrating T_{reg} cells have very high levels of different T_{reg} activation markers, such as T cell immunoglobulin mucin receptor 3 (HAVCR2), glucocorticoid-induced TNFR-related protein (GIRT), lymphocyte-activation gene 3 protein (LAG3), and inducible T cell co-stimulator (ICOS). Interestingly, this phenotype was not observed in peripheral blood samples from the same patients, whose expression levels in the blood remained the same. This could indicate that T_{reg} cells become activated in TME, where they exert their immune-suppressive functions [105].

3.5. Cross-Talk between T_{reg} Cells and the Tumor Microenvironment

It has recently been shown that adenosine produced by apoptotic T_{reg} cells present within the TME exerts higher immunosuppressive effects compared to live T_{reg} cells [21,106]. A weak NRF2-associated antioxidant pathway leads to a vulnerable system against reactive oxygen species in TME, possibly causing apoptosis in T_{reg} cells, which is a process that has been shown to convert high ATP levels into adenosine through T_{reg} cell-expressed ectoenzymes CD39 and CD73. In turn, the resulting abundance in adenosine engages purinergic adenosine A2A receptors (also known as ARORA2A), which is a family of G protein-coupled receptors with seven transmembrane alpha helices whose function is to regulate the oxygen demand and increase vasodilatation, as well as suppress immune cells. Apoptotic T_{reg} cells use the A2A pathway to suppress immune cells [21,106]. The mechanism postulated to explain the activation of T_{reg} cells in TME is that proliferating and dying tumor cells have loads of self-antigens, which are best recognized through T_{reg} cells and thereby become activated in TME [107]. Another explanation comes from results from mice experiments of two research groups showing that immune dendritic cells expressed in mice tumors activate T_{reg} cells in a TGFβ-dependent manner [107,108]. T_{reg} cells recognize specific self-antigens and can become clonally expanded in TME [109,110]. T_{reg} cells typically have higher affinity TCRs for self-antigens than T_{conv} cells and therefore, should be predominantly activated, even when in competition with T_{conv} cells. It must be stated, however, that these data come from animal studies and T_{reg} cells induced by TFGβ have not yet been fully demonstrated in humans. As for the epigenetic profile of tumor-infiltrating T_{reg} cells, very little is understood [111–113]. Epigenetic studies of T_{reg} cells are limited and future studies could shed more light on the subject, in order to better understand the origin and mechanisms of activation of T_{reg} cells. T_{reg} cells move to the TME by chemotaxis via chemokines and their receptors, such as CXCL12-CXCR4, CCL5-CCR5, CCL22-CCR4, and CCL1-CCR8 [83,105,114–118]. Blocking such chemotactic signals can reduce the accumulation of T_{reg} cells inside tumors [119]. Such chemokines are produced in the TME by the tumor and/or macrophages [83,105,114–116]. Additionally, some chemokines, such as CCL1 and CCL22, can be produced within tumors by exhausted or dysfunctional CD8+ T cells [119,120]. Therapies targeting chemokines could be considered to lower the T_{reg}:T_{conv} ratio in the tumor microenvironment, in order to produce more T_{conv} and less T_{reg} cells. Cancers engage various immune escape mechanisms that can be dependent on specific tumor intrinsic factors. In fact, alterations in tumor suppressor PTEN; Liver Kinase B1 (LKB1); or oncogenes WNT/β-catenin, KRAS, or basic leucine zipper transcriptional factor ATF-like 3 (BATF3), affect effector T-cell recruitment to the tumors [121–125]. On the contrary, tumor hyper-activation of FAK leads to a recruitment of T_{reg} cells, together with chemokine-driven CD8+ T cell exhaustion or poor infiltration within the tumor [126,127]. In fact, Jiang et al., using tissues from pancreatic ductal adenocarcinoma (PDAC) patients, observed that FAK was elevated and correlated with high levels of fibrosis and poor CD8+ cytotoxic T cell infiltration, which are signs of an immune-suppressive TME. The use of a FAK inhibitor (VS-4718) substantially limited tumor progression and doubled the survival of a humanized mice model of PDAC [126]. In squamous cell carcinoma (SCC) cells, FAK was shown to drive the exhaustion of CD8+ T cells and recruitment of T_{reg} cells in TME through the regulation of chemokines/cytokines and ligand-receptor networks (such as Ccl5/Ccr5), ultimately permitting tumor growth. FAK kinase inhibitor VS-4718 drove T_{reg} cell depletion and promoted the anti-tumor response of CD8+ T cells [127].

3.6. Treg Cells and Nonself Antigens

At the location of tumor cells, there are two types of antigens recognized by T_{reg} cells: Shared antigens and neoantigens. The first ones arise from highly or aberrantly expressed endogenous proteins encoded by the germ line. The second ones derive from either abnormal self-proteins formed from somatic genetic alterations or from oncogenic viral proteins. Experiments with animals have shown that T_{reg} cells primed to nonself antigens increased the affinity of CD8+ T cells, most likely by the inhibition of T cells carrying TCRs with low-avidity to antigens [128]. APCs can render CD8+ T cells targeting self-antigens self-tolerant through the control of T_{reg} cells [129]. In fact, using human T cells in vitro, the authors showed that T_{reg} cells were able to make the self-reactive human CD8+ T cells anergic upon antigen stimulation. In addition, they observed the proliferative activity of self-antigen-specific T cells in CTLA-4+ and CTLA-4- fractions. The CTLA-4+ fraction was highly proliferative, had a low expression level of BCL2, and was prone to death upon self-antigen stimulation. On the contrary, T_{reg} cells did not suppress non-self-specific CD8+ T cells [129]. Therefore, T_{reg} cell-mediated immunosuppression could be more effective in shared antigen-expressing tumors compared to those expressing neoantigens. This could be a reason why tumors expressing neoantigens respond better to immune checkpoint blocking than tumors with a low mutational burden [130,131]. One of the major aims of immunotherapy research is to understand why some cancer patients respond very well to immune checkpoint inhibitions while others do not, as well as discovering new biomarkers useful for just-in-time determination of treatment-responsive patients, before administrating immunotherapies.

4. Correlation between Anti-CTLA-4 Treatment and Its Effect on T_{reg} Cells

The anti-CTLA-4 monoclonal antibody ipilimumab (Yervoy, Bristol-Meyers Squibb) gained FDA approval in March 2011 for the treatment of advanced melanoma, which is the most dangerous type of skin cancer, after a large randomized phase III clinical trial consisting of 676 patients demonstrated that ipilimumab improved the overall survival (OS) of melanoma patients who did not respond to standard therapy. In fact, the median OS in 403 patients randomly assigned to receive 3 mg/kg ipilimumab with an investigational vaccine made of HLA-A*01201-restricted glycoprotein 100 with incomplete Freund' adjuvant was 10.0 months (gp100, 95% Confidence Interval [CI], 8.5–11.5) vs. 6.4 months observed for 136 patients treated with gp100 only (Hazard Ration [HR] for death = 0.68; p = 0.001). In total, 137 patients were treated with ipilimumab alone and had an OS of 10.1 months vs. 6.4 months for the gp100 alone group (95% CI, 9.0–13.8; HR for death = 0.66, p = 0.003) [132]. After its approval, the drug was added as a category 1 recommendation in the National Comprehensive Cancer Network (NCCN) guidelines for the systemic treatment of advanced or metastatic melanoma.

This clinical evidence shows that the antibody enhanced the ability of the immune system to attack cancer through CTLA-4 inhibition. It must be mentioned that adverse events occurred in 10–15% of patients treated with ipilimumab alone compared to patients treated with gp100 only [132].

In 2014, another pivotal phase III clinical trial (CA184-024) including 502 metastatic melanoma patients tested ipilimumab. The current standard of care treatment for the disease is chemotherapy (decarbazine), which has not been shown to increase OS. Interestingly, the treatment of patients with 850 mg/m^2 decarbazine with 10 mg/kg ipilimumab improved OS compared to an arm with only chemotherapy with the placebo. The OS of patients treated with ipilimumab plus decarbazine vs. decarbazine plus placebo was 47.3% vs. 36.3% at the first year, 28.5% vs. 17.9% at the second year, and 20.8% vs. 12.2% at the third year (HR for death with ipilimumab/decarbazine, 0.72; p < 0.001). The risk of progressing through the disease decreased by 24% when using ipilimumab/decarbazine vs. decarbazine/placebo (HR for progression, 0.76; p = 0.006). The ratios of the disease to control were similar for the two groups (33.2% for ipilimumab/decarbazine and 30.2% for decarbazome/placebo; p = 0.41). This study was important because it showed how ipilimuamb could be used as the first line treatment for metastatic melanoma [133]. The study tested a higher concentration (10 mg/kg) of ipilimuab than the approved 3 mg/kg [134]. Consequently, more adverse events were observed using higher doses of anti-CTLA-4, possibly because of CTLA-4 molecular degradation. In fact, CTLA-4 is needed to prevent immune-related adverse reactions and its degradation can be deleterious.

Interestingly, a recent report demonstrates that the immune-related Adverse Events (irAEs) of ipilimumab and alike result from the lysosomal degradation of CTLA-4 in T_{reg} cells. The study used the CTLA-4 mutant (Y201V), which is incapable of being recycled because it lacks interaction with the lipopolysaccharide (LPS)-responsive and beige-like anchor protein (LRBA). This indicates that the specific region of CTLA-4 is an essential mediator of CTLA-4 recycling. The investigators made antibodies targeting CTLA-4 (HL12 and HL32) that were not able to degrade the CTLA-4 of T_{reg} cells. In fact, in contrast to ipilimumab or TremeIgG1, the use of novel anti-CTLA-4 antibodies had no effect on the CTLA-4 level of T_{reg} cells in the same model. Additionally, HL12 and HL32 could more effectively lead to tumor rejection, with fewer irAEs in mice [135]. Such knowledge is useful for the generation of novel antibodies or molecules that could inhibit CLTA-4 without eliciting its degradation and could therefore be used in combination with other PD-1 or PD-L1 inhibitors with less toxicity.

Various studies show that consolidated or novel types of CTLA-4 therapies correlate with different expression levels of T_{reg} cells. Ji et al. showed that the treatment of mice with 0.25 mg anti-CTLA-4 monoclonal antibody correlated with a lower level of the CD25+Foxp3+ T_{reg} cell population ($p < 0.05$) [136]. Qu et al. observed that anti-CTLA-4 monoclonal antibodies enhanced IL36-stimulated antitumor activity by depleting T_{regs} in the tumor [137]. Mihic-Probst et al. showed that anti-CTLA-4 antibody ipilimumab, anti PD-1 antibody nivolumab, or pembrolizumab decreased the number of CD25+ T_{reg} cells [138]. Sun et al. observed that the number of T_{reg} cells decreased after treating mice with anti-CTLA-4 or anti PD-1 antibodies in an HPV16 E6/E7^{+} syngeneic mouse tumor model [139]. Kvarnhammar et al. showed that new IgG1 bispecific anti-CTLA-4 and anti-OX40 induced the activation of T cells and T_{reg} cell depletion in vitro and in vivo in the tumor [140]. Sharma et al., using samples from 19 melanoma, 17 prostate, and 9 bladder cancer patients treated with ipilimumab and 18 samples from melanoma cancer patients treated with tremelimumab, observed that the monoclonal antibodies depleted intratumoral FOXP3 T_{reg} cells in tumors [48]. Pai et al. devised a dual variable domain immunoglobulin of the anti-CTLA-4 antibody (anti-CTLA-4 DVD) possessing an outer tumor-specific antigen-binding site engineered to shield the inner anti-CTLA-4-binding domain. The latter only became available upon reaching the tumor after cleavage of the construct by proteases present in the tumor. In a preclinical tumor model, treatment with the anti-CTLA-4 DVD led to the depletion of tumor-resident T_{reg} cells, while preserving tissue-resident T_{reg} cells, resulting in an efficient antitumor response with a reduced multi-organ immune toxicity [141]. Morris et al. observed that anti-CTLA-4 antibodies IgG2a and IgG2b isotypes of 9D9 clone decreased the number of T_{reg} cells in syngeneic murine tumors of B78 melanoma and/or Panc02 pancreatic cancer [142]. Duperret et al. observed that, upon treatment with anti CTLA-4 in combination with a TERT DNA vaccine administered once a week for four rounds of immunization in C57BL/6 mice, the level of T_{reg} cells decreased within the tumors, while it remained unchanged within the peripheral blood [143]. Tang et al. observed, through IHC and quantitative real-time PCR, that the anti-CTLA-4 monoclonal antibody decreased the presence of T_{reg} cells in the mice tumor microenvironment, but not in peripheral lymphoid organs [144]. Son et al. showed that the anti-CTLA-4 antibody and radiotherapy suppressed CD25 T_{reg} cells in C57BL mice injected with lung cancer [145]. Schwarz et al. investigated the effect of using different doses of anti-CTLA-4 in the presence of T_{reg} cells in mice. They used a low dose of 0.25 mg CTLA-Ig antibody (LD, 10 mg/kg body weight), high dose of 1.25 mg CTLA-Ig antibody (HD, 50 mg/kg body weight), and very high dose of 6.25 mg CTLA-Ig antibody (VHD, 250 mg/kg body weight). T_{reg} cell levels decreased, independently of the doses [146]. Marabelle et al., using a combination of anti-CTLA-4 and anti-OX40 with CpG therapy, observed a reduction of T_{reg} cells in tumors [147].

Interestingly, Du et al. observed that anti-CTLA-4 antibodies are capable of efficiently inducing T_{reg} cell depletion and tumor regression in mice [148].

In contrast, several other groups reported an increase of T_{reg} cells in cancers after anti-CTLA-4 treatment. In fact, Sandi et al. observed that high dose treatment of anti-CTLA-4 increased the accumulation of T_{reg} cells in secondary lymphoid organs [149]. Kavanagh et al. observed that the anti-CTLA-4 antibody ipilimumab in four cohorts of patients increased T_{reg} cell levels in a dose-dependent manner. The drug was administered every 28 days [150]. Quezada et al. observed that a CTLA-4 blockade with GM-CSF combination immunotherapy in an in vivo B16/BL6 mouse model of melanoma led to a self-expansion of T_{reg} cells in tumors [47]. The reason for such discrepancies between the last four studies and the majority of studies described in the previous paragraphs remains unknown. A possible explanation could be that different subpopulations of T_{reg} cells were detected by the groups, such as bona fide and naïve Treg cells, or that the organisms' TMEs of either animals or humans were different across the different experimental settings.

Of note, CTLA-4 has two opposing and crucial properties in cancer and autoimmunity. For self-tolerance it is important to have functional CTLA-4. Current antibodies developed against CTLA-4 have the property of reducing the levels of CTLA-4 by 50% by lysosomal degradation, which is directly responsible for their toxicity [135]. Therefore, since CTLA-4 is crucial for preventing autoimmunity, which is the major cause of irAE triggered by monoclonal antibodies such as ipilimumab and tremelimumab [135], new drugs should be developed considering such a gap. Encouraging results have already been produced by Zhang et al. HL12 and HL32 anti-CTLA-4 antibodies did not change the CTLA-4 level total or that in the T_{reg} cell fraction, while exerting powerful anti-CTLA-4-induced tumor inhibition [135]. Table 1 summarizes all the studies investigating anti-CTLA-4 therapies' effect on T_{reg} cell levels.

Table 1. Effects of anti-CTLA-4 therapy on T_{reg} cells.

Reference	Anti-CTLA-4 Therapy and Samples	Effect on the Presence of T_{reg} Cells
Ji et al. 2020 [136]	In vivo investigated effect of administration of 0.25 mg anti-CTLA-4 monoclonal antibody on the CD25+Foxp3+ population in spleens and tumor tissues.	Decreased T_{reg} cells ($p < 0.05$) in tumor. It did not in spleen.
Qu et al. 2020 [137]	CTLA-4 monoclonal antibodies.	Decreased T_{reg} cells in tumors.
Probst et al. 2020 [138]	All patients received anti-CTLA-4 therapy and four received additional anti-PD1 therapy.	Decreased T_{reg} cells in tumors.
Zhang et al. 2019 [135]	In vivo anti-CTLA-4 therapy ipililumab and TremeIgG1 standard and HL12 and HL32 experimental antibodies.	Ipilimumab and TremeIgG1 downregulated cell-surface and total CTLA-4 levels in T_{reg} cells from spleen and lung. In contrast, HL12 and HL32 had no effect on CTLA-4 level of T_{reg} cells in the same model.
Sun et al. 2019 [139]	In vivo anti–CTLA-4 antibody.	Downregulation of T_{reg} cells in tumors of mice.
Kvarnhammar et al. 2019 [140]	CTLA-4 x OX40 bispecific antibody. ATOR-1015 was used in vivo.	Reduced the frequency of T_{reg} cells in vitro and at the tumor site in vivo.
Sharma et al. 2019 [48]	Nineteen melanoma patient, 17 prostate cancer patient, and 9 bladder cancer patient samples were treated with ipilimumab. Eighteen melanoma tumors were treated with tremelimumab.	mAbs depleted intratumoral FOXP3+ T_{reg} cells in tumors via Fc-dependent mechanisms.
Pai et al. 2019 [141]	Anti CTLA-4 DVD Ig tetravalent bispecific antibody-like antibody containing an Fc region and two pairs of variable domains joined in tandem by a short flexible linker.	Decreased T_{reg} cells in mouse tumors, but not in tissues.
Tang et al. 2019 [144]	Anti-CTLA-4 monoclonal antibody.	Increase of T_{reg} cells in tumors.
Morris et al. 2018 [142]	Anti-CTLA-4 (IgG2a and IgG2b isotypes of the 9D9 clone)	Decreased T_{reg} cells in tumors.
Duperret et al. 2018 [143]	Anti-CTLA-4 with a TERT DNA vaccine in vivo in C57BL/6 mice. Mice were immunized at 1-week intervals for a total of four immunizations.	Decreased T_{reg} cell frequency within the tumor. No decrease in peripheral blood.
Du et al. 2018 [148]	In vivo anti-CTLA-4 antibodies binding to human-like ipilimumab.	T_{reg} cell depletion.
Son et al. 2017 [145]	Anti-CTLA-4 antibody therapy and radiotherapy in vivo.	Suppression of T_{reg} cells in tumors.
Schwarz et al. 2016 [146]	In vivo anti-CTLA-4 low dose (0.25 mg), high dose (1.25 mg), and very high dose (6.25 mg) were given to mice.	CD25 T_{reg} cells were reduced independently of the doses.
Sandin et al. 2014 [149]	In vivo comparison of low-dose peritumoral and high-dose systemic CTLA-4 blockade therapy.	As opposed to low-dose therapy, high-dose systemic therapy stimulated accumulation of T_{reg} cells in secondary lymphoid organs. This could counteract immunotherapeutic benefit of CTLA-4 blockade.

Table 1. *Cont.*

Reference	Anti-CTLA-4 Therapy and Samples	Effect on the Presence of T_{reg} Cells
Marabelle et al. 2013 [147]	In vivo anti-CTLA-4 and anti-OX40 with CpG.	Depleted T_{reg} cells in tumors.
Sandin et al. 2010 [149]	In vivo anti-CTLA-4 or anti-PD-1 with CpG therapy.	The combinations reduced numbers of T_{reg} cells at tumor site.
Kavanagh et al. 2007 [150]	In vivo anti-CTLA-4 antibody dose escalation.	Increased T_{reg} cells in tumors in a dose-dependent manner.
Quezada et al. 2006 [47]	In vivo CTLA-4 blockade and GM-CSF combination immunotherapy mice model B16/BL6 melanoma.	Led to self-expansion of T_{reg} cells in tumors.

Moreover, in clinical routines, it should also be considered that T cells are made of multiple subpopulations with their own peculiar effects. The modulation of T_{reg} cells and/or T_{eff} cells and pro-inflammatory responses is critical for cancer. An immunosuppressive state (increased T_{reg} and/or decreased T_{eff}) may facilitate the growth and spread of abnormal cancer cells. Therefore, the $T_{reg}:T_{eff}$ ratio could be used in a clinical setting. The new checkpoint inhibitors attempt to pharmacologically modulate the $T_{reg}:T_{eff}$ ratio in the treatment of cancer therapy. However, in cancer progression, the expression of co-inhibitory molecules by tumors favors an imbalance in the tumor microenvironment toward an immune suppression status by increasing T_{reg} infiltration and decreasing T_{eff} activity [151]. On the contrary, the ratio of $T_{reg}:T_{eff}$ should be in favor of T_{reg} depletion and an increase of activated effector T cells, in order to potentiate an anti-tumor response [152]. Tremelimumab was shown to improve the proliferative response of T_{eff} and to abrogate the T_{reg} suppressive ability, suggesting that monitoring these populations may allow for the proper selection of responsive patients from those who would not obtain a benefit from immunotherapy [153]. With regards to the patients' management, it seems to be crucial to understand and monitor the "ping-pong" effect produced by treatment of the $T_{reg}:T_{eff}$ ratio in the regulation of autoimmunity and anti-tumor immunity. Clinicians should pay attention to monitoring this effect in order to maintain an effective anti-tumor response and immune homeostasis preventing the onset of IRAEs [154].

5. Conclusive Remarks and Future Directions

In conclusion, most studies have shown that anti-CTLA-4 antibodies mainly depleted T_{reg} cells in cancers, whereas very few have observed that the number of T_{reg} cells increased or remained the same because of different experimental settings or in some cases, the design of their therapeutic agents. It is generally known that T_{reg} cells inhibit anti-cancer immunity, blocking the immune surveillance of tumors, ultimately leading to cancer growth. In our opinion, antibodies or small molecules that inhibit CTLA-4, but do not alter CTLA-4 levels in T_{reg} cells, could be innovative and ultimately more effective in eradicating cancer cells. In fact, such drugs would not cause the degradation of CTLA-4 and consequently, do not interfere with T_{reg} cells' function in preventing autoimmunity. Consequently, the inhibition of CTLA-4 could be achieved without the degradation of CTLA-4 and adverse related events caused by toxicity. Testing their efficiency, together with other checkpoint inhibitors, such as anti-PD1 and anti-PD-L1, could further improve the therapy efficacy.

Author Contributions: N.S. conceptualized and drafted the initial version of the manuscript, researched the literature, and edited the manuscript; D.R.T.-C. contributed to the draft and revised the manuscript; A.D. contributed as a chemist, providing insights on the mechanisms of CTLA-4 inhibition in immunotherapies; D.G. contributed as an oncologist to the relevance of immunotherapies for cancer and revised the manuscript; R.R. revised the literature and edited the manuscript; Y.L. improved the idea, and revised and finalized the manuscript. All authors have read and agreed to the published version of the manuscript.

Funding: This research was financed by the seed fund provided to Y.L. from Baylor College of Medicine and a Recruitment of Established Investigator Award from Cancer Prevention & Research Institute of Texas (grant number RR190043). The funds had no influence on the content of this review.

Institutional Review Board Statement: "Not applicable" for studies not involving humans or animals.

Informed Consent Statement: "Not applicable" for studies not involving humans.

Data Availability Statement: Data sharing not applicable. No new data were created or analyzed in this study. Data sharing is not applicable to this article.

Conflicts of Interest: The authors declare no conflict of interest.

References

1. Miller, K.D.D.; Fidler-Benaoudia, M.; Keegan, T.H.H.; Hipp, H.S.S.; Jemal, A.; Siegel, R.L.L. Cancer statistics for adolescents and young adults, 2020. *CA Cancer J. Clin.* **2020**, *70*, 443–459. [CrossRef]
2. Siegel, R.L.; Miller, K.D.; Jemal, A. Cancer statistics, 2019. *CA Cancer J. Clin.* **2019**, *69*, 7–34. [CrossRef]
3. Speiser, D.E.E.; Ho, P.C.C.; Verdeil, G. Regulatory circuits of T cell function in cancer. *Nat. Rev. Immunol.* **2016**, *16*, 599–611.
4. Zou, W. Regulatory T cells, tumour immunity and immunotherapy. *Nat. Rev. Immunol.* **2006**, *6*, 295–307.
5. Kennedy, R.; Celis, E. Multiple roles for CD4+ T cells in anti-tumor immune responses. *Immunol. Rev.* **2008**, *222*, 129–144.
6. Melssen, M.; Slingluff, C.L. Vaccines targeting helper T cells for cancer immunotherapy. *Curr. Opin. Immunol.* **2017**, *47*, 85–92.
7. Borst, J.; Ahrends, T.; Bąbała, N.; Melief, C.J.M.; Kastenmüller, W. CD4+ T cell help in cancer immunology and immunotherapy. *Nat. Rev. Immunol.* **2018**, *18*, 635–647.
8. Raskov, H.; Orhan, A.; Christensen, J.P.; Gögenur, I. Cytotoxic CD8+ T cells in cancer and cancer immunotherapy. *Br. J. Cancer* **2021**, *124*, 359–367.
9. Raverdeau, M.; Cunningham, S.P.; Harmon, C.; Lynch, L. γδ T cells in cancer: A small population of lymphocytes with big implications. *Clin. Transl. Immunol.* **2019**, *8*. [CrossRef]
10. Bennstein, S.B. Unraveling natural killer T-cells development. *Front. Immunol.* **2018**, *8*, 1950.
11. Toubal, A.; Nel, I.; Lotersztajn, S.; Lehuen, A. Mucosal-associated invariant T cells and disease. *Nat. Rev. Immunol.* **2019**, *19*, 643–657.
12. Bedoui, S.; Gebhardt, T.; Gasteiger, G.; Kastenmüller, W. Parallels and differences between innate and adaptive lymphocytes. *Nat. Immunol.* **2016**, *17*, 490–494.
13. Kronenberg, M. Toward an understanding of NKT cell biology: Progress and paradoxes. *Annu. Rev. Immunol.* **2005**, *23*, 877–900.
14. Van Kaer, L.; Postoak, J.L.; Wang, C.; Yang, G.; Wu, L. Innate, innate-like and adaptive lymphocytes in the pathogenesis of MS and EAE. *Cell. Mol. Immunol.* **2019**, *16*, 531–539.
15. Vantourout, P.; Hayday, A. Six-of-the-best: Unique contributions of γδ T cells to immunology. *Nat. Rev. Immunol.* **2013**, *13*, 88–100.
16. Chien, Y.H.; Meyer, C.; Bonneville, M. γδ T cells: First line of defense and beyond. *Annu. Rev. Immunol.* **2014**, *32*, 121–155.
17. Lantz, O.; Legoux, F. MAIT cells: An historical and evolutionary perspective. *Immunol. Cell Biol.* **2018**, *96*, 564–572. [CrossRef]
18. Keller, A.N.; Corbett, A.J.; Wubben, J.M.; McCluskey, J.; Rossjohn, J. MAIT cells and MR1-antigen recognition. *Curr. Opin. Immunol.* **2017**, *46*, 66–74.
19. Mougiakakos, D.; Johansson, C.C.C.; Trocme, E.; All-Ericsson, C.; Economou, M.A.A.; Larsson, O.; Seregard, S.; Kiessling, R. Intratumoral forkhead box p3-positive regulatory T cells predict poor survival in cyclooxygenase-2-positive uveal melanoma. *Cancer* **2010**, *116*, 2224–2233. [CrossRef]
20. Piccirillo, C.A.A. Regulatory T cells in health and disease. *Cytokine* **2008**, *43*, 395–401.
21. Maj, T.; Wang, W.; Crespo, J.; Zhang, H.; Wang, W.; Wei, S.; Zhao, L.; Vatan, L.; Shao, I.; Szeliga, W.; et al. Oxidative stress controls regulatory T cell apoptosis and suppressor activity and PD-L1-blockade resistance in tumor. *Nat. Immunol.* **2017**, *18*, 1332–1341. [CrossRef]
22. Vanamee, É.S.S.; Faustman, D.L.L. TNFR2: A Novel Target for Cancer Immunotherapy. *Trends Mol. Med.* **2017**, *23*, 1037–1046.
23. Zeng, G.; Jin, L.; Ying, Q.; Chen, H.; Thembinkosi, M.C.C.; Yang, C.; Zhao, J.; Ji, H.; Lin, S.; Peng, R.; et al. Regulatory T cells in cancer immunotherapy: Basic research outcomes and clinical directions. *Cancer Manag. Res.* **2020**, *12*, 10411–10421.
24. Togashi, Y.; Shitara, K.; Nishikawa, H. Regulatory T cells in cancer immunosuppression—Implications for anticancer therapy. *Nat. Rev. Clin. Oncol.* **2019**, *16*, 356–371.
25. Atkins, M.B.B.; Clark, J.I.I.; Quinn, D.I.I. Immune checkpoint inhibitors in advanced renal cell carcinoma: Experience to date and future directions. *Ann. Oncol. Off. J. Eur. Soc. Med. Oncol.* **2017**, *28*, 1484–1494.
26. Foell, J.; Hewes, B. T Cell Costimulatory and Inhibitory Receptors as Therapeutic Targets for Inducing Anti-Tumor Immunity. *Curr. Cancer Drug Targets* **2007**, *7*, 55–70. [CrossRef]
27. Wei, S.C.; Duffy, C.R.; Allison, J.P. Fundamental mechanisms of immune checkpoint blockade therapy. *Cancer Discov.* **2018**, *8*, 1069–1086. [CrossRef]
28. Pentcheva-Hoang, T.; Egen, J.G.; Wojnoonski, K.; Allison, J.P. B7-1 and B7-2 selectively recruit CTLA-4 and CD28 to the immunological synapse. *Immunity* **2004**, *21*, 401–413. [CrossRef]

29. Walunas, T.L.; Lenschow, D.J.; Bakker, C.Y.; Linsley, P.S.; Freeman, G.J.; Green, J.M.; Thompson, C.B.; Bluestone, J.A. CTLA-4 can function as a negative regulator of T cell activation. *Immunity* **1994**, *1*, 405–413. [CrossRef]
30. Brunner, M.C.; Chambers, C.A.; Chan, F.K.; Hanke, J.; Winoto, A.; Allison, J.P. CTLA-4-Mediated inhibition of early events of T cell proliferation—PubMed. Available online: https://pubmed.ncbi.nlm.nih.gov/10229815/ (accessed on 18 November 2020).
31. Linsley, P.S.; Brady, W.; Urnes, M.; Grosmaire, L.S.; Damle, N.K.; Ledbetter, J.A. CTLA4 is a second receptor for the b cell activation antigen B7. *J. Exp. Med.* **1991**, *174*, 561–569. [CrossRef]
32. Linsley, P.S.; Greene, J.A.L.; Brady, W.; Bajorath, J.; Ledbetter, J.A.; Peach, R. Human B7-1 (CD80) and B7-2 (CD86) bind with similar avidities but distinct kinetics to CD28 and CTLA-4 receptors. *Immunity* **1994**, *1*, 793–801. [CrossRef]
33. Lanier, L.L.; O'Fallon, S.; Somoza, C.; Phillips, J.H.; Linsley, P.S.; Okumura, K.; Ito, D.; Azuma, M. CD80 (B7) and CD86 (B70) provide similar costimulatory signals for T cell proliferation, cytokine production, and generation of CTL. *J. Immunol.* **1995**, *154*, 97–105.
34. Van Der Merwe, P.A.; Bodian, D.L.; Daenke, S.; Linsley, P.; Davis, S.J. CD80 (B7-1) binds both CD28 and CTLA-4 with a low affinity and very fast kinetics. *J. Exp. Med.* **1997**, *185*, 393–403. [CrossRef]
35. Yi, J.; Kawabe, T.; Sprent, J. New insights on T-cell self-tolerance. *Curr. Opin. Immunol.* **2020**, *63*, 14–20.
36. Qureshi, O.S.; Zheng, Y.; Nakamura, K.; Attridge, K.; Manzotti, C.; Schmidt, E.M.; Baker, J.; Jeffery, L.E.; Kaur, S.; Briggs, Z.; et al. Trans-endocytosis of CD80 and CD86: A molecular basis for the cell-extrinsic function of CTLA-4. *Science* **2011**, *332*, 600–603. [CrossRef]
37. Bachmann, M.F.; Köhler, G.; Ecabert, B.; Mak, T.W.; Kopf, M. Cutting Edge: Lymphoproliferative Disease in the Absence of CTLA-4 is Not T Cell Autonomous—PubMed. Available online: https://pubmed.ncbi.nlm.nih.gov/10415006/ (accessed on 19 November 2020).
38. Wing, K.; Onishi, Y.; Prieto-Martin, P.; Yamaguchi, T.; Miyara, M.; Fehervari, Z.; Nomura, T.; Sakaguchi, S. CTLA-4 control over Foxp3+ regulatory T cell function. *Science* **2008**, *322*, 271–275. [CrossRef]
39. Yang, Y.; Li, X.; Ma, Z.; Wang, C.; Yang, Q.; Byrne-Steele, M.; Hong, R.; Min, Q.; Zhou, G.; Cheng, Y.; et al. CTLA-4 expression by B-1a B cells is essential for immune tolerance. *Nat. Commun.* **2021**, *12*, 1–17. [CrossRef]
40. Egen, J.G.; Allison, J.P. Cytotoxic T lymphocyte antigen-4 accumulation in the immunological synapse is regulated by TCR signal strength. *Immunity* **2002**, *16*, 23–35. [CrossRef]
41. Pagès, F.; Ragueneau, M.; Rottapel, R.; Truneh, A.; Nunes, J.; Imbert, J.; Olive, D. Binding of phosphatidyl-inositol-3-OH kinase to CD28 is required for T-cell signalling. *Nature* **1994**, *369*, 327–329. [CrossRef]
42. Kane, L.P.; Andres, P.G.; Howland, K.C.; Abbas, A.K.; Weiss, A. Akt provides the CD28 costimulatory signal for up-regulation of IL-2 and IFN-γ but not TH2 cytokines. *Nat. Immunol.* **2001**, *2*, 37–44. [CrossRef]
43. Chambers, C.A.; Cado, D.; Truong, T.; Allison, J.P. Thymocyte development is normal in CTLA-4-deficient mice. *Proc. Natl. Acad. Sci. USA* **1997**, *94*, 9296–9301. [CrossRef]
44. Waterhouse, P.; Penninger, J.M.; Timms, E.; Wakeham, A.; Shahinian, A.; Lee, K.P.; Thompson, C.B.; Griesser, H.; Mak, T.W. Lymphoproliferative disorders with early lethality in mice deficient in Ctla-4. *Science* **1995**, *270*, 985–988. [CrossRef]
45. Tivol, E.A.; Borriello, F.; Schweitzer, A.N.; Lynch, W.P.; Bluestone, J.A.; Sharpe, A.H. Loss of CTLA-4 leads to massive lymphoproliferation and fatal multiorgan tissue destruction, revealing a critical negative regulatory role of CTLA-4. *Immunity* **1995**, *3*, 541–547. [CrossRef]
46. Friedline, R.H.; Brown, D.S.; Nguyen, H.; Kornfeld, H.; Lee, J.; Zhang, Y.; Appleby, M.; Der, S.D.; Kang, J.; Chambers, C.A. CD4+ regulatory T cells require CTLA-4 for the maintenance of systemic tolerance. *J. Exp. Med.* **2009**, *206*, 421–434. [CrossRef]
47. Quezada, S.A.; Peggs, K.S.; Curran, M.A.; Allison, J.P. CTLA4 blockade and GM-CSF combination immunotherapy alters the intratumor balance of effector and regulatory T cells. *J. Clin. Investig.* **2006**, *116*, 1935–1945. [CrossRef]
48. Sharma, A.; Subudhi, S.K.; Blando, J.; Scutti, J.; Vence, L.; Wargo, J.; Allison, J.P.; Ribas, A.; Sharma, P. Anti-CTLA-4 immunotherapy does not deplete Foxp3 þ regulatory T cells (Tregs) in human cancers. *Clin. Cancer Res.* **2019**, *25*, 1233–1238. [CrossRef]
49. Jain, N.; Nguyen, H.; Chambers, C.; Kang, J. Dual function of CTLA-4 in regulatory T cells and conventional T cells to prevent multiorgan autoimmunity. *Proc. Natl. Acad. Sci. USA* **2010**, *107*, 1524–1528. [CrossRef]
50. Corse, E.; Allison, J.P. Cutting Edge: CTLA-4 on Effector T Cells Inhibits In Trans. *J. Immunol.* **2012**, *189*, 1123–1127. [CrossRef]
51. Paterson, A.M.; Lovitch, S.B.; Sage, P.T.; Juneja, V.R.; Lee, Y.; Trombley, J.D.; Arancibia-Cárcamo, C.V.; Sobel, R.A.; Rudensky, A.Y.; Kuchroo, V.K.; et al. Deletion of CTLA-4 on regulatory T cells during adulthood leads to resistance to autoimmunity. *J. Exp. Med.* **2015**, *212*, 1603–1621. [CrossRef]
52. Doyle, A.M.; Mullen, A.C.; Villarino, A.V.; Hutchins, A.S.; High, F.A.; Lee, H.W.; Thompson, C.B.; Reiner, S.L. Induction of cytotoxic T lymphocyte antigen 4 (CTLA-4) restricts clonal expansion of helper T cells. *J. Exp. Med.* **2001**, *194*, 893–902. [CrossRef]
53. Egen, J.G.; Kuhns, M.S.; Allison, J.P. CTLA-4: New insights into its biological function and use in tumor immunotherapy. *Nat. Immunol.* **2002**, *3*, 611–618.
54. Busch, D.H.; Pamer, E.G. T cell affinity maturation by selective expansion during infection. *J. Exp. Med.* **1999**, *189*, 701–709. [CrossRef]
55. Savage, P.A.; Boniface, J.J.; Davis, M.M. A kinetic basis for T cell receptor repertoire selection during an immune response. *Immunity* **1999**, *10*, 485–492. [CrossRef]

56. Metzler, W.J.; Bajorath, J.; Fenderson, W.; Shaw, S.Y.; Constantine, K.L.; Naemura, J.; Leytze, G.; Peach, R.J.; Lavoie, T.B.; Mueller, L.; et al. Solution structure of human CTLA-4 and delineation of a CD80/CD86 binding site conserved in CD28. *Nat. Struct. Biol.* **1997**, *4*, 527–531. [CrossRef]
57. Yu, C.; Sonnen, A.F.P.; George, R.; Dessailly, B.H.; Stagg, L.J.; Evans, E.J.; Orengo, C.A.; Stuart, D.I.; Ladbury, J.E.; Ikemizu, S.; et al. Rigid-body ligand recognition drives cytotoxic T-lymphocyte antigen 4 (CTLA-4) receptor triggering. *J. Biol. Chem.* **2011**, *286*, 6685–6696. [CrossRef]
58. Stamper, C.C.; Zhang, Y.; Tobin, J.F.; Erbe, D.V.; Ikemizu, S.; Davis, S.J.; Stahl, M.L.; Seehra, J.; Somers, W.S.; Mosyak, L. Crystal structure of the B7-1/CTLA-4 complex that inhibits human immune responses. *Nature* **2001**, *410*, 608–611. [CrossRef]
59. Schwartz, J.C.D.; Zhang, X.; Fedorov, A.A.; Nathenson, S.G.; Almo, S.C. Structural basis for co-stimulation by the human CTLA-4/B7-2 complex. *Nature* **2001**, *410*, 604–608. [CrossRef]
60. Zhang, F.; Qi, X.; Wang, X.; Wei, D.; Wu, J.; Feng, L.; Cai, H.; Wang, Y.; Zeng, N.; Xu, T.; et al. Structural basis of the therapeutic anti-PD-L1 antibody atezolizumab. *Oncotarget* **2017**, *8*, 90215–90224. [CrossRef]
61. Ramagopal, U.A.; Liu, W.; Garrett-Thomson, S.C.; Bonanno, J.B.; Yan, Q.; Srinivasan, M.; Wong, S.C.; Bell, A.; Mankikar, S.; Rangan, V.S.; et al. Structural basis for cancer immunotherapy by the first-in-class checkpoint inhibitor ipilimumab. *Proc. Natl. Acad. Sci. USA* **2017**, *114*, E4223–E4232. [CrossRef]
62. He, M.; Chai, Y.; Qi, J.; Zhang, C.W.H.; Tong, Z.; Shi, Y.; Yan, J.; Tan, S.; Gao, G.F. Remarkably similar CTLA-4 binding properties of therapeutic ipilimumab and tremelimumab antibodies. *Oncotarget* **2017**, *8*, 67129–67139. [CrossRef]
63. Hueber, A.J.; Matzkies, F.G.; Rahmeh, M.; Manger, B.; Kalden, J.R.; Nagel, T. CTLA-4 lacking the cytoplasmic domain costimulates IL-2 production in T-cell hybridomas. *Immunol. Cell Biol.* **2006**, *84*, 51–58. [CrossRef]
64. Rudd, C.E. The reverse stop-signal model for CTLA4 function. *Nat. Rev. Immunol.* **2008**, *8*, 153–160.
65. Thompson, C.B.; Allison, J.P. The emerging role of CTLA-4 as an immune attenuator. *Immunity* **1997**, *7*, 445–450.
66. Waterhouse, P.; Marengère, L.E.M.; Mittrücker, H.W.; Mak, T.W. CTLA-4, a negative regulator of T-Lymphocyte activation. *Immunol. Rev.* **1996**, *153*, 183–207.
67. Marengère, L.E.M.; Waterhouse, P.; Duncan, G.S.; Mittrücker, H.W.; Feng, G.S.; Mak, T.W. Regulation of T cell receptor signaling by tyrosine phosphatase SYP association with CTLA-4. *Science* **1996**, *272*, 1170–1173. [CrossRef]
68. Wang, H.; Sun, Y.; Zhou, X.; Chen, C.; Jiao, L.; Li, W.; Gou, S.; Li, Y.; Du, J.; Chen, G.; et al. CD47/SIRPα blocking peptide identification and synergistic effect with irradiation for cancer immunotherapy. *J. Immunother. Cancer* **2020**, *8*, 905. [CrossRef]
69. Kaspar, A.A.; Reichert, J.M. Future directions for peptide therapeutics development. *Drug Discov. Today* **2013**, *18*, 807–817.
70. Xie, J.; Si, X.; Gu, S.; Wang, M.; Shen, J.; Li, H.; Li, D.; Fang, Y.; Liu, C.; Zhu, J. Allosteric Inhibitors of SHP2 with Therapeutic Potential for Cancer Treatment. *J. Med. Chem.* **2017**, *60*, 10205–10219. [CrossRef]
71. Song, K.; Liu, X.; Huang, W.; Lu, S.; Shen, Q.; Zhang, L.; Zhang, J. Improved Method for the Identification and Validation of Allosteric Sites. *J. Chem. Inf. Model.* **2017**, *57*, 2358–2363. [CrossRef]
72. Roca, C.; Requena, C.; Sebastián-Pérez, V.; Malhotra, S.; Radoux, C.; Pérez, C.; Martinez, A.; Antonio Páez, J.; Blundell, T.L.; Campillo, N.E. Identification of new allosteric sites and modulators of AChE through computational and experimental tools. *J. Enzyme Inhib. Med. Chem.* **2018**, *33*, 1034–1047. [CrossRef]
73. Sakaguchi, S.; Sakaguchi, N.; Asano, M.; Itoh, M.; Toda, M. Immunologic self-tolerance maintained by activated T cells expressing IL-2 receptor alpha-chains (CD25). Breakdown of a single mechanism of self-tolerance causes various autoimmune diseases. *J. Immunol.* **1995**, *155*, 1151–1164.
74. Bennett, C.L.; Christie, J.; Ramsdell, F.; Brunkow, M.E.; Ferguson, P.J.; Whitesell, L.; Kelly, T.E.; Saulsbury, F.T.; Chance, P.F.; Ochs, H.D. The immune dysregulation, polyendocrinopathy, enteropathy, X-linked syndrome (IPEX) is caused by mutations of FOXP3. *Nat. Genet.* **2001**, *27*, 20–21. [CrossRef]
75. Fontenot, J.D.; Gavin, M.A.; Rudensky, A.Y. Foxp3 programs the development and function of CD4+CD25+ regulatory T cells. *J. Immunol.* **2017**, *198*, 986–992. [CrossRef]
76. Brunkow, M.E.; Jeffery, E.W.; Hjerrild, K.A.; Paeper, B.; Clark, L.B.; Yasayko, S.A.; Wilkinson, J.E.; Galas, D.; Ziegler, S.F.; Ramsdell, F. Disruption of a new forkhead/winged-helix protein, scurfin, results in the fatal lymphoproliferative disorder of the scurfy mouse. *Nat. Genet.* **2001**, *27*, 68–73. [CrossRef]
77. Linterman, M.A.; Pierson, W.; Lee, S.K.; Kallies, A.; Kawamoto, S.; Rayner, T.F.; Srivastava, M.; Divekar, D.P.; Beaton, L.; Hogan, J.J.; et al. Foxp3+ follicular regulatory T cells control the germinal center response. *Nat. Med.* **2011**, *17*, 975–982. [CrossRef]
78. Koch, M.A.; Tucker-Heard, G.; Perdue, N.R.; Killebrew, J.R.; Urdahl, K.B.; Campbell, D.J. The transcription factor T-bet controls regulatory T cell homeostasis and function during type 1 inflammation. *Nat. Immunol.* **2009**, *10*, 595–602. [CrossRef]
79. Chung, Y.; Tanaka, S.; Chu, F.; Nurieva, R.I.; Martinez, G.J.; Rawal, S.; Wang, Y.H.; Lim, H.; Reynolds, J.M.; Zhou, X.H.; et al. Follicular regulatory T cells expressing Foxp3 and Bcl-6 suppress germinal center reactions. *Nat. Med.* **2011**, *17*, 983–988. [CrossRef]
80. Chaudhry, A.; Rudra, D.; Treuting, P.; Samstein, R.M.; Liang, Y.; Kas, A.; Rudensky, A.Y. CD4+ regulatory T cells control TH17 responses in a stat3-dependent manner. *Science* **2009**, *326*, 986–991. [CrossRef]
81. Wing, K.; Sakaguchi, S. Regulatory T cells exert checks and balances on self tolerance and autoimmunity. *Nat. Immunol.* **2010**, *11*, 7–13.
82. Sakaguchi, S.; Miyara, M.; Costantino, C.M.; Hafler, D.A. FOXP3 + regulatory T cells in the human immune system. *Nat. Rev. Immunol.* **2010**, *10*, 490–500.

83. Togashi, Y.; Nishikawa, H. Regulatory T cells: Molecular and cellular basis for immunoregulation. In *Current Topics in Microbiology and Immunology*; Springer: Berlin, Germany, 2017; Volume 410, pp. 3–27.
84. Hori, S.; Nomura, T.; Sakaguchi, S. Control of regulatory T cell development by the transcription factor Foxp3. *J. Immunol.* **2017**, *198*, 981–985. [CrossRef]
85. Khattri, R.; Cox, T.; Yasayko, S.A.; Ramsdell, F. An essential role for Scurfin in CD4+CD25+T regulatory cells. *J. Immunol.* **2017**, *198*, 993–998. [CrossRef]
86. Steinbrink, K.; Wölfl, M.; Jonuleit, H.; Knop, J.; Enk, A.H.H. Induction of tolerance by IL-10-treated dendritic cells. *J. Immunol.* **1997**, *159*, 4772–4780.
87. Collison, L.W.W.; Workman, C.J.J.; Kuo, T.T.T.; Boyd, K.; Wang, Y.; Vignali, K.M.M.; Cross, R.; Sehy, D.; Blumberg, R.S.S.; Vignali, D.A.A.A.A. The inhibitory cytokine IL-35 contributes to regulatory T-cell function. *Nature* **2007**, *450*, 566–569. [CrossRef]
88. Turnis, M.E.E.; Sawant, D.V.V.; Szymczak-Workman, A.L.L.; Andrews, L.P.P.; Delgoffe, G.M.M.; Yano, H.; Beres, A.J.J.; Vogel, P.; Workman, C.J.J.; Vignali, D.A. Interleukin-35 Limits Anti-Tumor Immunity. *Immunity* **2016**, *44*, 316–329. [CrossRef]
89. Jarnicki, A.G.G.; Lysaght, J.; Todryk, S.; Mills, K.H.G.H.G. Suppression of Antitumor Immunity by IL-10 and TGF-β-Producing T Cells Infiltrating the Growing Tumor: Influence of Tumor Environment on the Induction of CD4+ and CD8+ Regulatory T Cells. *J. Immunol.* **2006**, *177*, 896–904. [CrossRef]
90. Takahashi, T.; Kuniyasu, Y.; Toda, M.; Sakaguchi, N.; Itoh, M.; Iwata, M.; Shimizu, J.; Sakaguchi, S. Immunologic self-tolerance maintained by CD25+CD4+ naturally anergic and suppressive T cells: Induction of autoimmune disease by breaking their anergic/suppressive state. *Int. Immunol.* **1998**, *10*, 1969–1980. [CrossRef]
91. Thornton, A.M.M.; Shevach, E.M.M. CD4+CD25+ immunoregulatory T cells suppress polyclonal T cell activation in vitro by inhibiting interleukin 2 production. *J. Exp. Med.* **1998**, *188*, 287–296. [CrossRef]
92. Perez, V.L.L.; Van Parijs, L.; Biuckians, A.; Zheng, X.X.X.; Strom, T.B.B.; Abbas, A.K.K. Induction of peripheral T cell tolerance in vivo requires CTLA-4 engagement. *Immunity* **1997**, *6*, 411–417. [CrossRef]
93. Onizuka, S.; Tawara, I.; Shimizu, J.; Sakaguchi, S.; Fujita, T.; Nakayama, E. Tumor rejection by in vivo administration of anti-CD25 (interleukin-2 receptor α) monoclonal antibody. *Cancer Res.* **1999**, *59*, 3128–3133.
94. Shimizu, J.; Yamazaki, S.; Sakaguchi, S. Induction of tumor immunity by removing CD25+CD4+ T cells: A common basis between tumor immunity and autoimmunity—PubMed. *J. Immunol.* **1999**, *163*, 5211–5218.
95. Lodish, H.; Berk, A.; Zipursky, S.L.; Matsudaira, P.; Baltimore, D.; Darnell, J. Proto-Oncogenes and Tumor-Suppressor Genes. In *Molecular Cell Biology*, 4th ed.; W. H. Freeman: New York, NY, USA, 2000.
96. Saito, T.; Nishikawa, H.; Wada, H.; Nagano, Y.; Sugiyama, D.; Atarashi, K.; Maeda, Y.; Hamaguchi, M.; Ohkura, N.; Sato, E.; et al. Two FOXP3 + CD4 + T cell subpopulations distinctly control the prognosis of colorectal cancers. *Nat. Med.* **2016**, *22*, 679–684. [CrossRef]
97. Tada, Y.; Togashi, Y.; Kotani, D.; Kuwata, T.; Sato, E.; Kawazoe, A.; Doi, T.; Wada, H.; Nishikawa, H.; Shitara, K. Targeting VEGFR2 with Ramucirumab strongly impacts effector/activated regulatory T cells and CD8+ T cells in the tumor microenvironment. *J. Immunother. Cancer* **2018**, *6*, 106. [CrossRef]
98. Fridman, W.H.; Pagès, F.; Sautès-Fridman, C.; Galon, J. The immune contexture in human tumours: Impact on clinical outcome. *Nat. Rev. Cancer* **2012**, *12*, 298–306.
99. Shang, B.; Liu, Y.; Jiang, S.J.; Liu, Y. Prognostic value of tumor-infiltrating FoxP3+ regulatory T cells in cancers: A systematic review and meta-analysis. *Sci. Rep.* **2015**, *5*. [CrossRef]
100. Kryczek, I.; Wu, K.; Zhao, E.; Wei, S.; Vatan, L.; Szeliga, W.; Huang, E.; Greenson, J.; Chang, A.; Roliński, J.; et al. IL-17 + Regulatory T Cells in the Microenvironments of Chronic Inflammation and Cancer. *J. Immunol.* **2011**, *186*, 4388–4395. [CrossRef]
101. Kryczek, I.; Liu, R.; Wang, G.; Wu, K.; Shu, X.; Szeliga, W.; Vatan, L.; Finlayson, E.; Huang, E.; Simeone, D.; et al. FOXP3 defines regulatory T cells in human tumor and autoimmune disease. *Cancer Res.* **2009**, *69*, 3995–4000. [CrossRef]
102. Colbeck, E.J.; Jones, E.; Hindley, J.P.; Smart, K.; Schulz, R.; Browne, M.; Cutting, S.; Williams, A.; Parry, L.; Godkin, A.; et al. Treg depletion licenses T cell–driven HEV neogenesis and promotes tumor destruction. *Cancer Immunol. Res.* **2017**, *5*, 1005–1015. [CrossRef]
103. Hindley, J.P.; Jones, E.; Smart, K.; Bridgeman, H.; Lauder, S.N.; Ondondo, B.; Cutting, S.; Ladell, K.; Wynn, K.K.; Withers, D.; et al. T-cell trafficking facilitated by high endothelial venules is required for tumor control after regulatory T-cell depletion. *Cancer Res.* **2012**, *72*, 5473–5482. [CrossRef]
104. Sugiyama, D.; Nishikawa, H.; Maeda, Y.; Nishioka, M.; Tanemura, A.; Katayama, I.; Ezoe, S.; Kanakura, Y.; Sato, E.; Fukumori, Y.; et al. Anti-CCR4 mAb selectively depletes effector-Type FoxP3+CD4+ regulatory T cells, evoking antitumor immune responses in humans. *Proc. Natl. Acad. Sci. USA* **2013**, *110*, 17945–17950. [CrossRef]
105. Facciabene, A.; Peng, X.; Hagemann, I.S.; Balint, K.; Barchetti, A.; Wang, L.P.; Gimotty, P.A.; Gilks, C.B.; Lal, P.; Zhang, L.; et al. Tumour hypoxia promotes tolerance and angiogenesis via CCL28 and T reg cells. *Nature* **2011**, *475*, 226–230. [CrossRef]
106. Togashi, Y.; Nishikawa, H. Suppression from beyond the grave. *Nat. Immunol.* **2017**, *18*, 1285–1286.
107. Nishikawa, H.; Kato, T.; Tawara, I.; Saito, K.; Ikeda, H.; Kuribayashi, K.; Allen, P.M.; Schreiber, R.D.; Sakaguchi, S.; Old, L.J.; et al. Definition of target antigens for naturally occurring CD4+ CD25+ regulatory T cells. *J. Exp. Med.* **2005**, *201*, 681–686. [CrossRef]
108. Ghiringhelli, F.; Puig, P.E.; Roux, S.; Parcellier, A.; Schmitt, E.; Solary, E.; Kroemer, G.; Martin, F.; Chauffert, B.; Zitvogel, L. Tumor cells convert immature myeloid dendritic cells into TGF-β-secreting cells inducing CD4 +CD25 + regulatory T cell proliferation. *J. Exp. Med.* **2005**, *202*, 919–929. [CrossRef]

109. Hindley, J.P.; Ferreira, C.; Jones, E.; Lauder, S.N.; Ladell, K.; Wynn, K.K.; Betts, G.J.; Singh, Y.; Price, D.A.; Godkin, A.J.; et al. Analysis of the T-cell receptor repertoires of tumor-infiltrating conventional and regulatory T cells reveals no evidence for conversion in carcinogen-induced tumors. *Cancer Res.* **2011**, *71*, 736–746. [CrossRef]
110. Sainz-Perez, A.; Lim, A.; Lemercier, B.; Leclerc, C. The T-cell receptor repertoire of tumor-infiltrating regulatory T lymphocytes is skewed toward public sequences. *Cancer Res.* **2012**, *72*, 3557–3569. [CrossRef]
111. Morikawa, H.; Ohkura, N.; Vandenbon, A.; Itoh, M.; Nagao-Sato, S.; Kawaji, H.; Lassmann, T.; Carninci, P.; Hayashizaki, Y.; Forrest, A.R.R.; et al. Differential roles of epigenetic changes and Foxp3 expression in regulatory T cell-specific transcriptional regulation. *Proc. Natl. Acad. Sci. USA* **2014**, *111*, 5289–5294. [CrossRef]
112. Wei, G.; Wei, L.; Zhu, J.; Zang, C.; Hu-Li, J.; Yao, Z.; Cui, K.; Kanno, Y.; Roh, T.Y.; Watford, W.T.; et al. Global Mapping of H3K4me3 and H3K27me3 Reveals Specificity and Plasticity in Lineage Fate Determination of Differentiating CD4+ T Cells. *Immunity* **2009**, *30*, 155–167. [CrossRef]
113. Ohkura, N.; Hamaguchi, M.; Morikawa, H.; Sugimura, K.; Tanaka, A.; Ito, Y.; Osaki, M.; Tanaka, Y.; Yamashita, R.; Nakano, N.; et al. T Cell Receptor Stimulation-Induced Epigenetic Changes and Foxp3 Expression Are Independent and Complementary Events Required for Treg Cell Development. *Immunity* **2012**, *37*, 785–799. [CrossRef]
114. Curiel, T.J.; Coukos, G.; Zou, L.; Alvarez, X.; Cheng, P.; Mottram, P.; Evdemon-Hogan, M.; Conejo-Garcia, J.R.; Zhang, L.; Burow, M.; et al. Specific recruitment of regulatory T cells in ovarian carcinoma fosters immune privilege and predicts reduced survival. *Nat. Med.* **2004**, *10*, 942–949. [CrossRef]
115. Wei, S.; Kryczek, I.; Edwards, R.P.; Zou, L.; Szeliga, W.; Banerjee, M.; Cost, M.; Cheng, P.; Chang, A.; Redman, B.; et al. Interleukin-2 administration alters the CD4+FOXP3+ T-cell pool and tumor trafficking in patients with ovarian carcinoma. *Cancer Res.* **2007**, *67*, 7487–7494. [CrossRef]
116. Tan, M.C.B.; Goedegebuure, P.S.; Belt, B.A.; Flaherty, B.; Sankpal, N.; Gillanders, W.E.; Eberlein, T.J.; Hsieh, C.-S.; Linehan, D.C. Disruption of CCR5-Dependent Homing of Regulatory T Cells Inhibits Tumor Growth in a Murine Model of Pancreatic Cancer. *J. Immunol.* **2009**, *182*, 1746–1755. [CrossRef]
117. Hoelzinger, D.B.; Smith, S.E.; Mirza, N.; Dominguez, A.L.; Manrique, S.Z.; Lustgarten, J. Blockade of CCL1 Inhibits T Regulatory Cell Suppressive Function Enhancing Tumor Immunity without Affecting T Effector Responses. *J. Immunol.* **2010**, *184*, 6833–6842. [CrossRef]
118. Sen, D.R.; Kaminski, J.; Barnitz, R.A.; Kurachi, M.; Gerdemann, U.; Yates, K.B.; Tsao, H.W.; Godec, J.; LaFleur, M.W.; Brown, F.D.; et al. The epigenetic landscape of T cell exhaustion. *Science* **2016**, *354*, 1165–1169. [CrossRef]
119. Spranger, S.; Spaapen, R.M.; Zha, Y.; Williams, J.; Meng, Y.; Ha, T.T.; Gajewski, T.F. Up-regulation of PD-L1, IDO, and Tregs in the melanoma tumor microenvironment is driven by CD8+ T cells. *Sci. Transl. Med.* **2013**, *5*, 200ra116.
120. Williams, J.B.; Horton, B.L.; Zheng, Y.; Duan, Y.; Powell, J.D.; Gajewski, T.F. The EGR2 targets LAG-3 and 4-1BB describe and regulate dysfunctional antigen-specific CD8+ T cells in the tumor microenvironment. *J. Exp. Med.* **2017**, *214*, 381–400. [CrossRef]
121. Zakiryanova, G.K.; Wheeler, S.; Shurin, M.R. Oncogenes in immune cells as potential therapeutic targets. *ImmunoTargets Ther.* **2018**, *7*, 21–28. [CrossRef]
122. Spranger, S.; Bao, R.; Gajewski, T.F. Melanoma-intrinsic β-catenin signalling prevents anti-tumour immunity. *Nature* **2015**, *523*, 231–235. [CrossRef]
123. Peng, W.; Chen, J.Q.; Liu, C.; Malu, S.; Creasy, C.; Tetzlaff, M.T.; Xu, C.; McKenzie, J.A.; Zhang, C.; Liang, X.; et al. Loss of PTEN promotes resistance to T cell–mediated immunotherapy. *Cancer Discov.* **2016**, *6*, 202–216. [CrossRef]
124. Spranger, S.; Gajewski, T.F. Impact of oncogenic pathways on evasion of antitumour immune responses. *Nat. Rev. Cancer* **2018**, *18*, 139–147.
125. Hamarsheh, S.; Groß, O.; Brummer, T.; Zeiser, R. Immune modulatory effects of oncogenic KRAS in cancer. *Nat. Commun.* **2020**, *11*. [CrossRef]
126. Jiang, H.; Hegde, S.; Knolhoff, B.L.; Zhu, Y.; Herndon, J.M.; Meyer, M.A.; Nywening, T.M.; Hawkins, W.G.; Shapiro, I.M.; Weaver, D.T.; et al. Targeting focal adhesion kinase renders pancreatic cancers responsive to checkpoint immunotherapy. *Nat. Med.* **2016**, *22*, 851–860. [CrossRef]
127. Serrels, A.; Lund, T.; Serrels, B.; Byron, A.; McPherson, R.C.; Von Kriegsheim, A.; Gómez-Cuadrado, L.; Canel, M.; Muir, M.; Ring, J.E.; et al. Nuclear FAK Controls Chemokine Transcription, Tregs, and Evasion of Anti-tumor Immunity. *Cell* **2015**, *163*, 160–173. [CrossRef]
128. Pace, L.; Tempez, A.; Arnold-Schrauf, C.; Lemaitre, F.; Bousso, P.; Fetler, L.; Sparwasser, T.; Amigorena, S. Regulatory T cells increase the avidity of primary CD8+ T cell responses and promote memory. *Science* **2012**, *338*, 532–536. [CrossRef]
129. Maeda, Y.; Nishikawa, H.; Sugiyama, D.; Ha, D.; Hamaguchi, M.; Saito, T.; Nishioka, M.; Wing, J.B.; Adeegbe, D.; Katayama, I.; et al. Detection of self-reactive CD8+ T cells with an anergic phenotype in healthy individuals. *Science* **2014**, *346*, 1536–1540. [CrossRef]
130. Snyder, A.; Makarov, V.; Merghoub, T.; Yuan, J.; Zaretsky, J.M.; Desrichard, A.; Walsh, L.A.; Postow, M.A.; Wong, P.; Ho, T.S.; et al. Genetic Basis for Clinical Response to CTLA-4 Blockade in Melanoma. *N. Engl. J. Med.* **2014**, *371*, 2189–2199. [CrossRef]
131. Rizvi, N.A.; Hellmann, M.D.; Snyder, A.; Kvistborg, P.; Makarov, V.; Havel, J.J.; Lee, W.; Yuan, J.; Wong, P.; Ho, T.S.; et al. Mutational landscape determines sensitivity to PD-1 blockade in non-small cell lung cancer. *Science* **2015**, *348*, 124–128. [CrossRef]

132. Hodi, F.S.; O'Day, S.J.; McDermott, D.F.; Weber, R.W.; Sosman, J.A.; Haanen, J.B.; Gonzalez, R.; Robert, C.; Schadendorf, D.; Hassel, J.C.; et al. Improved Survival with Ipilimumab in Patients with Metastatic Melanoma. *N. Engl. J. Med.* **2010**, *363*, 711–723. [CrossRef]
133. Robert, C.; Thomas, L.; Bondarenko, I.; O'Day, S.; Weber, J.; Garbe, C.; Lebbe, C.; Baurain, J.-F.; Testori, A.; Grob, J.-J.; et al. Ipilimumab plus Dacarbazine for Previously Untreated Metastatic Melanoma. *N. Engl. J. Med.* **2011**, *364*, 2517–2526. [CrossRef]
134. Fellne, C. Ipilimumab (Yervoy) prolongs survival in advanced melanoma: Serious side effects and a hefty price tag may limit its use. *Pharm. Ther.* **2012**, *37*, 503.
135. Zhang, Y.; Du, X.; Liu, M.; Tang, F.; Zhang, P.; Ai, C.; Fields, J.K.; Sundberg, E.J.; Latinovic, O.S.; Devenport, M.; et al. Hijacking antibody-induced CTLA-4 lysosomal degradation for safer and more effective cancer immunotherapy. *Cell Res.* **2019**, *29*, 609–627. [CrossRef]
136. Ji, D.; Song, C.; Li, Y.; Xia, J.; Wu, Y.; Jia, J.; Cui, X.; Yu, S.; Gu, J. Combination of radiotherapy and suppression of Tregs enhances abscopal antitumor effect and inhibits metastasis in rectal cancer. *J. Immunother. Cancer* **2020**, *8*, 826. [CrossRef]
137. Qu, Q.; Zhai, Z.; Xu, J.; Li, S.; Chen, C.; Lu, B. IL36 Cooperates with Anti-CTLA-4 mAbs to Facilitate Antitumor Immune Responses. *Front. Immunol.* **2020**, *11*. [CrossRef]
138. Mihic-Probst, D.; Reinehr, M.; Dettwiler, S.; Kolm, I.; Britschgi, C.; Kudura, K.; Maggio, E.M.; Lenggenhager, D.; Rushing, E.J. The role of macrophages type 2 and T-regs in immune checkpoint inhibitor related adverse events. *Immunobiology* **2020**, *225*, 152009. [CrossRef]
139. Sun, N.Y.; Chen, Y.L.; Lin, H.W.; Chiang, Y.C.; Chang, C.F.; Tai, Y.J.; Chen, C.A.; Sun, W.Z.; Chien, C.L.; Cheng, W.F. Immune checkpoint Ab enhances the antigen-specific anti-tumor effects by modulating both dendritic cells and regulatory T lymphocytes. *Cancer Lett.* **2019**, *444*, 20–34. [CrossRef]
140. Kvarnhammar, A.M.; Veitonmäki, N.; Hägerbrand, K.; Dahlman, A.; Smith, K.E.; Fritzell, S.; Von Schantz, L.; Thagesson, M.; Werchau, D.; Smedenfors, K.; et al. The CTLA-4 x OX40 bispecific antibody ATOR-1015 induces anti-tumor effects through tumor-directed immune activation. *J. Immunother. Cancer* **2019**, *7*, 103. [CrossRef]
141. Pai, C.C.S.; Simons, D.M.; Lu, X.; Evans, M.; Wei, J.; Wang, Y.H.; Chen, M.; Huang, J.; Park, C.; Chang, A.; et al. Tumor-conditional anti-CTLA4 uncouples antitumor efficacy from immunotherapy-related toxicity. *J. Clin. Investig.* **2019**, *129*, 349. [CrossRef]
142. Morris, Z.S.; Guy, E.I.; Werner, L.R.; Carlson, P.M.; Heinze, C.M.; Kler, J.S.; Busche, S.M.; Jaquish, A.A.; Sriramaneni, R.N.; Carmichael, L.L.; et al. Tumor-specific inhibition of in situ vaccination by distant untreated tumor sites. *Cancer Immunol. Res.* **2018**, *6*, 825–834. [CrossRef]
143. Duperret, E.K.; Wise, M.C.; Trautz, A.; Villarreal, D.O.; Ferraro, B.; Walters, J.; Yan, J.; Khan, A.; Masteller, E.; Humeau, L.; et al. Synergy of Immune Checkpoint Blockade with a Novel Synthetic Consensus DNA Vaccine Targeting TERT. *Mol. Ther.* **2018**, *26*, 435–445. [CrossRef]
144. Tang, F.; Du, X.; Liu, M.; Zheng, P.; Liu, Y. Anti-CTLA-4 antibodies in cancer immunotherapy: Selective depletion of intratumoral regulatory T cells or checkpoint blockade? *Cell Biosci.* **2018**, *8*, 30. [CrossRef]
145. Son, C.H.; Bae, J.; Lee, H.R.; Yang, K.; Park, Y.S. Enhancement of antitumor immunity by combination of anti-CTLA-4 antibody and radioimmunotherapy through the suppression of Tregs. *Oncol. Lett.* **2017**, *13*, 3781–3786. [CrossRef]
146. Schwarz, C.; Unger, L.; Mahr, B.; Aumayr, K.; Regele, H.; Farkas, A.M.; Hock, K.; Pilat, N.; Wekerle, T. The Immunosuppressive Effect of CTLA-4 Immunoglobulin Is Dependent on Regulatory T Cells at Low But Not High Doses. *Am. J. Transplant.* **2016**, *16*, 3404–3415. [CrossRef]
147. Marabelle, A.; Kohrt, H.; Sagiv-Barfi, I.; Ajami, B.; Axtell, R.C.; Zhou, G.; Rajapaksa, R.; Green, M.R.; Torchia, J.; Brody, J.; et al. Depleting tumor-specific Tregs at a single site eradicates disseminated tumors. *J. Clin. Invest.* **2013**, *123*, 2447–2463. [CrossRef]
148. Du, X.; Tang, F.; Liu, M.; Su, J.; Zhang, Y.; Wu, W.; Devenport, M.; Lazarski, C.A.; Zhang, P.; Wang, X.; et al. A reappraisal of CTLA-4 checkpoint blockade in cancer immunotherapy. *Cell Res.* **2018**, *28*, 416–432. [CrossRef]
149. Sandin, L.C.; Eriksson, F.; Ellmark, P.; Loskog, A.S.I.; Tötterman, T.H.; Mangsbo, S.M. Local CTLA4 blockade effectively restrains experimental pancreatic adenocarcinoma growth in vivo. *Oncoimmunology* **2014**, *3*. [CrossRef]
150. Kavanagh, B.; O'Brien, S.; Lee, D.; Hou, Y.; Weinberg, V.; Rini, B.; Allison, J.P.; Small, E.J.; Fong, L. CTLA4 blockade expands FoxP3+ regulatory and activated effector CD4 + T cells in a dose-dependent fashion. *Blood* **2008**, *112*, 1175–1183. [CrossRef]
151. Francisco, L.M.; Salinas, V.H.; Brown, K.E.; Vanguri, V.K.; Freeman, G.J.; Kuchroo, V.K.; Sharpe, A.H. PD-L1 regulates the development, maintenance, and function of induced regulatory T cells. *J. Exp. Med.* **2009**, *206*, 3015–3029. [CrossRef]
152. Li, C.; Jiang, P.; Wei, S.; Xu, X.; Wang, J. Regulatory T cells in tumor microenvironment: New mechanisms, potential therapeutic strategies and future prospects. *Mol. Cancer* **2020**, *19*, 1–23. [CrossRef]
153. Khan, S.; Burt, D.J.; Ralph, C.; Thistlethwaite, F.C.; Hawkins, R.E.; Elkord, E. Tremelimumab (anti-CTLA4) mediates immune responses mainly by direct activation of T effector cells rather than by affecting T regulatory cells. *Clin. Immunol.* **2011**, *138*, 85–96. [CrossRef]
154. Kumar, P.; Saini, S.; Prabhakar, B.S. Cancer immunotherapy with check point inhibitor can cause autoimmune adverse events due to loss of Treg homeostasis. *Semin. Cancer Biol.* **2020**, *64*, 29–35. [CrossRef]

Review

Role of Methylation in Pro- and Anti-Cancer Immunity

Ali Mehdi [1,2] and Shafaat A. Rabbani [1,2,*]

[1] Department of Human Genetics, McGill University, Montreal, QC H3A 2B4, Canada; ali.mehdi@mail.mcgill.ca
[2] Department of Medicine, Research Institute of the McGill University Health Centre, Montreal, QC H4A 3J1, Canada
* Correspondence: shafaat.rabbani@mcgill.ca; Tel.: +1-514-843-1632

Simple Summary: Epigenetic mechanisms including methylation play an essential role in regulating gene expression not only in cancer cells but also in immune cells. Although role of DNA methylation has been extensively studied in tumor cells in tumor microenvironment (TME), the understanding of transcriptional regulation of pro- and anti-cancer immune cells in TME is beginning to unfold. This review focuses on the role of DNA and RNA methylation in regulating immune responses in innate and adaptive immune cells during their activation, differentiation, and function phase in cancer and in non-cancer pathologies. Uncovering these crucial regulatory mechanisms can trigger discovery of novel therapeutic targets which could enhance immunity against cancer to decrease cancer associated morbidity and mortality.

Abstract: DNA and RNA methylation play a vital role in the transcriptional regulation of various cell types including the differentiation and function of immune cells involved in pro- and anti-cancer immunity. Interactions of tumor and immune cells in the tumor microenvironment (TME) are complex. TME shapes the fate of tumors by modulating the dynamic DNA (and RNA) methylation patterns of these immune cells to alter their differentiation into pro-cancer (e.g., regulatory T cells) or anti-cancer (e.g., CD8+ T cells) cell types. This review considers the role of DNA and RNA methylation in myeloid and lymphoid cells in the activation, differentiation, and function that control the innate and adaptive immune responses in cancer and non-cancer contexts. Understanding the complex transcriptional regulation modulating differentiation and function of immune cells can help identify and validate therapeutic targets aimed at targeting DNA and RNA methylation to reduce cancer-associated morbidity and mortality.

Keywords: DNA methylation; RNA methylation; S-adenosylmethionine (SAM); cancer; tumor microenvironment; innate immunity; adaptive immunity; T cells; m^{6}A

1. Introduction

Epigenetic modifications are heritable changes regulating the cellular gene expression patterns required for the normal development and maintenance of various tissue functions [1–3]. Whereas genetic mutations result in the activation/inactivation of certain genes playing a pivotal role in carcinogenesis, abnormalities in the epigenetic landscape can lead to altered gene expression and function, genomic instability, and malignant cellular transformation (Figure 1) [3,4]. The three most studied epigenetic mechanisms that result in cancer are alterations in DNA methylation, histone modification, and non-coding RNA (ncRNA) expression.

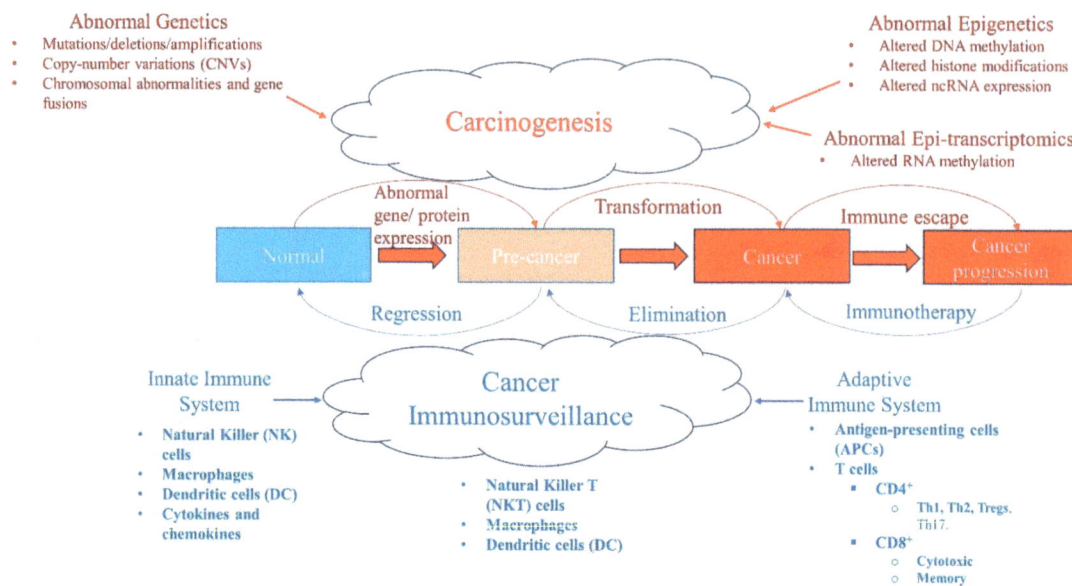

Figure 1. A balance between carcinogenesis and cancer immunosurveillance system. Abnormal genetic modifications such as gene mutations, deletions, amplifications, copy-number variations (CNVs), chromosomal abnormalities, or instability and gene fusions can all result in abnormal expression of genes and proteins leading to transformation of a normal cell into a pre-cancer state and/or cancer stage. Similarly, abnormal epigenetics, such as aberrant DNA methylation patterns, histone modifications, and ncRNA expression (e.g., miRNA) levels, also cause tumorigenesis. Recently, abnormal RNA methylation patterns, such as m^{6}A RNA post-transcriptional modifications (epi-transcriptomics), have been shown to result in the initiation and progression of cancer. Although these abnormalities in malignancy promote tumorigenesis, the cancer immunosurveillance system acts as a tumor suppressor working against the formation of pre-malignant and cancer cells. The cancer immunosurveillance system comprises the innate and adaptive immune systems that have various components that help to regress or eliminate tumor cells. However, some immune cells can be pro-tumor, which paradoxically help tumor progression in the tumor microenvironment. Cancer can evolve and escape the immune system by developing immunosuppressive escape mechanisms (such as high expression of PD-L1) that allow it to progress. This state can be reversed with immunotherapy, such as immune checkpoint inhibitors (ICPi).

1.1. DNA Methylation: Writers, Readers, Erasers, and Co-Factors

DNA methylation is the most well-characterized epigenetic mechanism, and was linked to cancer as early as the 1980s [5]. Specific DNA methylation patterns are crucial for parental imprinting, genomic stability, and importantly, regulation of gene expression [6,7]. DNA methylation is the covalent addition of a methyl (-CH3) group at the cytosine (C) base adjacent to 5′ of a guanosine (G) [8,9]. The methyl donor for this methylation reaction is s-adenosylmethionine (SAM). In the human genome, more than 28 million CpG dinucleotides exist, and 60–80% show methylation in any given cell [10]. In contrast, there are specific regions where CpG dinucleotides are enriched, called CpG islands, which are primarily located near gene promoters [10]. Increased methylation at CpG islands is typically associated with gene silencing. However, varying levels of DNA methylation at other regions, including gene bodies, enhancers, 5′ and 3′ UTRs, and partially methylated domains (PMDs), can also differentially affect gene expression to regulate dynamic biological processes [11–14].

In mammals, the addition of methyl groups to DNA is carried out by "writers", DNA methyltransferase (DNMT) 1, DNMT3A, and DNMT3B, converting unmodified C into 5-methyl-cytosine (5mC) [15]. DNMT3A and DNMT3B add methyl groups to DNA without template DNA and hence, undertake de novo methylation, whereas DNMT1, maintenance DNMT, adds methyl groups to hemi-methylated DNA by copying DNA methylation patterns from the parental strand to the daughter strand during cell division. DNMTs utilize methyl groups from SAM, which is a universal methyl donor and acts as a co-factor in this reaction [16].

DNA methylation can be recognized by readers including methyl-CpG-binding domain (MBD) proteins, certain transcription factors, and zinc finger (ZNF) proteins [17]. Generally, methylation of the CpG can directly affect gene transcription by interference with the binding of the transcription factors at a regulatory site leading to transcriptional silencing. In addition, DNMTs and MBD proteins can recruit histone modifiers to the methylated promoter region, and stimulate chromatin condensation and gene silencing [15,18–21].

Methyl groups from DNA can be removed either passively or actively. Active DNA demethylation is performed by "erasers", called ten-eleven translocation (TET), which remove methyl groups from DNA by oxidizing 5mC into 5hmC (5-hydroxymethyl-cytosine), 5fC (5-formylcytosine), and 5caC (5-carboxylcytosine) [22]. The 5fC and 5caC marks are later identified by thymine DNA glycosylase (TDG), and repaired and replaced by unmodified C. Passive DNA demethylation occurs when DNA methylation maintenance proteins are altered or the DNMT1/UHRF1 complex is unable to read 5hmC, 5fC, or 5caC, leaving C on a newly formed strand unmethylated and, due to multiple rounds of cell division, the original DNA methylation patterns are lost [22].

1.2. m^6A RNA Methylation: Writers, Readers, and Erasers

An emerging crucial layer of post-transcriptional gene regulation, N6-methyladenosine (m^6A) RNA methylation, plays an essential role in gene expression regulation and development, and human diseases [23–30]. m^6A is the most common and characterized modification in RNA amongst 150 other post-transcriptional modifications in eukaryotes [23–30]. Alterations in m^6A RNA methylation and its regulators target different genes in various cancers, including melanoma, acute myeloid leukemia (AML), liver cancer, glioblastoma, and breast and pancreatic cancer (Figure 1) [24,26–30]. m^6A RNA regulators include writers/methyltransferases, erasers/demethylases, and readers that can add/methylate, remove/demethylate, and read/recognize m^6A modified sites on RNA, respectively [23,25,26,28]. The major methyltransferases of m^6A are methyltransferase-like (METTL) 3 and METTL14 complexes that add a methyl group donated from SAM to the 6th Adenosine of the RNAs [23,25,26,28]. In contrast, active demethylation of m^6A is performed by demethylases AlkB homolog 5 (ALKBH5) or fat mass and obesity-associated (FTO), which remove the methyl groups from the RNA [23,25,26,28]. Readers recognize the m^6A modification either directly using the YTH domain (e.g., YTH-domain containing reader; YTHDF1/2/3; or YTHDC1/2) or indirectly, which leads to either RNA degradation or enhanced translation of the mRNA [23,25,26].

1.3. Immune System: Pro- and Anti-Cancer Immunity

Humans have evolved their immune system, including the innate and adaptive immune systems, to combat a broad range of diseases, including cancer (Figure 1) [31–33]. The innate immune system consists of immune cells including natural killer (NK) cells, dendritic cells (DC), macrophages, and neutrophils. The innate immune system is typically the first line of defense, has a nonspecific and immediate response against pathogens, and exhibits germline inheritance [31–33]. Innate immune cells use pattern recognition receptors (PRRs), such as toll-like receptors (TLRs), and identify pathogens based on non-specific molecular patterns including single-stranded RNAs or lipopolysaccharide. The adaptive immune system, by comparison, is highly specific and forms the immunological memory. Adaptive immunity comprises lymphocytes, and T and B cells, which produce cytokines and antibodies to counter pathogens [31–33]. A large number of extremely diverse but highly specific receptors on T cells—T cell receptors (TCRs)—and B cells—B cell receptors (BCR)—which recognize and differentiate self from non-self antigens are extremely useful in response to foreign pathogens. Long-lasting memory cells generated after pathogen clearance provide a rapid and robust pathogen control upon re-exposure to the same pathogen.

After a century of controversy, it has now been established that a functional cancer immunosurveillance system indeed exists, and acts as a tumor suppressor or killer (Figure 1) [31–35]. Interestingly, both innate and adaptive immune systems can recognize and eliminate malignant cells. Components of the immune system in the tumor microenvironment (TME) can be either anti-tumor, regressing or killing tumor cells; or pro-tumor, helping tumor progression. TME is a complex interaction of tumor cells, immune cells, and stromal cells, and is influenced by various factors including cytokines, chemokines, the extracellular matrix, tissue-specific factors, and inflammation [31,36]. Tumor inhibition or progression depends on TME factors, which can be anti- or pro-tumorigenic. Tumor progression is suppressed or eliminated by the cancer immunosurveillance system; however, tumor cells can evolve and develop mechanisms that allow them to evade or escape the immune system (Figures 1 and 2) [31,36,37]. There are three main immune escape mechanisms: (1) loss of antigenicity—tumor cells increase defects in antigen processing and presentation machinery resulting in lower presentation of antigens to immune cells; (2) loss of immunogenicity—tumor cells produce low levels of immunogenic tumor antigens and high levels of immunosuppressive ligands (e.g., PD-L1); and (3) creating an immunosuppressive TME—tumor cells transform to cause alterations in oncogenes and tumor suppressor genes to increase inflammation and recruitment of pro-tumor immune cells in TME.

Solid tumors typically have immune cells that can be anti-tumor or pro-tumor as a result of factors including differentiation (Figure 2). In summary, pro-tumor factors include high type II M2 macrophages; high CD4$^+$ regulatory T cells (Tregs); high type II CD4$^+$ Th2 cells; typically low or exhausted tumor infiltrating lymphocytes (TILs) (cold tumor); and low antigenicity and immunogenicity of the tumor cells. In contrast, anti-tumor factors include high NK cells; high type I M1 macrophages; high type I CD4$^+$ Th1 cells; low Tregs; high tumor infiltrating CD8$^+$ T cells (memory, cytotoxic); high type I cellular immune response (e.g., IFN-g, IL-2, granzyme B); more and functional TILs (hot tumor); and high antigenicity and immunogenicity of the tumor cells (Figure 2) [31–33,36–44].

Epigenetic mechanisms including miRNAs and histone modifications are crucial for the regulation of the immune system in the TME and has been extensively reviewed [45–49]. DNA methylation also plays an essential role in the differentiation and function of immune cells into various subtypes, and the manner in which these immune cells influence each other in the TME, which ultimately results in tumor progression or suppression. Schuyler et al. [50] carried out analysis of large whole-genome bisulfite sequencing datasets (112 datasets from the BLUEPRINT Epigenome Project) to delineate trends of changes in DNA methylation in different lineages of immune cells, including myeloid and lymphoid cells in TME of various cancer models. Global methylation, in general, increases during macrophage differentiation and activation, whereas it reduces during lymphocyte differentiation (T and B). Numerous studies have also shown methylation changes in the differentiation and activation of pro- or anti-cancer myeloid and lymphoid cells [22,51,52].

The role of methylation in hematopoiesis and in immune disorders is now well established [22,51,52]. The focus of this review is to discuss the role of DNA and RNA methylation (m^{6}A) and its regulators in key pro- or anti-cancer immune cells of innate and adaptive immune systems. Examples from other non-cancer immune triggering pathologies are also included. Additionally, the translational potential of targeting methylation with DNA methyltransferase inhibitors (DNMTi), methylating agents such as SAM, and m^{6}A RNA demethylase inhibitors in the treatment of liquid and solid cancers is also discussed.

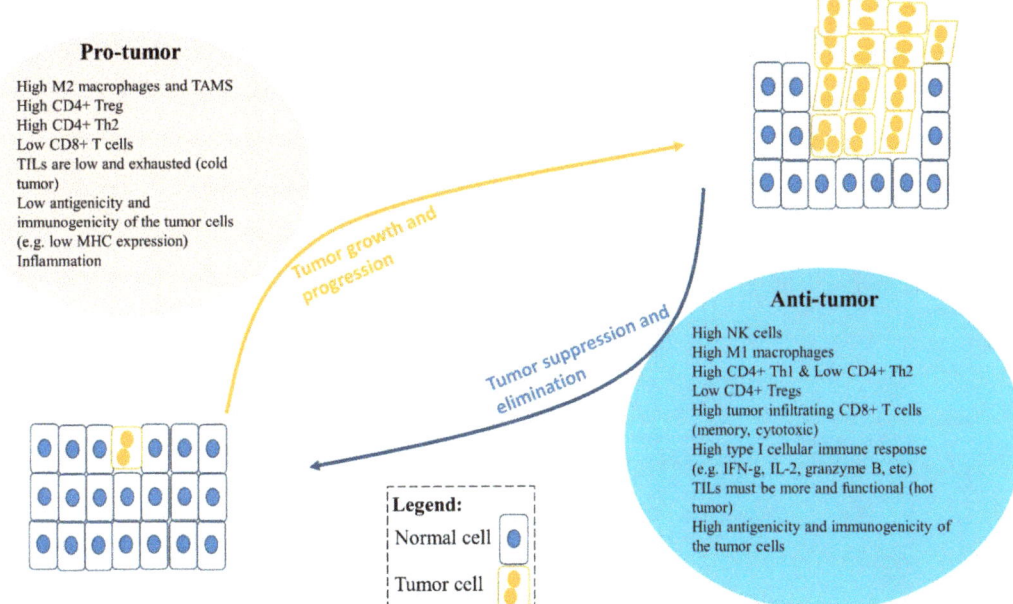

Figure 2. An imbalance between pro-tumor and anti-tumor immune cells and factors in the tumor microenvironment (TME) can lead to tumor growth and progression or tumor suppression and elimination. Pro-tumor immune cells can promote tumor progression, including type II M2 or TAMs (tumor-associated macrophages), regulatory T cells (Tregs), and type II Th2 cells. Moreover, factors that influence tumor progression are low tumor infiltrating lymphocytes (TILs) in the TME, low antigenicity and immunogenicity of tumor cells, and inflammation. Anti-tumor immune cells can reduce tumor growth and suppress tumor progression in the TME. These include CD8+ T cells, type I Th1 cells, NK cells, and type I M1 cells and their type I cytokines such as IFNγ, TNFα, IL-2, and granzyme B. Furthermore, anti-tumor immune factors can also influence tumor suppression, including high infiltration of functional TILs, and greater antigenicity and immunogenicity of the tumor cells, such as high MHC-I expression and tumor-associated antigen expression.

2. Role of DNA Methylation in Innate and Adaptive Immunity

2.1. Innate Immunity

2.1.1. Dendritic Cells (DCs)

DCs and macrophages are the first innate immunity cell types which are triggered for defense against pathogen invasion. DCs are professional antigen presenting cells (APCs) that are essential for triggering adaptive T cell responses in an antigen-specific manner. DCs can undergo marked changes in their phenotype and function under various stimuli and inflammatory conditions [53]. For instance, DCs can be polarized towards producing specific type of cytokines (e.g., IL-12, IL-23) and Notch ligands (e.g., DLL1/4) to induce different effector CD4 (Th1, Th2, Th17) and CD8 (cytotoxic) T cells [53].

The role of DNA methylation is crucial for regulating differentiation and activation of DCs; however, this has not been fully elucidated, particularly in the TME. Nevertheless, DNA methylation changes have been reported during differentiation of monocytes into DCs and immature DCs (iDCs) into mature DCs (mDCs) [54–57]. Bullwinkel et al. investigated epigenetic changes occurring at *CD14* and *CD209* gene loci, which are essential for the function of monocytes and DCs, respectively, and found CD14 expression was lost, whereas CD209 expression was elevated, upon differentiation from monocytes to DCs [54]. The reciprocal expression changes in CD14 and CD209 were associated with histone modifications at the *CD14* locus leading to *CD14* silencing, whereas loss of "repressive" histone marks and DNA demethylation at the *CD209* locus resulted in CD209 transcriptional activation. Zhang et al. carried out a comprehensive study of DNA methylation changes at single nucleotide-resolution for human monocytes and monocyte-derived iDCs and mDCs [56]. Several known genes and pathways regulating DC differentiation and maturation were identified. A total of 1608 differentially methylated positions (DMPs) from monocytes to iDCs and 156 DMPs from iDC to mDCs were identified. Major DNA demethylation occurred at the binding sites of the transcription factors of genes involved in DC differentiation and function that ultimately increased transcription of these genes. Moreover, the demethylation was locus-specific, and is associated with changes in DNA methylation regulators, including DNMT1, DNMT3A, DNMT3B, and TET2 [56]. Interestingly, DNA methylation reader, MBD2, in DCs was previously shown to have a dominant role in inducing CD4+ T cells differentiation into the Th2 cell type. Specifically, loss of Mbd2, resulted in reduced phenotypic activation of DCs and capability to initiate Th2 immunity against helminths or allergens [58]. In addition, during IL-4-mediated differentiation from human monocytes to DCs and macrophages, TET2 was identified as the main regulator of DNA demethylation of dendritic cell-specific or macrophage-specific gene sets mostly in intergenic regions and gene bodies [57]. Essentially, the IL-4-JAK3-STAT6 pathway is required for dendritic cell-specific demethylation and expression signature, and STAT6 also prevents demethylation of macrophage-specific genes required for monocyte to macrophage differentiation. Pacis et al. performed a comprehensive epigenome and transcriptome analysis of DCs infected with a live pathogenic bacterium (*Mycobacterium tuberculosis*) [59]. A rapid and active DNA demethylation at distal enhancers was identified that activates master immune transcription factors such as NF-κB and IFN regulatory families [59]. Although the above studies provide strong evidence of DNA methylation regulating monocyte to DC differentiation, and activation of DCs, the role of DNA methylation in the TME needs further characterization.

2.1.2. Macrophages

Macrophages are myeloid cells that have a spectrum of phenotypes in which M1 or M2 subtypes are the extreme ends. M1 cells are "classically activated" by IFNγ, and destroy tumor cells through their production of nitric oxide and type 1 cytokines and chemokines [31,60]. Moreover, M1 act as APCs to activate cytotoxic CD8+ T cells in an antigen (Ag)-specific manner. M2 cells are activated by "alternative" pathways via IL-4, IL-13, and/or TGFβ [31,60]. M2 secrete type II chemokines and cytokines, thereby promoting tumor growth and progression. Stromal and tumor-associated factors in the TME can shift macrophages to M2 types, specifically the tumor-associated macrophages (TAMs) type that promotes angiogenesis, tumor progression, and metastasis [60–62]. The differentiation from monocyte into macrophages and between the M1 or M2 (or TAMS) phenotype is regulated by DNA methylation at lineage-specific promoter and enhancer regions.

Upon examining global DNA methylation between human monocytes, naïve macrophages, and activated macrophages, Dekkers et al. reported major DNA methylation changes during monocyte to macrophage differentiation [63]. Differential methylation was generally fixed to short regions or single CpGs, and was prevalent at lineage-specific enhancers. The differential methylation was either gain (e.g., *IRF8, CEBPB*) or loss (e.g., *PPARG*) of methylation at specific transcription factor binding sites involved in monocyte to macrophage transition. Authors also analyzed different types of activated macrophages and found some genes for lipopolysaccharide (LPS)/IFNγ macrophage-specific activation (e.g., *CCL5*). In another study, the transcriptome and epigenome of human monocytes differentiated into macrophages with colony-stimulating factor 1 (CSF1) identifying several RNAs (mRNA and miRNAs) that are differentially expressed [64]. In addition, 100 differentially methylated regions (DMRs) between monocytes and macrophages were identified in enhancer regions that were uniquely demethylated in macrophages and repressed in monocytes, and were linked to actin cytoskeleton, phagocytosis, and innate immune response [64]. Evidence has shown that both methyltransferases DNMT1 and DNMT3A/B play a vital role in differentiation and macrophage polarization [51]. For instance, knock-down (KD) of

DNMT3B in RAW264.7 cells showed a higher polarization towards the M2 macrophage phenotype compared to M1, and leads to suppressed inflammation; the opposite pattern was observed for overexpression of DNMT3B [65]. During chronic inflammation, DNMT1 expression is elevated and has been associated with DNA hypermethylation. A study examined the role of TAMS in DNA methylation of a tumor suppressor gene gelsolin (*GSN*) during gastric cancer progression. Firstly, DNMT1 overexpression was shown to methylate and silence the *GSN* gene, and secondly, DNMT1 overexpression was associated with higher TAMs infiltration in the TME of gastric cancer [66]. Further analysis revealed that TAMs secreted CCL5 that triggered DNMT1 overexpression by activating the JAK2/STAT3 pathway in gastric cells, resulting in GSN silencing and tumorigenesis. In another study, DNMT1 was associated with M1 polarization by silencing the *SOCS1* gene and a subsequent increase in tumor necrosis factor (TNF) and IL-6 production [67]. Furthermore, DNMT1 overexpression was shown to promote M1 activation induced by LPS and IFNγ [67].

In contrast, TET proteins appear to have a role in the downregulation of inflammatory gene expression in normal myeloid cells [22]. In a model of TET2-deficient macrophages and DC, a higher expression of IL-6 was observed upon stimulation [68]. TET2 was shown to reduce IL-6 expression by interacting with IκbζІ (a member of the nuclear IκB family) and binding to the IL-6 promoter region in addition to recruitment of histone deacetylase 2 (HDAC2) [69]. Furthermore, Tet2-deficient mice are more susceptible to septic shock and colitis induced by endotoxin and dextran sulfate sodium (DSS), respectively, both due to elevated IL-6 expression [69]. TET2 expression is elevated in tumor infiltrating myeloid cells of both melanoma patients and mouse models via the IL-1R-MyD88 pathway. Moreover, TET2 acts as an oncogene in melanoma tumorigenesis by suppressing anti-cancer immune cells [70]. This is consistent with the TET protein acting as anti-inflammatory to myeloid cells [22]. Overall, these studies show the role of DNA methylation in regulating monocyte to macrophage differentiation and macrophage polarization.

2.1.3. Natural Killer (NK) Cells

NK cells can directly lyse MHC class I-deficient tumor cells [31,35]. NK cells have activating receptors that identify malignant cells expressing stress-induced ligands (e.g., MICA) [31,35]. NK cells kill the tumor cells by making them undergo apoptosis through either expressing death ligands (e.g., Fas ligand) or by releasing granzymes and perforin [31,35].

The role of DNA methylation in NK cells' activation or differentiation has not been fully elucidated. However, it was reported that the MHC-I cytotoxicity of NK cells, which is mediated by the KIR (killer cell Ig-like receptor) family, is regulated via methylation. In progenitor cells, KIR genes are silenced via hypermethylation and histone modifications, whereas in KIR-expressing cells, such as NK cells, KIR genes are demethylated and expressed [71]. Furthermore, work with human cytomegalovirus (HCMV) viral infection has shown that, upon infection, subjects have elevated levels of a "memory-like" subtype of NK cells which survive long term and have increased response upon re-exposure of the same pathogen. These memory-like NK cells are characterized by activation of NKG2C, which is in turn epigenetically regulated. In addition, in some HCMV-infected patients, memory-like NK cells were reported to lack B-cell and myeloid signaling proteins such as tyrosine kinase SYK. Further analysis showed that the gene promoter of *SYK* was hypermethylated and SYK expression was downregulated [72]. HCMV-associated NK cells also have low expression of signaling adaptors, including EAT-2, FCER1G, and transcription factor PLZF due to hypermethylation at their DNA [73]. Wiencke et al. examined human naïve vs. activated NK cells' DNA methylome and found reproducible genome-wide DNA methylation changes [74]. Methylation analysis showed primarily CpG hypomethylation (81% of significant loci) during activation of NK cells. Several previously reported and novel genes or pathways associated with activation of NK cells were identified. The high priority gene *BHLHE40* had high demethylation in activated NK cells, whereas it had low demethylation in naïve NK cells and was shown to be a potential biomarker for NK activation in peripheral blood. Interestingly, increased NK cells and CD8+ T cells tumor infiltration was reported using the DNA methyltransferase inhibitor (DNMTi), AzaC, through type I IFN signaling while reducing the tumor burden of the murine epithelial ovarian cancer model [75]. Histone deacetylase inhibitors (HDACi) lead to further activation of these anti-tumor immune cells and reduction in pro-tumor macrophages in the TME. Furthermore, ligands (such as ULBPs and MICA) of NK cells activating receptor NKG2D, which are essential for NK cell lytic activity, are downregulated in gliomas and hepatocellular carcinoma (HCC) cells via DNA methylation and histone methylation, respectively [76,77]. Indeed, treatment with DNMTi and Enhancer of zeste homolog 2 (EZH2) inhibitor was shown to upregulate NKG2D ligand expression, resulting in the lysis of glioma and HCC cells by NK cells, respectively. These studies show that DNA methylation not only controls the critical gene expression in NK cells that regulates differentiation and activation of NK cells but also genes in cancer cells that regulate NK cell tumor lytic activity.

2.2. Adaptive Immunity

Binding of the T cell receptor (TCR) present on T cells to the antigen/MHC complex (signal 1) expressed on APCs is essential for the activation of naive T cells [78]. Additional binding of positive co-stimulatory molecules present on activated APCs, called signal 2 (e.g., CD80/86 and B7RP1 on APCs onto CD28 and ICOS on T cells, respectively), helps in further activation. TCR activation is a multistep process that leads to an intracellular signaling cascade that results in activation, differentiation, and proliferation (clonal expansion) of T cells, and transforms them into effector cells producing cytokines [78]. DNA methylation has a key role in regulating these processes. For instance, upon TCR stimulation of T cells, IL-2 is highly expressed and is required for T cell activation and clonal expansion in mouse [79]. The increase in IL-2 cytokine results from active demethylation at a promoter-enhancer region of the *IL-2* locus upon T cell activation and remains demethylated afterwards [79]. In addition to IL-2 cytokine, DNA methylation also plays an important role in the activation, proliferation, and effector functions of CD4 and CD8 T cells as discussed below.

2.2.1. CD4+ T Cells

CD4+ T cells are unique T cells that can, depending on the nature of the Ag signal and type of cytokine stimulation, differentiate into various subtypes including helper T cell 1, 2, and 17 (Th1, Th2, and Th17) and Tregs (Figure 3). Th1 produce type I cytokines, including IL-2 and IFNγ, facilitating optimal expansion, trafficking, and effector functions of CD8+ T cells, thereby reducing tumor growth and progression [31,36,37]. In contrast, Th2 produce type II cytokines (IL-4, IL-5, and IL-13) and polarize immunity towards tumor progression [31,36,37]. This differentiation of CD4+ T cells into various subtypes is regulated by DNA methylation (Figure 3) [31,36,37]. The differentiated CD4 T cells then regulate downstream immune functions, such as enhancement of CD8 T cells, macrophages, and B cell effector functions, and immunological memory.

Numerous studies have analyzed the methylation status of immune genes and correlated it with immune responses in the TME (Figure 3). Upon antigenic stimulation, naïve CD4+ T cells differentiate into Th1 and Th2 by epigenetically activating or silencing a certain set of genes, usually by DNA demethylation and hypermethylation, respectively [80–82]. By analyzing the methylation status of a key gene, *IFNG* or *IFNγ*, essential for anti-tumor activity, Janson et al. reported demethylation of the *IFNγ* gene promoter and enhancer, and upregulation of IFNγ in Th1 cells [83]. In contrast, Th2 cells had hypermethylation at the *IFNγ* gene promoter and had low IFNγ expression. Studies show that naïve T cells that develop in the thymus have hypermethylated DNA at enhancer regions of the *IFNγ* and *IL-4* cluster (IL-4, IL-5, IL-13), and methylated H3K27me3 marks [80,81]. These marks limit chromatin accessibility and inhibit transcription of these genes and hence, naïve T cells minimally transcribe these genes. Interestingly, these regions become demethylated in T cell lineages that require expression of these cytokines—for instance, the demethylated promoter of the *IFNγ* gene in Th1 and CD8⁺ T cells [81]. These CpGs are maintained by Dmnt1 as deletion of *Dnmt1* results in global hypomethylation in naïve precursors, including DNA regions which are normally hypermethylated at these cytokine regulatory regions [84]. For instance, in Dnmt1-deficient mice, naïve T cells produce effector cytokines such as IFNγ immediately after activation. This shows that Dmnt1 is required to maintain these hypermethylated regions during T cell development to suppress and induce cytokine gene expression in naïve and active T cells, respectively [84,85]. Indeed, Th1 cells produce 100-times more IFNγ transcripts than naïve T cells but the *IL-4* gene loci are silenced [81].

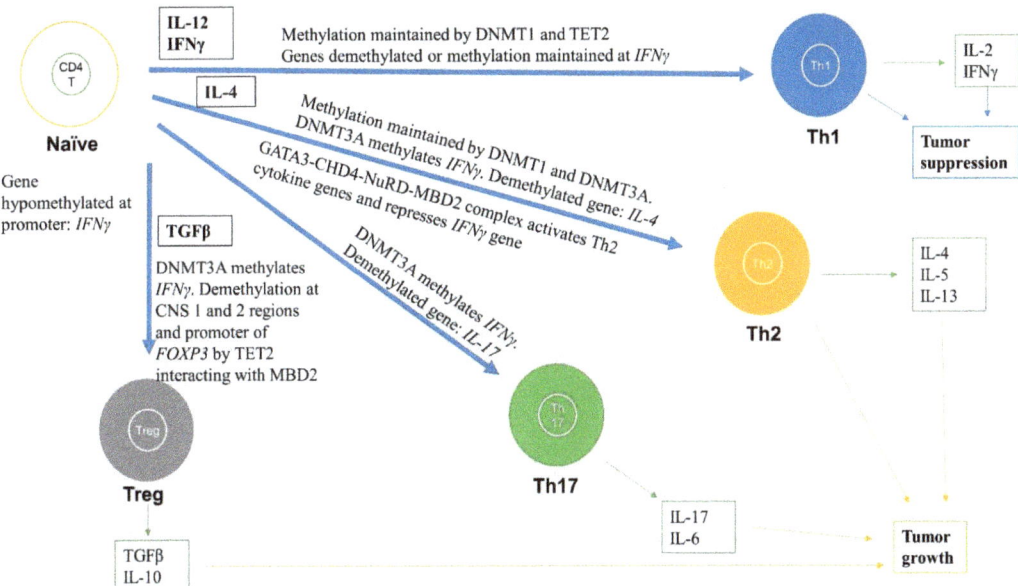

Figure 3. Role of DNA methylation in regulating differentiation and activation of naïve CD4⁺ T cells into effector cells including Th1, Th2, Th17, and Tregs subtypes. DNA methylation changes during differentiation can lead to formation of subtypes of CD4⁺ T cells. The black boxes are cytokines that help in the differentiation and activation process for each subtype. For instance, Th1 are formed when naïve CD4⁺ T cells are stimulated by IL-12 and IFNγ cytokines and the *IFNγ* gene promoter remains hypomethylated and IFNγ is highly expressed. For the Th2 subtype, the *IL-4* gene is demethylated and is highly expressed, whereas *IFNγ* is methylated and repressed. For Th17 cells, the *IL-17* gene is demethylated and highly expressed. For Tregs, *FOXP3* is demethylated at various regions, including promoter and enhancer, thereby markedly increasing FOXP3 expression. These methylation levels are maintained by DNMT1, DNMT3A, and TET2. The green boxes indicate the cytokines released from differentiated cells. These immune cells and released cytokines can further lead to tumor progression or suppression.

In contrast, some genes have the opposite pattern, i.e., they have hypomethylation in naïve cells but hypermethylation in differentiated T cells. For example, the *IFNγ* promoter region is unmethylated in naïve CD4⁺ T cells and continues to be hypomethylated upon Th1 cell differentiation; however, upon Th2 cell differentiation, which do not produce IFNγ, the *IFNγ* promoter is methylated via *de novo* DNA methylation by Dnmt3a [86,87]. Moreover, in mouse, *Dnmt3a* deletion in T cells can lead to a complete failure of naïve T cell differentiation into Th2, Th17, and iTreg lineage cells, due to their inability to methylate DNA (de novo) by Dnmt3a at the *Ifnγ* promoter region [88]. Indeed, *Dnmt3a* expression is stimulated upon TCR activation and is recruited to the *Ifnγ* promoter region to carry out methylation in Th2 cells [89]. In addition, deregulated de novo methylation patterns resulted in reduced histone silencing mark (H3K27me3) and increased transcriptionally active histone mark (H3K4me3) upon re-stimulation in the presence of IL-12 [81,88]. Furthermore, Th2 cells produce high amounts of IL-4 as a result of DNA hypomethylation at the *IL4* gene loci and transcriptional activation, whereas in naïve T cells, the *IL4* gene loci are hypermethylated [88]. Finally, during differentiation of naïve CD4$^+$ T cells into memory CD4$^+$ T cells a global loss of DNA methylation was observed, suggesting a role of DNA methylation in memory CD4$^+$ T cell formation [51]. These data suggest that CD4$^+$ T cells differentiation into Th1, Th2, Th17, and memory subtypes require DNA methylation changes at gene promoters and enhancers of critical genes such as *IFNγ* and *IL-4* (Figure 3) [36,81–83,88].

Strong evidence suggests that the MBD proteins together with the nucleosome remodeling deacetylase (NuRD) complex are essential in regulating DNA methylation-dependent differentiation of T cells [90–92]. For instance, loss of either MBD2 or NuRD complex can result in polarization of CD4+ T cells to Th2 cell type. Aoki et al. suggested that the NuRD–MBD2 complex may be required for the demethylation of gene loci encoding cytokines specific for Th2 differentiation [91]. Mechanistically, the chromodomain-helicase-DNA-binding protein 4 (Chd4) subunit of the NuRD–Mbd2 complex forms a complex with Gata3 that both activates Th2 cytokine transcription and represses the Th1 cytokine, IFN-γ, by forming a transcriptional activation complex at Th2 cytokine gene loci and a transcriptional repressive complex at the Tbx21 (encoding T-bet) gene locus in Th2 cells, respectively (Figure 3) [90]. TET proteins have also been linked to the differentiation and function of CD4$^+$ T cells (Figure 3). A study analyzing 5-hydroxymethyl-cytosine (5hmC) patterns in CD4$^+$ peripheral T cells found a positive correlation between 5hmC alterations at gene bodies of transcription factors, including *Tbx21* and *Gata3*, which drive differentiation into Th1 and Th2 subtypes and their expression levels, respectively [93–95]. Similarly, another study suggested similar Th1/2-specific 5hmC alterations during differentiation of human CD4$^+$ T cells [93]. In addition, a Tet2 knock-out (KO) mouse model was reported to have Th1 and Th17 cells producing low IFNγ and IL-17, respectively [94]. Overall, these studies suggest that not only DNA methyltransferases (DNMT1 and DNMT3A/B) are required for regulating differentiation of CD4$^+$ T cells into various subtypes but also DNA readers and DNA demethylases such as MBD2 and TET proteins, respectively [22,93–95].

Regulatory T Cells (Tregs)

Tregs can be natural (nTreg), i.e., derived from the thymus, or Ag-induced (iTreg), i.e., differentiated from naïve T cells by TGF-β and IL-2 in the periphery (Figure 3) [31,36,37]. Tregs typically act as pro-tumor, are immunosuppressive, and are associated with poorer prognosis in several cancer types [35,96]. Tregs block the activation of CD8$^+$ T cells through expressing cytotoxic T lymphocyte antigen 4 (CTLA4), which is an inhibitory molecule for CD8$^+$ T cells [31,96]. In addition, inflammation enhances Treg function because prostaglandin E2 (PGE2) causes differentiation of Tregs. Tregs were also reported to block killing by NK cells, and thus downregulate both adaptive and innate anti-tumor immunity [31,97].

A master regulator switch for Tregs is FOXP3, which is required for its functions (Figure 3). DNA methylation of *FOXP3* together with intergenic CD3G/CD3D regions were utilized as a biomarker for TILs and Treg quantification in several tumor tissues [98]. This DNA methylation-based quantification of immune cells was even comparable to flow cell cytometry and outperformed IHC techniques. Using differential methylation analysis between nTreg, naive CD4$^+$ T cells, activated CD4$^+$ T cells, and iTreg, Lal et al. found a unique CpG site at the enhancer of *Foxp3* that was unmethylated in nTreg compared to other Tregs that were heavily methylated at this locus [99]. Demethylation by DNMTi (Aza) promoted acetylation of histone 3, and interaction with TIEG1 and Sp1, which ultimately led to upregulation of *Foxp3*. To study Tregs in non-small-cell lung cancer (NSCLC) using a co-culture system, Ke et al. showed demethylation of *FOXP3* in the promoter region increased FOXP3 expression in Tregs, which led to downregulation of immune response in the TME (Figure 3) [100].

Treg-specific demethylated region (TSDR) is a CpG dinucleotide dense region which is within the conserved non-coding sequences 2 (CNS2) located in the first intron of the *FOXP3* gene [101]. DNA demethylation at the TSDR region can discriminate between Tregs and other cell types [102]. Interestingly, using ChIP analysis, Wang et al. showed that MBD2 binds to the TSDR site of the *FOXP3* locus in Tregs [103]. Knocking down Mbd2, in vitro and in vivo, reduced the number of Tregs and impaired Treg-suppressive function (Figure 3). Surprisingly, this was due to increased methylation (>75%) of the TSDR in the Mbd2-/- Tregs because: (i) WT Tregs had a complete TSDR demethylation; and (ii) expressing Mbd2 in Mbd2-/- Tregs rescued the TSDR demethylation. TET proteins are essential for stable Foxp3 expression because they were shown to demethylate the CNS2 region as well as another non-coding sequence, CNS 1, in the *Foxp3* gene (Figure 3) [104,105]. Deletion of Tet2/3 in $CD4^+$ T cells of mice led to hypermethylation of CNS1 and 2 in Tregs. Moreover, deletion of *Tet1/2* also resulted in hypermethylation of CNS2 [104,105]. Overexpression of the TET1 catalytic domain in $CD4^+$ T cells also resulted in partial demethylation of CNS2 and differentiation of $CD4^+$ into iTregs in vitro [106]. TET2 protein may function via interacting with the MBD2 protein because loss of MBD2 resulted in hypermethylation of TSDR in CNS2 [103]. In TME, higher demethylation at the TSDR FOXP3 locus in adjacent normal tissues in colon cancer patient samples were associated with distant metastases and worse recurrence-free survival. The poor survival rates could be due to abnormal recruitment of nTregs in TME [101]. Collectively, these studies show a potential role of DNA methylation in controlling the effector function of Tregs through regulating the expression of the master switch FOXP3 of Tregs.

2.2.2. $CD8^+$ T Cells

$CD8^+$ T cells control tumor growth and kill tumor cells directly in an Ag-specific manner within the TME [31,36,37]. The $CD8^+$ T cells, upon recognizing an Ag, can undergo activation and clonal expansion, thereby carrying out effector functions, such as cytokine production (IFNγ, TNFα), and these processes are regulated by DNA methylation (Figure 4) [31,36,37,78].

Epigenetic mechanisms that govern these processes are largely unknown. A study was conducted to delineate these mechanisms and compared Ag-specific naive and effector $CD8^+$ T cells after stimulating them with an acute CMV viral infection [107]. The DNA methylome was rewired globally upon effector differentiation of $CD8^+$ T cells, and a negative correlation between DNA methylation at gene promoters and gene expression was observed. The DMRs were associated with transcription binding sites and promoters of genes that control effector $CD8^+$ T cell function. For instance, DMR at promoters of *Gzmb*, which encodes a serine protease granzyme B essential for cytolytic function, and *Zbtb32*, which encodes a transcription factor induced in activated lymphocytes, was demethylated and had high expression in the effector $CD8^+$ T cells compared to naïve cells. In contrast, *Ccr7*, *Ccr2*, *Ccr9*, and *Tcf7*, essential for naïve T cell development and homeostasis, were methylated and had reduced expression. Another study examined *Dnmt3a* KO $CD8^+$ T cells and found effector functions to be normal; however, *Dnmt3a* KO T cells developed into fewer terminal effector cells and more memory precursors in a T-cell intrinsic manner. This was due to ineffective repression of *Tcf1* expression by Dnmt3a in *Dnmt3a* KO T cells [108]. The role of Dnmt1 in regulating T cell activation and production of Ag-specific effector and memory $CD8^+$ T cells after a viral infection was also investigated. Dnmt1 was knocked-out at the time of activation and *Dnmt1*-/- had marked reduction (>80%) in Ag-specific clonal expansion in effector $CD8^+$ T cells but only moderately affected memory $CD8^+$ T cells. Even in reduced T cell expansion, the infection was effectively controlled. Thus, Dnmt1 may be required for proliferation of Ag-specific $CD8^+$ T cells but not differentiation into effector and memory $CD8^+$ T cells [109].

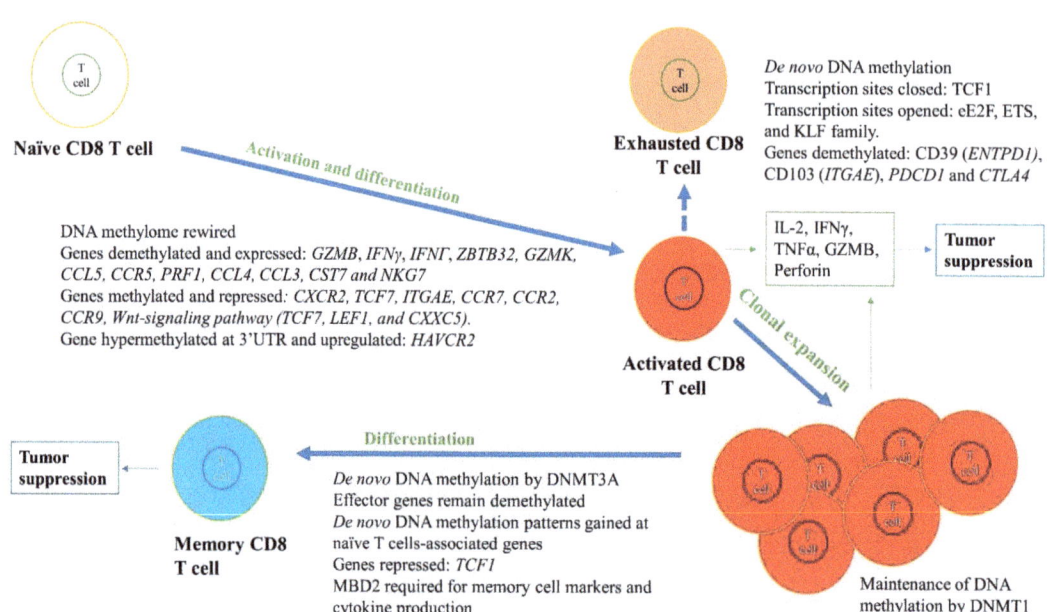

Figure 4. Role of DNA methylation in regulating differentiation and activation of naïve CD8+ T cells into effector cells, including cytotoxic and memory T cell subtypes. DNA methylation changes during differentiation and activation can lead to formation of subtypes of CD8+ T cells. For instance, cytotoxic CD8+ T cells are formed due to whole genome remodeling and expression, and repression of various genes in naïve CD8+ T cells. The genes that are essential for activation, proliferation, and effector functions are demethylated and highly expressed, such as *IL-2*, *IFNG* or *IFNγ*, and *GZMB*, whereas genes that are not required are methylated and repressed (e.g., *TCF7*). Although methylation and gene silencing are usually positively correlated, there are examples of genes that could be methylated and expressed, such as *HAVCR2*, depending upon the precise location of the methylation. In memory CD8+ T cell differentiation, effector genes remain demethylated, whereas methylation at naïve T cell-associated genes are gained and repressed, such as in the case of *TCF1*. These methylation levels are maintained by DNMT1, DNMT3A, and TET2. The green boxes indicate the cytokines released from differentiated cells. These immune cells and released cytokines can further lead to tumor suppression and elimination. However, CD8+ T cells can become exhausted in the TME, highly expressing exhaustive markers such as CD39, CD103, PD-1, and CTLA-4. The exhausted CD8+ T cells are non-functional and produce low amounts of effector cytokines (e.g., IFNγ).

Memory CD8+ T cells, which are formed from a subset of effector CD8+ T cells after Ag/pathogen clearance, remain in the blood and lymphoid organs for a long time, giving long-term immunity. These memory CD8+ T cells also resemble naïve T cells as they have pluripotency and can travel to lymph nodes and the spleen. A study comparing memory CD8+ T cells with terminal effector cells found that memory cells formed from effector cells gain de novo DNA methylation patterns at naïve CD8+ T cells-associated genes while becoming demethylated at the loci that are effector-specific genes [110]. *Dnmt3a* KO in effector T cells resulted in reduced DNA methylation and quicker re-expression of naïve T cell genes, decreasing the time for memory T cell development. Therefore, in memory CD8+ T cells, DNA methylation repression at the naïve-related genes can be reversed and effector genes remain demethylated without the need for memory cells to differentiate, allowing them to become faster effector CD8+ T cells upon Ag/pathogen re-exposure.

Long-lived memory CD8+ T cells can be identified with a few markers, such as CD127^{hi} and KLRG1^{low}. CD127^{low} and KLRG1^{hi} are typically markers for short-lived effector CD8+ T cells. Moreover, transcription factors, including T-bet, Eomes, Blimp-1, Bcl-6, Irf4, and Runx3, define the fate of activated CD8+ T cells and these are further regulated by DNA methylation. In a mouse model with Tet2-deficient CD8+ T cells infected with lymphocytic choriomeningitis virus (LCMV), CD8+ T cells differentiated more into long-lived memory cells having gp33-specific memory markers, KLRG1^{low} CD127^{hi}, and less into effector short-lived effector cells (CD127^{low} and KLRG1^{hi}) [111]. These memory-like cells had markers of central memory cells expressing CD27, CD62L, and CXCR3, and high expression of transcription factor Eomes compared to wild-type Tet2. Furthermore, these memory cells also had superior pathogen control upon re-challenge. Global methylation analysis revealed several DMRs that gained 5mC/5hmC in Tet2-deficient cells versus WT CD8$^+$ T cells. These DMRs were present in transcriptional regulator genes known to be vital for effector and memory CD8$^+$ T cell differentiation. Pharmacological inhibition of TET2 by 2-HG also showed similar results to genetic *Tet2* KO, such as a decrease in 5hmC and an increase in Eomes and CD62L expression [112]. The role of MBD2 in the differentiation of naïve CD8$^+$ T cells into effector and memory cells was determined following LCMV infection. In contrast to Tet2-deficient CD8$^+$ T cells, Mbd2-deficient mice had a reduced number of Ag-specific memory CD8$^+$ T cells and an effective primary effector CD8$^+$ T cell response leading to a rapid viral clearance. Essentially, generation of precursor memory CD8$^+$ T cells (IL-7Rα^{high}) was delayed and the MBD2 KO memory cells were phenotypically defective with altered memory cell markers (e.g., IL-7Rα, KLRG-1, CD27) and cytokine production, and were unprotective against re-challenge (Figure 4) [113]. These studies suggest a key role of MBD2 and TET proteins in regulating the differentiation of CD8$^+$ T cells into memory versus effector cells. Together, the above studies show the crucial role of DNA methylation in differentiation of naïve CD8$^+$ T cells into effector cytotoxic CD8$^+$ T cells and memory CD8$^+$ T cells (Figure 4).

3. Role of DNA Methylation in Regulating T Cell Exhaustion

If an Ag is exposed to CD8$^+$ T cells for a long time, CD8$^+$ T cells can become non-functional or exhausted, resulting in reduced effector functions, such as decreased cytokine production (IFNγ and TNF-α) and/or loss of cytotoxicity (e.g., low granzyme B production). Exhausted T cells generally have high surface expression of multiple inhibitory molecules, such as PD-1, TIM3, LAG3, TIGIT, and 2B4, and transcription factors associated with high PD-1 expression are T-bet, Eomes, and YY1 [114–116]. YY1 is a key transcription factor that can regulate the inhibitory molecules PD-1, LAG3, and TIM3 expression, and was shown to have downregulated IL-2 via EZH2 activation, features characteristic of exhausted T cells [114–116]. In human patient tumors treated with immune checkpoint inhibitors (ICPi), around 72% of TILs were found to be dysfunctional. These TILs showed different stages of differentiation and interestingly, had higher proliferation rates compared to effector T cells, ruling out the possibility that exhausted T cells have low proliferation rates [114–116].

CD8$^+$ TILs become exhausted and lose their effector functions in the TME due to numerous factors, such as immunosuppressive mechanisms by tumor cells. Analyzing the transcriptome and methylome of CD8$^+$ TILs in the TME of colorectal cancer simultaneously, Yang et al. confirmed tumor-reactive TILs have an exhausted tissue-resident memory signature [117]. They showed tumor-reactive markers CD39 and CD103 of CD8$^+$ TILs were demethylated and CD8$^+$ TILs had an exhausted phenotype, including high expression of CTLA4, HAVCR2, LAYN, and TOX [117,118]. To delineate changes in methylation from naïve to cytotoxic CD8$^+$ T cell phenotype and cytotoxic to exhausted CD8$^+$ T cell phenotype, promoter methylation of these cells was compared. Naïve CD8$^+$ T cells showed the most promoter demethylation compared to cytotoxic and exhausted T cells; however, essential cytotoxic CD8$^+$ T cell effector genes, including *PRF1, GZMB, IFNG, CCL4, CCL3, CST7*, and *NKG7*, went through hypermethylation to hypomethylation from naïve to cytotoxic CD8$^+$ T cell differentiation, respectively [117]. For exhausted T cells, two inhibitory checkpoint markers, *PDCD1* (encoding PD-1) and *CTLA4*, were demethylated within cytotoxic CD8$^+$ T cells. Moreover, *LAG3* and *LAYN* were also differentially methylated from naïve to cytotoxic CD8$^+$ T cell transition [117]. Therefore, these studies determined that aberrant DNA methylation at certain gene loci could result in T cell exhaustion (Figure 4) [116–118].

Interestingly, DNA methylation could determine if T cell exhaustion can be reversed. In chronic LCMV infection, the PD-1 gene promoter of the effector CD8$^+$ T cells remained unmethylated, whereas the exhausted T cells showed complete demethylation [116,119]. Furthermore, studies analyzing the chromatin states using transposase-accessible chromatin using sequencing (ATAC-seq) have determined two chromatin states that define exhaustion: one in which T cell factor 1 (TCF1) transcription sites are closed and another in which transcription sites for eE2F, ETS, and KLF family proteins are opened (Figure 4) [120]. Low TCF1 expression is associated with the low effector function of CD8$^+$ T cells and non-renewal of CD8$^+$ effector T cells [121]. DNA methylation can, therefore, regulate the state of exhaustion of CD8+ T cells, which, due to the reversable nature of DNA methylation patterns, provides new opportunities for therapeutic intervention.

4. Role of m^{6}A RNA Methylation in Immunity

m^{6}A has various functions, including mRNA stability, translation, splicing, and phase separation, and also takes part in cell differentiation and development [23–30]. These essential functions indicate that m^{6}A RNA methylation can potentially regulate immunity. Although the role of m^{6}A RNA methylation in immunity has not been fully elucidated, few studies have reported its role in both innate and adaptive immune response [122–131].

4.1. Role of m^{6}A RNA Methylation in Innate Immune Response

Certain DNA and RNA molecules can be detected by the innate immune system as non-self entities via PPRs, such as TLRs. For instance, a study investigated the mammalian innate immune response of DCs through stimulation with DNA, RNA, and modified RNAs, including m^{6}A-modified RNA [128]. Although DNA containing methylated CpG were not stimulatory, RNA could be stimulatory or not stimulatory depending upon modification on RNA [123,129,130]. Modified RNA, including m^{6}A modification exposed to DCs, did not activate their TLR3, TLR7, and TLR8, and led to lower cytokines and activation markers, compared to DC stimulated with unmodified RNA that activated TLRs [123]. Unmodified RNA that is present in bacteria could trigger innate immune response to bacterial infection, whereas highly modified RNA, such as mammalian RNA, would not, indicating a role of RNA modifications in selectively triggering the immune system against pathogens. Indeed, DC are activated via m^{6}A RNA modifications and lack of METTL3 can result in lack of DC maturation [123,128,129]. Regulators of m^{6}A RNA, METTL14, and ALKBH5 were reported to regulate type I IFN production triggered by dsDNA or HCMV [125,129,130]. Depletion of METTL14 decreased viral replication and induced IFNβ1 mRNA production and stability upon dsDNA and HCMV infection, whereas ALKBH5 depletion had an opposing effect (with the exception of affecting IFNβ1 mRNA stability). This control of IFNβ1 mRNA was due to m^{6}A modification at the coding sequence and the 3' UTR of the *IFNβ1* gene. Another study reported increased interferon-stimulated genes upon METTL3 (m^{6}A writer) or YTHDF2 (m^{6}A reader) deletion. Specifically, following deletion of METTL3 or YTHDF2, mRNA of IFNβ was modified at m^{6}A, increasing its stability [125,129,130]. These studies indicate that m^{6}A can play a role in the negative regulation of anti-viral response by dictating increased turnover of IFN mRNAs. One study established a key link of m^{6}A to cellular antiviral response by showing that m^{6}A induces antiviral immunity as it regulates crucial proteins of innate immunity [131]. Mechanistically, m^{6}A demethylase ALKBH5 is recruited by RNA helicase DDX46 to remove m^{6}A from 3' UTRs of genes encoding TRAF3, TRAF6, and MAVS, thereby reducing export of their transcript out of the nucleus and subsequently preventing production of type I IFNs.

4.2. Role of m^{6}A RNA Methylation in Adaptive Immune Response

m^{6}A RNA methylation has also been shown to regulate adaptive immune responses. Similar to DNA methylation regulating differentiation of CD4$^+$ T cells into various subtypes, m^{6}A RNA methylation was shown to regulate differentiation of CD4$^+$ T cells [124]. The authors utilized a conditional KO mouse model (CD4$^+$-CRE conditional Mettl3 $^{flox/flox}$) to delete Mettl3 in CD4$^+$ T cells [124]. After validating Mettl3 deletion, they checked for thymocyte differentiation or cellularity and found no difference compared to WT mouse.

However, the proportion of naïve T cells (CD44^{lo} CD62L^{hi}) was higher in spleens and lymph nodes compared to WT. When the function of Mettl3-/- CD4$^+$ T cells was compared to WT, they observed normal sensitivity to TCR signaling; however, T helper polarization had abnormalities. For instance, the KO CD4$^+$ T cells had a significant reduction in differentiation into Th1 and Th17 cells, but increased differentiation into Th2 cells. In-depth analysis showed that m^{6}A targets the mRNA of the IL-7 protein, which regulates T cell homeostatic proliferation and differentiation to various subtypes upon numerous external stimuli. SOCS proteins are adaptors which bind to cytokine receptors, such as the IL-7 receptor, thereby preventing STAT5 and downstream signaling [126,129]. SOCS proteins are produced immediately in response to acute stimuli but are degraded quickly and have short half-lives [126,129]. The m^{6}A modification was shown to regulate the degradation of the *Socs* genes, via the IL-7-JAK1/STAT5 signaling pathway, and without m^{6}A, Socs mRNA persists, leading to high levels of SOCS proteins and reduced sensitivity to IL-7. This study indicates that m^{6}A not only regulates CD4$^+$ T cells differentiation but also T cell homeostasis [124]. Using a similar Mettl3 conditional KO mouse model, the authors analyzed the Tregs subset (Mettl3-/- and WT) of CD4$^+$ T cells and found that Mettl3 -/- Tregs mice developed severe autoimmune disorders compared to WT, suggesting loss of m^{6}A modification can lead to loss of Treg immune suppressive functions [127]. In addition to the writer of m^{6}A, readers have shown potential in regulating immune response. As such, compared to WT, a direct reader of m^{6}A, Ythdf1 KO mice showed better cross-presentation of tumor antigens in DC and better cross-priming with CD8$^+$ T cells, leading to high Ag-specific CD8$^+$ T cells in response to tumors [122]. Specifically, binding of Ythdf1 at the m^{6}A of transcripts encoding lysosomal proteases lead to increased translation of these lysosomal proteases' (cathepsins) transcripts in DCs, whereas inhibition of Ythdf1 led to inhibition of these cathepsins, resulting in enhanced cross-presentation by DCs and cross-priming of CD8$^+$ T cells by DCs. Indeed, mature DCs were reported to have higher expression of writer complex, including METTL3, than naïve DCs [128]. In addition, patient tumor samples that had low YTHDF1 expression had higher tumor-infiltrating CD8$^+$ T cells [122]. Interestingly, mice with Ythdf1 KO showed a better response to ICPi (anti-PD-L1) therapy than the Ythdf1 WT [122].

Collectively, the above studies show the essential role of m^{6}A RNA methylation in regulating innate and adaptive immune responses. The role of RNA methylation in immunity is still at its infancy and requires further research for discovery of novel therapeutic targets for its translational potential.

5. Targeting Methylation in the Treatment of Human Disease

Alterations in methylation have been strongly associated with the initiation and progression of cancer [132]. Compared to normal control tissues in tumors, DNA hypomethylation occurs at global and gene-specific levels, which results in genomic instability and activation of silenced oncogenes [133]. In contrast, DNA hypermethylation occurs at the promoter regions of tumor suppressor genes (TSGs), which leads to their silencing [133]. With our increasing understanding of the role of methylation in cancer and immunity, further efforts are now aimed at its translational potential to develop new therapeutic strategies that can alter the methylation landscape. Towards these goals, both DNA hypo- and hyper-methylation can serve as viable targets which, unlike genetic changes, are both dynamic and reversible.

5.1. Targeting DNA Hypermethylation

Several DNA hypomethylating agents have been developed that target DNA hypermethylation. However, among these DNA methyltransferase inhibitors (DNMTi), 5-azacytidine (Vidaza®) and 5-aza-2′deoxycytidine (Decitabine, Dacogen®) have been approved by the Food and Drug Administration (FDA) [16]. Because multiple hematologic malignancies are linked to abnormal DNA methylation patterns, DNMTi were first tested in these cancers. Among these, myelodysplastic syndromes (MDS) comprising a group of hematologic disorders derived from abnormal progenitor cells were the first to be evaluated. Patients with MDS have hypoproliferative bone marrow and a risk of developing different forms of acute leukemia [51]. The inhibitor 5-azacytidine was first tested on MDS patients, where it showed improved response rates, lower transformation to acute leukemia, and prolonged survival [134], and 5-aza-2′deoxycytidine showed similar clinical outcomes [135]. Both 5-azacytidine and 5-aza-2′deoxycytidine have also shown success in a clinical setting for acute myeloid leukemia (AML) and chronic myelomonocytic leukemia (CMML) [16].

Following the clinical success of DNMTi with hematologic malignancies, DNMTi were also tested in solid tumors [136–138]. Although DNMTi showed a good response in patients with ovarian cancer and non-small cell lung cancer, the response was highly variable and less effective in other solid tumors [136–138]. DNMTi has shown the greatest potential in combination with cytotoxic agents or immunotherapies. With cytotoxic agents, DNMTi appear to sensitize tumors and increase the efficacy of conventional cytotoxic agents, even for patients who were previously resistant to the cytotoxic agents alone [139]. Recently, studies have established that malignant cells escape host immune recognition by acquiring an immune evasive phenotype through epigenetically downregulating essential molecules for cancer and immune interactions [35]. For instance, these mechanisms include suppression of tumor associated antigens (TAAs), reducing the expression of many components of antigen processing and presentation machinery (APM), and decreasing co-stimulatory molecules, stress-induced ligands, and death receptors [35]. DNMTi and histone deacetylase inhibitors (HDACi) reverse the immune evasive phenotype, for example, by upregulating the expression of TAAs and APM components on tumor cells, which helps the immune system to recognize and eliminate tumor cells [35,140–142]. Additionally, T cell exhaustion can also be reversed using DNMTi in mouse models, resulting in enhanced anti-cancer immunity [143,144]. DNMTi can also trigger a state of "viral mimicry" by activating dsRNAs, thereby increasing type I interferon responses [35,145]. In addition, DNMTi and HDACi increased cytotoxic activity of CD8 T cells and NK cells, and increased these anti-tumor cells' immune infiltration in the TME while reducing pro-tumor macrophage infiltration in a murine ovarian cancer model [75]. These anti-cancer effects were further elevated in triple combination with ICPi (anti-PD-1), which reduced the tumor burden and provided longest overall survival. Collectively, the above studies indicate priming of the immune system by DNMTi (and HDACi), thereby increasing the efficacy of ICPi therapy.

5.2. Targeting DNA Hypomethylation

In cancer, promoter hypermethylation of TSGs and silencing of TSGs resulting in tumorigenesis have been the focus of the last few decades, resulting in the discovery of DNMTi [146–149]. By comparison, a phenomenon that is relatively underestimated is genome-wide DNA hypomethylation, which occurs in various solid tumors [133,150]. Several studies have also demonstrated that gene-specific and global hypomethylation play a crucial role in the initiation and progression of cancer [7,133]. However, there is still no approved agent that targets DNA hypomethylation. Currently, the most studied approach to target DNA hypomethylation uses SAM. SAM is a natural and universal methyl donor of all methylation reactions [151,152]. As such, SAM donates its methyl group to key cellular components including proteins, nucleic acids (RNA and DNA), lipids, and secondary metabolites to modulate several physiological functions [151–153].

Although studies investigating the effect of SAM on the immune system are still lacking, SAM has been shown to modulate the immune system [154–167]. SAM manipulates methylation levels, which further modulates T cell functions by regulating the TCR signaling pathway, impairing Th1/Th2 cytokines release, and decreasing T cell proliferation and activation in autoimmunity [154]. Moreover, SAM reduces IL-1 levels in rats with cecal ligation and puncture. In macrophages, SAM inhibited LPS-induced gene expression via modulation of H3K4 methylation [155]. Similarly, deregulation of SAM levels can result in immune disorders, such as in liver inflammatory diseases. Molecular links between SAM and innate immune functions were reported in which low levels of SAM were shown to affect hepatic PC synthesis and may limit stress-induced protective gene expression upon infection [156]. In addition, SAM prevented upregulation of TLR signaling by blocking the overexpression of TLR2/4 and their downstream partners MyD88 and TRAF-6 in the Mallory–Denk body, forming hepatocytes [157].

Interestingly, studies have shown that SAM is essential for T cell activation and proliferation [154–167]. In activated T cells, both the SAM quantity and the rate of SAM utilization increase dramatically via increased transcription of *MAT2A*, which encodes the catalytic subunit of MATII and is vital for SAM biosynthesis [161,162,164,165]. Blockage of SAM synthesis resulted in blocked T cell proliferation [160]. Furthermore, SAM was shown to be indispensable for T cell proliferation and activation by decreasing both caspase-3 activity and apoptosis in ethanol-related activation-induced cell death (AICD) [159]. Furthermore, SAM was shown to lower the suppressive capacity of Tregs (nTreg cells) by methylating the *FOXP3* gene, thereby reducing its protein and mRNA expression in a dose-dependent manner. SAM was also found to decrease expression of an immunosuppressive cytokine, IL-10, and increase expression of IFNγ [168].

Aberrant methylome is a common consequence of a disrupted SAM cycle associated with transformation of cells towards tumorigenesis [152,169,170]. SAM, which increases DNA methylation, has been shown to cause significant anti-tumor effects in breast, osteosarcoma, prostate, hepatocellular, gastric, colon, and other cancers [151,152,169–174]. In addition, SAM levels are depleted by cancer cells through various mechanisms, such as increased conversion of SAM to by-products, which reduces the methylation potential of cancer cells [175,176]. A recent study has shown that an essential immune evasive mechanism used by tumor cells is depriving the CD8+ T cells of SAM and methionine (the pre-cursor of SAM) in the TME. This makes CD8+ T cells non-functional and unresponsive to ICPi [175]. Indeed, we showed that SAM in combination with ICPi (anti-PD-1) significantly reduced tumor volume and weight compared to monotherapy in a syngeneic mouse model of advanced melanoma [177]. This effect was partially due to the elevated activation, proliferation, and cytokine production of CD8 T cells. We also observed increased tumor infiltration of CD8 T cells, a higher number of polyfunctional CD8 T cells, and a lower number of exhausted CD8 T cells in the TME. The above studies show a potential of SAM, a co-factor of methylation, in targeting aberrant DNA methylation patterns in the TME as a novel anti-cancer approach that also enhances anti-cancer immunity. Therefore, the effect of SAM on anti-cancer immunity should be studied comprehensively in future studies.

5.3. Targeting m^6A RNA Methylation

The role of DNA methylation in regulating the immune system and cancer has been the focus of research for more than three decades. Regulation of immunity and cancer by m^6A RNA methylation is still at its infancy. However, novel studies have shown the potential of targeting RNA methylation in cancer. For instance, FTO inhibition through selective inhibitors, such as Meclofenamic acid (MA), MA2, and R-2-hydroxyglutarate (R-2HG), have shown potent anti-cancer activity in several cancers including AML, glioblastoma multiforme (GBM), and colorectal cancer (CRC) [26,30,178]. In contrast to other RNA demethylase inhibitors, Rhein was identified to be reversibly bound to the FTO catalytic domain via a crystal structure approach and shown to increase m^6A RNA methylation levels [178,179]. Rhein is attractive as it is a natural compound and selective against FTO and not ALKBH5 [179]. Rhein has shown significant anti-cancer activity in various cancers; however, comprehensive in vivo evidence is still lacking and would require further in-depth studies [180]. Citrate was identified as an ALKBH5 inhibitor via a crystal structure approach; however, the effect of citrate on ALKBH5 demethylase activity in reducing cancer growth and progression is yet to be determined [181].

Although the inhibitors for RNA methylation regulators have been identified, none of them have been tested in a clinical setting. Furthermore, the effect of these pharmacological inhibitors of RNA methylation on the immune system is yet to be determined. Along this line, recently, RNA demethylase FTO was reported to promote tumorigenesis in melanoma and knockdown of FTO-reduced resistance to ICPi (anti-PD-1) therapy [182]. FTO regulates important immune genes (including PD-1, CXCR4, and SOX10 genes) and KD of FTO led to increased mRNA decay of these genes through the m6A reader YTHDF2. Furthermore, KD of FTO sensitized melanoma cells to IFNγ, thereby reducing resistance to anti-PD-1 therapy. Similarly, RNA demethylase ALKBH5 KO showed significant reduction in tumor growth and prolonged mouse survival during ICPi therapy in B16 melanoma and CT26 colon cancer mouse models [183]. This was due to ALKBH5 altering gene expression and splicing that leads to changes in lactate levels in the TME. These metabolic changes result in decreased Treg and MDSCs infiltration in the TME. Interestingly, the authors also tested an ALKBH5 inhibitor and showed similar phenotype to the ALKBH5 KO model. These studies not only show the inhibition of m^6A demethylases as a potential anti-cancer target but also their potential in anti-cancer immunity within the TME.

6. Conclusions

The role of DNA and RNA methylation in regulating the differentiation and activity of immune cells within the TME is key to determining the fate of tumor growth or suppression (Figure 5). A pro-cancer TME has immune cells expressing pro-tumor cytokines that lead to tumor growth and progression, whereas the reverse is seen in the anti-cancer TME. Precise methylation patterns change gene expression, leading to specific immune cell subtypes. For instance, DNA demethylation and high expression of *IL4* and *FOXP3* genes occur in Th2 and Tregs, respectively. In contrast, DNA demethylation and high expression of *IFNγ* and *IL2* genes occur in both Th1 and CD8 T cells, which results in a better anti-cancer immune response. Studies should further investigate the effect of DNA and RNA methylation on transcriptional regulation of immune cells along with tumor cells in a time-dependent manner in order to uncover the complexity of the TME at various stages of cancer growth and progression. As explained earlier, the balance between pro- and anti-cancer immune cells within the TME is key to tumor progression or suppression. However, most studies investigating the role of methylation have focused only on one immune cell subtype. Future studies should investigate various immune subtypes simultaneously. These comprehensive studies will provide deeper insights into the interplay between the immune system and cancer, and allow discovery of novel epi-therapies that can enhance the immune system against cancer and other pathologies. Targeting methylation is a particularly attractive anti-cancer strategy because it is dynamic and reversible. For instance, DNMTi that target DNA hypermethylation can also enhance the efficacy of immunotherapies. Similarly, SAM, targeting DNA hypomethylation, has shown profound effects in combination with ICPi. Along the same line, inhibitors of m^6A RNA demethylases have shown potential in enhancing anti-cancer immunity. However, further comprehensive studies are required to delineate the mechanism of action before these inhibitors can be tested in a clinical setting. In addition, SAM, which donates methyl groups to RNA, has shown significant anti-cancer activity in numerous cancer models by regulating DNA methylation. It is yet to be determined if SAM causes inhibition of tumor growth and metastasis through modulating m^6A RNA methylation levels. Although the efficacy of epigenetic-based therapeutic strategies targeting tumor and immune cells needs further elucidation, the current state of knowledge provides compelling evidence to suggest that it will be effective in blocking cancer progression and reducing cancer associated morbidity and mortality.

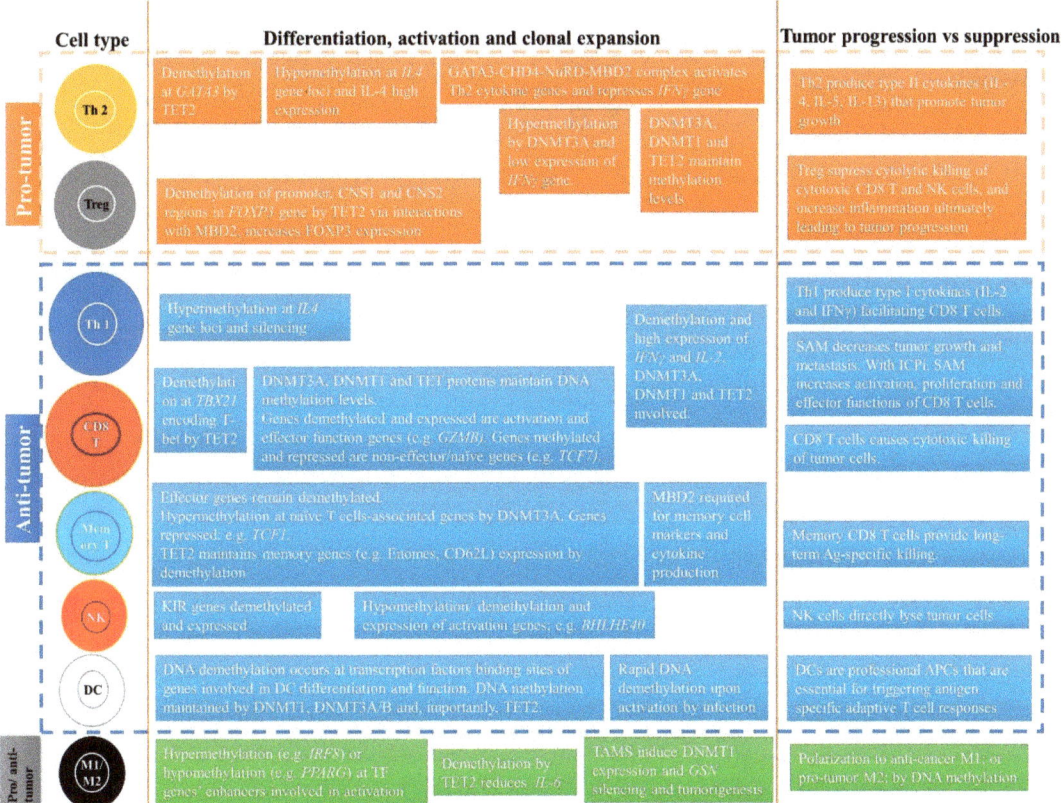

Figure 5. Summary of the role of DNA methylation and co-factor (s-adenosylmethionine, SAM) in regulating differentiation, activation, and proliferation of pro- and anti-cancer immune cells. The pro- or anti-tumor effect of the immune cells in the TME is also described. Abbreviations: Th2, CD4+ helper T cell 2; Tregs, regulatory T cell; Th1, CD4+ helper T cell 1; CD8 T, CD8 cytotoxic T cells; Memory T cells, CD8 memory T cells; NK, natural killer cell; DC, dendritic cell; M1, macrophage M1 subtype; M2, macrophage M2 subtype; TAMS, Tumor associated macrophages; ICPi, Immune checkpoint inhibitors; s-adenosylmethionine, SAM.

Author Contributions: Writing—research, review, and editing, A.M.; review, editing, and supervision, S.A.R. All authors have read and agreed to the published version of the manuscript.

Funding: This work was supported by grants from the Canadian Institutes for Health Research, PJT-156225 to SAR.

Institutional Review Board Statement: Not applicable.

Informed Consent Statement: Not applicable.

Data Availability Statement: Not applicable.

Conflicts of Interest: The authors declare no conflict of interest.

References

1. Allis, C.D.; Jenuwein, T. The molecular hallmarks of epigenetic control. *Nat. Rev. Genet.* **2016**, *17*, 487–500. [CrossRef] [PubMed]
2. Bird, A. DNA methylation patterns and epigenetic memory. *Genes Dev.* **2002**, *16*, 6–21. [CrossRef] [PubMed]
3. Sharma, S.; Kelly, T.K.; Jones, P.A. Epigenetics in cancer. *Carcinogenesis* **2010**, *31*, 27–36. [CrossRef] [PubMed]
4. Sandoval, J.; Esteller, M. Cancer epigenomics: Beyond genomics. *Curr. Opin. Genet. Dev.* **2012**, *22*, 50–55. [CrossRef] [PubMed]

5. Feinberg, A.P.; Vogelstein, B. Hypomethylation distinguishes genes of some human cancers from their normal counterparts. *Nature* **1983**, *301*, 89–92. [CrossRef]
6. Hsieh, J.; Gage, F.H. Epigenetic control of neural stem cell fate. *Curr. Opin. Genet. Dev.* **2004**, *14*, 461–469. [CrossRef]
7. Robertson, K.D. DNA methylation and human disease. *Nat. Rev. Genet.* **2005**, *6*, 597–610. [CrossRef]
8. Jones, P.A.; Baylin, S.B. The fundamental role of epigenetic events in cancer. *Nat. Rev. Genet.* **2002**, *3*, 415–428. [CrossRef]
9. Patil, V.; Ward, R.L.; Hesson, L.B. The evidence for functional non-CpG methylation in mammalian cells. *Epigenetics* **2014**, *9*, 823–828. [CrossRef]
10. Saxonov, S.; Berg, P.; Brutlag, D.L. A genome-wide analysis of CpG dinucleotides in the human genome distinguishes two distinct classes of promoters. *Proc. Natl. Acad. Sci. USA* **2006**, *103*, 1412–1417. [CrossRef]
11. Luo, C.; Hajkova, P.; Ecker, J.R. Dynamic DNA methylation: In the right place at the right time. *Science (N. Y.)* **2018**, *361*, 1336–1340. [CrossRef] [PubMed]
12. Neri, F.; Rapelli, S.; Krepelova, A.; Incarnato, D.; Parlato, C.; Basile, G.; Maldotti, M.; Anselmi, F.; Oliviero, S. Intragenic DNA methylation prevents spurious transcription initiation. *Nature* **2017**, *543*, 72–77. [CrossRef] [PubMed]
13. Schultz, M.D.; He, Y.; Whitaker, J.W.; Hariharan, M.; Mukamel, E.A.; Leung, D.; Rajagopal, N.; Nery, J.R.; Urich, M.A.; Chen, H.; et al. Human body epigenome maps reveal noncanonical DNA methylation variation. *Nature* **2015**, *523*, 212–216. [CrossRef] [PubMed]
14. McGuire, M.H.; Herbrich, S.M.; Dasari, S.K.; Wu, S.Y.; Wang, Y.; Rupaimoole, R.; Lopez-Berestein, G.; Baggerly, K.A.; Sood, A.K. Pan-cancer genomic analysis links 3′UTR DNA methylation with increased gene expression in T cells. *EBioMedicine* **2019**, *43*, 127–137. [CrossRef]
15. Lin, R.K.; Wang, Y.C. Dysregulated transcriptional and post-translational control of DNA methyltransferases in cancer. *Cell Biosci.* **2014**, *4*, 46. [CrossRef]
16. Mahmood, N.; Rabbani, S.A. Targeting DNA hypomethylation in malignancy by epigenetic therapies. *Adv. Exp. Med. Biol.* **2019**, *1164*, 179–196. [CrossRef]
17. Mahmood, N.; Rabbani, S.A. DNA methylation readers and cancer: Mechanistic and therapeutic applications. *Front. Oncol.* **2019**, *9*, 489. [CrossRef]
18. Deaton, A.M.; Bird, A. CpG islands and the regulation of transcription. *Genes Dev.* **2011**, *25*, 1010–1022. [CrossRef]
19. Hermann, A.; Gowher, H.; Jeltsch, A. Biochemistry and biology of mammalian DNA methyltransferases. *Cell. Mol. Life Sci.* **2004**, *61*, 2571–2587. [CrossRef]
20. Ropero, S.; Esteller, M. The role of histone deacetylases (HDACs) in human cancer. *Mol. Oncol.* **2007**, *1*, 19–25. [CrossRef]
21. Morris, M.R.; Latif, F. The epigenetic landscape of renal cancer. *Nat. Rev. Nephrol.* **2016**, *13*, 47–60. [CrossRef] [PubMed]
22. Lio, C.J.; Rao, A. TET Enzymes and 5hmC in adaptive and innate immune systems. *Front. Immunol.* **2019**, *10*, 210. [CrossRef] [PubMed]
23. Zaccara, S.; Ries, R.J.; Jaffrey, S.R. Reading, writing and erasing mRNA methylation. *Nat. Rev. Mol. Cell. Biol.* **2019**, *20*, 608–624. [CrossRef] [PubMed]
24. Sun, T.; Wu, R.; Ming, L. The role of m6A RNA methylation in cancer. *Biomed. Pharmacother.* **2019**, *112*, 108613. [CrossRef]
25. Shi, H.; Wei, J.; He, C. Where, when, and how: Context-dependent functions of RNA methylation writers, readers, and erasers. *Mol. Cell* **2019**, *74*, 640–650. [CrossRef]
26. Pan, Y.; Ma, P.; Liu, Y.; Li, W.; Shu, Y. Multiple functions of m(6)A RNA methylation in cancer. *J. Hematol. Oncol.* **2018**, *11*, 48. [CrossRef]
27. Ianniello, Z.; Fatica, A. N6-Methyladenosine role in acute myeloid leukaemia. *Int. J. Mol. Sci.* **2018**, *19*, 2345. [CrossRef]
28. Hsu, P.J.; Shi, H.; He, C. Epitranscriptomic influences on development and disease. *Genome Biol.* **2017**, *18*, 197. [CrossRef]
29. Deng, X.; Su, R.; Weng, H.; Huang, H.; Li, Z.; Chen, J. RNA N6-methyladenosine modification in cancers: Current status and perspectives. *Cell Res.* **2018**, *28*, 507–517. [CrossRef]
30. Chen, X.Y.; Zhang, J.; Zhu, J.S. The role of m(6)A RNA methylation in human cancer. *Mol. Cancer* **2019**, *18*, 103. [CrossRef]
31. Ostrand-Rosenberg, S. Immune surveillance: A balance between pro- and anti-tumor immunity. *Curr. Opin. Genet. Dev.* **2008**, *18*, 11–18. [CrossRef] [PubMed]
32. Page, D.B.; Bourla, A.B.; Daniyan, A.; Naidoo, J.; Smith, E.; Smith, M.; Friedman, C.; Khalil, D.N.; Funt, S.; Shoushtari, A.N.; et al. Tumor immunology and cancer immunotherapy: Summary of the 2014 SITC primer. *J. Immunother. Cancer* **2015**, *3*, 1–10. [CrossRef]
33. Pandya, P.H.; Murray, M.E.; Pollok, K.E.; Renbarger, J.L. The immune system in cancer pathogenesis: Potential therapeutic approaches. *J. Immunol. Res.* **2016**, *2016*. [CrossRef] [PubMed]
34. Dunn, G.P.; Old, L.J.; Schreiber, R.D. The immunobiology of cancer immunosurveillance and immunoediting. *Immunity* **2004**, *21*, 137–148. [CrossRef] [PubMed]
35. Dunn, J.; Rao, S. Epigenetics and immunotherapy: The current state of play. *Mol. Immunol.* **2017**, *87*, 227–239. [CrossRef]
36. Zhang, M.; Fujiwara, K.; Che, X.; Zheng, S.; Zheng, L. DNA methylation in the tumor microenvironment. *J. Zhejiang Univ. Sci. B* **2017**, *18*, 365–372. [CrossRef]
37. Sukari, A.; Nagasaka, M.; Al-Hadidi, A.; Lum, L.G. Cancer immunology and immunotherapy. *Anticancer Res.* **2016**, *36*, 5593–5606. [CrossRef]
38. Raval, R.R.; Sharabi, A.B.; Walker, A.J.; Drake, C.G.; Sharma, P. Tumor immunology and cancer immunotherapy: Summary of the 2013 SITC primer. *J. Immunother. Cancer* **2014**, *2*, 14. [CrossRef]

39. Demaria, O.; Cornen, S.; Daeron, M.; Morel, Y.; Medzhitov, R.; Vivier, E. Harnessing innate immunity in cancer therapy. *Nature* **2019**, *574*, 45–56. [CrossRef]
40. Gorentla, B.K.; Zhong, X.P. T cell receptor signal transduction in T lymphocytes. *J. Clin. Cell. Immunol.* **2012**. [CrossRef]
41. Galluzzi, L.; Buque, A.; Kepp, O.; Zitvogel, L.; Kroemer, G. Immunological effects of conventional chemotherapy and targeted anticancer agents. *Cancer Cell.* **2015**, *28*, 690–714. [CrossRef] [PubMed]
42. Wargo, J.A.; Reuben, A.; Cooper, Z.A.; Oh, K.S.; Sullivan, R.J. Immune effects of chemotherapy, radiation, and targeted therapy and opportunities for combination with immunotherapy. *Semin. Oncol.* **2015**, *42*, 601–616. [CrossRef] [PubMed]
43. Beatty, G.L.; Gladney, W.L. Immune escape mechanisms as a guide for cancer immunotherapy. *Clin. Cancer Res.* **2015**, *21*, 687–692. [CrossRef] [PubMed]
44. Galon, J.; Bruni, D. Approaches to treat immune hot, altered and cold tumours with combination immunotherapies. *Nat. Rev. Drug Discov.* **2019**, *18*, 197–218. [CrossRef] [PubMed]
45. Audia, J.E.; Campbell, R.M. Histone modifications and cancer. *Cold Spring Harb. Perspect. Biol.* **2016**, *8*, a019521. [CrossRef]
46. Hirschberger, S.; Hinske, L.C.; Kreth, S. MiRNAs: Dynamic regulators of immune cell functions in inflammation and cancer. *Cancer Lett.* **2018**, *431*, 11–21. [CrossRef]
47. Paladini, L.; Fabris, L.; Bottai, G.; Raschioni, C.; Calin, G.A.; Santarpia, L. Targeting microRNAs as key modulators of tumor immune response. *J. Exp. Clin. Cancer Res.* **2016**, *35*, 103. [CrossRef]
48. Sun, W.; Lv, S.; Li, H.; Cui, W.; Wang, L. Enhancing the anticancer efficacy of immunotherapy through combination with histone modification inhibitors. *Genes (Basel)* **2018**, *9*, 633. [CrossRef]
49. Yi, M.; Xu, L.; Jiao, Y.; Luo, S.; Li, A.; Wu, K. The role of cancer-derived microRNAs in cancer immune escape. *J. Hematol. Oncol.* **2020**, *13*, 25. [CrossRef]
50. Schuyler, R.P.; Merkel, A.; Raineri, E.; Altucci, L.; Vellenga, E.; Martens, J.H.A.; Pourfarzad, F.; Kuijpers, T.W.; Burden, F.; Farrow, S.; et al. Distinct trends of DNA methylation patterning in the innate and adaptive immune systems. *Cell Rep.* **2016**, *17*, 2101–2111. [CrossRef]
51. Morales-Nebreda, L.; McLafferty, F.S.; Singer, B.D. DNA methylation as a transcriptional regulator of the immune system. *Transl. Res.* **2019**, *204*, 1–18. [CrossRef] [PubMed]
52. Calle-Fabregat, C.; Morante-Palacios, O.; Ballestar, E. Understanding the relevance of DNA methylation changes in immune differentiation and disease. *Genes (Basel)* **2020**, *11*, 110. [CrossRef] [PubMed]
53. Tian, Y.; Meng, L.; Zhang, Y. Epigenetic regulation of dendritic cell development and function. *Cancer J.* **2017**, *23*, 302–307. [CrossRef] [PubMed]
54. Bullwinkel, J.; Ludemann, A.; Debarry, J.; Singh, P.B. Epigenotype switching at the CD14 and CD209 genes during differentiation of human monocytes to dendritic cells. *Epigenetics* **2011**, *6*, 45–51. [CrossRef] [PubMed]
55. Klug, M.; Heinz, S.; Gebhard, C.; Schwarzfischer, L.; Krause, S.W.; Andreesen, R.; Rehli, M. Active DNA demethylation in human postmitotic cells correlates with activating histone modifications, but not transcription levels. *Genome Biol.* **2010**, *11*, R63. [CrossRef]
56. Zhang, X.; Ulm, A.; Somineni, H.K.; Oh, S.; Weirauch, M.T.; Zhang, H.X.; Chen, X.; Lehn, M.A.; Janssen, E.M.; Ji, H. DNA methylation dynamics during ex vivo differentiation and maturation of human dendritic cells. *Epigenetics Chromatin* **2014**, *7*, 21. [CrossRef]
57. Vento-Tormo, R.; Company, C.; Rodriguez-Ubreva, J.; de la Rica, L.; Urquiza, J.M.; Javierre, B.M.; Sabarinathan, R.; Luque, A.; Esteller, M.; Aran, J.M.; et al. IL-4 orchestrates STAT6-mediated DNA demethylation leading to dendritic cell differentiation. *Genome Biol.* **2016**, *17*, 4. [CrossRef]
58. Cook, P.C.; Owen, H.; Deaton, A.M.; Borger, J.G.; Brown, S.L.; Clouaire, T.; Jones, G.R.; Jones, L.H.; Lundie, R.J.; Marley, A.K.; et al. A dominant role for the methyl-CpG-binding protein Mbd2 in controlling Th2 induction by dendritic cells. *Nat. Commun.* **2015**, *6*, 6920. [CrossRef]
59. Pacis, A.; Tailleux, L.; Morin, A.M.; Lambourne, J.; MacIsaac, J.L.; Yotova, V.; Dumaine, A.; Danckaert, A.; Luca, F.; Grenier, J.C.; et al. Bacterial infection remodels the DNA methylation landscape of human dendritic cells. *Genome Res.* **2015**, *25*, 1801–1811. [CrossRef]
60. Gordon, S. Alternative activation of macrophages. *Nat. Rev. Immunol.* **2003**, *3*, 23–35. [CrossRef]
61. Wyckoff, J.B.; Wang, Y.; Lin, E.Y.; Li, J.F.; Goswami, S.; Stanley, E.R.; Segall, J.E.; Pollard, J.W.; Condeelis, J. Direct visualization of macrophage-assisted tumor cell intravasation in mammary tumors. *Cancer Res.* **2007**, *67*, 2649–2656. [CrossRef] [PubMed]
62. Lin, E.Y.; Li, J.F.; Gnatovskiy, L.; Deng, Y.; Zhu, L.; Grzesik, D.A.; Qian, H.; Xue, X.N.; Pollard, J.W. Macrophages regulate the angiogenic switch in a mouse model of breast cancer. *Cancer Res.* **2006**, *66*, 11238–11246. [CrossRef] [PubMed]
63. Dekkers, K.F.; Neele, A.E.; Jukema, J.W.; Heijmans, B.T.; de Winther, M.P.J. Human monocyte-to-macrophage differentiation involves highly localized gain and loss of DNA methylation at transcription factor binding sites. *Epigenetics Chromatin* **2019**, *12*, 34. [CrossRef] [PubMed]
64. Wallner, S.; Schroder, C.; Leitao, E.; Berulava, T.; Haak, C.; Beisser, D.; Rahmann, S.; Richter, A.S.; Manke, T.; Bonisch, U.; et al. Epigenetic dynamics of monocyte-to-macrophage differentiation. *Epigenetics Chromatin* **2016**, *9*, 33. [CrossRef]
65. Yang, X.; Wang, X.; Liu, D.; Yu, L.; Xue, B.; Shi, H. Epigenetic regulation of macrophage polarization by DNA methyltransferase 3b. *Mol. Endocrinol.* **2014**, *28*, 565–574. [CrossRef]

66. Wang, H.C.; Chen, C.W.; Yang, C.L.; Tsai, I.M.; Hou, Y.C.; Chen, C.J.; Shan, Y.S. Tumor-associated macrophages promote epigenetic silencing of gelsolin through DNA methyltransferase 1 in gastric cancer cells. *Cancer Immunol. Res.* **2017**, *5*, 885–897. [CrossRef]
67. Cheng, C.; Huang, C.; Ma, T.T.; Bian, E.B.; He, Y.; Zhang, L.; Li, J. SOCS1 hypermethylation mediated by DNMT1 is associated with lipopolysaccharide-induced inflammatory cytokines in macrophages. *Toxicol. Lett.* **2014**, *225*, 488–497. [CrossRef]
68. Cull, A.H.; Snetsinger, B.; Buckstein, R.; Wells, R.A.; Rauh, M.J. Tet2 restrains inflammatory gene expression in macrophages. *Exp. Hematol.* **2017**, *55*, 56e13–70e13. [CrossRef]
69. Zhang, Q.; Zhao, K.; Shen, Q.; Han, Y.; Gu, Y.; Li, X.; Zhao, D.; Liu, Y.; Wang, C.; Zhang, X.; et al. Tet2 is required to resolve inflammation by recruiting Hdac2 to specifically repress IL-6. *Nature* **2015**, *525*, 389–393. [CrossRef]
70. Pan, W.; Zhu, S.; Qu, K.; Meeth, K.; Cheng, J.; He, K.; Ma, H.; Liao, Y.; Wen, X.; Roden, C.; et al. The DNA methylcytosine dioxygenase Tet2 sustains immunosuppressive function of tumor-infiltrating myeloid cells to promote melanoma progression. *Immunity* **2017**, *47*, 284–297e285. [CrossRef]
71. Santourlidis, S.; Graffmann, N.; Christ, J.; Uhrberg, M. Lineage-specific transition of histone signatures in the killer cell Ig-like receptor locus from hematopoietic progenitor to NK cells. *J. Immunol. (Baltim.)* **2008**, *180*, 418–425. [CrossRef] [PubMed]
72. Lee, J.; Zhang, T.; Hwang, I.; Kim, A.; Nitschke, L.; Kim, M.; Scott, J.M.; Kamimura, Y.; Lanier, L.L.; Kim, S. Epigenetic modification and antibody-dependent expansion of memory-like NK cells in human cytomegalovirus-infected individuals. *Immunity* **2015**, *42*, 431–442. [CrossRef] [PubMed]
73. Schlums, H.; Cichocki, F.; Tesi, B.; Theorell, J.; Beziat, V.; Holmes, T.D.; Han, H.; Chiang, S.C.; Foley, B.; Mattsson, K.; et al. Cytomegalovirus infection drives adaptive epigenetic diversification of NK cells with altered signaling and effector function. *Immunity* **2015**, *42*, 443–456. [CrossRef] [PubMed]
74. Wiencke, J.K.; Butler, R.; Hsuang, G.; Eliot, M.; Kim, S.; Sepulveda, M.A.; Siegel, D.; Houseman, E.A.; Kelsey, K.T. The DNA methylation profile of activated human natural killer cells. *Epigenetics* **2016**, *11*, 363–380. [CrossRef]
75. Stone, M.L.; Chiappinelli, K.B.; Li, H.; Murphy, L.M.; Travers, M.E.; Topper, M.J.; Mathios, D.; Lim, M.; Shih, I.M.; Wang, T.L.; et al. Epigenetic therapy activates type I interferon signaling in murine ovarian cancer to reduce immunosuppression and tumor burden. *Proc. Natl. Acad. Sci. USA* **2017**, *114*, E10981–E10990. [CrossRef]
76. Figueroa, M.E.; Abdel-Wahab, O.; Lu, C.; Ward, P.S.; Patel, J.; Shih, A.; Li, Y.; Bhagwat, N.; Vasanthakumar, A.; Fernandez, H.F.; et al. Leukemic IDH1 and IDH2 mutations result in a hypermethylation phenotype, disrupt TET2 function, and impair hematopoietic differentiation. *Cancer Cell.* **2010**, *18*, 553–567. [CrossRef]
77. Bugide, S.; Green, M.R.; Wajapeyee, N. Inhibition of Enhancer of zeste homolog 2 (EZH2) induces natural killer cell-mediated eradication of hepatocellular carcinoma cells. *Proc. Natl. Acad. Sci. USA* **2018**, *115*, E3509–E3518. [CrossRef]
78. Gaud, G.; Lesourne, R.; Love, P.E. Regulatory mechanisms in T cell receptor signalling. *Nat. Rev. Immunol.* **2018**, *18*, 485–497. [CrossRef]
79. Bruniquel, D.; Schwartz, R.H. Selective, stable demethylation of the interleukin-2 gene enhances transcription by an active process. *Nat. Immunol.* **2003**, *4*, 235–240. [CrossRef]
80. Wei, G.; Wei, L.; Zhu, J.; Zang, C.; Hu-Li, J.; Yao, Z.; Cui, K.; Kanno, Y.; Roh, T.Y.; Watford, W.T.; et al. Global mapping of H3K4me3 and H3K27me3 reveals specificity and plasticity in lineage fate determination of differentiating CD4+ T cells. *Immunity* **2009**, *30*, 155–167. [CrossRef]
81. Ansel, K.M.; Lee, D.U.; Rao, A. An epigenetic view of helper T cell differentiation. *Nat. Immunol.* **2003**, *4*, 616–623. [CrossRef] [PubMed]
82. Hirahara, K.; Vahedi, G.; Ghoreschi, K.; Yang, X.P.; Nakayamada, S.; Kanno, Y.; O'Shea, J.J.; Laurence, A. Helper T-cell differentiation and plasticity: Insights from epigenetics. *Immunology* **2011**, *134*, 235–245. [CrossRef] [PubMed]
83. Janson, P.C.; Marits, P.; Thorn, M.; Ohlsson, R.; Winqvist, O. CpG methylation of the IFNG gene as a mechanism to induce immunosuppression [correction of immunosupression] in tumor-infiltrating lymphocytes. *J. Immunol. (Baltim.)* **2008**, *181*, 2878–2886. [CrossRef]
84. Lee, P.P.; Fitzpatrick, D.R.; Beard, C.; Jessup, H.K.; Lehar, S.; Makar, K.W.; Perez-Melgosa, M.; Sweetser, M.T.; Schlissel, M.S.; Nguyen, S.; et al. A critical role for Dnmt1 and DNA methylation in T cell development, function, and survival. *Immunity* **2001**, *15*, 763–774. [CrossRef]
85. Makar, K.W.; Wilson, C.B. DNA methylation is a nonredundant repressor of the Th2 effector program. *J. Immunol. (Baltim.)* **2004**, *173*, 4402–4406. [CrossRef]
86. Young, H.A.; Ghosh, P.; Ye, J.; Lederer, J.; Lichtman, A.; Gerard, J.R.; Penix, L.; Wilson, C.B.; Melvin, A.J.; McGurn, M.E.; et al. Differentiation of the T helper phenotypes by analysis of the methylation state of the IFN-gamma gene. *J. Immunol. (Baltim.)* **1994**, *153*, 3603–3610.
87. Winders, B.R.; Schwartz, R.H.; Bruniquel, D. A distinct region of the murine IFN-gamma promoter is hypomethylated from early T cell development through mature naive and Th1 cell differentiation, but is hypermethylated in Th2 cells. *J. Immunol. (Baltim.)* **2004**, *173*, 7377–7384. [CrossRef]
88. Thomas, R.M.; Gamper, C.J.; Ladle, B.H.; Powell, J.D.; Wells, A.D. De novo DNA methylation is required to restrict T helper lineage plasticity. *J. Biol. Chem.* **2012**, *287*, 22900–22909. [CrossRef]
89. Jones, B.; Chen, J. Inhibition of IFN-gamma transcription by site-specific methylation during T helper cell development. *EMBO J.* **2006**, *25*, 2443–2452. [CrossRef]

90. Hosokawa, H.; Tanaka, T.; Suzuki, Y.; Iwamura, C.; Ohkubo, S.; Endoh, K.; Kato, M.; Endo, Y.; Onodera, A.; Tumes, D.J.; et al. Functionally distinct Gata3/Chd4 complexes coordinately establish T helper 2 (Th2) cell identity. *Proc. Natl. Acad. Sci. USA* **2013**, *110*, 4691–4696. [CrossRef]
91. Aoki, K.; Sato, N.; Yamaguchi, A.; Kaminuma, O.; Hosozawa, T.; Miyatake, S. Regulation of DNA demethylation during maturation of CD4+ naive T cells by the conserved noncoding sequence 1. *J. Immunol. (Baltim.)* **2009**, *182*, 7698–7707. [CrossRef] [PubMed]
92. Wood, K.H.; Zhou, Z. Emerging molecular and biological functions of MBD2, a reader of DNA methylation. *Front. Genet.* **2016**, *7*, 93. [CrossRef] [PubMed]
93. Nestor, C.E.; Lentini, A.; Hagg Nilsson, C.; Gawel, D.R.; Gustafsson, M.; Mattson, L.; Wang, H.; Rundquist, O.; Meehan, R.R.; Klocke, B.; et al. 5-Hydroxymethylcytosine remodeling precedes lineage specification during differentiation of human CD4(+) T cells. *Cell Rep.* **2016**, *16*, 559–570. [CrossRef] [PubMed]
94. Ichiyama, K.; Chen, T.; Wang, X.; Yan, X.; Kim, B.S.; Tanaka, S.; Ndiaye-Lobry, D.; Deng, Y.; Zou, Y.; Zheng, P.; et al. The methylcytosine dioxygenase Tet2 promotes DNA demethylation and activation of cytokine gene expression in T cells. *Immunity* **2015**, *42*, 613–626. [CrossRef] [PubMed]
95. Tsagaratou, A.; Aijo, T.; Lio, C.W.; Yue, X.; Huang, Y.; Jacobsen, S.E.; Lahdesmaki, H.; Rao, A. Dissecting the dynamic changes of 5-hydroxymethylcytosine in T-cell development and differentiation. *Proc. Natl. Acad. Sci. USA* **2014**, *111*, E3306–3315. [CrossRef]
96. Zou, W. Regulatory T cells, tumour immunity and immunotherapy. *Nat. Rev. Immunol.* **2006**, *6*, 295–307. [CrossRef]
97. Ralainirina, N.; Poli, A.; Michel, T.; Poos, L.; Andres, E.; Hentges, F.; Zimmer, J. Control of NK cell functions by CD4+CD25+ regulatory T cells. *J. Leukoc. Biol.* **2007**, *81*, 144–153. [CrossRef]
98. Sehouli, J.; Loddenkemper, C.; Cornu, T.; Schwachula, T.; Hoffmuller, U.; Grutzkau, A.; Lohneis, P.; Dickhaus, T.; Grone, J.; Kruschewski, M.; et al. Epigenetic quantification of tumor-infiltrating T-lymphocytes. *Epigenetics* **2011**, *6*, 236–246. [CrossRef]
99. Lal, G.; Zhang, N.; van der Touw, W.; Ding, Y.; Ju, W.; Bottinger, E.P.; Reid, S.P.; Levy, D.E.; Bromberg, J.S. Epigenetic regulation of Foxp3 expression in regulatory T cells by DNA methylation. *J. Immunol. (Baltim.)* **2009**, *182*, 259–273. [CrossRef]
100. Ke, X.; Zhang, S.; Xu, J.; Liu, G.; Zhang, L.; Xie, E.; Gao, L.; Li, D.; Sun, R.; Wang, F.; et al. Non-small-cell lung cancer-induced immunosuppression by increased human regulatory T cells via Foxp3 promoter demethylation. *Cancer Immunol. Immunother.* **2016**, *65*, 587–599. [CrossRef]
101. Zhuo, C.; Li, Z.; Xu, Y.; Wang, Y.; Li, Q.; Peng, J.; Zheng, H.; Wu, P.; Li, B.; Cai, S. Higher FOXP3-TSDR demethylation rates in adjacent normal tissues in patients with colon cancer were associated with worse survival. *Mol. Cancer* **2014**, *13*, 153. [CrossRef] [PubMed]
102. Baron, U.; Floess, S.; Wieczorek, G.; Baumann, K.; Grutzkau, A.; Dong, J.; Thiel, A.; Boeld, T.J.; Hoffmann, P.; Edinger, M.; et al. DNA demethylation in the human FOXP3 locus discriminates regulatory T cells from activated FOXP3(+) conventional T cells. *Eur. J. Immunol.* **2007**, *37*, 2378–2389. [CrossRef] [PubMed]
103. Wang, L.; Liu, Y.; Han, R.; Beier, U.H.; Thomas, R.M.; Wells, A.D.; Hancock, W.W. Mbd2 promotes foxp3 demethylation and T-regulatory-cell function. *Mol. Cell. Biol.* **2013**, *33*, 4106–4115. [CrossRef] [PubMed]
104. Zheng, Y.; Josefowicz, S.; Chaudhry, A.; Peng, X.P.; Forbush, K.; Rudensky, A.Y. Role of conserved non-coding DNA elements in the Foxp3 gene in regulatory T-cell fate. *Nature* **2010**, *463*, 808–812. [CrossRef] [PubMed]
105. Yue, X.; Trifari, S.; Aijo, T.; Tsagaratou, A.; Pastor, W.A.; Zepeda-Martinez, J.A.; Lio, C.W.; Li, X.; Huang, Y.; Vijayanand, P.; et al. Control of Foxp3 stability through modulation of TET activity. *J. Exp. Med.* **2016**, *213*, 377–397. [CrossRef]
106. Someya, K.; Nakatsukasa, H.; Ito, M.; Kondo, T.; Tateda, K.I.; Akanuma, T.; Koya, I.; Sanosaka, T.; Kohyama, J.; Tsukada, Y.I.; et al. Improvement of Foxp3 stability through CNS2 demethylation by TET enzyme induction and activation. *Int. Immunol.* **2017**, *29*, 365–375. [CrossRef]
107. Scharer, C.D.; Barwick, B.G.; Youngblood, B.A.; Ahmed, R.; Boss, J.M. Global DNA methylation remodeling accompanies CD8 T cell effector function. *J. Immunol. (Baltim.)* **2013**, *191*, 3419–3429. [CrossRef]
108. Ladle, B.H.; Li, K.P.; Phillips, M.J.; Pucsek, A.B.; Haile, A.; Powell, J.D.; Jaffee, E.M.; Hildeman, D.A.; Gamper, C.J. De novo DNA methylation by DNA methyltransferase 3a controls early effector CD8+ T-cell fate decisions following activation. *Proc. Natl. Acad. Sci. USA* **2016**, *113*, 10631–10636. [CrossRef]
109. Chappell, C.; Beard, C.; Altman, J.; Jaenisch, R.; Jacob, J. DNA methylation by DNA methyltransferase 1 is critical for effector CD8 T cell expansion. *J. Immunol. (Baltim.)* **2006**, *176*, 4562–4572. [CrossRef]
110. Youngblood, B.; Hale, J.S.; Kissick, H.T.; Ahn, E.; Xu, X.; Wieland, A.; Araki, K.; West, E.E.; Ghoneim, H.E.; Fan, Y.; et al. Effector CD8 T cells dedifferentiate into long-lived memory cells. *Nature* **2017**, *552*, 404–409. [CrossRef]
111. Carty, S.A.; Gohil, M.; Banks, L.B.; Cotton, R.M.; Johnson, M.E.; Stelekati, E.; Wells, A.D.; Wherry, E.J.; Koretzky, G.A.; Jordan, M.S. The loss of TET2 promotes CD8(+) T cell memory differentiation. *J. Immunol. (Baltim.)* **2018**, *200*, 82–91. [CrossRef] [PubMed]
112. Tyrakis, P.A.; Palazon, A.; Macias, D.; Lee, K.L.; Phan, A.T.; Velica, P.; You, J.; Chia, G.S.; Sim, J.; Doedens, A.; et al. S-2-hydroxyglutarate regulates CD8(+) T-lymphocyte fate. *Nature* **2016**, *540*, 236–241. [CrossRef] [PubMed]
113. Kersh, E.N. Impaired memory CD8 T cell development in the absence of methyl-CpG-binding domain protein 2. *J. Immunol. (Baltim.)* **2006**, *177*, 3821–3826. [CrossRef] [PubMed]
114. McKinney, E.F.; Smith, K.G. T cell exhaustion and immune-mediated disease-the potential for therapeutic exhaustion. *Curr. Opin. Immunol.* **2016**, *43*, 74–80. [CrossRef]

115. Emran, A.A.; Chatterjee, A.; Rodger, E.J.; Tiffen, J.C.; Gallagher, S.J.; Eccles, M.R.; Hersey, P. Targeting DNA methylation and EZH2 activity to overcome melanoma resistance to immunotherapy. *Trends Immunol.* **2019**, *40*, 328–344. [CrossRef]
116. Wherry, E.J.; Kurachi, M. Molecular and cellular insights into T cell exhaustion. *Nat. Rev. Immunol.* **2015**, *15*, 486–499. [CrossRef]
117. Yang, R.; Cheng, S.; Luo, N.; Gao, R.; Yu, K.; Kang, B.; Wang, L.; Zhang, Q.; Fang, Q.; Zhang, L.; et al. Distinct epigenetic features of tumor-reactive CD8+ T cells in colorectal cancer patients revealed by genome-wide DNA methylation analysis. *Genome Biol.* **2019**, *21*, 2. [CrossRef]
118. Duhen, T.; Duhen, R.; Montler, R.; Moses, J.; Moudgil, T.; de Miranda, N.F.; Goodall, C.P.; Blair, T.C.; Fox, B.A.; McDermott, J.E.; et al. Co-expression of CD39 and CD103 identifies tumor-reactive CD8 T cells in human solid tumors. *Nat. Commun.* **2018**, *9*, 2724. [CrossRef]
119. Youngblood, B.; Oestreich, K.J.; Ha, S.J.; Duraiswamy, J.; Akondy, R.S.; West, E.E.; Wei, Z.; Lu, P.; Austin, J.W.; Riley, J.L.; et al. Chronic virus infection enforces demethylation of the locus that encodes PD-1 in antigen-specific CD8(+) T cells. *Immunity* **2011**, *35*, 400–412. [CrossRef]
120. Philip, M.; Fairchild, L.; Sun, L.; Horste, E.L.; Camara, S.; Shakiba, M.; Scott, A.C.; Viale, A.; Lauer, P.; Merghoub, T.; et al. Chromatin states define tumour-specific T cell dysfunction and reprogramming. *Nature* **2017**, *545*, 452–456. [CrossRef]
121. Kratchmarov, R.; Magun, A.M.; Reiner, S.L. TCF1 expression marks self-renewing human CD8(+) T cells. *Blood Adv.* **2018**, *2*, 1685–1690. [CrossRef] [PubMed]
122. Han, D.; Liu, J.; Chen, C.; Dong, L.; Liu, Y.; Chang, R.; Huang, X.; Liu, Y.; Wang, J.; Dougherty, U.; et al. Anti-tumour immunity controlled through mRNA m(6)A methylation and YTHDF1 in dendritic cells. *Nature* **2019**, *566*, 270–274. [CrossRef]
123. Kariko, K.; Buckstein, M.; Ni, H.; Weissman, D. Suppression of RNA recognition by Toll-like receptors: The impact of nucleoside modification and the evolutionary origin of RNA. *Immunity* **2005**, *23*, 165–175. [CrossRef] [PubMed]
124. Li, H.B.; Tong, J.; Zhu, S.; Batista, P.J.; Duffy, E.E.; Zhao, J.; Bailis, W.; Cao, G.; Kroehling, L.; Chen, Y.; et al. m(6)A mRNA methylation controls T cell homeostasis by targeting the IL-7/STAT5/SOCS pathways. *Nature* **2017**, *548*, 338–342. [CrossRef]
125. Rubio, R.M.; Depledge, D.P.; Bianco, C.; Thompson, L.; Mohr, I. RNA m(6) A modification enzymes shape innate responses to DNA by regulating interferon beta. *Genes Dev.* **2018**, *32*, 1472–1484. [CrossRef] [PubMed]
126. Shulman, Z.; Stern-Ginossar, N. The RNA modification N(6)-methyladenosine as a novel regulator of the immune system. *Nat. Immunol.* **2020**, *21*, 501–512. [CrossRef]
127. Tong, J.; Cao, G.; Zhang, T.; Sefik, E.; Amezcua Vesely, M.C.; Broughton, J.P.; Zhu, S.; Li, H.; Li, B.; Chen, L.; et al. m(6)A mRNA methylation sustains Treg suppressive functions. *Cell Res.* **2018**, *28*, 253–256. [CrossRef]
128. Wang, H.; Hu, X.; Huang, M.; Liu, J.; Gu, Y.; Ma, L.; Zhou, Q.; Cao, X. Mettl3-mediated mRNA m(6)A methylation promotes dendritic cell activation. *Nat. Commun.* **2019**, *10*, 1898. [CrossRef]
129. Wang, Y.N.; Yu, C.Y.; Jin, H.Z. RNA N(6)-methyladenosine modifications and the immune response. *J. Immunol. Res.* **2020**, *2020*, 6327614. [CrossRef]
130. Winkler, R.; Gillis, E.; Lasman, L.; Safra, M.; Geula, S.; Soyris, C.; Nachshon, A.; Tai-Schmiedel, J.; Friedman, N.; Le-Trilling, V.T.K.; et al. m(6)A modification controls the innate immune response to infection by targeting type I interferons. *Nat. Immunol.* **2019**, *20*, 173–182. [CrossRef]
131. Zheng, Q.; Hou, J.; Zhou, Y.; Li, Z.; Cao, X. The RNA helicase DDX46 inhibits innate immunity by entrapping m(6)A-demethylated antiviral transcripts in the nucleus. *Nat. Immunol.* **2017**, *18*, 1094–1103. [CrossRef]
132. Biswas, S.; Rao, C.M. Epigenetics in cancer: Fundamentals and beyond. *Pharmacol. Ther.* **2017**, *173*, 118–134. [CrossRef]
133. Ehrlich, M. DNA hypomethylation in cancer cells. *Epigenomics* **2009**, *1*, 239–259. [CrossRef]
134. Silverman, L.R.; Demakos, E.P.; Peterson, B.L.; Kornblith, A.B.; Holland, J.C.; Odchimar-Reissig, R.; Stone, R.M.; Nelson, D.; Powell, B.L.; DeCastro, C.M.; et al. Randomized controlled trial of azacitidine in patients with the myelodysplastic syndrome: A study of the cancer and leukemia group B. *J. Clin. Oncol.* **2002**, *20*, 2429–2440. [CrossRef]
135. Kantarjian, H.; Issa, J.P.; Rosenfeld, C.S.; Bennett, J.M.; Albitar, M.; DiPersio, J.; Klimek, V.; Slack, J.; de Castro, C.; Ravandi, F.; et al. Decitabine improves patient outcomes in myelodysplastic syndromes: Results of a phase III randomized study. *Cancer* **2006**, *106*, 1794–1803. [CrossRef]
136. Derissen, E.J.; Beijnen, J.H.; Schellens, J.H. Concise drug review: Azacitidine and decitabine. *Oncologist* **2013**, *18*, 619–624. [CrossRef]
137. Nervi, C.; De Marinis, E.; Codacci-Pisanelli, G. Epigenetic treatment of solid tumours: A review of clinical trials. *Clin. Epigenetics* **2015**, *7*, 127. [CrossRef]
138. Koch, A.; Joosten, S.C.; Feng, Z.; de Ruijter, T.C.; Draht, M.X.; Melotte, V.; Smits, K.M.; Veeck, J.; Herman, J.G.; Van Neste, L.; et al. Analysis of DNA methylation in cancer: Location revisited. *Nat. Rev. Clin. Oncol.* **2018**, *15*, 459–466. [CrossRef]
139. Azad, N.; Zahnow, C.A.; Rudin, C.M.; Baylin, S.B. The future of epigenetic therapy in solid tumours—Lessons from the past. *Nat. Rev. Clin. Oncol.* **2013**, *10*, 256–266. [CrossRef]
140. Sigalotti, L.; Fratta, E.; Coral, S.; Maio, M. Epigenetic drugs as immunomodulators for combination therapies in solid tumors. *Pharmacol. Ther.* **2014**, *142*, 339–350. [CrossRef]
141. Chiappinelli, K.B.; Zahnow, C.A.; Ahuja, N.; Baylin, S.B. Combining epigenetic and immunotherapy to combat cancer. *Cancer Res.* **2016**, *76*, 1683–1689. [CrossRef] [PubMed]
142. Larkin, J.; Hodi, F.S.; Wolchok, J.D. Combined nivolumab and ipilimumab or monotherapy in untreated melanoma. *N Engl. J. Med.* **2015**, *373*, 1270–1271. [CrossRef] [PubMed]

143. Zhang, F.; Zhou, X.; DiSpirito, J.R.; Wang, C.; Wang, Y.; Shen, H. Epigenetic manipulation restores functions of defective CD8(+) T cells from chronic viral infection. *Mol. Ther.* **2014**, *22*, 1698–1706. [CrossRef]
144. Ghoneim, H.E.; Fan, Y.; Moustaki, A.; Abdelsamed, H.A.; Dash, P.; Dogra, P.; Carter, R.; Awad, W.; Neale, G.; Thomas, P.G.; et al. De novo epigenetic programs inhibit PD-1 blockade-mediated T cell rejuvenation. *Cell* **2017**, *170*, 142–157e119. [CrossRef]
145. Chiappinelli, K.B.; Strissel, P.L.; Desrichard, A.; Li, H.; Henke, C.; Akman, B.; Hein, A.; Rote, N.S.; Cope, L.M.; Snyder, A.; et al. Inhibiting DNA methylation causes an interferon response in cancer via dsRNA including endogenous retroviruses. *Cell* **2015**, *162*, 974–986. [CrossRef]
146. Subramaniam, D.; Thombre, R.; Dhar, A.; Anant, S. DNA methyltransferases: A novel target for prevention and therapy. *Front. Oncol.* **2014**, *4*, 80. [CrossRef]
147. Giri, A.K.; Aittokallio, T. DNMT inhibitors increase methylation in the cancer genome. *Front. Pharmacol.* **2019**, *10*, 385. [CrossRef]
148. Gnyszka, A.; Jastrzebski, Z.; Flis, S. DNA methyltransferase inhibitors and their emerging role in epigenetic therapy of cancer. *Anticancer Res.* **2013**, *33*, 2989–2996.
149. Ehrlich, M. DNA hypermethylation in disease: Mechanisms and clinical relevance. *Epigenetics* **2019**, *14*, 1141–1163. [CrossRef]
150. Ehrlich, M. DNA methylation in cancer: Too much, but also too little. *Oncogene* **2002**, *21*, 5400–5413. [CrossRef]
151. Mahmood, N.; Cheishvili, D.; Arakelian, A.; Tanvir, I.; Khan, H.A.; Pepin, A.S.; Szyf, M.; Rabbani, S.A. Methyl donor S-adenosylmethionine (SAM) supplementation attenuates breast cancer growth, invasion, and metastasis in vivo; therapeutic and chemopreventive applications. *Oncotarget* **2018**, *9*, 5169–5183. [CrossRef] [PubMed]
152. Lu, S.C.; Mato, J.M. S-Adenosylmethionine in cell growth, apoptosis and liver cancer. *J. Gastroenterol. Hepatol.* **2008**, *23*, S73–S77. [CrossRef] [PubMed]
153. Bottiglieri, T. S-Adenosyl-L-methionine (SAMe): From the bench to the bedside–molecular basis of a pleiotrophic molecule. *Am. J. Clin. Nutr.* **2002**, *76*, 1151s–1157s. [CrossRef] [PubMed]
154. Yang, M.L.; Gee, A.J.; Gee, R.J.; Zurita-Lopez, C.I.; Khare, S.; Clarke, S.G.; Mamula, M.J. Lupus autoimmunity altered by cellular methylation metabolism. *Autoimmunity* **2018**, *46*, 21–31. [CrossRef] [PubMed]
155. Ara, A.I.; Xia, M.; Ramani, K.; Mato, J.M.; Lu, S.C. S-adenosylmethionine inhibits lipopolysaccharide-induced gene expression via modulation of histone methylation. *Hepatology* **2008**, *47*, 1655–1666. [CrossRef]
156. Ding, W.; Smulan, L.J.; Hou, N.S.; Taubert, S.; Watts, J.L.; Walker, A.K. S-adenosylmethionine levels govern innate immunity through distinct methylation-dependent pathways. *Cell Metab.* **2015**, *22*, 633–645. [CrossRef]
157. Bardag-Gorce, F.; Oliva, J.; Lin, A.; Li, J.; French, B.A.; French, S.W. SAMe prevents the up regulation of toll-like receptor signaling in mallory-denk body forming hepatocytes. *Exp. Mol. Pathol.* **2010**, *88*, 376–379. [CrossRef]
158. Gomez-Santos, L.; Luka, Z.; Wagner, C.; Fernandez-Alvarez, S.; Lu, S.C.; Mato, J.M.; Martinez-Chantar, M.L. Inhibition of natural killer cells protects the liver against acute injury in the absence of glycine N-methyltransferase. *Hepatology* **2012**, *56*, 747–759. [CrossRef]
159. Hote, P.T.; Sahoo, R.; Jani, T.S.; Ghare, S.S.; Chen, T.; Joshi-Barve, S.; McClain, C.J.; Barve, S.S. Ethanol inhibits methionine adenosyltransferase II activity and S-adenosylmethionine biosynthesis and enhances caspase-3-dependent cell death in T lymphocytes: Relevance to alcohol-induced immunosuppression. *J. Nutr. Biochem.* **2008**, *19*, 384–391. [CrossRef]
160. Tobena, R.; Horikawa, S.; Calvo, V.; Alemany, S. Interleukin-2 induces gamma-S-adenosyl-L-methionine synthetase gene expression during T-lymphocyte activation. *Biochem. J.* **1996**, *319*, 929–933. [CrossRef]
161. LeGros, H.L., Jr.; Geller, A.M.; Kotb, M. Differential regulation of methionine adenosyltransferase in superantigen and mitogen stimulated human T lymphocytes. *J. Biol. Chem.* **1997**, *272*, 16040–16047. [CrossRef] [PubMed]
162. Kotb, M.; Dale, J.B.; Beachey, E.H. Stimulation of S-Adenosylmethionine synthetase in human lymphocytes by streptococcal M protein. *J. Immunol. (Baltim.)* **1987**, *139*, 202–206.
163. De La Rosa, J.; Geller, A.M.; LeGros, H.L., Jr.; Kotb, M. Induction of interleukin 2 production but not methionine adenosyltransferase activity or S-adenosylmethionine turnover in Jurkat T-cells. *Cancer Res.* **1992**, *52*, 3361–3366.
164. De La Rosa, J.; Kotb, M.; Kredich, N.M. Regulation of S-adenosylmethionine synthetase activity in cultured human lymphocytes. *Biochim. Biophys. Acta* **1991**, *1077*, 225–232. [CrossRef]
165. German, D.C.; Bloch, C.A.; Kredich, N.M. Measurements of S-adenosylmethionine and L-homocysteine metabolism in cultured human lymphoid cells. *J. Biol. Chem.* **1983**, *258*, 10997–11003. [CrossRef]
166. Zeng, Z.; Yang, H.; Huang, Z.Z.; Chen, C.; Wang, J.; Lu, S.C. The role of c-Myb in the up-regulation of methionine adenosyltransferase 2A expression in activated Jurkat cells. *Biochem. J.* **2001**, *353*, 163–168. [CrossRef]
167. Kotb, M.; Kredich, N.M. S-Adenosylmethionine synthetase from human lymphocytes. Purification and characterization. *J. Biol. Chem.* **1985**, *260*, 3923–3930. [CrossRef]
168. Sahin, E.; Sahin, M. Epigenetical Targeting of the FOXP3 Gene by S-adenosylmethionine diminishes the suppressive capacity of regulatory T cells ex vivo and alters the expression profiles. *J. Immunother. (Hagerstown)* **2019**, *42*, 11–22. [CrossRef]
169. Shukeir, N.; Stefanska, B.; Parashar, S.; Chik, F.; Arakelian, A.; Szyf, M.; Rabbani, S.A. Pharmacological methyl group donors block skeletal metastasis in vitro and in vivo. *Br. J. Pharmacol.* **2015**, *172*, 2769–2781. [CrossRef]
170. Murín, R.; Vidomanová, E.; Bhavani, S. Kowtharapu; Jozef Hatok; Dobrota, D. Role of S-adenosylmethionine cycle in carcinogenesis. *Gen. Physiol. Biophys.* **2017**, *36*, 513–520. [CrossRef]
171. Pakneshan, P.; Szyf, M.; Farias-Eisner, R.; Rabbani, S.A. Reversal of the hypomethylation status of urokinase (uPA) promoter blocks breast cancer growth and metastasis. *J. Biol. Chem.* **2004**, *279*, 31735–31744. [CrossRef] [PubMed]

172. Parashar, S.; Cheishvili, D.; Arakelian, A.; Hussain, Z.; Tanvir, I.; Khan, H.A.; Szyf, M.; Rabbani, S.A. S-adenosylmethionine blocks osteosarcoma cells proliferation and invasion in vitro and tumor metastasis in vivo: Therapeutic and diagnostic clinical applications. *Cancer Med.* **2015**, *4*, 732–744. [CrossRef]
173. Chik, F.; Machnes, Z.; Szyf, M. Synergistic anti-breast cancer effect of a combined treatment with the methyl donor S-adenosyl methionine and the DNA methylation inhibitor 5-aza-2′-deoxycytidine. *Carcinogenesis* **2014**, *35*, 138–144. [CrossRef] [PubMed]
174. Luo, J.; Li, Y.N.; Wang, F.; Zhang, W.M.; Geng, X. S-adenosylmethionine inhibits the growth of cancer cells by reversing the hypomethylation status of c-myc and H-ras in human gastric cancer and colon cancer. *Int. J. Biol. Sci.* **2010**, *6*, 784–795. [CrossRef] [PubMed]
175. Bian, Y.; Li, W.; Kremer, D.M.; Sajjakulnukit, P.; Li, S.; Crespo, J.; Nwosu, Z.C.; Zhang, L.; Czerwonka, A.; Pawlowska, A.; et al. Cancer SLC43A2 alters T cell methionine metabolism and histone methylation. *Nature* **2020**, *585*, 277–282. [CrossRef]
176. Ulanovskaya, O.A.; Zuhl, A.M.; Cravatt, B.F. NNMT promotes epigenetic remodeling in cancer by creating a metabolic methylation sink. *Nat. Chem. Biol.* **2013**, *9*, 300–306. [CrossRef]
177. Mehdi, A.; Attias, M.; Mahmood, N.; Arakelian, A.; Mihalcioiu, C.; Piccirillo, C.A.; Szyf, M.; Rabbani, S.A. Enhanced anticancer effect of a combination of S-adenosylmethionine (SAM) and Immune Checkpoint Inhibitor (ICPi) in a syngeneic mouse model of advanced melanoma. *Front. Oncol.* **2020**, *10*, 1361. [CrossRef]
178. Niu, Y.; Wan, A.; Lin, Z.; Lu, X.; Wan, G. N (6)-Methyladenosine modification: A novel pharmacological target for anti-cancer drug development. *Acta Pharm. Sin. B* **2018**, *8*, 833–843. [CrossRef]
179. Chen, B.; Ye, F.; Yu, L.; Jia, G.; Huang, X.; Zhang, X.; Peng, S.; Chen, K.; Wang, M.; Gong, S.; et al. Development of cell-active N6-methyladenosine RNA demethylase FTO inhibitor. *J. Am. Chem. Soc.* **2012**, *134*, 17963–17971. [CrossRef]
180. Sun, H.; Luo, G.; Chen, D.; Xiang, Z. A Comprehensive and system review for the pharmacological mechanism of action of Rhein, an active anthraquinone ingredient. *Front. Pharmacol.* **2016**, *7*, 247. [CrossRef]
181. Xu, C.; Liu, K.; Tempel, W.; Demetriades, M.; Aik, W.; Schofield, C.J.; Min, J. Structures of human ALKBH5 demethylase reveal a unique binding mode for specific single-stranded N6-methyladenosine RNA demethylation. *J. Biol. Chem.* **2014**, *289*, 17299–17311. [CrossRef] [PubMed]
182. Yang, S.; Wei, J.; Cui, Y.H.; Park, G.; Shah, P.; Deng, Y.; Aplin, A.E.; Lu, Z.; Hwang, S.; He, C.; et al. m(6)A mRNA demethylase FTO regulates melanoma tumorigenicity and response to anti-PD-1 blockade. *Nat. Commun.* **2019**, *10*, 2782. [CrossRef] [PubMed]
183. Li, N.; Kang, Y.; Wang, L.; Huff, S.; Tang, R.; Hui, H.; Agrawal, K.; Gonzalez, G.M.; Wang, Y.; Patel, S.P.; et al. ALKBH5 regulates anti-PD-1 therapy response by modulating lactate and suppressive immune cell accumulation in tumor microenvironment. *Proc. Natl. Acad. Sci. USA* **2020**, *117*, 20159–20170. [CrossRef] [PubMed]

Review

Involvement of Kynurenine Pathway in Hepatocellular Carcinoma

Shivani Krishnamurthy [1], David Gilot [2], Seong Beom Ahn [1], Vincent Lam [1], Joo-Shik Shin [3], Gilles Jackie Guillemin [1,†] and Benjamin Heng [1,*,†]

1. Faculty of Medicine, Health and Human Sciences, Macquarie University, Sydney 2109, Australia; shivani.krishnamurthy@hdr.mq.edu.au (S.K.); charlie.ahn@mq.edu.au (S.B.A.); vincent.lam@mq.edu.au (V.L.); gilles.guillemin@mq.edu.au (G.J.G.)
2. INSERM U1242, University of Rennes, 35000 Rennes, France; david.gilot@univ-rennes1.fr
3. Tissue Pathology and Diagnostic Oncology, Royal Prince Alfred Hospital, Faculty of Medicine, University of Sydney, Sydney 2006, Australia; JooShik.Shin@health.nsw.gov.au
* Correspondence: benjamin.heng@mq.edu.au; Tel.: +61-9850-2733; Fax: +61-9850-2701
† Co-senior author.

Simple Summary: The kynurenine pathway (KP) is a biochemical pathway that synthesizes the vital coenzyme, nicotinamide adenine dinucleotide (NAD^+). In cancer, the KP is significantly activated, leading to tryptophan depletion and the production of downstream metabolites, which skews the immune response towards tumour tolerance. More specifically, advanced stage cancers that readily metastasize evidence the most dysregulation in KP enzymes, providing a clear link between the KP and cancer morbidity. Consequently, this provides the rationale for an attractive new drug discovery opportunity for adjuvant therapeutics targeting KP-mediated immune tolerance, which would greatly complement current pharmacological interventions. In this review, we summarize recent developments in the roles of the KP and clinical trials examining KP inhibition in liver cancer.

Abstract: As the second and third leading cancer-related death in men and the world, respectively, primary liver cancer remains a major concern to human health. Despite advances in diagnostic technology, patients with primary liver cancer are often diagnosed at an advanced stage. Treatment options for patients with advanced hepatocarcinoma (HCC) are limited to systemic treatment with multikinase inhibitors and immunotherapy. Furthermore, the 5-year survival rate for these late-stage HCC patients is approximately 12% worldwide. There is an unmet need to identify novel treatment options and/or sensitive blood-based biomarker(s) to detect this cancer at an early stage. Given that the liver harbours the largest proportion of immune cells in the human body, understanding the tumour–immune microenvironment has gained increasing attention as a potential target to treat cancer. The kynurenine pathway (KP) has been proposed to be one of the key mechanisms used by the tumour cells to escape immune surveillance for proliferation and metastasis. In an inflammatory environment such as cancer, the KP is elevated, suppressing local immune cell populations and enhancing tumour growth. In this review, we collectively describe the roles of the KP in cancer and provide information on the latest research into the KP in primary liver cancer.

Keywords: primary liver cancer; kynurenine pathway; immune evasion; indoleamine 2,3 dioxygenase 1; tryptophan 2,3 dioxygenase 2; IDO inhibitor

1. Primary Liver Cancer

Primary liver cancer is the second leading cause of cancer mortality in men and the sixth most commonly occurring cancer worldwide, with an estimated 905,677 cases and 830,180 deaths in 2020 [1]. It is a tumour that develops in the liver and is known to be highly invasive and spread to other organs such as the lungs, bone marrow, lymph nodes, and brain [2–4]. Hepatocellular carcinoma (HCC), accounting for more than 75% of all primary liver cancer cases, and intrahepatic cholangiocarcinoma (ICC), which accounts for a lesser proportion, approximately 12–15% of all liver cancer cases, are the two main histological types of this malignancy [5]. HCC arises from hepatocytes in the liver and is the most common cause of death in people with a history of chronic liver disease [6] or cirrhosis [7].

The global burden of liver cancer-related mortality is increasing worldwide, with an estimation of >1 million diagnosed with this cancer annually by 2025 [8,9]. The highest HCC incidence and mortality rates are observed in Africa and East Asia, although a growing trend in incidence rates has been observed in western countries, including the USA and parts of Europe [10]. In Australia, the incidence rate of primary liver cancer has increased 5-fold from 2003 to 2011. According to the Australian Institute of Health and Welfare's burden of cancer report, this cancer is a significant health threat and a burden to the Australian community [11]. A recent study showed that the age-adjusted incidence of HCC increased from 1.38/100,000 persons in 1982 to 4.96/100,000 in 2014 [12]. Incidence of HCC is up to four times higher in men compared to women and is projected to be the fifth and sixth most common cause of cancer death in Australian men and women, respectively in 2020. The gender discrepancy in primary liver cancer incidence can be attributed to biological and behavioural risk factors [13].

Important risk factors are chronic hepatitis B virus (HBV) or hepatitis C virus (HCV) infections, liver cirrhosis, chronic alcohol consumption, metabolic-associated fatty liver disease (MAFLD), and non-alcoholic steatohepatitis (NASH) [14]. HCC predominantly develops in the setting of cirrhosis and chronic liver diseases. Cirrhosis of the liver caused by any liver disease is a major risk factor, and HCC is the primary cause of death in hepatic cirrhosis patients [15]. The most common risk factor is chronic viral hepatitis [16–18], with HBV infection accounting for approximately 50% of the HCC cases. However, HBV vaccinations have reduced the risk associated with HBV-induced HCC [19,20]. Chronic HCV patients with cirrhosis or chronic liver damage are at higher risk of developing HCC [21]. However, a significant decrease in the risk of HCC attributed to HCV infections has been observed because of effective antiviral drugs [22]. Additionally, higher prevalence of obesity- or diabetes-related MAFLD and NASH (the most severe form of MAFLD) is also driving the increase in HCC incidence rates [23–26]. Studies suggest that older age is another important risk factor that increases the risk of developing primary liver cancer [27–30]. Statistical epidemiology shows that primary liver cancer patients mostly comprise individuals above 50 years, with mean onset age increasing from 58.2 years in 1990 to 62.5 years in 2017 [31].

HCC Stages and Its Prognosis

Overall survival for HCC patients is poor, with a 5-year relative survival rate of 34% for patients diagnosed with localized tumour mass, 12% for patients with regional cancer that has spread outside the liver to surrounding tissues or lymph nodes, and 3% for patients diagnosed with distant or metastasized liver cancer [32]. The Barcelona Clinical Liver Cancer (BCLC) staging system is widely accepted and used to identify the stage of HCC based on tumour characteristics and burden, the Child–Pugh score of hepatic function, and patient performance status [33]. The median survival time for HCC patients can vary according to the stage of cancer diagnosed. Based on the BCLC staging system, these values are more than 6 years for early stage (0 and A), 26 to 30 months for intermediate stage (B), 12 to 19 months for advanced stage (C), and nearly 3 months for end-stage (D) HCC after receiving treatment (Figure 1) [8].

Surgical resection or partial hepatectomy [34], laparoscopic liver resection [35], and liver transplantation [36] are the most common treatments used for early stage HCC patients (i.e., BCLC stage A), when the tumour mass is more than 2 cm but less than 5 cm in size and is confined to the liver, with no evidence of vascular invasion. Radiofrequency ablation is the primary treatment of choice for single tumours less than 2 cm in size (BCLC stage 0) and is also an alternative for early stage HCC patients unsuitable for surgery or liver transplantation due to the presence of multiple tumour nodules and liver dysfunction [37].

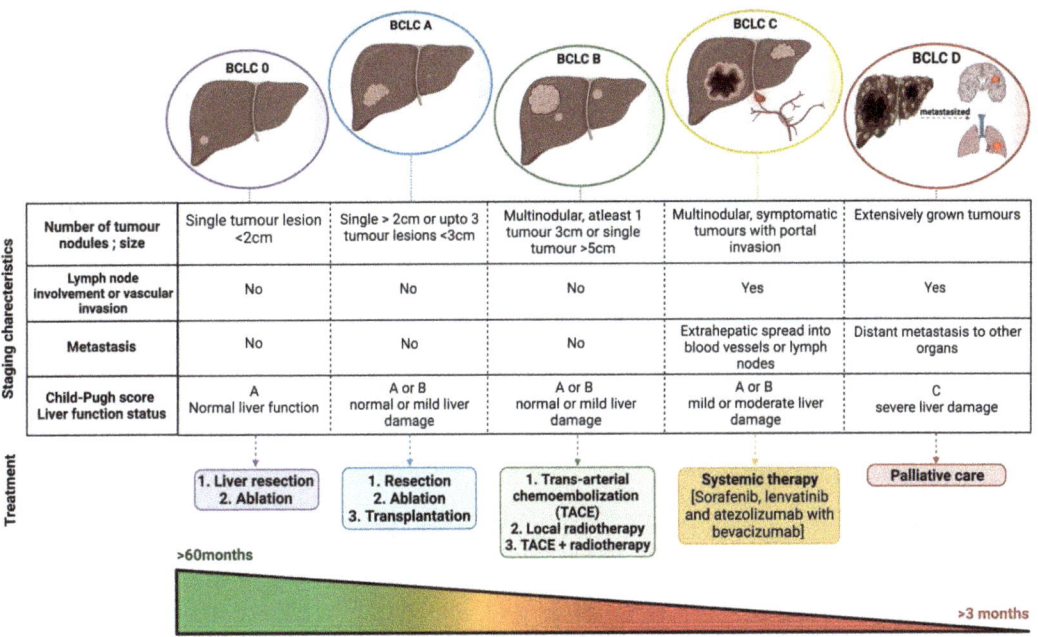

Figure 1. Classification of HCC and its characteristics: Based on the BCLC staging system, HCC can be classified as stages 0, A, B, C, and D. Stage A has the highest median survival time of more than 60 months while stage D has less than 4 months. Localised surgery and radiotherapy are the choice of treatments for stage 0 to B, while systemic treatment with palliative care is usually recommended for stages C to D.

Unfortunately, patients with HCC are often asymptomatic in the early stages; hence, detecting early stages of cancer in patients remains a challenge. A combined diagnostic approach consisting of ultrasound imaging, magnetic resonance imaging, computed tomography, and detecting alpha-fetoprotein (AFP) levels in patient sera is used to diagnose cancer and predict HCC prognosis [38]. AFP is a type of glycoprotein that is produced by embryonic endoderm tissue cells and is usually in high concentrations in maternal serum during foetal development [39,40]. This concentration of AFP drops during adulthood due to the inability of mature hepatocytes to synthesize this glycoprotein [40]. Transformed cancer cells including hepatocytes can regain this ability to synthesize AFP and have therefore been used as blood-based biomarkers for HCC diagnosis [41]. However, this biomarker is not effective in detecting patients with a low concentration of AFP (AFP <20 ng/mL), such as during early stage HCC, and a portion of advanced HCC, where AFP remains low throughout disease progression [42]. A promising alternate blood biomarker is glypican-3 (GPC-3). GPC-3 is a cell-surface proteoglycan that is highly expressed in embryonic tissues and is involved in cell proliferation and survival during foetal development [43]. In adults, GPC-3 expression is only limited to lung, ovary, mesothelium, mammary glands, and kidney [44,45]. However, high levels of GPC-3 expression are observed in HCC tissues but not in healthy adult liver, and it is a commonly used immunohistochemical marker to detect the degree of HCC tumour differentiation [46,47]. Although studies have shown 83.4% sensitivity in HCC [48], the diagnostic use of GPC-3 as an HCC biomarker remains controversial due to conflicting results [49–51]. A delay of as little as three months in diagnosis can result in the cancer progressing to later stages and, more importantly, it reduces patient survival rate. Focusing on early diagnosis is important to increase patients' survival rate rather than treatment options. [52]. Other locoregional treatment strategies for some early and intermediate HCC patients (BCLC stage B) who are not fit to undergo surgery or transplantation include trans-arterial chemoembolization (TACE) [53], local radiotherapy, or a combination approach of laparoscopy with TACE or radiotherapy is used to prevent from further cancer progression [34].

Most HCC cases are diagnosed at advanced stages (BCLC stage C and D) when the tumours are too aggressive for surgical resection and have metastasized to other organ sites. Systemic treatment, which includes molecular-targeted therapy, remains a recommended treatment for locally advanced or metastatic unresectable HCC tumours [34]. To date, the first-line drug treatments for advanced HCC patients include sorafenib [54], lenvatinib [55], and atezolizumab (anti-PDL1 antibody) in combination with bevacizumab (anti-VEGF antibody) [56]. The recent IMbrave150 trial reported that patients treated with the combination regimen of atezolizumab and bevacizumab showed improved overall survival and progression-free survival compared to sorafenib. The most common treatment-related adverse events observed with combination immunotherapy are fatigue, pain, loss of appetite, and diarrhoea [57]. On the basis of these positive findings from the trial, the Therapeutic Goods Administration (TGA)-approved regimen has now been extensively used to treat patients with unresectable HCC and was added to the Australian Pharmaceutical Benefits Scheme (PBS) program in 2020 [58]. While there has been significant improvement in treatment opportunities over the last decade, this malignancy is associated with a high recurrence rate and poor overall survival. Clinical trials evaluating the efficacy and safety of immune-therapeutic drugs such as pembrolizumab or nivolumab for advanced liver cancer treatment failed to improve overall survival of patients and significant immune-related adverse side effects were observed, resulting in failure of the clinical trials [59,60].

Although the understanding of the disease and treatment opportunities for HCC have drastically improved over the last decade, this malignancy remains a fatal disease worldwide. There is an urgent need to identify a specific set of biomarkers to (1) detect early stage HCC with high accuracy in patients and (2) to effectively allow the assessment of response to treatment to rapidly estimate whether a patient responds to treatment. Identification of novel and specific diagnostic set of biomarkers to detect patients who may be at risk and with early stage HCC, prognostic predictors that can effectively distinguish between patients with favourable or unfavourable prognosis in the same tumour stage, and more specific treatment targets are all critical. An important aspect to consider is the unique relationship between the liver and the immune system. The liver is a critical immunological frontline of the body, where complex immunological activity occurs to prevent infection in the body [61,62]. Interestingly, some biochemical pathways promote tumour tolerance by decreasing the recognition of cancer antigen, inducing immune suppression and chronic inflammation. Notably, an interesting biochemical pathway that mediates tumour tolerance is the kynurenine pathway (KP) of tryptophan (TRP) metabolism. Elevation of KP activity by tumour cells suppresses the local immune response and enhances tumour survival and invasion [63,64]. This review will examine the role of KP in HCC progression. Understanding how HCC manipulates immune-suppressive KP may lead to the identification of potential therapeutic targets for HCC.

2. The KP

TRP is one of the eight essential amino acids that are only obtainable through the diet [65]. TRP and its metabolites play a critical role in various cellular growth and maintenance processes. Up to 90% of the TRP is catabolized by the KP to produce nicotinamide adenine dinucleotide (NAD^+), an important enzyme co-factor involved in the regulation of important cellular processes (Figure 2) [66]. KP is tightly regulated under a healthy physiological state and produces various metabolites with immune-suppressive and redox activity. These metabolites include kynurenine (KYN), kynurenic acid (KYNA), 3-hydroxykynurenine (3-HK), anthranilic acid, 3-hydroxyanthranilic acid

(3-HAA), picolinic acid, and quinolinic acid (QUIN) [67]. The pathway begins with three rate-limiting enzymes, indoleamine 2,3-dioxygenase (IDO1) [68], indoleamine 2,3 dioxygenase 2 (IDO2) [69,70], and tryptophan 2,3-dioxygenase (TDO2) [71] that catabolise the substrate TRP to KYN.

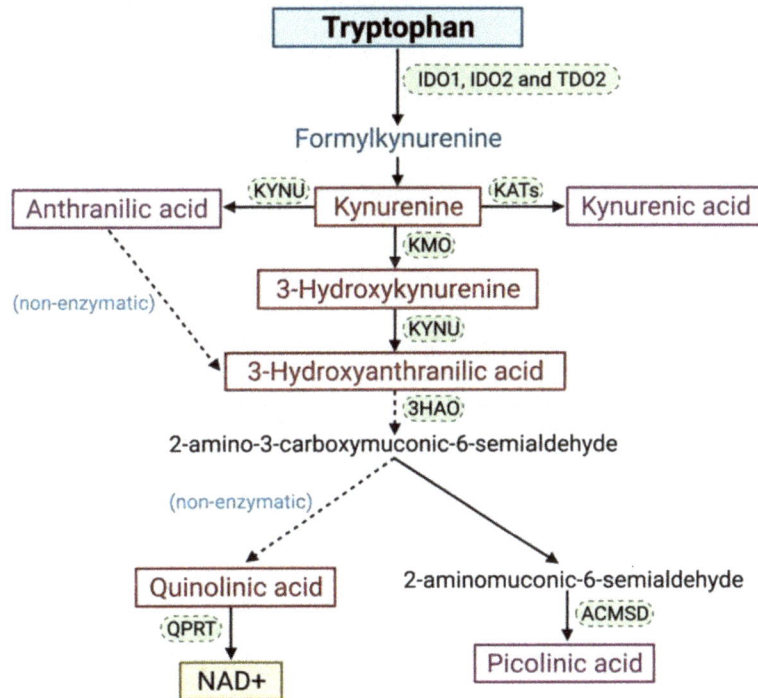

Figure 2. A simplified diagram of the KP: majority of TRP is catabolized through the KP to synthesize the vital energy cofactor, NAD^+.

Although the three rate-limiting enzymes catabolise the same substrate, TRP, they each have different inducers and regions of expression. In normal physiological conditions, IDO1 enzyme expression is limited to endothelial cells in the lungs and placenta, epithelial cells scattered in the female genital tract and mature dendritic cells in secondary lymphoid organs, and is known to be induced by interferon-gamma (IFN-γ) [72]. Compared to IDO1, IDO2 enzyme expression is restricted and confined to hepatocytes, bile duct, neuronal cells of the cerebral cortex, and kidneys [73]. While IDO1 and IDO2 share 43% gene similarity, IDO1 remains the dominant enzyme [69]. Interestingly, the activity of IDO2 elevates when the IDO1 gene is deleted [74]. The third rate-limiting enzyme, TDO2, is primarily expressed in liver, and is the major enzyme to regulate systemic TRP levels in the liver [75,76]. TDO2 enzyme expression is known to be induced partly by glucocorticoids and its substrate TRP [77]. Though these rate-limiting enzymes are cytosolic, their enzymatic activity induces TRP metabolism and accumulation of KP metabolites in the extracellular space, which is facilitated by specific amino-acid transporters [78]. In an inflammatory environment such as cancer, KP is highly activated, resulting in depletion of local TRP in the tumour micro-environment. This process facilitates tumour cells to evade immune detection by reducing the proliferation of effector T lymphocytes and favouring the differentiation of regulatory T (T_{regs}) cells [79].

Involvement of the KP in Cancer

After the discovery that placental IDO1 was the key enzyme mediating immune suppression in maternal–foetal tolerance in 1998, the research focus was expanded to examine whether the KP was involved in immune evasion and cancer [80,81]. Indeed, the KP is frequently dysregulated in cancer and suppresses tumour surveillance in two different mechanisms. The first mechanism involves the overexpression of the rate-limiting enzymes IDO1 and TDO2 to deplete TRP within the tumour microenvironment. TRP is one of the amino acids required for the survival and proliferation of immune T-cells such as T helper (T_h) and cytotoxic T-cells (T_c). Therefore, immune surveillance will be strongly suppressed in a TRP-deprived tumour microenvironment driven by an overactive IDO1/TDO2 tumour [82]. A study by Uyttenhove et al. confirmed overexpression of IDO1 in various human cancer tissues and cell lines, suggesting that was involved in protecting tumours from immune detection [83]. The overexpression of IDO1 in tumours has been suggested to be induced by the IFN-γ generated by tumour-infiltrating T-cells as an adaptive resistance mechanism [84]. Syngeneic animal studies showed that treatment of the IDO1 inhibitor 1-methyltryptophan (1-MT) limited the growth of IDO1-overexpressed tumours [83,85]. A subsequent breast cancer animal model study by Muller et al. demonstrated that combined treatment with 1-MT and cancer chemotherapeutic drug paclitaxel slowed down the tumour growth progression by 30% [86]. Importantly, they observed that the efficacy of this combination therapy was highly dependent on the presence of T-cells, and the inhibition of IDO1 could potentiate the efficacy of chemotherapy.

Apart from IDO1, overexpression of TDO2 in tumour cells has been shown to facilitate immune escape. TDO2 mRNA expression was detected in different types of tumours including hepatocarcinoma [87], glioblastoma [88], breast cancer [89], and colorectal cancer [90,91]. These studies also demonstrated that TDO2 was responsible for the depletion of TRP in IDO-negative tumours to evade immune surveillance [63,88,92]. This notion was supported by an animal model study by Pilotte et al., who showed that treatment using TDO2 inhibitor in an animal model reversed the TDO2-mediated immune evasion mechanism and prevented the growth of TDO2-overexpressing tumours [92]. Consequently, this led to further studies exploring new TDO2 inhibitors for use in the treatment of TDO2-overexpressing cancer [93–95].

Though the role of the IDO2 enzyme in cancer remains less understood, studies have shown that IDO2 expression is upregulated in certain malignancies such as colon cancer, gastric and renal cancer [96], pancreatic cancer [97], non-small cell lung cancer [98], and may have roles in tumour immune escape, facilitating cancer cell proliferation and metastasis. Sorensen et al. described the immunogenic role of IDO2 by demonstrating the presence of spontaneous T_c reactivity against IDO2 in healthy and cancer patient blood samples, and reported that IDO2 supported T_{regs} cells generation that was induced by human dendritic cells [99].

The second mechanism of KP-mediated tumour evasion involves the bioactive KP metabolites KYN, 3HK, 3-HAA, and QUIN. Studies have shown that these metabolites can promote tumour proliferation and modulate the immune cell population. KYN, the first metabolite of KP, can function as an endogenous ligand to activate the aryl hydrocarbon receptor (AhR) in an autocrine/paracrine fashion, and emerging evidence points toward the tumour-promoting role of KYN-mediated activation of the AhR [100,101]. AhR is a ligand-activated transcription factor of the basic helix–loop–helix (bHLH) Per–Arnt–Sim (PAS) family [102]. It is expressed in many immune cells and plays a vital role in regulating various immune functions in a wide range of physical and pathological processes [103–105]. Activation of AhR may facilitate cancer cell proliferation, tissue invasion, metastasis, and angiogenesis [106]. The KYN-AhR signalling pathway can suppress the differentiation and activity of immune cells, resulting in an impaired immune response against tumours, leading to tumour immune tolerance [107]. Various studies have demonstrated the importance of KYN-AhR activation in IDO1- or TDO2-expressing tumour cells and its role in enhancing cancer cell survival and motility. These studies suggested that TDO2-expressing cancer cells escape immune surveillance by activating AhR in various immune cells including dendritic cells, macrophages, natural killer cells, innate lymphoid cells, T_c cells, and T_{regs} cells [108,109]. Opitz et al., found that murine tumours in AhR-proficient mice expressing high AhR and TDO2 expression levels had an enhanced tumour growth rate by suppressing the infiltration of antitumour immune cells, increasing levels of inflammatory cytokines. Furthermore, the study suggested that the TDO2-Kyn-AhR signalling pathway might also be involved in other malignancies, including sarcoma, bladder cancer, cervix cancer, colorectal cancer, lung, and ovarian cancer [88]. Moreover, Ulrike et al. revealed that IDO1 enzyme expression was induced by inflammatory cytokines such as Interleukin 6 (IL-6), and could activate an autocrine-positive inflammatory feedback loop (IDO-AhR-IL-6-STAT3 signalling pathway) that could promote tumour growth and survival [110].

In addition to KYN, kynurenic acid (KYNA) is also an endogenous AhR ligand [111]. In the presence of IL-1β, KYNA binds to AhR and induces production of IL-6, which may also contribute to the IDO-AhR-IL-6-STAT3 autocrine-positive inflammatory feedback loop mentioned earlier. Interestingly, the production of KYNA may not be limited to just via KP but rather through an alternate TRP metabolism mediated by Interleukin-4-induced gene 1 (IL4I1) in a cancer setting. Sadik et al. revealed that IL4I1 was elevated in cancers such as melanoma. An IL4I1-driven AhR activity though KYNA increases tumour cell motility and T-cell proliferation [112]. Given that the activity of IL4I1 is independent of the KP and can limit antitumor immune cell response [113], inhibiting the formation of KYNA metabolite either via the KP or through IL4I1 gene reaction may be necessary to block the activation of AhR in cancer.

The KP metabolites downstream of KYN, including 3-HK, 3-HAA, and QUIN have been shown to inhibit T-cell proliferation and activation. A study by Fallarino et al. showed that 3-3-HAA and QUIN could induce selective apoptosis in T_h1 cells and thymocytes of effector T-cell population in vitro by the activation of caspase-8 activity and the release of cytochrome c from mitochondria [114]. The 3-HAA also significantly inhibits $CD8^+$ T-cell proliferation stimulated through cytokines by driving the T-cells to a proliferative arrest and directly inhibiting the phosphorylation of phosphoinositide-dependent kinase 1 and preventing the activation of nuclear factors after T-cell receptor stimulation [115]. A study by Favre et al. showed that 3HAA also disturbed the balance between T_h and T_{reg} cell populations, driving them towards an immunosuppressive T_{reg} pathway in vitro [116]. Furthermore, a later study by Zaher et al. confirmed that 3HK and 3HAA suppressed $CD4^+$ T-cell proliferation along with significant T-cell death [117].

3. Involvement of the KP in Chronic Liver Disease and HCC

The role of KP in liver diseases has been gaining interest in the recent years. A number of studies have measured high KP activity in chronic liver diseases such as primary biliary cirrhosis, HCV-associated chronic hepatitis, and liver cirrhosis [118,119]. Claria et al. [120] reported that KP activity was elevated in patients with acute decompensation and acute-on-chronic liver failure, and was associated with pathogenesis and mortality in cirrhotic patients. The study concluded that elevated KP activity may be used as an independent prognostic predictor of poor clinical outcomes in cirrhotic patients. In contrast, elevated IDO1 activity during early stages of the HBV infection in hepatocytes was reported to significantly reduce viral replication and enhance the protective immune response [121].

Although the liver is a site of robust immunological activity, liver cancer cells can remain undetected and proliferate. This suggests that these cancer cells can evade local immune surveillance, possibly by using the KP, as observed in various malignancies. Although the research on KP and HCC is limited, the activity of the three upstream enzymes of the pathway, including IDO1, TDO2, and KMO enzymes, has been extensively studied in HCC cells and tissue specimens. These study findings revealed that IDO1, TDO2, and KMO enzyme activity was upregulated in HCC (Table 1).

Table 1. Summary of all KP research carried out on HCC.

Author (Year) [Ref]	Sample Type	Sample Size	Enzyme/Metabolite Studied	Technique Used	Finding
Ishio et al. (2004) [122]	Cell lines HCC tumour specimens	4 HCC cell lines 21 HCC	IDO1	RT-PCR IHC	IDO1 expression may play a role in antitumour immune response.
Pan et al. (2008) [123]	Cell lines Tumour and distant normal liver tissue	6 HCC and human normal hepatocytes 138 HCC	IDO1	RT-PCR IHC	HCC cancer cells and surrounding noncancerous tissue express high IDO1 expression. High IDO1 expression is confined to the tumour creating an immune suppressive microenvironment.
Li et al. (2018) [124]	Cell lines Tumour tissue	2 HCC cell lines 112 HCC (HBV) *	IDO1	RT-PCR WB IHC and IF	IDO1 is expressed in HCC cells on stimulation by IFN-γ via the JAK2-STAT1 signalling pathway. High IDO1 expression indicates antitumour immune response. IDO1 is a favourable prognostic indicator.
Brown et al. (2018) [125]	Cell lines	2 HCC cell lines	IDO1	RT-PCR	IDO1 inhibitors in combination with immune checkpoint inhibitors might be an effective treatment option for HCC patients.
Hoffman et al. (2020) [126]	Cell lines Tumour and normal tissue	1 HCC cell line 171 tissue specimens	TDO2	RT-PCR WB, HPLC IHC and IF	High TDO2 expression observed in HCC tumour cells. TDO2 may be a novel immunotherapeutic target for HCC.
Li et al. (2020) [127]	Cell lines Paired tumour and adjacent normal tissues	5 HCC cell lines and 1 normal liver cell line 93 HCC	TDO2	RT-PCR WB RT-PCR WB, IHC	TDO2 is overexpressed in HCC and may be facilitating HCC progression and invasion. TDO2 enzyme can be a novel prognostic biomarker for HCC patients.
Lei et al. (2021) [87]	Cell lines Paired tumour and adjacent normal tissue	6 HCC cell lines and 1 normal liver cell 23 HCC	TDO2	RT-PCR WB Knockdown using shRNAs HPLC IHC and IF	TDO2 supports EMT of HCC cells via the KYN-AhR pathway, facilitating HCC metastasis and invasion.
Bekki et al. (2020) [128]	Serum	604 HCC * (HCV) 288 Control **	KYN	ELISA	A high level of serum KYN correlated with poor prognosis of HCC.
Jin et al. (2015) [129]	Tumour and adjacent noncancerous liver tissue Cell lines	120 matched HCC and adjacent tissue 205 HCC 5 HCC and 2 human normal liver cells	KMO	IHC RT-PCR WB Knockdown using siRNAs	High KMO expression correlated with HCC tumour aggression, recurrence, and shorter survival rate. KMO knockdown suppressed HCC progression in vitro. KMO overexpression enhanced HCC cell proliferation, migration, and invasion.

* HCC patients with chronic hepatitis B (HBV) or hepatitis C (HCV) virus infection. ** Patients with chronic hepatitis C virus infection without HCC. RT-PCR: reverse transcription-polymerase chain reaction, WB: Western blot, IHC: immunohistochemistry, IF: immunofluorescence, HPLC: high-performance liquid chromatography, shRNA: short hairpin or small hairpin RNA, siRNA: small interfering RNA.

3.1. IDO1

The immunological and prognostic roles of IDO1 in HCC were first investigated by Ishio et al. in 2004 [122]. The results showed that IDO1 mRNA expression was strongly induced in tumour-infiltrating cells of the HCC tumour, which might facilitate an antitumour immune reaction and the expression of IDO in tissue specimens of HCC patients significantly correlated with better recurrence-free survival rates. A later study by Ke Pan et al. observed elevated IDO1 enzyme mRNA and protein expressions in liver tumour and its adjacent normal tissues compared to distant non-involved normal tissues, suggesting that IDO1 overexpression was confined to the tumour microenvironment [123]. A potential explanation for the confined IDO1 expression could be due to the presence of inflammatory cytokine(s) in the tumour microenvironment that activate IDO1 activity. Indeed, a later study by Li et al. demonstrated that IDO1 enzyme expression was observed only in IFN-γ-stimulated HCC cells through the IFN-γ-JAK2-STAT1-signalling pathway. Moreover, high IDO1 expression in HCC positively correlated with abundance of CD8+ T-cells, thus reflecting an antitumour immune response and suggesting that IDO1 could be used as a favourable prognostic indicator for HCC patients [124]. Lastly, Brown et al. suggested that IDO1 enzyme inhibitors in combination with immune checkpoint inhibitors could be a novel treatment approach for liver cancer treatment [125].

3.2. TDO2

A recent study conducted by Hoffman et al., showed that the majority of the tumour cells in HCC tissues expressed TDO2 in HCC [126]. This study demonstrated the immune-regulatory role of the TDO2 enzyme in HCC tumour cells, and suggested that the TDO2 enzyme was a promising immunotherapy treatment target for HCC. Another study by Li et al. characterized the overexpression of TDO2 enzyme in HCC cancer cells and suggested that it might play a vital role in promoting HCC cancer cell growth, migration, and invasion in vitro and in vivo [127]. Additionally, TDO2 expression was correlated with the development of the tumour, such as size, tumour differentiation, and vascular invasion. Based on these strong correlation data, the authors suggested that TDO2 expression could be used as an effective biomarker to predict overall or disease-free survival of HCC patients. Activation of AhR is associated with the loss of cell contact inhibition and changes to the extracellular matrix, and extensive studies have demonstrated that this activation induces epithelial to mesenchymal transition (EMT) in various cancers [130–132]. Overexpression of AhR in HCC has been shown to be associated with its tumour proliferation and invasion [133,134]. A recent study by Lei Li et al. showed that upregulated expression of the TDO2 enzyme promotes the migration and invasion capabilities of HCC cells by the KYN-AhR-mediated induction of epithelial to mesenchymal transition, a process that is vital for cancer metastasis [87].

3.3. KYN Levels in Patient Sera

A recent retrospective study on a cohort of HCC patients with chronic HCV infection revealed that KYN levels were elevated in HCV-mediated HCC patient sera in comparison to healthy controls (non-HCC patients). Bekki et al. observed that KYN production gradually increased when chronic HCV progressed to HCC, and suggested the potential of using serum KYN levels as a biomarker for predicting survival and prognosis in early stage HCV-mediated HCC patients [128].

3.4. KMO

Kynurenine 3-monooxygenase (KMO) is the immediate KP enzyme after the rate-limiting step, and it is widely distributed in the peripheral tissues of the liver and kidney, astrocytes and microglial cells situated in the brain, central nervous system [135,136], and phagocytes, including macrophages and monocytes [137]. KMO localizes to the outer membrane of mitochondria and catabolizes KYN to 3-HK. The role of KMO enzyme expression in cancer has rarely been studied in comparison to IDO and TDO2 enzymes. Liu et al. identified the oncogenic role of KMO in triple-negative breast cancer progression [138]. Moreover, high surface expression of KMO was detected in cytosol and on the cell membranes of breast cancer tissue specimens, indicating its potential as a treatment target for TNBC [139]. A recent study investigated the correlation between upregulated KMO activity and poor clinical outcomes in colorectal cancer (CRC) patients and demonstrated that KMO inhibition suppressed CRC cell proliferation in vitro [140]. On analysing KMO enzyme expression in 120 matched HCC tissue samples, Jin et al. showed that the expression of the KMO enzyme is significantly elevated in HCC tumour tissue compared to adjacent normal liver tissue. High KMO expression correlated with poor patient outcomes, which indicates that the KMO enzyme may be a significant prognostic marker in HCC patients [129]. Results from the in vitro experiment comparing KMO enzyme levels in human normal liver cells and HCC cell lines showed that KMO enzyme was upregulated in HCC cells and might play a role in promoting tumour proliferation, metastasis, and invasion. The study also demonstrated that KMO knockdown in HCC cell lines by small interfering RNA (siRNA) transfection decreased cancer cell proliferation, thus suggesting that KMO could be a novel target for HCC treatment.

3.5. Clinical Trials: IDO1 Inhibitors as HCC Treatment

IDO1 inhibitors are small molecule drugs that competitively block the activity of the IDO1 enzyme without inhibiting IDO2 or TDO2 [141]; several of these drugs are in clinical development. The safety and efficacy of many IDO1 inhibitors, including Indoximod, Epacadostat, Navoximod, BMS-986205, and others, have been tested in combination with other immunotherapy drugs such as pembrolizumab and nivolumab for the treatment of various metastatic cancers. Currently, two small molecule IDO1 inhibitors, BMS-986205/NCT03695250 and INCB024360 (Epacadostat)/NCT02178722, are in phase I/II clinical trial to evaluate their safety and efficacy in HCC patients [142,143]. The clinical trial NCT03695250 is a single-group assignment that examines the safety, tolerability, and efficacy of BMS-986205 with nivolumab in unresectable/metastatic HCC. It is still active but not recruiting patients; hence, the results have not been published yet. The expected treatment-related adverse events of BMS-986205 would be at grade 1–2 such as fatigue and nausea, as reported in the other trials examining the efficacy of BMS-986205 in cancer patients. Clinical trial NCT02178722 evaluated the safety, tolerability, and efficacy of Epacadostat in combination with pembrolizumab. This trial concluded that the combination regime has an acceptable safety profile in patients with advanced cancers, achieving an objective response rate in 12 of 22 cancer patients [144,145]. Treatment-related adverse events observed in 84% of the patients enrolled were of grade 1–2. The most common events were fatigue, rash, arthralgia pruritus, and nausea. This result supports additional phase 3 studies in other malignancies but not in HCC.

4. Conclusions

HCC is one of the few malignancies for which the risk factors have been well-established. Although patients with early stage HCC have the best median survival time and can usually be cured by resection, liver transplant, or ablation, they are often asymptomatic. Hence, most patients present with late-stage HCC and have a poor prognosis. The approved first-line treatment of late-stage HCC is multikinase inhibitors such as sorafenib, which confers a slightly longer survival time. However, this treatment is associated with substantial side effects that have a negative impact on quality of life. This therefore changes the treatment focus by combining current antitumoral drugs with immunotherapy, and this approach has significantly benefited HCC patients. A recently concluded trial examining combination therapy of atezolizumab with bevacizumab showed a significant improvement in overall survival and progression-free survival as compared to sorafenib. Since this study, it has been adopted as the first-line treatment for late-stage HCC. Considering the strong evidence of its ability to mediate immune suppression, the KP might be an alternative immunotherapy target and play a role in the progression of liver cancer, as summarized in Figure 3.

This notion is supported by clinical studies that showed an elevated KP enzyme profile in HCC cells and tumour tissue specimens, with elevated expressions associated with disease aggressiveness. Although current IDO1 inhibitor clinical trials are still in phase I/II evaluation, it is possible to suggest that the use of KP inhibitors in combination regimens may improve the survival mark of early and advanced HCC.

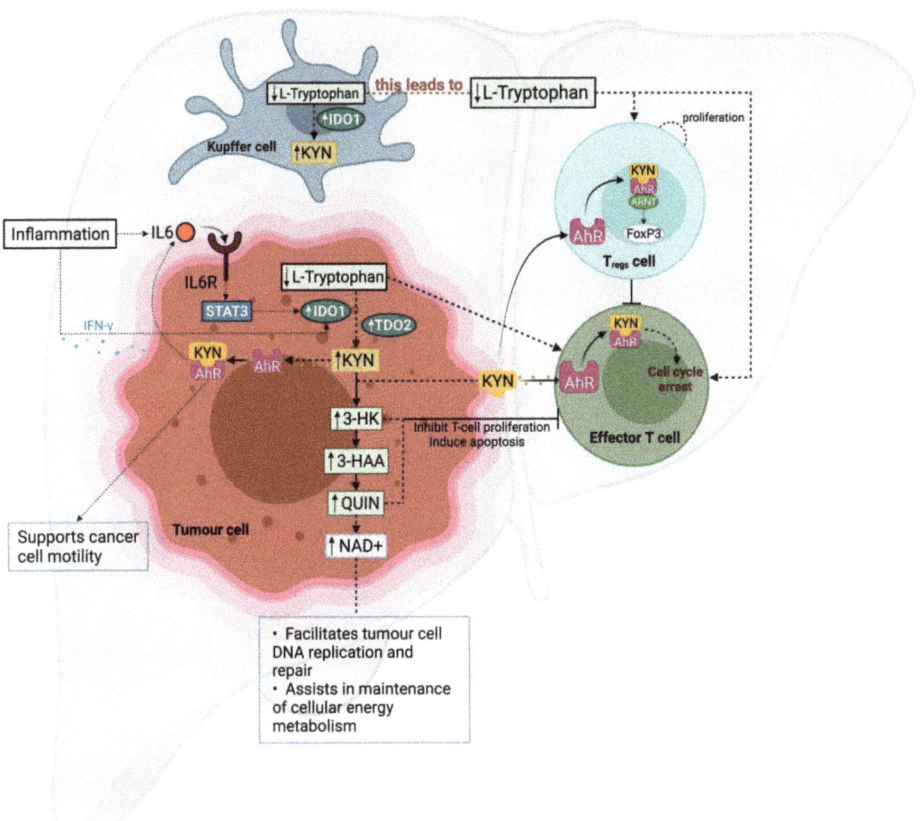

Figure 3. The KP-mediated immune tolerance and cancer invasion: KP promotes immune tolerance by two different mechanisms. Firstly, elevated IDO1/TDO2 enzyme activity in either tumour or immune cell depletes TRP concentration in its local tumour environment. A TRP-stripped environment induces cell arrest in T-cells while inducing differentiation and proliferation of T_{reg} cells. Secondly, downstream KP metabolites induce cell arrest in T-cells and T_{reg} proliferation by either interaction with AhR or by direct interaction with immune cells themselves. In addition to KP-mediated immune tolerance, elevated KP promotes cancer cell motility and proliferation by either overproduction of NAD^+ for cellular repair or by activation of AhR.

Author Contributions: Conceptualization, S.K., G.J.G. and B.H.; writing—original draft preparation, S.K., G.J.G. and B.H.; Writing—review and editing, D.G., S.B.A., V.L. and J.-S.S.; funding acquisition, G.J.G. and B.H. All authors have read and agreed to the published version of the manuscript.

Funding: S.K. is supported by International Macquarie University Research Excellence Scholarship—Master of Research scholarship; D.G. is supported by Aviesan Plan Cancer and Cancer pole Grand Ouest; S.B.A. is supported by Cancer Council NSW funding RG19-04; G.J.G. is supported by the National Health and Medical Research Council (NHMRC) APP1176660, PANDIS.org and Macquarie University; B.H. is supported by NHMRC.

Acknowledgments: The figures were created with BioRender.com.

Conflicts of Interest: The authors declare that they have no competing interests.

Abbreviations

1-MT	1-methyltryptophan
3-HAA	3-hydroxyanthranilic acid
3HAO	3-hydroxyanthranilate 3,4-dioxygenase
3-HK	3-hydroxykynurenine
ACMSD	2-amino-3-carboxymuconate semialdehyde decarboxylase
AFP	Alpha-fetoprotein
AhR	Aryl hydrocarbon receptor
BCLC	Barcelona clinic liver cancer
bHLH	basic helix–loop–helix
CRC	Colorectal cancer
GPC-3	Glypican-3
HCC	Hepatocarcinoma
HBV	Hepatitis B virus
HCV	Hepatitis C virus
ICC	Intrahepatic cholangiocarcinoma
IDO1	Indoleamine 2,3 dioxygenase 1
IDO2	Indoleamine 2,3 dioxygenase 2
IL4I1	Interleukin-4 induced gene 1
IL-6	Interleukin 6
KMO	Kynurenine-3-monooxygenase
KP	Kynurenine pathway
KYN	Kynurenine
KYNA	Kynurenic acid
KYNU	Kynureninase
NAD^+	Nicotinamide adenine dinucleotide
NAFLD	Non-alcoholic fatty liver disease
NASH	Non-alcoholic steatohepatitis
QPRT	Quinolate phosphoribosyltransferase
QUIN	Quinolinic acid
PAS	Per–Arnt–Sim
TACE	Trans-arterial chemoembolization
T_c	Cytotoxic T-cells
TDO2	Tryptophan 2,3 dioxygenase 2
T_h	T helper cells
T_{regs}	CD4+CD25+Foxp3+ Regulatory T-cells
TRP	Tryptophan

References

1. Sung, H.; Ferlay, J.; Siegel, R.L.; Laversanne, M.; Soerjomataram, I.; Jemal, A.; Bray, F. Global Cancer Statistics 2020: GLOBOCAN Estimates of Incidence and Mortality Worldwide for 36 Cancers in 185 Countries. *CA Cancer J. Clin.* **2021**, *71*, 209–249. [CrossRef] [PubMed]
2. Olubuyide, I. Pattern of metastasis of primary liver cancer at autopsy: An African series. *Trop. Gastroenterol.* **1991**, *12*, 67–72. [PubMed]
3. Lee, Y.-T.M.; Geer, D.A. Primary liver cancer: Pattern of metastasis. *J. Surg. Oncol.* **1987**, *36*, 26–31. [CrossRef]
4. Jiang, X.-B.; Ke, C.; Zhang, G.-H.; Zhang, X.-H.; Sai, K.; Chen, Z.-P.; Mou, Y.-G. Brain metastases from hepatocellular carcinoma: Clinical features and prognostic factors. *BMC Cancer* **2012**, *12*, 49. [CrossRef]
5. Tang, D.; Nagano, H.; Nakamura, M.; Wada, H.; Marubashi, S.; Miyamoto, A.; Takeda, Y.; Umeshita, K.; Dono, K.; Monden, M. Clinical and Pathological Features of Allen's Type C Classification of Resected Combined Hepatocellular and Cholangiocarcinoma: A Comparative Study with Hepatocellular Carcinoma and Cholangiocellular Carcinoma. *J. Gastrointest. Surg.* **2006**, *10*, 987–998. [CrossRef]
6. Chen, G.; Lin, W.; Shen, F.; Iloeje, U.H.; London, W.T.; Evans, A.A. Past HBV Viral Load as Predictor of Mortality and Morbidity from HCC and Chronic Liver Disease in a Prospective Study. *Am. J. Gastroenterol.* **2006**, *101*, 1797–1803. [CrossRef]
7. Fattovich, G.; Stroffolini, T.; Zagni, I.; Donato, F. Hepatocellular carcinoma in cirrhosis: Incidence and risk factors. *Gastroenterology* **2004**, *127*, S35–S50. [CrossRef]
8. Llovet, J.M.; Kelley, R.K.; Villanueva, A.; Singal, A.G.; Pikarsky, E.; Roayaie, S.; Lencioni, R.; Koike, K.; Zucman-Rossi, J.; Finn, R.S. Hepatocellular carcinoma. *Nat. Rev. Dis. Prim.* **2021**, *7*, 6. [CrossRef] [PubMed]

9. International Agency for Research on Cancer. GLOBOCAN 2018. IARC 2020. Available online: https://gco.iarc.fr/today/online-analysis-map?v=2020&mode=population&mode_population=continents&population=900&populations=900&key=asr&sex=0&cancer=11&type=0&statistic=5&prevalence=0&population_group=earth&color_palette=default&map_scale=quantile&map_nb_colors=5&continent=0&rotate=%255B10%252C0%255D (accessed on 30 August 2021).
10. McGlynn, K.A.; Petrick, J.L.; El-Serag, H.B. Epidemiology of hepatocellular carcinoma. *Hepatology* **2020**, *73*, 4–13. [CrossRef] [PubMed]
11. Australian Institute of Health and Welfare. Cancer in Australia: Actual incidence data from 1982 to 2013 and mortality data from 1982 to 2014 with projections to 2017. *Asia-Pac. J. Clin. Oncol.* **2018**, *14*, 5–15. [CrossRef]
12. Wallace, M.C.; Preen, D.; Short, M.W.; Adams, L.A.; Jeffrey, G.P. Hepatocellular carcinoma in Australia 1982–2014: Increasing incidence and improving survival. *Liver Int.* **2018**, *39*, 522–530. [CrossRef]
13. Wu, E.M.; Wong, L.L.; Hernandez, B.Y.; Ji, J.-F.; Jia, W.; Kwee, S.A.; Kalathil, S. Gender differences in hepatocellular cancer: Disparities in nonalcoholic fatty liver disease/steatohepatitis and liver transplantation. *Hepatoma Res.* **2018**, *4*, 66. [CrossRef]
14. Yang, J.D.; Hainaut, P.; Gores, G.J.; Amadou, A.; Plymoth, A.; Roberts, L.R. A global view of hepatocellular carcinoma: Trends, risk, prevention and management. *Nat. Rev. Gastroenterol. Hepatol.* **2019**, *16*, 589–604. [CrossRef] [PubMed]
15. Trinchet, J.-C.; Bourcier, V.; Chaffaut, C.; Ahmed, M.A.; Allam, S.; Marcellin, P.; Guyader, D.; Pol, S.; Larrey, D.; De Lédinghen, V.; et al. Complications and competing risks of death in compensated viral cirrhosis (ANRS CO12 CirVir prospective cohort). *Hepatology* **2015**, *62*, 737–750. [CrossRef] [PubMed]
16. Ringelhan, M.; McKeating, J.A.; Protzer, U. Viral hepatitis and liver cancer. *Philos. Trans. R. Soc. Lond. Ser. B Biol. Sci.* **2017**, *372*, 20160274. [CrossRef]
17. Yang, J.D.; Kim, W.R.; Coelho, R.; Mettler, T.A.; Benson, J.T.; Sanderson, S.O.; Therneau, T.M.; Kim, B.; Roberts, L. Cirrhosis Is Present in Most Patients with Hepatitis B and Hepatocellular Carcinoma. *Clin. Gastroenterol. Hepatol.* **2011**, *9*, 64–70. [CrossRef] [PubMed]
18. Zamor, P.J.; Delemos, A.S.; Russo, M.W. Viral hepatitis and hepatocellular carcinoma: Etiology and management. *J. Gastrointest. Oncol.* **2017**, *8*, 229–242. [CrossRef] [PubMed]
19. Global Burden of Disease Liver Cancer Collaboration; Akinyemiju, T.; Abera, S.; Ahmed, M.; Alam, N.; Alemayohu, M.A.; Allen, C.; Al-Raddadi, R.; Alvis-Guzman, N.; Amoako, Y.; et al. The Burden of Primary Liver Cancer and Underlying Etiologies from 1990 to 2015 at the Global, Regional, and National Level. *JAMA Oncol.* **2017**, *3*, 1683–1691. [CrossRef]
20. Chang, M.-H. Hepatitis B virus and cancer prevention. *Recent Results Cancer Res.* **2011**, *188*, 75–84. [CrossRef]
21. Hoshida, Y.; Fuchs, B.C.; Bardeesy, N.; Baumert, T.F.; Chung, R.T. Pathogenesis and prevention of hepatitis C virus-induced hepatocellular carcinoma. *J. Hepatol.* **2014**, *61*, S79–S90. [CrossRef]
22. Kanwal, F.; Kramer, J.; Asch, S.M.; Chayanupatkul, M.; Cao, Y.; El-Serag, H.B. Risk of Hepatocellular Cancer in HCV Patients Treated with Direct-Acting Antiviral Agents. *Gastroenterology* **2017**, *153*, 996–1005.e1. [CrossRef]
23. Blonski, W.; Kotlyar, D.S.; Forde, K.A. Non-viral causes of hepatocellular carcinoma. *World J. Gastroenterol.* **2010**, *16*, 3603–3615. [CrossRef]
24. Michelotti, G.A.; Machado, M.V.; Diehl, A.M. NAFLD, NASH and liver cancer. *Nat. Rev. Gastroenterol. Hepatol.* **2013**, *10*, 656–665. [CrossRef]
25. Said, A.; Ghufran, A. Epidemic of non-alcoholic fatty liver disease and hepatocellular carcinoma. *World J. Clin. Oncol.* **2017**, *8*, 429. [CrossRef] [PubMed]
26. Ascha, M.S.; Hanouneh, I.A.; Lopez, R.; Tamimi, T.A.-R.; Feldstein, A.F.; Zein, N.N. The incidence and risk factors of hepatocellular carcinoma in patients with nonalcoholic steatohepatitis. *Hepatology* **2010**, *51*, 1972–1978. [CrossRef]
27. El-Serag, H.B.; Kanwal, F. Epidemiology of hepatocellular carcinoma in the United States: Where are we? Where do we go? *Hepatology* **2014**, *60*, 1767. [CrossRef]
28. Yu, M.-L.; Chuang, W.-L. Treatment of chronic hepatitis C in Asia: When East meets West. *J. Gastroenterol. Hepatol.* **2009**, *24*, 336–345. [CrossRef]
29. Altekruse, S.F.; Henley, S.J.; Cucinelli, J.E.; McGlynn, K.A. Changing Hepatocellular Carcinoma Incidence and Liver Cancer Mortality Rates in the United States. *Am. J. Gastroenterol.* **2014**, *109*, 542–553. [CrossRef]
30. Lan, X.; Li, L. Association between hepatitis B virus/hepatitis C virus infection and primary hepatocellular carcinoma risk: A meta-analysis based on Chinese population. *J. Cancer Res. Ther.* **2016**, *12*, 284. [CrossRef] [PubMed]
31. Lin, L.; Yan, L.; Liu, Y.; Qu, C.; Ni, J.; Li, H. The Burden and Trends of Primary Liver Cancer Caused by Specific Etiologies from 1990 to 2017 at the Global, Regional, National, Age, and Sex Level Results from the Global Burden of Disease Study 2017. *Liver Cancer* **2020**, *9*, 563–582. [CrossRef] [PubMed]
32. Liver Cancer Survival Rates. American Cancer Society. Available online: https://www.cancer.org/cancer/liver-cancer/detection-diagnosis-staging/survival-rates.html (accessed on 30 August 2021).
33. Pons, F.; Varela, M.; Llovet, J.M. Staging systems in hepatocellular carcinoma. *HPB* **2005**, *7*, 35–41. [CrossRef]
34. Wege, H.; Li, J.; Ittrich, H. Treatment Lines in Hepatocellular Carcinoma. *Visc. Med.* **2019**, *35*, 266–272. [CrossRef]
35. Hibi, T.; Cherqui, D.; Geller, D.A.; Itano, O.; Kitagawa, Y.; Wakabayashi, G. Expanding indications and regional diversity in laparoscopic liver resection unveiled by the International Survey on Technical Aspects of Laparoscopic Liver Resection (INSTALL) study. *Surg. Endosc.* **2015**, *30*, 2975–2983. [CrossRef]

36. Fuks, D.; Dokmak, S.; Paradis, V.; Diouf, M.; Durand, F.; Belghiti, J. Benefit of initial resection of hepatocellular carcinoma followed by transplantation in case of recurrence: An intention-to-treat analysis. *Hepatology* **2011**, *55*, 132–140. [CrossRef] [PubMed]
37. Rhim, H.; Lim, H.K. Radiofrequency Ablation of Hepatocellular Carcinoma: Pros and Cons. *Gut Liver* **2010**, *4*, S113–S118. [CrossRef] [PubMed]
38. Gomaa, A.I.; Khan, S.A.; Leen, E.L.S.; Waked, I.; Taylor-Robinson, S.D. Diagnosis of hepatocellular carcinoma. *World J. Gastroenterol.* **2009**, *15*, 1301–1314. [CrossRef] [PubMed]
39. Andrews, G.K.; Dziadek, M.; Tamaoki, T. Expression and methylation of the mouse alpha-fetoprotein gene in embryonic, adult, and neoplastic tissues. *J. Biol. Chem.* **1982**, *257*, 5148–5153. [CrossRef]
40. Tilghman, S.M.; Belayew, A. Transcriptional control of the murine albumin/alpha-fetoprotein locus during development. *Proc. Natl. Acad. Sci. USA* **1982**, *79*, 5254–5257. [CrossRef]
41. Kelly, S.L.; Bird, T.G. The Evolution of the Use of Serum Alpha-fetoprotein in Clinical Liver Cancer Surveillance. *J. Immunobiol.* **2016**, *1*. [CrossRef]
42. Colombo, M. Screening for Cancer in Viral Hepatitis. *Clin. Liver Dis.* **2001**, *5*, 109–122. [CrossRef]
43. Filmus, J.; Selleck, S.B. Glypicans: Proteoglycans with a surprise. *J. Clin. Investig.* **2001**, *108*, 497–501. [CrossRef]
44. Iglesias, B.V.; Centeno, G.; Pascuccelli, H.; Ward, F.; Peters, M.G.; Filmus, J.; Puricelli, L.; Joffé, E.B.D.K. Expression pattern of glypican-3 (GPC3) during human embryonic and fetal development. *Histol. Histopathol.* **2008**, *23*, 1333–1340.
45. Filmus, J. Glypicans in growth control and cancer. *Glycobiology* **2001**, *11*, 19–23. [CrossRef] [PubMed]
46. Yamauchi, N.; Watanabe, A.; Hishinuma, M.; Ohashi, K.-I.; Midorikawa, Y.; Morishita, Y.; Niki, T.; Shibahara, J.; Mori, M.; Makuuchi, M.; et al. The glypican 3 oncofetal protein is a promising diagnostic marker for hepatocellular carcinoma. *Mod. Pathol.* **2005**, *18*, 1591–1598. [CrossRef] [PubMed]
47. Capurro, M.; Wanless, I.R.; Sherman, M.; Deboer, G.; Shi, W.; Miyoshi, E.; Filmus, J. Glypican-3: A novel serum and histochemical marker for hepatocellular carcinoma. *Gastroenterology* **2003**, *125*, 89–97. [CrossRef]
48. Wang, F.H.; Yip, Y.C.; Zhang, M.; Vong, H.T.; Chan, K.I.; Wai, K.C.; Wen, J.M. Diagnostic utility of glypican-3 for hepatocellular carcinoma on liver needle biopsy. *J. Clin. Pathol.* **2010**, *63*, 599–603. [CrossRef]
49. Hippo, Y.; Watanabe, K.; Watanabe, A.; Midorikawa, Y.; Yamamoto, S.; Ihara, S.; Tokita, S.; Iwanari, H.; Ito, Y.; Nakano, K.; et al. Identification of Soluble NH2-Terminal Fragment of Glypican-3 as a Serological Marker for Early-Stage Hepatocellular Carcinoma. *Cancer Res.* **2004**, *64*, 2418–2423. [CrossRef] [PubMed]
50. Jia, X.; Gao, Y.; Zhai, D.; Liu, J.; Cai, J.; Wang, Y.; Jing, L.; Du, Z. Assessment of the Clinical Utility of Glypican 3 as a Serum Marker for the Diagnosis of Hepatocellular Carcinoma. *Technol. Cancer Res. Treat.* **2016**, *15*, 780–786. [CrossRef]
51. Yasuda, E.; Kumada, T.; Toyoda, H.; Kaneoka, Y.; Maeda, A.; Okuda, S.; Yoshimi, N.; Kozawa, O. Evaluation for clinical utility of GPC3, measured by a commercially available ELISA kit with Glypican-3 (GPC3) antibody, as a serological and histological marker for hepatocellular carcinoma. *Hepatol. Res.* **2010**, *40*, 477–485. [CrossRef]
52. Kubota, K.; Ina, H.; Okada, Y.; Irie, T. Growth Rate of Primary Single Hepatocellular Carcinoma: Determining Optimal Screening Interval with Contrast Enhanced Computed Tomography. *Dig. Dis. Sci.* **2003**, *48*, 581–586. [CrossRef]
53. Lewandowski, R.J.; Geschwind, J.-F.; Liapi, E.; Salem, R. Transcatheter Intraarterial Therapies: Rationale and Overview. *Radiology* **2011**, *259*, 641–657. [CrossRef] [PubMed]
54. Llovet, J.M.; Ricci, S.; Mazzaferro, V.; Hilgard, P.; Gane, E.; Blanc, J.-F.; de Oliveira, A.C.; Santoro, A.; Raoul, J.-L.; Forner, A.; et al. Sorafenib in advanced hepatocellular carcinoma. *N. Engl. J. Med.* **2008**, *359*, 378–390. [CrossRef] [PubMed]
55. Kudo, M.; Finn, R.S.; Qin, S.; Han, K.-H.; Ikeda, K.; Piscaglia, F.; Baron, A.; Park, J.-W.; Han, G.; Jassem, J.; et al. Lenvatinib versus sorafenib in first-line treatment of patients with unresectable hepatocellular carcinoma: A randomised phase 3 non-inferiority trial. *Lancet* **2018**, *391*, 1163–1173. [CrossRef]
56. Finn, R.S.; Qin, S.; Ikeda, M.; Galle, P.R.; Ducreux, M.; Kim, T.-Y.; Lim, H.Y.; Kudo, M.; Breder, V.V.; Merle, P.; et al. IMbrave150: Updated overall survival (OS) data from a global, randomized, open-label phase III study of atezolizumab (atezo) + bevacizumab (bev) versus sorafenib (sor) in patients (pts) with unresectable hepatocellular carcinoma (HCC). *J. Clin. Oncol.* **2021**, *39*, 267. [CrossRef]
57. Galle, P.R.; Finn, R.S.; Qin, S.; Ikeda, M.; Zhu, A.X.; Kim, T.-Y.; Kudo, M.; Breder, V.; Merle, P.; Kaseb, A.; et al. Patient-reported outcomes with atezolizumab plus bevacizumab versus sorafenib in patients with unresectable hepatocellular carcinoma (IMbrave150): An open-label, randomised, phase 3 trial. *Lancet Oncol.* **2021**, *22*, 991–1001. [CrossRef]
58. Lubel, J.S.; Roberts, S.K.; Strasser, S.I.; Thompson, A.J.; Philip, J.; Goodwin, M.; Clarke, S.; Crawford, D.H.; Levy, M.T.; Shackel, N. Australian recommendations for the management of hepatocellular carcinoma: A consensus statement. *Med. J. Aust.* **2020**, *214*, 475–483. [CrossRef] [PubMed]
59. Pembrolizumab (Keytruda) in Advanced Hepatocellular Carcinoma. ClinicalTrials.gov U.S. National Library of Medicine. Available online: https://clinicaltrials.gov/ct2/show/NCT02658019 (accessed on 30 August 2021).
60. Nivolumab and Bevacizumab in Patients with Advanced and or Metastatic Hepatocellular Carcinoma (NUANCE). ClinicalTrials.gov U.S. National Library of Medicine. Available online: https://clinicaltrials.gov/ct2/show/NCT03382886?term=Nivolumab&cond=Hepatocellular+Carcinoma&draw=2&rank=3 (accessed on 30 August 2021).
61. Kubes, P.; Jenne, C. Immune Responses in the Liver. *Annu. Rev. Immunol.* **2018**, *36*, 247–277. [CrossRef]
62. Robinson, M.; Harmon, C.; O'Farrelly, C. Liver immunology and its role in inflammation and homeostasis. *Cell. Mol. Immunol.* **2016**, *13*, 267–276. [CrossRef]

63. Platten, M.; Wick, W.; Eynde, B.J.V.D. Tryptophan Catabolism in Cancer: Beyond IDO and Tryptophan Depletion. *Cancer Res.* **2012**, *72*, 5435–5440. [CrossRef]
64. Munn, D.H.; Mellor, A.L. IDO in the Tumor Microenvironment: Inflammation, Counter-Regulation, and Tolerance. *Trends Immunol.* **2016**, *37*, 193–207. [CrossRef]
65. Moffett, J.R.; Namboodiri, M.A. Tryptophan and the immune response. *Immunol. Cell Biol.* **2003**, *81*, 247–265. [CrossRef] [PubMed]
66. Colabroy, K.L.; Begley, T.P. Tryptophan Catabolism: Identification and Characterization of a New Degradative Pathway. *J. Bacteriol.* **2005**, *187*, 7866–7869. [CrossRef] [PubMed]
67. Heng, R.B.; Lim, E.; Lovejoy, D.B.; Bessede, A.; Gluch, L.; Guillemin, G.J. Understanding the role of the kynurenine pathway in human breast cancer immunobiology. *Oncotarget* **2015**, *7*, 6506–6520. [CrossRef]
68. Takikawa, O.; Yoshida, R.; Kido, R.; Hayaishi, O. Tryptophan degradation in mice initiated by indoleamine 2,3-dioxygenase. *J. Biol. Chem.* **1986**, *261*, 3648–3653. [CrossRef]
69. Ball, H.J.; Sanchez-Perez, A.; Weiser, S.; Austin, C.J.; Astelbauer, F.; Miu, J.; McQuillan, J.A.; Stocker, R.; Jermiin, L.; Hunt, N.H. Characterization of an indoleamine 2,3-dioxygenase-like protein found in humans and mice. *Gene* **2007**, *396*, 203–213. [CrossRef]
70. Ball, H.J.; Yuasa, H.J.; Austin, C.J.; Weiser, S.; Hunt, N.H. Indoleamine 2, 3-dioxygenase-2; a new enzyme in the kynurenine pathway. *Int. J. Biochem. Cell Biol.* **2009**, *41*, 467–471. [CrossRef]
71. Ren, S.; Liu, H.; Licad, E.; Correia, M.A. Expression of Rat Liver Tryptophan 2, 3-Dioxygenase inEscherichia coli: Structural and Functional Characterization of the Purified Enzyme. *Arch. Biochem. Biophys.* **1996**, *333*, 96–102. [CrossRef]
72. Théate, I.; Van Baren, N.; Pilotte, L.; Moulin, P.; Larrieu, P.; Renauld, J.-C.; Hervé, C.; Gutierrez-Roelens, I.; Marbaix, E.; Sempoux, C.; et al. Extensive Profiling of the Expression of the Indoleamine 2,3-Dioxygenase 1 Protein in Normal and Tumoral Human Tissues. *Cancer Immunol. Res.* **2014**, *3*, 161–172. [CrossRef]
73. Fukunaga, M.; Yamamoto, Y.; Kawasoe, M.; Arioka, Y.; Murakami, Y.; Hoshi, M.; Saito, K. Studies on tissue and cellular distribution of indoleamine 2, 3-dioxygenase 2: The absence of IDO1 upregulates IDO2 expression in the epididymis. *J. Histochem. Cytochem.* **2012**, *60*, 854–860. [CrossRef]
74. Metz, R.; Smith, C.; DuHadaway, J.B.; Chandler, P.; Baban, B.; Merlo, L.; Pigott, E.; Keough, M.P.; Rust, S.; Mellor, A.L.; et al. IDO2 is critical for IDO1-mediated T-cell regulation and exerts a non-redundant function in inflammation. *Int. Immunol.* **2014**, *26*, 357–367. [CrossRef] [PubMed]
75. Salter, M.; Hazelwood, R.; Pogson, C.I.; Iyer, R.; Madge, D.J. The effects of a novel and selective inhibitor of tryptophan 2,3-dioxygenase on tryptophan and serotonin metabolism in the rat. *Biochem. Pharmacol.* **1995**, *49*, 1435–1442. [CrossRef]
76. Kanai, M.; Funakoshi, H.; Takahashi, H.; Hayakawa, T.; Mizuno, S.; Matsumoto, K.; Nakamura, T. Tryptophan 2,3-dioxygenase is a key modulator of physiological neurogenesis and anxiety-related behavior in mice. *Mol. Brain* **2009**, *2*, 1–16. [CrossRef]
77. Ren, S.; Correia, M.A. Heme: A regulator of rat hepatic tryptophan 2, 3-dioxygenase? *Arch. Biochem. Biophys.* **2000**, *377*, 195–203. [CrossRef]
78. Kudo, Y.; Boyd, C.A.R. Characterisation of L-tryptophan transporters in human placenta: A comparison of brush border and basal membrane vesicles. *J. Physiol.* **2001**, *531*, 405–416. [CrossRef]
79. Mellor, A.L.; Munn, D.H. Tryptophan catabolism and T-cell tolerance: Immunosuppression by starvation? *Immunol. Today* **1999**, *20*, 469–473. [CrossRef]
80. Van Baren, N.; van den Eynde, B.J. Tryptophan-degrading enzymes in tumoral immune resistance. *Front. Immunol.* **2016**, *6*, 34. [CrossRef] [PubMed]
81. Munn, D.H.; Zhou, M.; Attwood, J.T.; Bondarev, I.; Conway, S.J.; Marshall, B.; Brown, C.; Mellor, A.L. Prevention of Allogeneic Fetal Rejection by Tryptophan Catabolism. *Science* **1998**, *281*, 1191–1193. [CrossRef] [PubMed]
82. Van Baren, N.; Van den Eynde, B.J. Tumoral immune resistance mediated by enzymes that degrade tryptophan. *Cancer Immunol. Res.* **2015**, *3*, 978–985. [CrossRef] [PubMed]
83. Uyttenhove, C.; Pilotte, L.; Théate, I.; Stroobant, V.; Colau, D.; Parmentier, N.; Boon, T.; Eynde, B.J.V.D. Evidence for a tumoral immune resistance mechanism based on tryptophan degradation by indoleamine 2,3-dioxygenase. *Nat. Med.* **2003**, *9*, 1269–1274. [CrossRef]
84. Spranger, S.; Spaapen, R.; Zha, Y.; Williams, J.; Meng, Y.; Ha, T.T.; Gajewski, T.F. Up-Regulation of PD-L1, IDO, and Tregs in the Melanoma Tumor Microenvironment Is Driven by CD8+ T Cells. *Sci. Transl. Med.* **2013**, *5*, 200ra116. [CrossRef] [PubMed]
85. Friberg, M.; Jennings, R.; Alsarraj, M.; Dessureault, S.; Cantor, A.; Extermann, M.; Mellor, A.L.; Munn, D.H.; Antonia, S.J. Indoleamine 2,3-dioxygenase contributes to tumor cell evasion of T cell-mediated rejection. *Int. J. Cancer* **2002**, *101*, 151–155. [CrossRef]
86. Muller, A.; DuHadaway, J.B.; Donover, P.S.; Sutanto-Ward, E.; Prendergast, G.C. Inhibition of indoleamine 2,3-dioxygenase, an immunoregulatory target of the cancer suppression gene Bin1, potentiates cancer chemotherapy. *Nat. Med.* **2005**, *11*, 312–319. [CrossRef]
87. Li, L.; Wang, T.; Li, S.; Chen, Z.; Wu, J.; Cao, W.; Wo, Q.; Qin, X.; Xu, J. TDO2 Promotes the EMT of Hepatocellular Carcinoma Through Kyn-AhR Pathway. *Front. Oncol.* **2021**, *10*, 3008.
88. Opitz, C.; Litzenburger, U.M.; Sahm, F.; Ott, M.; Tritschler, I.; Trump, S.; Schumacher, T.; Jestaedt, L.; Schrenk, D.; Weller, M.; et al. An endogenous tumour-promoting ligand of the human aryl hydrocarbon receptor. *Nature* **2011**, *478*, 197–203. [CrossRef]
89. Liu, Q.; Zhai, J.; Kong, X.; Wang, X.; Wang, Z.; Fang, Y.; Wang, J. Comprehensive Analysis of the Expression and Prognosis for TDO2 in Breast Cancer. *Mol. Ther. Oncolytics* **2020**, *17*, 153–168. [CrossRef]

90. Zhao, L.; Wang, B.; Yang, C.; Lin, Y.; Zhang, Z.; Wang, S.; Ye, Y.; Shen, Z. TDO2 knockdown inhibits colorectal cancer progression via TDO2-KYNU-AhR pathway. *Gene* **2021**, *792*, 145736. [CrossRef]
91. Chen, I.-C.; Lee, K.-H.; Hsu, Y.-H.; Wang, W.-R.; Chen, C.-M.; Cheng, Y.-W. Expression Pattern and Clinicopathological Relevance of the Indoleamine 2,3-Dioxygenase 1/Tryptophan 2,3-Dioxygenase Protein in Colorectal Cancer. *Dis. Markers* **2016**, *2016*, 1–9. [CrossRef] [PubMed]
92. Pilotte, L.; Larrieu, P.; Stroobant, V.; Colau, D.; Dolušić, E.; Frédérick, R.; de Plaen, E.; Uyttenhove, C.; Wouters, J.; Masereel, B.; et al. Reversal of tumoral immune resistance by inhibition of tryptophan 2, 3-dioxygenase. *Proc. Natl. Acad. Sci. USA* **2012**, *109*, 2497–2502. [CrossRef] [PubMed]
93. Dolusic, E.; Larrieu, P.; Moineaux, L.; Stroobant, V.; Pilotte, L.; Colau, D.; Pochet, L.; van den Eynde, B.; Masereel, B.; Wouters, J.; et al. Tryptophan 2, 3-dioxygenase (TDO) inhibitors. 3-(2-(pyridyl) ethenyl) indoles as potential anticancer immunomodulators. *J. Med. Chem.* **2011**, *54*, 5320–5334. [CrossRef] [PubMed]
94. Abdel-Magid, A.F. Targeting the Inhibition of Tryptophan 2,3-Dioxygenase (TDO-2) for Cancer Treatment. *ACS Med. Chem. Lett.* **2016**, *8*, 11–13. [CrossRef]
95. Cheong, J.E.; Sun, L. Targeting the IDO1/TDO2–KYN–AhR pathway for cancer immunotherapy–challenges and opportunities. *Trends Pharmacol. Sci.* **2018**, *39*, 307–325. [CrossRef]
96. Löb, S.; Königsrainer, A.; Zieker, D.; Brücher, B.L.D.M.; Rammensee, H.-G.; Opelz, G.; Terness, P. IDO1 and IDO2 are expressed in human tumors: Levo- but not dextro-1-methyl tryptophan inhibits tryptophan catabolism. *Cancer Immunol. Immunother.* **2008**, *58*, 153–157. [CrossRef] [PubMed]
97. Nevler, A.; Muller, A.J.; Sutanto-Ward, E.; DuHadaway, J.B.; Nagatomo, K.; Londin, E.; O'Hayer, K.; Cozzitorto, J.A.; Lavu, H.; Yeo, T.P.; et al. Host IDO2 Gene Status Influences Tumor Progression and Radiotherapy Response in KRAS-Driven Sporadic Pancreatic Cancers. *Clin. Cancer Res.* **2018**, *25*, 724–734. [CrossRef] [PubMed]
98. Mandarano, M.; Bellezza, G.; Belladonna, M.L.; Vannucci, J.; Gili, A.; Ferri, I.; Lupi, C.; Ludovini, V.; Falabella, G.; Metro, G.; et al. Indoleamine 2,3-Dioxygenase 2 Immunohistochemical Expression in Resected Human Non-small Cell Lung Cancer: A Potential New Prognostic Tool. *Front. Immunol.* **2020**, *11*, 839. [CrossRef] [PubMed]
99. Sørensen, R.B.; Køllgaard, T.; Andersen, R.S.; Berg, J.H.V.D.; Svane, I.M.; Straten, P.T.; Andersen, M.H. Spontaneous Cytotoxic T-Cell Reactivity against Indoleamine 2,3-Dioxygenase-2. *Cancer Res.* **2011**, *71*, 2038–2044. [CrossRef]
100. Zhou, L. AHR Function in Lymphocytes: Emerging Concepts. *Trends Immunol.* **2015**, *37*, 17–31. [CrossRef]
101. Leclerc, D.; Pires, A.C.S.; Guillemin, G.J.; Gilot, D. Detrimental activation of AhR pathway in cancer: An overview of therapeutic strategies. *Curr. Opin. Immunol.* **2021**, *70*, 15–26. [CrossRef]
102. Kewley, R.J.; Whitelaw, M.L.; Chapman-Smith, A. The mammalian basic helix–loop–helix/PAS family of transcriptional regulators. *Int. J. Biochem. Cell Biol.* **2004**, *36*, 189–204. [CrossRef]
103. Kimura, A.; Naka, T.; Nakahama, T.; Chinen, I.; Masuda, K.; Nohara, K.; Fujii-Kuriyama, Y.; Kishimoto, T. Aryl hydrocarbon receptor in combination with Stat1 regulates LPS-induced inflammatory responses. *J. Exp. Med.* **2009**, *206*, 2027–2035. [CrossRef] [PubMed]
104. Wang, C.; Ye, Z.; Kijlstra, A.; Zhou, Y.; Yang, P. Activation of the aryl hydrocarbon receptor affects activation and function of human monocyte-derived dendritic cells. *Clin. Exp. Immunol.* **2014**, *177*, 521–530. [CrossRef]
105. Liu, Y.; Liang, X.; Dong, W.; Fang, Y.; Lv, J.; Zhang, T.; Fiskesund, R.; Xie, J.; Liu, J.; Yin, X.; et al. Tumor-Repopulating Cells Induce PD-1 Expression in CD8+ T Cells by Transferring Kynurenine and AhR Activation. *Cancer Cell* **2018**, *33*, 480–494.e7. [CrossRef]
106. Safe, S.; Lee, S.-O.; Jin, U.-H. Role of the Aryl Hydrocarbon Receptor in Carcinogenesis and Potential as a Drug Target. *Toxicol. Sci.* **2013**, *135*, 1–16. [CrossRef]
107. Xue, P.; Fu, J.; Zhou, Y. The Aryl Hydrocarbon Receptor and Tumor Immunity. *Front. Immunol.* **2018**, *9*, 286. [CrossRef]
108. Routy, J.-P.; Routy, B.; Graziani, G.M.; Mehraj, V. The Kynurenine Pathway is a Double-Edged Sword in Immune-Privileged Sites and in Cancer: Implications for Immunotherapy. *Int. J. Tryptophan Res.* **2016**, *9*, 67–77. [CrossRef] [PubMed]
109. Hjortsø, M.D.; Larsen, S.K.; Kongsted, P.; Met, Ö.; Frøsig, T.M.; Andersen, G.H.; Ahmad, S.M.; Svane, I.M.; Becker, J.C.; Straten, P.T.; et al. Tryptophan 2,3-dioxygenase (TDO)-reactive T cells differ in their functional characteristics in health and cancer. *OncoImmunology* **2015**, *4*, e968480. [CrossRef] [PubMed]
110. Litzenburger, U.M.; Opitz, C.A.; Sahm, F.; Rauschenbach, K.J.; Trump, S.; Winter, M.; Ott, M.; Ochs, K.; Lutz, C.; Liu, X.; et al. Constitutive IDO expression in human cancer is sustained by an autocrine signaling loop involving IL-6, STAT3 and the AHR. *Oncotarget* **2014**, *5*, 1038–1051. [CrossRef] [PubMed]
111. DiNatale, B.C.; Murray, I.A.; Schroeder, J.C.; Flaveny, C.A.; Lahoti, T.S.; Laurenzana, E.M.; Omiecinski, C.J.; Perdew, G.H. Kynurenic Acid Is a Potent Endogenous Aryl Hydrocarbon Receptor Ligand that Synergistically Induces Interleukin-6 in the Presence of Inflammatory Signaling. *Toxicol. Sci.* **2010**, *115*, 89–97. [CrossRef]
112. Sadik, A.; Patterson, L.F.S.; Öztürk, S.; Mohapatra, S.R.; Panitz, V.; Secker, P.F.; Pfänder, P.; Loth, S.; Salem, H.; Prentzell, M.T.; et al. IL4I1 Is a Metabolic Immune Checkpoint that Activates the AHR and Promotes Tumor Progression. *Cell* **2020**, *182*, 1252–1270.e34. [CrossRef]
113. Castellano, F.; Prevost-Blondel, A.; Cohen, J.L.; Molinier-Frenkel, V. What role for AHR activation in IL4I1-mediated immunosuppression? *Oncoimmunology* **2021**, *10*, 1924500. [CrossRef]
114. Fallarino, F.; Grohmann, U.; Vacca, C.; Bianchi, R.; Orabona, C.; Spreca, A.; Fioretti, M.C.; Puccetti, P. T cell apoptosis by tryptophan catabolism. *Cell Death Differ.* **2002**, *9*, 1069–1077. [CrossRef]

115. Hayashi, T.; Mo, J.-H.; Gong, X.; Rossetto, C.; Jang, A.; Beck, L.; Elliott, G.I.; Kufareva, I.; Abagyan, R.; Broide, D.H.; et al. 3-Hydroxyanthranilic acid inhibits PDK1 activation and suppresses experimental asthma by inducing T cell apoptosis. *Proc. Natl. Acad. Sci. USA* **2007**, *104*, 18619–18624. [CrossRef]
116. Favre, D.; Mold, J.; Hunt, P.W.; Kanwar, B.; Loke, P.; Seu, L.; Barbour, J.D.; Lowe, M.M.; Jayawardene, A.; Aweeka, F.; et al. Tryptophan catabolism by indoleamine 2, 3-dioxygenase 1 alters the balance of TH17 to regulatory T cells in HIV disease. *Sci. Transl. Med.* **2010**, *2*, 32ra36. [CrossRef]
117. Zaher, S.S.; Germain, C.; Fu, H.; Larkin, D.F.; George, A.J. 3-hydroxykynurenine suppresses CD4+ T-cell proliferation, induces T-regulatory-cell development, and prolongs corneal allograft survival. *Investig. Ophthalmol. Vis. Sci.* **2011**, *52*, 2640–2648. [CrossRef]
118. Asghar, K.; Brain, J.; Palmer, J.M.; Douglass, S.; Naemi, F.M.A.; O'Boyle, G.; Kirby, J.; Ali, S. Potential role of indoleamine 2,3-dioxygenase in primary biliary cirrhosis. *Oncol. Lett.* **2017**, *14*, 5497–5504. [CrossRef] [PubMed]
119. Yang, R.; Gao, N.; Chang, Q.; Meng, X.; Wang, W. The role of IDO, IL-10, and TGF-β in the HCV-associated chronic hepatitis, liver cirrhosis, and hepatocellular carcinoma. *J. Med. Virol.* **2019**, *91*, 265–271. [CrossRef] [PubMed]
120. Clària, J.; Moreau, R.; Fenaille, F.; Amorós, A.; Junot, C.; Gronbaek, H.; Coenraad, M.J.; Pruvost, A.; Ghettas, A.; Chu-Van, E.; et al. Orchestration of Tryptophan-Kynurenine pathway, acute decompensation, and Acute-on-Chronic liver failure in cirrhosis. *Hepatology* **2019**, *69*, 1686–1701. [CrossRef]
121. Yoshio, S.; Sugiyama, M.; Shoji, H.; Mano, Y.; Mita, E.; Okamoto, T.; Matsuura, Y.; Okuno, A.; Takikawa, O.; Mizokami, M.; et al. Indoleamine-2,3-dioxygenase as an effector and an indicator of protective immune responses in patients with acute hepatitis B. *Hepatology* **2015**, *63*, 83–94. [CrossRef]
122. Ishio, T.; Goto, S.; Tahara, K.; Tone, S.; Kawano, K.; Kitano, S. Immunoactivative role of indoleamine 2, 3-dioxygenase in human hepatocellular carcinoma. *J. Gastroenterol. Hepatol.* **2004**, *19*, 319–326. [CrossRef]
123. Pan, K.; Wang, H.; Chen, M.; Zhang, H.; Weng, D.; Zhou, J.; Huang, W.; Li, J.; Song, H.; Xia, J. Expression and prognosis role of indoleamine 2,3-dioxygenase in hepatocellular carcinoma. *J. Cancer Res. Clin. Oncol.* **2008**, *134*, 1247–1253. [CrossRef] [PubMed]
124. Li, S.; Han, X.; Lyu, N.; Xie, Q.; Deng, H.; Mu, L.; Pan, T.; Huang, X.; Wang, X.; Shi, Y.; et al. Mechanism and prognostic value of indoleamine 2,3-dioxygenase 1 expressed in hepatocellular carcinoma. *Cancer Sci.* **2018**, *109*, 3726–3736. [CrossRef] [PubMed]
125. Brown, Z.J.; Yu, S.J.; Heinrich, B.; Ma, C.; Fu, Q.; Sandhu, M.; Agdashian, D.; Zhang, Q.; Korangy, F.; Greten, T.F. Indoleamine 2,3-dioxygenase provides adaptive resistance to immune checkpoint inhibitors in hepatocellular carcinoma. *Cancer Immunol. Immunother.* **2018**, *67*, 1305–1315. [CrossRef]
126. Hoffmann, D.; Dvorakova, T.; Stroobant, V.; Bouzin, C.; Daumerie, A.; Solvay, M.; Klaessens, S.; Letellier, M.-C.; Renauld, J.C.; Van Baren, N.; et al. Tryptophan 2,3-Dioxygenase Expression Identified in Human Hepatocellular Carcinoma Cells and in Intratumoral Pericytes of Most Cancers. *Cancer Immunol. Res.* **2019**, *8*, 19–31. [CrossRef]
127. Li, S.; Li, L.; Wu, J.; Song, F.; Qin, Z.; Hou, L.; Xiao, C.; Weng, J.; Qin, X.; Xu, J. TDO Promotes Hepatocellular Carcinoma Progression. *OncoTargets Ther.* **2020**, *13*, 5845–5855. [CrossRef]
128. Bekki, S.; Hashimoto, S.; Yamasaki, K.; Komori, A.; Abiru, S.; Nagaoka, S.; Saeki, A.; Suehiro, T.; Kugiyama, Y.; Beppu, A.; et al. Serum kynurenine levels are a novel biomarker to predict the prognosis of patients with hepatocellular carcinoma. *PLoS ONE* **2020**, *15*, e0241002. [CrossRef]
129. Jin, H.; Zhang, Y.; You, H.; Tao, X.; Wang, C.; Jin, G.; Wang, N.; Ruan, H.; Gu, D.; Huo, X.; et al. Prognostic significance of kynurenine 3-monooxygenase and effects on proliferation, migration and invasion of human hepatocellular carcinoma. *Sci. Rep.* **2015**, *5*, srep10466. [CrossRef] [PubMed]
130. Moretti, S.; Nucci, N.; Menicali, E.; Morelli, S.; Bini, V.; Colella, R.; Mandarano, M.; Sidoni, A.; Puxeddu, E. The Aryl Hydrocarbon Receptor Is Expressed in Thyroid Carcinoma and Appears to Mediate Epithelial-Mesenchymal-Transition. *Cancers* **2020**, *12*, 145. [CrossRef] [PubMed]
131. Song, L.; Guo, L.; Li, Z. Molecular mechanisms of 3,3′4,4′,5-pentachlorobiphenyl-induced epithelial-mesenchymal transition in human hepatocellular carcinoma cells. *Toxicol. Appl. Pharmacol.* **2017**, *322*, 75–88. [CrossRef] [PubMed]
132. Pierre, S.; Chevallier, A.; Teixeira-Clerc, F.; Ambolet-Camoit, A.; Bui, L.-C.; Bats, A.-S.; Fournet, J.-C.; Fernandez-Salguero, P.M.; Aggerbeck, M.; Lotersztajn, S.; et al. Aryl Hydrocarbon Receptor–Dependent Induction of Liver Fibrosis by Dioxin. *Toxicol. Sci.* **2013**, *137*, 114–124. [CrossRef]
133. Liu, Z.; Wu, X.; Zhang, F.; Han, L.; Bao, G.; He, X.; Xu, Z. AhR expression is increased in hepatocellular carcinoma. *J. Mol. Histol.* **2013**, *44*, 455–461. [CrossRef]
134. Hsu, S.-H.; Wang, L.-T.; Chai, C.-Y.; Wu, C.-C.; Hsi, E.; Chiou, S.-S.; Wang, S.-N. Aryl hydrocarbon receptor promotes hepatocellular carcinoma tumorigenesis by targeting intestine-specific homeobox expression. *Mol. Carcinog.* **2017**, *56*, 2167–2177. [CrossRef]
135. Giorgini, F.; Möller, T.; Kwan, W.; Zwilling, D.; Wacker, J.L.; Hong, S.; Tsai, L.-C.L.; Cheah, C.S.; Schwarcz, R.; Guidetti, P.; et al. Histone Deacetylase Inhibition Modulates Kynurenine Pathway Activation in Yeast, Microglia, and Mice Expressing a Mutant Huntingtin Fragment. *J. Biol. Chem.* **2008**, *283*, 7390–7400. [CrossRef] [PubMed]
136. Guillemin, G.J.; Kerr, S.; Smythe, G.A.; Smith, D.G.; Kapoor, V.; Armati, P.J.; Croitoru, J.; Brew, B.J. Kynurenine pathway metabolism in human astrocytes: A paradox for neuronal protection. *J. Neurochem.* **2001**, *78*, 842–853. [CrossRef] [PubMed]
137. Heyes, M.P.; Saito, K.; Markey, S.P. Human macrophages convert l-tryptophan into the neurotoxin quinolinic acid. *Biochem. J.* **1992**, *283*, 633–635. [CrossRef]

138. Liu, C.-Y.; Huang, T.-T.; Lee, C.-H.; Wang, W.-L.; Lee, H.-C.; Tseng, L.-M. Kynurenine-3-monooxygenase (KMO) protein promotes triple negative breast cancer progression. *Ann. Oncol.* **2017**, *28*, v3. [CrossRef]
139. Lai, M.-H.; Liao, C.-H.; Tsai, N.-M.; Chang, K.-F.; Liu, C.-C.; Chiu, Y.-H.; Huang, K.-C.; Lin, C.-S. Surface Expression of Kynurenine 3-Monooxygenase Promotes Proliferation and Metastasis in Triple-Negative Breast Cancers. *Cancer Control* **2021**, *28*. [CrossRef]
140. Liu, C.-Y.; Huang, T.-T.; Chen, J.-L.; Chu, P.-Y.; Lee, C.-H.; Lee, H.-C.; Lee, Y.-H.; Chang, Y.-Y.; Yang, S.-H.; Jiang, J.-K.; et al. Significance of Kynurenine 3-Monooxygenase Expression in Colorectal Cancer. *Front. Oncol.* **2021**, *11*, 620361. [CrossRef]
141. Liu, X.; Shin, N.; Koblish, H.K.; Yang, G.; Wang, Q.; Wang, K.; Leffet, L.; Hansbury, M.J.; Thomas, B.; Rupar, M.; et al. Selective inhibition of IDO1 effectively regulates mediators of antitumor immunity. *Blood* **2010**, *115*, 3520–3530. [CrossRef] [PubMed]
142. Study to Explore the Safety, Tolerability and Efficacy of MK-3475 in Combination with INCB024360 in Participants with Selected Cancers. ClinicalTrials.gov U.S. National Library of Medicine. Available online: https://clinicaltrials.gov/ct2/show/NCT02178722?term=Epacadostat&cond=Hepatocarcinoma&draw=2&rank=1 (accessed on 30 August 2021).
143. BMS-986205 and Nivolumab as First or Second Line Therapy in Treating Patients with Liver Cancer. ClinicalTrials.gov U.S. National Library of Medicine. Available online: https://clinicaltrials.gov/ct2/results?recrs=&cond=Hepatocarcinoma&term=BMS-986205+&cntry=&state=&city=&dist= (accessed on 30 August 2021).
144. Hamid, O.; Bauer, T.M.; Spira, A.I.; Smith, D.C.; Olszanski, A.J.; Tarhini, A.A.; Lara, P.; Gajewski, T.; Wasser, J.S.; Patel, S.P.; et al. Safety of epacadostat 100 mg bid plus pembrolizumab 200 mg Q3W in advanced solid tumors: Phase 2 data from ECHO-202/KEYNOTE-037. *J. Clin. Oncol.* **2017**, *35*, 3012. [CrossRef]
145. Mitchell, T.C.; Hamid, O.; Smith, D.C.; Bauer, T.M.; Wasser, J.S.; Olszanski, A.; Luke, J.J.; Balmanoukian, A.S.; Schmidt, E.V.; Zhao, Y.; et al. Epacadostat Plus Pembrolizumab in Patients with Advanced Solid Tumors: Phase I Results from a Multicenter, Open-Label Phase I/II Trial (ECHO-202/KEYNOTE-037). *J. Clin. Oncol.* **2018**, *36*, 3223–3230. [CrossRef] [PubMed]

Review

Tumor Microenvironment: Key Players in Triple Negative Breast Cancer Immunomodulation

Hongmei Zheng [1,2,*], Sumit Siddharth [2], Sheetal Parida [2], Xinhong Wu [1,*] and Dipali Sharma [2]

1. Hubei Provincial Clinical Research Center for Breast Cancer, Department of Breast Surgery, Hubei Cancer Hospital, Tongji Medical College, Huazhong University of Science and Technology, Wuhan 430079, China
2. The Sidney Kimmel Comprehensive Cancer Center, Department of Oncology, Johns Hopkins University School of Medicine, Baltimore, MD 21218, USA; ssiddha2@jhmi.edu (S.S.); sparida1@jhu.edu (S.P.); dsharma7@jhmi.edu (D.S.)
* Correspondence: zhenghongmeicj@163.com (H.Z.); wuxinhong_9@sina.com (X.W.)

Simple Summary: The tumor microenvironment (TME) is a complicated network composed of various cells, signaling molecules, and extra cellular matrix. TME plays a crucial role in triple negative breast cancer (TNBC) immunomodulation and tumor progression, paradoxically, acting as an immunosuppressive as well as immunoreactive factor. Research regarding tumor immune microenvironment has contributed to a better understanding of TNBC subtype classification. Shall we treat patients precisely according to specific subtype classification? Moving beyond traditional chemotherapy, multiple clinical trials have recently implied the potential benefits of immunotherapy combined with chemotherapy. In this review, we aimed to elucidate the paradoxical role of TME in TNBC immunomodulation, summarize the subtype classification methods for TNBC, and explore the synergistic mechanism of chemotherapy plus immunotherapy. Our study may provide a new direction for the development of combined treatment strategies for TNBC.

Citation: Zheng, H.; Siddharth, S.; Parida, S.; Wu, X.; Sharma, D. Tumor Microenvironment: Key Players in Triple Negative Breast Cancer Immunomodulation. *Cancers* **2021**, *13*, 3357. https://doi.org/10.3390/cancers13133357

Academic Editor: Charles Theillet

Received: 16 June 2021
Accepted: 1 July 2021
Published: 4 July 2021

Publisher's Note: MDPI stays neutral with regard to jurisdictional claims in published maps and institutional affiliations.

Copyright: © 2021 by the authors. Licensee MDPI, Basel, Switzerland. This article is an open access article distributed under the terms and conditions of the Creative Commons Attribution (CC BY) license (https://creativecommons.org/licenses/by/4.0/).

Abstract: Triple negative breast cancer (TNBC) is a heterogeneous disease and is highly related to immunomodulation. As we know, the most effective approach to treat TNBC so far is still chemotherapy. Chemotherapy can induce immunogenic cell death, release of damage-associated molecular patterns (DAMPs), and tumor microenvironment (TME) remodeling; therefore, it will be interesting to investigate the relationship between chemotherapy-induced TME changes and TNBC immunomodulation. In this review, we focus on the immunosuppressive and immunoreactive role of TME in TNBC immunomodulation and the contribution of TME constituents to TNBC subtype classification. Further, we also discuss the role of chemotherapy-induced TME remodeling in modulating TNBC immune response and tumor progression with emphasis on DAMPs-associated molecules including high mobility group box1 (HMGB1), exosomes, and sphingosine-1-phosphate receptor 1 (S1PR1), which may provide us with new clues to explore effective combined treatment options for TNBC.

Keywords: triple negative breast cancer; tumor microenvironment; immunomodulation

1. Introduction

Triple negative breast cancer (TNBC), characterized by the absence of estrogen receptor (ER), progesterone receptor (PR), and human epidermal growth factor receptor 2 (HER2) expression, comprises 10–20% of all breast cancers [1]. Owing to the lack of ER/PR/Her2 protein expression/amplification, TNBCs do not respond to existing endocrine and Her2-targeted therapies and exhibit poor prognosis [2]. It has been proposed that TNBCs with a higher involvement of immune cells termed as 'hot tumors' have better prognosis and a greater response to immunotherapy while TNBCs with a lower involvement of immune cells termed as 'cold tumors' are marked with poor prognosis and poor response to immunotherapy [3]. From this point of view, TNBC patients have been further segregated into different subgroups [4–8]. The tumor microenvironment (TME) is an ensemble of endothelial cells, cells of the immune system, adipocytes, and fibroblasts, in addition to the soluble factors released from all the cellular components (including cancer cells) [9,10]. TME can be classified from different perspectives such as host and non-host origin, cellular origin and constituents [9,11–13]. TME presents a complex network that plays a crucial role in TNBC immunomodulation and tumor progression.

Cancer initiation and development is not just a biological process triggered by cancer cells in isolation; in fact, it has to be evaluated along with the complicated TME with an emphasis on the interaction between cancer cells and their surrounding extra-cellular matrix. Indeed, considering alterations in microenvironment as active players during cancer progression brings another dimension of complexity [14]. During TNBC progression, tumor immune microenvironment remodeling including the change of the ratio of immune cells and release of multiple immune inhibitory and reactive cytokines is a critical feature [15,16]. Based on the constituents of TME, TNBCs have been stratified into 'tumor immune microenvironment (TIME) subtypes' aiding in predicting outcomes and proposing potential treatments guided by the distinct phenotypes of TNBC [16,17]. Chemotherapy, the foremost treatment for TNBC, could induce immunogenic cell death (ICD) and promote the release of damage-associated molecular patterns (DAMPs) [18] including high mobility group box1 (HMGB1), exosomes and sphingosine-1-phosphate receptor 1 (S1PR1) by damaged or activated cells via the activation of TLR4 signal pathway [19] and stimulate the release of various immune molecules such as TGF-β, IK12p7, and IFN-γ [20].

In this review, we focus on immune TME and summarize its immunosuppressive and immunoreactive roles, discuss constituent immune cells involved in TNBC immunomodulation, and the contribution of TIME in stratification of TNBC. Further, we discuss the role of chemotherapy-induced TME changes in modulating TNBC immune response and tumor progression, with a focus on HMGB1, exosomes, and sphingosine-1-phosphate (S1P)/sphingosine kinase 1 (SPHK1)/S1PR1, an axis whose therapeutic modulation may result in neoteric combination therapy for TNBC patients.

2. Two Roles of TME in TNBC Immunomodulation

According to the contribution to immune response, the tumor microenvironment (TME) can be classified as immunosuppressive and immunoreactive. Tumor infiltrating lymphocytes (TILs), the major cell types in the microenvironment, are heterogeneous and mainly composed of lymphocytes in tumor nests and tumor stroma. TILs can be classified into several different subtypes, mainly $CD3^+$ T cells and $CD20^+$ B cells in solid tumors, though $CD20^+$ B cell infiltration is relatively less. $CD3^+$ T cells include $CD8^+$ cytotoxic T lymphocytes ($CD8^+$ TILs), $CD4^+$ helping T lymphocytes, and $Foxp3^+$ regulatory T lymphocytes ($Foxp3^+$ Tregs) [21,22]. Different subtypes of TILs take part in immunomodulation with distinct mechanisms and play various roles in breast cancer immunomodulation [22]. Figure 1 pictorially represents immunosuppressive and immunoreactive TMEs (Figure 1).

2.1. Immunosuppressive TME in TNBC
2.1.1. PD-1/PD-L1 Axis

Programmed death-ligand 1 (PD-L1) and programmed cell death protein-1 (PD-1) are important negative co-stimulating signaling molecules in immunoglobulin superfamily (IgSF) and play an important role in host immunomodulation [23]. PD-L1 is expressed in many solid tumors including breast cancer and is a negative prognosis indicator [24,25]. PD-1 is expressed in TILs [26]. Theoretically, PD-L1 expression on tumor cells combined with PD-1 expression on TILs play a negative role in immunomodulation, which inhibits the activation of TILs, causing the tumor cell to survive through immune escape.

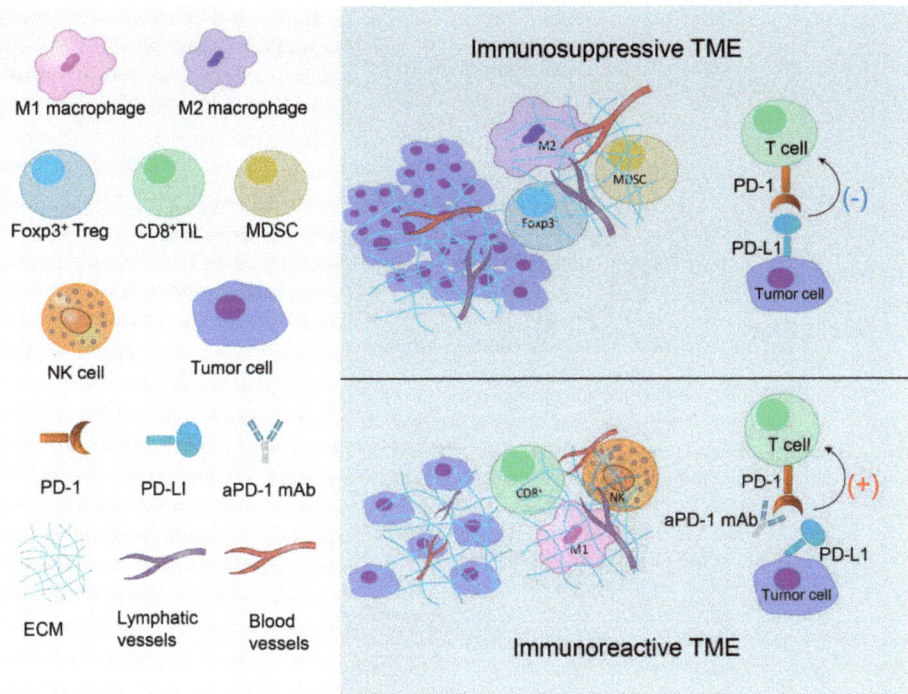

Figure 1. Immunosuppressive and immunoreactive TME. Immunosuppressive tumor microenvironment (TME) is mainly constituted of M2 macrophages, forkhead box P3+ (Foxp3+) regulatory T lymphocytes (Tregs), myeloid-derived suppressor cells (MDSCs), and PD-1/PD-L1 axis. Immunoreactive TME is mainly constituted of CD8+ T cells, natural killer (NK) cells, and M1 macrophages. PD-1/PD-L1 axis becomes immunoreactive in response to anti-PD1 or anti-PD-L1 monoclonal antibody (aPD-1/PD-L1 mAb) owing to the activation of CD8+ T cells. (Foxp3, forkhead box P3; Tregs, regulatory T lymphocytes; MDSC, myeloid-derived suppressor cell; NK, natural killer; PD-1, programmed cell death protein-1; PD-L1, programmed death-ligand 1; aPD-1 mAb, anti-PD-1 monoclonal antibody; ECM, extra cellular matrix; TME, tumor microenvironment).

The TME involves immune suppressing factors to support the progression of tumors which have escaped host immune surveillance [27–31]. Various immune check-point inhibitors have been developed that have shown efficacy in TNBC patients [32,33]. Clinical studies have shown a paradoxical role of PD-L1 regarding its prognostic value in patients with TNBC owing to the heterogeneity of PD-L1 expression in different tumor sites, non-standard detection methods, and distinct antibodies [31,34–41]. In the impassion 130 clinical trial, compared to TNBC patients receiving nab-paclitaxel plus placebo, a better median overall survival (OS) was observed in patients receiving atezolizumab (PD-L1 inhibitor) combined with nab-paclitaxel and most benefit was observed in PD-L1 positive subgroup [42]. However, in a phase 1b clinical trial (ClinicalTrials.gov Identifier: NCT01848834) which evaluated the safety and effectiveness of PD-1 inhibitor (pembrolizumab) in PD-L1 positive TNBC patients, the overall response rate was only 18.5% and the expression level of PD-L1 was not significantly related to the clinical response [43]. These disparate results might be related to multiple TME-related factors that can modulate the therapeutic effects of PD-1/PD-L1 inhibitors in TNBC. Preclinical studies have shown that PD-L1 expression is modulated by multiple signaling pathways including microRNA-200/ZEB1 axis, WNT, loss of PTEN, PI3K, and MUC1-C/MYC/NF-κB axis [31,44–46]. Voorwerk and colleagues reported that doxorubicin and cisplatin treatment caused an upregulation of inflammation-related genes JAK-STAT and TNF-α signaling, immune-related genes associated with PD-1/PD-L1, and T cell cytotoxicity pathways. Short-term and low-dose doxorubicin and cisplatin may create an immunoreactive TME and increase the response to PD-1 inhibitor in TNBC [47]. In conclusion, specifically designed clinical trials are needed to interrogate the involvement of various TME-related factors in order to enhance the efficacy of PD-1/PD-L1 inhibitors in TNBC.

2.1.2. Foxp3$^+$ Tregs

In TME, different classes of TILs exist, which have shown great prognostic value in patients with TNBC. Regulatory T lymphocytes (Tregs) are a lineage of lymphocytes involved in immunosuppression that are characterized by the expression of the forkhead box P3 (Foxp3) transcription factor [48,49]. Foxp3$^+$ Tregs are the major constituent of the TILs in claudin-low TNBC tumors and it has been speculated that the recruitment of Foxp3$^+$ Tregs to the TME inhibits an effective anti-tumor immune response of checkpoint inhibitors [50]. Jamiyan and colleagues detected the expression of stromal Foxp3$^+$ Tregs in 107 TNBC samples using IHC and found that a low stromal Foxp3$^+$ Tregs level was significantly associated with favorable recurrence free survival (RFS) and OS [51]. In contrast, high Foxp3$^+$ TILs expression in 43 TNBC tissues by IHC and Foxp3$^+$/CD25$^+$ TILs were positively correlated with better OS [52]. High densities of intra-tumoral Tregs and CD20$^+$ B cells represented a good prognostic panel in TNBCs [53]. However, mRNA expression of Foxp3 by qRT-PCR in 826 breast tumor tissue samples including 84 TNBC samples, was not significantly related to disease free survival (DFS), while none of the markers studied including CD3, CD8, and Foxp3 were of prognostic value for OS [54]. This phenomenon is somewhat explained by a study showing that activation of tumor antigen-specific Tregs in the bone marrow caused the accumulation of Tregs in breast cancer tissue leading to both antitumor immunity and local immune suppression in breast cancer [55]. The mechanisms underlying pro-tumor role of Foxp3$^+$ Tregs included (i) down-regulation of Notch pathway [56]; (ii) direct suppression via cell-cell contact and indirect suppression via secretion of anti-inflammatory mediators such as interleukins (IL-4, IL-5 and IL-10) [57–59]; (iii) decreased secretion of cytokine IFN-γ and IL-17 and activation of STAT1/STAT3 [59]. The prognostic significance of Tregs in TNBCs, therefore, remains controversial and warrants more careful investigations.

2.1.3. M2 Macrophages

M2 macrophages, the main tumor-associated macrophages, (TAMs), can promote breast cancer initiation, angiogenesis, invasion, and metastasis by generating an immunosuppressive TME via releasing cytokines, chemokines, and growth factors [60]. TAMs expressing CD163$^+$ (marker of M2 macrophages) positively correlate with tumor associated fibroblasts and epithelial-mesenchymal transition, which in turn are associated with aggressive behaviors and short DFS in 278 patients with histologically confirmed TNBC [61,62]. Another clinical study showed that high CD68$^+$ (marker of M2 macrophage) TAMs expression associates with poor distant metastasis free survival (DMFS), DFS and OS in 287 patients with TNBC [63]. Mechanistically, in vivo and in vitro studies showed that the presence of CD11b$^+$F4/80$^+$CD206$^+$ TAMs significantly associate with proliferating tumor cells in a TNBC mouse model. RNA sequencing analysis revealed that TAMs promote MAPK pathway activation in 4T1 cells [64]. Reactive oxygen species (ROS)-induced macrophages produce an immunosuppressive subtype (M2) and increase the expression of PD-L1 via activating NF-κB signaling, as well as release immunosuppressive chemokines such as interleukin-10 (IL-10), IL-17, IL-4, IL-1β, insulin-like growth factor-binding protein 3 (IGFBP-3), and chemokine (C-X-C motif) ligand 1 (CXCL1) [65]. The JAK2/STAT3 signaling pathway can up-regulate the expression of PD-L1 in CD169$^+$ macrophages, but cannot up-regulate the expression of PD-L1 in breast cancer cells, thus avoiding immune surveillance [66]. Metastasis- and inflammation-associated microenvironmental factor S100A4 activates the basal-like subtype of breast cancer cells to trigger monocyte-to-macrophage (M2) differentiation and polarization, and elevates secretion of pro-inflammatory cytokines such as IL-8, IL-6, CXCL10, CCL2 and CCL5 [67]. Further, macrophage colony-stimulating factor (M-CSF), the main stimulator of macrophage migration, caused aggregation of M2 macrophages through an increased elongation of pseudopodia [68]. Inhibitors of differentiation (ID) 4 significantly associates with M2 macrophage marker CD68 protein expression in a series of TNBC tissues. ID4 activates an angiogenic procedure at the molecular level in the macrophages through paracrine signaling including the decrease of constituents of the anti-angiogenic miR-15b/107 group and an increase of angiogenesis-associated mRNAs [69,70]. GM-CSF BRCA1-IRIS overexpressing TNBC cells secrete high quantities of GM-CSF in an NF-κB and a HIF-1α-dependent manner to induce macrophages to IRIS overexpressing cells and polarize them to pro-tumor TAMs (M2). GM-CSF triggers TGF-β1 expression on TAMs through activating STAT5, NF-κB and/or ERK signaling [71].

2.1.4. MDSCs

Myeloid-derived suppressor cells (MDSCs) are an important part of immunosuppressive network [72]. CD33+ MDSCs are a risk factor for progressive disease (PD) plus stable disease (SD) in breast cancer tissues prior to neoadjuvant chemotherapy [73]. Higher expression of MDSCs has been noted in TNBCs in comparison to non-TNBCs with their recruitment to the primary cancer and metastasis occurring via ΔNp63-dependent activation of the chemokines CCL22 and CXCL2 [74]. Glycolysis restriction reduces MDSCs through inhibiting cancer granulocyte G-CSF and GM-CSF expression [75] while hypoxia enhances the expansion of MDSCs and upregulates the expression of PD-L1 in the hypoxic TME of 4T1 tumor-bearing mice [76]. Studies have shown that the monoclonal antibody that neutralizes IL-8 (HuMax-IL8) and the traditional Chinese medicine Prim-O-glucosylcimifugin (POG) can inhibit the recruitment, proliferation, metabolism and immunosuppressive ability of MDSCs [77,78]. The 4T1 TNBC model effectively exhibits induction of immunosuppressive MDSCs accumulation by releasing inflammatory cytokines that produce permissive pro-metastatic TME [79]. Monocytic MDSCs (M-MDSC) and granulocytic MDSCs (G-MDSC) are two types of MDSCs in circulating peripheral blood. G-MDSC levels increase sharply and M-MDSCs decrease significantly after doxorubicin and cyclophosphamide treatment [80]. Investigations have shown that CCL5 is a key modulator of Rb1 activation and is associated with the immunosuppressive activity of MDSCs, especially the G-MDSC subset [81,82].

2.2. Immunoreactive TME in TNBC

2.2.1. NK Cells

Natural killer (NK) cells, a type of cytotoxic lymphocytes, are crucial constituents of the innate immune system whose function in enhancing the anti-tumor immunity in TNBC has been studied extensively. NK cells are abundant in early cancer tissue in human solid tumors; however, they dwindle in metastatic human cancers [83]. These findings show that NK cells play a key role in immune surveillance, but once tumorigenesis occurs, TME is suppressive for NK cells. Evasion of active immune suppression in the TME is an important consideration for enhancing the anti-tumor ability of tumor-infiltrating NK cells. Zhang and colleagues detected the expression of NKp46, Foxp3, CD8, CD163 or Gas6 in 278 TNBC tissues using IHC with an aim to develop a prognostic risk model for TNBC. Multivariate analysis showed that TNM stage, Foxp3 positive lymphocytes along with prognostic risk scores can be used as independent indicators of OS and DFS in TNBC [84]. Tumor-derived IL-18 upregulates PD-1 expression on $CD56^{dim}CD16^{dim/-}$ NK cells and relates to the bad/ prognosis of TNBC [85]. McArdle and colleagues examined the abundance of NK cells, MDSCs, monocyte subsets and Foxp3+ Tregs in the peripheral blood of 85 breast cancer patients and they found that chemotherapy had no effect on the percentage of these immune cells, but peripheral blood cells could distinguish TNBC patients that are at high risk of relapse after chemotherapy [86]. Tissue-infiltrating NK cells in solid tumors appear to have a less robust activity compared with circulating NK cells [87–90]. NK cells isolated from either breast cancer patients or healthy donors show high cytotoxicity against patient-derived tumor cells in vitro and prevent tumor initiation and growth in immunocompromised mice in vivo [91]. Expanded cord blood-NK cells show cytotoxicity towards primary breast tumor cells derived from TNBC and estrogen receptor-positive/progesterone receptor-positive breast cancer [92]. Baseline circulating tumor cells (CTCs) status is positively associated with peripheral NK cells among those receiving first-line treatment in 75 patients with TNBC. Baseline CTCs combined with peripheral NK enumeration (CTC-NK) can predict PFS of TNBC patients more precisely [93]. NK cells are the major effectors of antibody (Ab)-dependent cell-mediated cytotoxicity (ADCC) and thus play an important role in Ab-based therapies. In vivo and in vitro studies revealed that tissue factor (TF)-targeting antibody-like immunoconjugate (called L-ICON)-CAR-NK cells have direct killing effects against TNBC cells and also mediate L-ICON ADCC to acquire a stronger effect [94]. Avelumab, a human IgG anti-PD-L1 mAb, triggers ADCC against a panel of TNBC cells and enhances NK-cell mediated cytotoxicity, which is independent of the blockade of the PD-1/PD-L1 pathway but is involved with IL-2 and IL-15 [95]. CD85j, an inhibitory receptor which can recognize both classical and non-classical HLA-I molecules, is highly expressed in TNBC, and can impair the function of cetuximab through NK-cell functional deficiency even when stimulatory cytokines IL-2 or IL-15 are abundantly present [96]. More interestingly, NK cell infiltration and recruitment can be mediated by a bispecific Ab (MesobsFab) whose anti-tumor activity depend on mesothelin expression on the target cells and it can be a potential antibody-based immunotherapeutic for TNBC patients [97]. NK cell function is regulated by molecules from promoting and suppressing receptors interacting with ligands on target cells. Lectin-like Transcript-1 (OCIL, CLEC2D, LLT1) is a ligand that interacts with NK cell receptor NKRP1A and prevents NK cell activation. Inhibiting LLT1 on TNBCs with antibodies hinders the interaction with NKRP1A and increases lysis of TNBCs by primary NK cells [98].

2.2.2. CD8$^+$ TILs

CD8$^+$ TILs are the main kind of cytolytic lymphocytes in tumors. Kronqvist and group detected the expression of stromal TILs and CD8$^+$ TILs in 179 patients with TNBC using IHC and observed that the prognostic value of CD8$^+$ TILs and TILs varied when detected in various cancer compartments [99]. Presence of CD8$^+$ TILs in a large cohort of 12,439 breast cancer patients correlated with a significant decrease in the relative hazard of death in both the ER- positive and the ER- negative HER2-positive subtypes [100]. Ishida and colleagues assessed the CD8$^+$ TILs and Foxp3$^+$ Tregs status of the residual tumors in 131 patients with TNBC who received neoadjuvant chemotherapy (NAC) at three institutions and the rates of their changes before and after NAC were evaluated. They found that TNBC patients with a high CD8$^+$ TILs level or high CD8/Foxp3 ratio in residual tumors exhibit significantly favorable recurrence-free survival (RFS) and breast cancer-specific survival (BCSS) [101]. Another study also showed that CD8$^+$ TILs were related to favorable DMFS, DFS, and BCSS in the entire 207 breast cancer group and in 56 TNBC group [102]. BRCA1-IRIS overexpressing (IRISOE) TNBC carcinomas had more CD25$^+$/Foxp3$^+$ Tregs and few CD8$^+$/PD-1$^+$ cytotoxic T-cells, which showed that the interaction between macrophages and IRISOE cells initiated an immunosuppressive TME within TNBC tumors [71]. TOPOIIα and CD4$^+$ TILs were significantly positively associated with CD8$^+$ TILs and they exhibited a significantly good 5-year DFS but only a high infiltration of CD8$^+$ TILs showed significantly better 5-year OS in 52 TNBC patients that received taxane-anthracycline-based NAC [103,104]. Calcium/calmodulin-dependent kinase (CaMKK2), expressed in tumor-related stromal cells, could promote tumor growth. The inhibition of CaMKK2 within myeloid cells suppresses tumor growth by increasing immune-stimulatory myeloid subsets and intra-tumoral accumulation of CD8$^+$ T cells in TNBC [105]. PARP inhibitor Olaparib induced CD8$^+$ T cell activation and infiltration via activation of the cGAS/STING pathway, which provided rationale for combining the PARP inhibitors with immunotherapies for TNBC [106]. A recent study reported that CD8$^+$ TILs were crucial for infected cell vaccine (ICV) efficacy, which was composed of autologous tumor cells infected with an oncolytic Maraba MG1 virus in vitro in the BALB/c-4T1 model. Increased migration and proliferation ability of human CD8$^+$ TILs were observed following exposure to ICV [107]. A series of studies illuminated the mechanisms of different infiltration levels of CD8$^+$ TILs in immunomodulation and anti-tumor response of TNBC. By spatially modulating the diffusion/chemotactic coefficients of T cells via partial differential equations, Almohanad et al. found that a type of chemorepellent inside cancer cell clusters but not dense collagen fibers, prevents the infiltration of CD8$^+$ TILs into cancers and cancer cell clusters, which may imply a poor prognosis in TNBC [108]. Intra tumoral CD8$^+$ TILs enhance the efficacy of treatment through triple combined inhibition of PDGFRβ / MEK1/2/JAK2 signal pathway in vivo in TNBC [109]. Gruosso et al., found that there were many different kinds of CD8$^+$ TILs localization profiles with distinct meta-signatures, which were prognostic indicators in a cohort of TNBC [17]. Dong et al. investigated the genome-scale CD8$^+$ TILs CRISPR screen in the context of immunotherapy in vivo and in vitro and found that DHX37 interacts with PDCD11 and affects NF-κB activity to modulate CD8$^+$ TILs activation, cytokine production, and cytotoxicity [110].

2.2.3. M1 Macrophages

M1 phenotype macrophages, also called classical macrophages, are pro-inflammatory, and can activate the immune response and oppose tumorigenesis [111]. In vitro and in vivo studies have shown that M1 macrophage polarization decreases the expression of nuclear REST corepressor 1 (CoREST), LSD1 and the zinc finger protein SNAIL, and LSD1 inhibitors can target both CoREST and flavin adenine dinucleotide (FAD) binding domains of LSD1 to initiate macrophages toward M1 phenotype in TNBC successfully [112]. Another study revealed that exposure to infected cell vaccine (ICV) could induce the polarization of monocytes to M1 subtype [107].

Using the 4T1 TNBC murine model, Meyer and colleagues showed that in the early stages of disease, higher M1-related cytokines are released and decreased M2 macrophages infiltrate in the TME, while upon metastasis a dramatic enhancement in M2-related cytokine expression levels are detected and more immunosuppressive cells such as M2 macrophages infiltrate in the TME [113]. High level of CCL5 is related to recruitment of M1 macrophages, CD8$^+$ TILs, CD4 activated T lymphocytes, and NK activated cells in TNBC using CIBERSORT analysis [114]. The clinical significance and involved mechanisms of each constituent in TNBC microenvironment are included in Table 1.

Table 1. Clinical significance and involved mechanisms of immune cells and markers.

Items	Clinical Significance	Involved Mechanisms	References
PD-1/PD-L1	Paradoxical role in prognosis	microRNA-200/ZEB1 axis, WNT signaling, loss of PTEN, PI3K signaling, and MUC1-C/MYC/NF-κB pathway	[31,34–41,44–46]
Foxp3+ Tregs	Paradoxical role in prognosis	Notch pathway, IL-35/STAT1/STAT3, secretion of anti-inflammatory mediators such as interleukin	[50–54,56–59,115]
M2 macrophages	Adverse prognostic indicator	MAPK pathway, NF-κB/PD-L1, release of immunosuppressive chemokines, JAK2/STAT3 signaling pathway, S100A4 activation, angiogenic program, HIF-1α, STAT5, NF-κB and ERK signaling	[61–64,116]
MDSCs	Risk factor for PD plus SD	ΔNp63-dependent activation of the chemokines CXCL2 and CCL22, Glycolysis, hypoxia, secretion of inflammatory cytokines, Rb1 activation	[73–76,81,82]
NK cells	Positive prognostic indicator	ADCC, Lectin-like Transcript-1 activation, bispecific antibody (MesobsFab) modulating chemorepellent inside tumor cell clusters	[84,85,92,94–98,117]

Table 1. Cont.

Items	Clinical Significance	Involved Mechanisms	References
CD8+ TILs	Favorable prognostic indicator	Inhibition of PDGFRβ/MEK1/2/JAK2 signal pathway, distinct metasignatures of CD8+ TILs, DHX37/PDCD11/NF-κB	[17,99–101,108–110]
M1 macrophages	Favorable prognostic indicator	M1 polarization by FAD, CoREST and exposure to cell vaccine (ICV), release of CCL5	[112–114]

3. The Composition of TME Contributes to TNBC Subtype Classification

During TNBC progression, TME reconstruction including the ratio of immune cells and release of various immune cytokines play crucial roles, and the research focusing on stromal and immune composition of TME has contributed significantly to different subtype classification of TNBC [17]. Lehmann and colleagues distinguished six TNBC subtypes showing unique gene expression profiles and ontologies, comprised of two basal-like (BL1 and BL2), a mesenchymal stem-like (MSL), a mesenchymal (M), an immunomodulatory (IM), and a luminal androgen receptor (LAR) subtype. Interestingly, immune genes in IM subtype overlap with gene signatures in medullary breast cancer which is correlated with good prognosis despite its high-grade scores [118]. Park and colleagues distinguished four stromal axes abundant for T cells, B cells, epithelial markers and desmoplasia and assigned a score along with each marker and associated it with different TNBC subtypes. This classification method better depicted tumor heterogeneity and led to a superior evaluation of benefit from therapeutics and prognosis [119].

In addition, three subtypes of TNBC have been identified: an apocrine cluster (C1), which is more related to luminal, PIK3CA-mutated hallmarks and shows intermediate biological aggressiveness; and two basal-like clusters (C2 and C3), which show a major biological discrepancy related to immune response and are sensitive to drugs combating immunosuppression or stimulate adaptive immune response respectively [120]. Shao and colleagues analyzed genomic, clinical, and transcriptomic data of 465 primary TNBC patients, and also identified four subtypes of TNBC, including basal-like immune-suppressed (BLIS), immunomodulatory (IM), luminal androgen receptor (LAR) and mesenchymal-like (MES). They also showed that IM subtype is related to immune response and there are elevated immune cell signaling, TILs, high mRNA expression quantities of immune checkpoint blocking genes such as PD-L1, PD-1, CTLA4, and IDO1 [121]. Using the data of 465 Taiwanese with breast cancer, five TNBC subtypes were classified, namely, basal-like (BL), mesenchymal stem like (MSL), immunomodulatory (IM), mesenchymal (M), and luminal androgen receptor (LAR), and they observed the interaction between IM subtype and MSL subtype, which also implied the involvement of TME in TNBC subtype classification [122]. Distinguishing a four-gene decision tree signature (TP53BP2, EXO1, RSU1 and FOXM1) using transcriptomic and genomic data analysis established six subtypes of TNBC, named MC1 to MC6, comprised by five of varying sizes (MC1-MC5) and one large subtype MC6. Further study showed high level of CD8+ and CD4+ immune signatures and decreased expression of MAPK pathway related genes in MC6 subtype [123]. Another group identified three TNBC subtypes including Immunity High (Immunity H), Immunity Medium (Immunity M), and Immunity Low (Immunity L) based on the immunogenomic profiling of 29 immune signatures. In Immunity H subtype, greater anti-tumor immune response and immune cell infiltration, as well as favorable prognosis were detected compared to the other subtypes, which showed the close relationship between tumor immune microenvironment and TNBC classification [124]. TNBC tumors were classified into four subgroups (luminal-androgen receptor expressing, basal, claudin-high and claudin-low), in addition to two subgroups associated with immune activity using gene expression and clinical data and the latter two immune subgroups were defined as correlated to immune activity closely. Meanwhile, claudin-high subgroup had low response to neoadjuvant chemotherapy, and luminal immune-positive subgroup had favorable survival prognoses [125]. A recent study identified four TNBC epitopes, named as Epi-CL-A, Epi-CL-B, Epi-CLC, and Epi-CL-D using genome-wide DNA methylation properties and clinical and demographic variables, as well as gene mutation and gene expression data. Intriguingly, subtype Epi-CL-D showed a positive regulation of T lymphocyte-mediated cytotoxicity and associated molecules such as IL15RA and CCL18, which partially explained the favorable outcome and a positive immune response in this subtype [126]. Furthermore, a research group classified TNBC tumors into immune subtype A and B by the density of monocytes, γδ T cells, stromal CD4$^+$ T cells, M1 macrophages and M2 macrophages using CIBERSORT or IHC method and they proved that enriched immune-related pathways and higher levels of immune checkpoint cytokines such as PD-1, PD-L1 and CTLA-4 could be detected in phenotype A [127]. Romero-Cordoba et al. also identified three immuno-clusters in TNBC tumors using clustering analysis based on immune-related gene expression signatures and found that platelet to lymphocyte ratio (PLR) was associated with tumor immune infiltration [128].

We have included all the classification methods and the clinical significance (Table 2). Classification of TNBC has been developed extensively implying that a precision-treatment era has come in TNBC. Chemotherapy still remains the key treatment for TNBC but other targeted therapies including immunotherapy can be combined for better tailored treatments and are the focus of ongoing research efforts.

Table 2. TNBC subtype classification.

Subtype of TNBC	Subtype Number	Basis of Classification	Clinical Significance	References
BL1, BL2, IM, M, MSL, LAR	6	Gene expression profiles	IM subtype was associated with favorable prognosis.	[118]
4 stroma axes (T,B,E,D)	4	Transcriptome of stroma	Better evaluated patient benefit from therapeutics.	[119]
C1, C2, C3	3	Gene expression profiling	C2 and C3 subtypes were sensitive to drugs combating immunosuppression.	[120]
LAR, IM, BLIS, MES	4	Clinical, genomic, and transcriptomic data	Elevated immune cells and signaling in IM subtype.	[121]
BL, IM, M, MSL, LAR	5	Gene expression profiles	Interaction between IM and MSL subtype suggested involvement of TME.	[122]
MC1, MC2, MC3, MC4, MC5, MC6	6	Transcriptomic and genomic data	High level of CD8$^+$ and CD4$^+$ immune signatures in MC6 subtype.	[123]
Immunity_H, Immunity_M, Immunity_L	3	Immunogenomic profiling	Immunity_H subtype was correlated with immune cell expression and good prognosis	[124]
LAR, basal, claudin-low, claudin-high and two immune subtypes	6	Clinical and gene expression data	Claudin-h and immune-positive subtype was associated with low pCR and favorable prognosis separately.	[125]
Epi-CL-A, Epi-CL-B, Epi-CLC, Epi-CL-D	4	Genome-wide DNA methylation profiles	Positive regulation of T lymphocyte cytotoxicity and associated genes in Epi-CL-D subtype.	[126]
Immune phenotype A and B	2	Density of five prognosis-related immune cells	Enriched immune-related pathways and molecules in phenotype A.	[127]
ImA, ImB and ImC	3	Immune-related gene expression signatures	Platelet to lymphocyte ratio (PLR) was associated with tumor immune infiltration in TNBC.	[128]

4. Chemotherapy-Induced TME Remodeling Modulates TNBC Immune Response

It has been reported that cytotoxic drugs such as anthracycline and platinum agents, could induce immunogenic cell death (ICD), and stimulate anti-tumor immune response of T lymphocytes [18,129]. Damage-associated molecular patterns (DAMPs) are cytokines that are released by damaged or activated cells; have great immune stimulating response, and cause ICD [18]. ICD involves the cell surface exposure of calreticulin (CRT), release of DAMPs-related high mobility group box1 (HMGB1) and autophagy-dependent ATP release, which together, leads to the antigen uptake and presentation of DC cell, and then activates the CD8$^+$ TILs to play the anti-tumor role [130,131]. Carboplatin or paclitaxel combined with radiation generates both chemotherapeutic enhancement of ICD and a dose-dependent induction of ICD in TSA mammary carcinoma cells [132]. Doxorubicin and paclitaxel treatment results in the recruitment of innate immune cells and CSF1R-dependent macrophages infiltration in PyMT-MMTV mammary carcinoma through an increase of CCL2, CXCL2, CSF-1, interleukin-34 and vascular permeability [133,134]. Docetaxel polarizes MDSCs toward M1-like phenotype and upregulates macrophages markers (CD86, MHC class II, and CD11c) in vivo and in vitro partly through an inhibition of STAT-3 in 4T1-Neu mammary cancer implants [135]. All these studies emphasize that chemotherapy can induce TME remodeling through distinct signaling pathways. In this part, we have focused on three crucial factors related to chemotherapy-induced TME remodeling, which are HMGB1, exosomes and S1PR1. The clinical significance of HMGB1, exosomes and S1P/SPHK1/S1PR1 as well as their involvement in TNBC immunomodulation and tumor progression is shown in Figure 2.

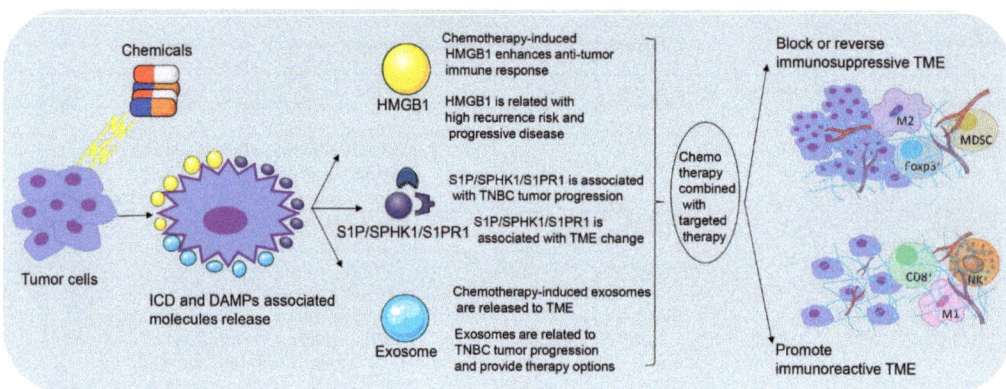

Figure 2. Chemotherapy-induced immunogenic cell death and immunomodulation in TNBC. Chemotherapy induces immunogenic cell death (ICD), and then promotes the release of damage-associated molecular patterns (DAMPs) including high mobility group box1 (HMGB1), exosomes and sphingosine-1-phosphate receptor 1 (S1PR1) by damaged or activated cells. Chemotherapy combined with targeted therapy could enhance anti-tumor immunity through promoting function of immunoreactive lymphocytes and blocking or reversing function of immunosuppressive cells. (ICD, immunogenic cell death; DAMPs, damage-associated molecular patterns; HMGB1, high mobility group box1; S1P, sphingosine-1-phosphate; SPHK1, sphingosine kinase 1; S1PR1, sphingosine-1-phosphate receptor 1; TNBC, triple negative breast cancer; TME, tumor microenvironment).

4.1. Chemotherapy-Induced HMGB1 Release Participates in TNBC Immunomodulation
4.1.1. Chemotherapy-Induced HMGB1 Enhances Anti-Tumor Immune Response

High mobility group box1 (HMGB1) is a highly conserved DNA-binding nuclear protein, involved in many kinds of diseases, including cancer, arthritis, and sepsis [136]. Extracellular HMGB1 in response to inflammation activates the host immune system. HMGB1 can combine with TLR-2, TLR-4, and TLR-9, and recruit the inflammatory cells to microenvironment. This activates the DCs, enhances the antigen presentation ability and anti-tumor immune response [137].

4.1.2. HMGB1 Is Related to High Recurrence Risk and Progressive Disease after Neoadjuvant Chemotherapy

A study indicated that the nuclear expression of HMGB1 in breast cancer cells negatively correlates with Tregs and TAMs [138], and could predict the recurrence risk of residual tumor [139]. HMGB1 expression in cytoplasm is higher in HER2-positive and TNBCs tumors than in hormone receptor (HR)-positive tumors. High cytoplasmic HMGB1 significantly correlates with advanced histologic grade, abundant TILs, and high expression of CD8$^+$ TILs but shows no prognostic significance in TNBC [140]. Intracellular HMGB1 expression has been detected in fibroblasts conditioned medium (CM) treated breast cancer cells and in doxorubicin-treated cells. Extracellular HMGB1 is upregulated in CM after doxorubicin-induced MDA-MB-231 cell death, which show the potential of fibroblasts in stroma to contribute to chemo-resistance partly by fibroblast-induced HMGB1 production [141]. It has been shown that low cytoplasmic HMGB1-positive breast tumor cells and high ASMA-positive fibroblasts predict adverse prognosis in TNBC [142]. Tanabe and colleagues reported that positive HMGB1 expressions are higher in the clinical progressive disease (cPD) than in control group during neoadjuvant chemotherapy in TNBC patients [143]. Some of HMGB1 single nucleotide polymorphisms (SNPs) have been related to tumor progression in T2 tumor, pathologic grade 3 disease, and distant metastasis in TNBC and HER2-enriched tumors compared with luminal tumors [144]. By targeting HMGB1-RAGE signaling pathway, miR-205 impairs the viability and epithelial-to-mesenchymal transition in TNBC cells [145]. HMGB1 released by breast cancer cells is N-glycosylated at Asn37, which promotes the transition from monocytes to MDSC-like cells and contributes to M-MDSC differentiation from bone marrow through the p38/NFκB/Erk1/2 signaling pathway [146].

4.2. Chemotherapy-Induced Exosomes Secretion Interconnects TME and TNBC Immune Response

4.2.1. Chemotherapy-Induced Exosomes Are Released to TME

Exosomes are tiny membrane vesicles (30–100 nm in diameter) synthesized in late endosomes and secreted into the extracellular milieu by various cells. They contain functional molecules (lipids, proteins, DNA, and RNA) that can be transferred to recipient cells, playing a key role in intercellular communication [147]. Apoptosis exosome vesicles (AEVs) are special exosomes overexpressing S1PR1 and S1PR3 released by the tumor cells in response to certain chemicals. These AEVs induce the expression of inflammatory chemokines and cytokines which participate in the pathological and physiological process of DAMPs [147].

4.2.2. Exosomes Are Related to TNBC Tumor Progression and Provide Therapy Options

Some investigations have explored connections between exosomes and TNBCs [148]. Hypoxia induces the production of exosomes and microvesicles (MVs) in breast cancer cells through HIF-dependent RAB22A expression, which can stimulate ECM invasion, focal adhesion formation, lung colonization and is associated with decreased OS and MFS in the mouse models [149]. Stevic and colleagues determined miRNA expression profiles of exosomes originated from the plasma of TNBC and HER2-positive breast cancer patients before neoadjuvant therapy. They found that exosomal miRNAs (miR-155 and miR-301) correlate with the risk factors and clinicopathological factors significantly and can predict pCR rate [150]. Extracellular vesicles (EVs) from HCC1806 but not from MDA-MB-231 cells exhibit enhanced drug resistance and alter the levels of genes involved in cell apoptosis and proliferation pathways in MCF10A cells [151]. Ni and colleagues quantified the levels of miRNAs expression in exosomes from plasma of 8 ductal carcinoma in situ (DCIS) patients, 32 breast cancer (BC) patients and 8 healthy women; they found that different levels of exosomal miRNAs had distinct prognostic value in different subtypes of BC and the expression of miR-16 was lower in TNBC than HR-positive counterparts [152]. Exosomes from TNBC tissues regulate cell apoptosis and TME changes. MiR-770 played its multi-functional role in TNBC by down-regulating gene STMN1 as follows: (i) was associated with favorable prognosis of TNBC, (ii) increased the sensitivity of TNBC cells to doxorubicin through induction of apoptosis, (iii) regulated TAMs-induced chemotherapy resistance, and (iv) inhibited invasion and migration ability of TNBC cells via EMT pathway [153]. Intriguingly, chemotherapy-induced senescent cells secreted more extracellular vesicles than non-senescent cells in TNBC [154]. Exosomes could facilitate co-delivery of cholesterol-modified miR-159 and therapeutic quantities of doxorubicin to TNBC cells both in vitro and in vivo [155]. A formulation of erastin (a low molecular weight chemotherapy drug that induces ferroptosis)-loaded exosome was labeled with special chemicals to target TNBC cells, which enhanced the uptake efficiency of drugs into MDA-MB-231 cells and had a better preventing effect on the migration and proliferation, revealing that the exosome-based therapy might serve as a novel and powerful delivery method for anti-cancer therapy [156].

4.3. S1P/SPHK1/S1PR1 Link TME Changes to TNBC Immunomodulation

4.3.1. S1P/SPHK1/S1PR1 Is Associated with TME Changes

Sphingosine-1-phosphate (S1P), a novel lipid signaling mediator with both intracellular and extracellular functions, is generated by sphingosine kinase 1 (SPHK1), an enzyme catalyzing phosphorylation of sphingosine. S1P/SPHK1 interacts with constituents in TME and modulate the progression and metastasis of breast cancer. Binding of S1P to sphingosine-1-phosphate receptor (S1PRs) on cell surface activates cytokines in the cytoplasm and gene activation in the nucleus in an autocrine and paracrine manner [157,158]. S1P, S1PRs, and SPHK1 expression are related to metastatic progression in breast cancers in vivo [159]. An investigation in melanoma suggested that S1PR1 causes immune functional change of T lymphocytes via PPARγ signal pathway [160]. A recent investigation in breast cancer showed that S1PR1 causes the change of TAMs phenotype, promotes neo-lymph vascularization, and the change of TME via activating inflammatory factors such as Nlrp3 and IL-1β [161]. Another team also showed that S1PR1 phosphorylates the complex of vasculogenic mimicry (VM), and the inhibition of S1PR1 decreases endothelium-dependent vessel (EDV), but causes the production of VM, invasion, and metastasis in vitro and in vivo [162]. Kim and colleagues showed that IL-22 induces S1PR1 and IL22R1 expression in myeloid cells and macrophages, and induce MCP1 expression in myeloid stem cells (MSCs), and then facilitate macrophage infiltration, implying a potential effect of IL-22 on promoting bone metastasis of breast cancers via IL22R1/S1PR1 pathway [158]. S1P1 is expressed in tumor antigen-specific bone marrow (BM) Tregs selectively in breast cancer, and can be induced by BM-resident antigen-presenting cells in conjunction with T cell receptor stimulation [163].

4.3.2. S1P/SPHK1/S1PR1 Is Associated with TNBC Tumor Progression

A preclinical study detected the function of S1PR1-antibody on the growth of breast cancer cell lines MDA-MB-231 and SK-BR-3. They found that S1PR1-antibody not only increases the cytotoxicity of carboplatin on MDA-MB-231 cells but also enhances the antiproliferative outcome of S1P on SK-BR-3 cells [164]. It has been reported that apoptotic tumor cells release S1P, and then stimulate the generation of lipocalin 2 (LCN2) in TAMs and is associated with breast cancer metastasis [165]. As the key kinase of S1P combination, SPHK1 has been found to be overexpressed in TNBC compared with other breast cancer subtypes, and promotes tumor metastasis. By targeting SPHK1 or its downstream signaling pathway (NF-κB pathway) with available inhibitors, TNBC metastasis is effectively inhibited [166]. Maiti and colleagues found that SPHKs/S1P axis is a crucial constituent of survival and growth of LM2-4 cells compared to parental MDA-MB-231 cells, and nuclear SPHK2 (in MDA-MB-231 cells) is also indispensable for LM2-4 cells survival and growth [167]. Obesity and high-fat diet are the main cause for increased expression of the S1P and SPHK1, and targeting the SPHK1/S1P/S1PR1 decreases key proinflammatory cytokines, macrophage infiltration, and tumor progression [168]. However, Lei and colleagues found that S1PRs expression inhibits tumor progression in breast cancer patients [169]. The clinical significance of DAMPs-associated molecules (HMGB1, exosomes, and S1P/SPHK1/S1PR1) and the mechanisms involved in TNBC immunomodulation and tumor progression are included in Table 3.

Table 3. Clinical significance and involved mechanisms of DAMPs-associated molecules.

Items	Clinical Significance	Involved Mechanisms	References
HMGB1	Predict recurrence risk of residual tumor after neoadjuvant chemotherapy	TLR4 signal pathway, immune molecules such as TGF-β, IK12p7, and IFN-γ, p38/NFκB/Erk1/2 pathway, RAGE/IRF3/NF-κB	[19,20,139,140,142,144–146,170]
Exosome	pCR prediction and distinct prognosis value in different subtype of breast cancer	HMGB1/TLR4/NF-κB signaling	[150,152,171,172]
S1P/SPHK1/S1PR1	Paradoxical role in tumor progression of TNBC	PPARγ signal pathway, STAT3/IL-6, IL-22, TCR activation	[158,160,163,169,173–175]

5. Conclusions

The role of the tumor microenvironment (TME) in triple negative breast cancer (TNBC) immunomodulation is vitally important. The deeper understanding of immunosuppressive and immunoreactive TME has contributed to specific subtype classification of TNBC. In future, we may be able to treat TNBC patients with more precision according to their subtype. Agents that remodel TME, promote function of immunoreactive lymphocytes, block function of immunosuppressive cells, and prevent inhibitory signaling pathways can all be considered. Furthermore, therapies targeting HMGB1, exosomal microRNAs, and S1P/SPHK1/S1PR1, can also be considered in combination with chemotherapy. In conclusion, immunosuppressive and immunoreactive role of TME, the contribution of TME in TNBC subtype classification, chemotherapy-induced TME changes and its role in TNBC immunomodulation are crucial for TNBC management. TME has provided a new direction to explore novel and effective combination regimens for precision treatment of TNBC.

Author Contributions: Conceptualization, H.Z., S.P., S.S., X.W. and D.S.; methodology, H.Z. and S.P.; software, S.S.; validation, S.S., D.S. and S.P.; formal analysis, D.S.; investigation, H.Z.; resources, H.Z., D.S.; data curation, H.Z.; writing—original draft preparation, H.Z.; writing—review and editing, D.S.; visualization, S.P.; supervision, D.S.; project administration, X.W.; funding acquisition, H.Z. and X.W. All authors have read and agreed to the published version of the manuscript.

Funding: This research was funded by Natural Science Foundation of Hubei Province, grant number 2020CFB874.

Conflicts of Interest: The authors declare no conflict of interest.

References

1. Perou, C.M.; Sørlie, T.; Eisen, M.B.; van de Rijn, M.; Jeffrey, S.S.; Rees, C.A.; Pollack, J.R.; Ross, D.T.; Johnsen, H.; Akslen, L.A.; et al. Molecular portraits of human breast tumours. *Nature* **2000**, *406*, 747–752. [CrossRef]
2. Prat, A.; Adamo, B.; Cheang, M.C.; Anders, C.K.; Carey, L.A.; Perou, C.M. Molecular characterization of basal-like and non-basal-like triple-negative breast cancer. *Oncologist* **2013**, *18*, 123–133. [CrossRef] [PubMed]
3. Xiao, Y.; Ma, D.; Zhao, S.; Suo, C.; Shi, J.; Xue, M.Z.; Ruan, M.; Wang, H.; Zhao, J.; Li, Q.; et al. Multi-Omics Profiling Reveals Distinct Microenvironment Characterization and Suggests Immune Escape Mechanisms of Triple-Negative Breast Cancer. *Clin. Cancer Res.* **2019**, *25*, 5002–5014. [CrossRef] [PubMed]
4. Adams, S.; Gray, R.J.; Demaria, S.; Goldstein, L.; Perez, E.A.; Shulman, L.N.; Martino, S.; Wang, M.; Jones, V.E.; Saphner, T.J.; et al. Prognostic value of tumor-infiltrating lymphocytes in triple-negative breast cancers from two phase III randomized adjuvant breast cancer trials: ECOG 2197 and ECOG 1199. *J. Clin. Oncol.* **2014**, *32*, 2959–2966. [CrossRef] [PubMed]
5. Dieci, M.V.; Mathieu, M.C.; Guarneri, V.; Conte, P.; Delaloge, S.; Andre, F.; Goubar, A. Prognostic and predictive value of tumor-infiltrating lymphocytes in two phase III randomized adjuvant breast cancer trials. *Ann. Oncol.* **2015**, *26*, 1698–1704. [CrossRef] [PubMed]

6. Callari, M.; Cappelletti, V.; D'Aiuto, F.; Musella, V.; Lembo, A.; Petel, F.; Karn, T.; Iwamoto, T.; Provero, P.; Daidone, M.G.; et al. Subtype-Specific Metagene-Based Prediction of Outcome after Neoadjuvant and Adjuvant Treatment in Breast Cancer. *Clin. Cancer Res.* **2016**, *22*, 337–345. [CrossRef] [PubMed]
7. Loi, S.; Sirtaine, N.; Piette, F.; Salgado, R.; Viale, G.; Van Eenoo, F.; Rouas, G.; Francis, P.; Crown, J.P.; Hitre, E.; et al. Prognostic and predictive value of tumor-infiltrating lymphocytes in a phase III randomized adjuvant breast cancer trial in node-positive breast cancer comparing the addition of docetaxel to doxorubicin with doxorubicin-based chemotherapy: BIG 02-98. *J. Clin. Oncol.* **2013**, *31*, 860–867. [CrossRef]
8. Loi, S.; Drubay, D.; Adams, S.; Pruneri, G.; Francis, P.A.; Lacroix-Triki, M.; Joensuu, H.; Dieci, M.V.; Badve, S.; Demaria, S.; et al. Tumor-Infiltrating Lymphocytes and Prognosis: A Pooled Individual Patient Analysis of Early-Stage Triple-Negative Breast Cancers. *J. Clin. Oncol.* **2019**, *37*, 559–569. [CrossRef]
9. Gupta, S.; Roy, A.; Dwarakanath, B.S. Metabolic Cooperation and Competition in the Tumor Microenvironment: Implications for Therapy. *Front. Oncol* **2017**, *7*, 68. [CrossRef]
10. Kim, I.S.; Zhang, X.H. One microenvironment does not fit all: Heterogeneity beyond cancer cells. *Cancer Metastasis Rev.* **2016**, *35*, 601–629. [CrossRef]
11. Hanahan, D.; Coussens, L.M. Accessories to the crime: Functions of cells recruited to the tumor microenvironment. *Cancer Cell* **2012**, *21*, 309–322. [CrossRef]
12. Varn, F.S.; Mullins, D.W.; Arias-Pulido, H.; Fiering, S.; Cheng, C. Adaptive immunity programmes in breast cancer. *Immunology* **2017**, *150*, 25–34. [CrossRef]
13. Quail, D.F.; Joyce, J.A. Microenvironmental regulation of tumor progression and metastasis. *Nat. Med.* **2013**, *19*, 1423–1437. [CrossRef]
14. Nguyen-Ngoc, K.V.; Cheung, K.J.; Brenot, A.; Shamir, E.R.; Gray, R.S.; Hines, W.C.; Yaswen, P.; Werb, Z.; Ewald, A.J. ECM microenvironment regulates collective migration and local dissemination in normal and malignant mammary epithelium. *Proc. Natl. Acad. Sci. USA* **2012**, *109*, E2595–E2604. [CrossRef] [PubMed]
15. Deng, W.; Lira, V.; Hudson, T.E.; Lemmens, E.E.; Hanson, W.G.; Flores, R.; Barajas, G.; Katibah, G.E.; Desbien, A.L.; Lauer, P.; et al. Recombinant Listeria promotes tumor rejection by CD8+ T cell-dependent remodeling of the tumor microenvironment. *Proc. Natl. Acad. Sci. USA* **2018**, *115*, 8179–8184. [CrossRef] [PubMed]
16. Marra, A.; Viale, G.; Curigliano, G. Recent advances in triple negative breast cancer: The immunotherapy era. *BMC Med.* **2019**, *17*, 90. [CrossRef] [PubMed]
17. Gruosso, T.; Gigoux, M.; VSK, M.; Bertos, N.; Zuo, D.; Perlitch, I.; SMI, S.; Zhao, H.; Souleimanova, M.; Johnson, R.M.; et al. Spatially distinct tumor immune microenvironments stratify triple-negative breast cancers. *J. Clin. Investig.* **2019**, *129*, 1785–1800. [CrossRef] [PubMed]
18. Galluzzi, L.; Senovilla, L.; Zitvogel, L.; Kroemer, G. The secret ally: Immunostimulation by anticancer drugs. *Nat. Rev. Drug Discov.* **2012**, *11*, 215–233. [CrossRef] [PubMed]
19. Huang, C.Y.; Chiang, S.F.; Ke, T.W.; Chen, T.W.; Lan, Y.C.; You, Y.S.; Shiau, A.C.; Chen, W.T.; KSC, C. Cytosolic high-mobility group box protein 1 (HMGB1) and/or PD-1+ TILs in the tumor microenvironment may be contributing prognostic biomarkers for patients with locally advanced rectal cancer who have undergone neoadjuvant chemoradiotherapy. *Cancer Immunol. Immunother.* **2018**, *67*, 551–562. [CrossRef]
20. Zhang, Y.; Yang, S.; Yang, Y.; Liu, T. Resveratrol induces immunogenic cell death of human and murine ovarian carcinoma cells. *Infect. Agents Cancer* **2019**, *14*, 27. [CrossRef]
21. Savas, P.; Salgado, R.; Denkert, C.; Sotiriou, C.; Darcy, P.K.; Smyth, M.J.; Loi, S. Clinical relevance of host immunity in breast cancer: From TILs to the clinic. *Nat. Rev. Clin. Oncol.* **2016**, *13*, 228–241. [CrossRef]
22. Bense, R.D.; Sotiriou, C.; Piccart-Gebhart, M.J.; Haanen, J.B.A.G.; van Vugt, M.A.T.M.; de Vries, E.G.E.; Schröder, C.P.; Fehrmann, R.S.N. Relevance of Tumor-Infiltrating Immune Cell Composition and Functionality for Disease Outcome in Breast Cancer. *J. Natl. Cancer Inst.* **2017**, *109*. [CrossRef]
23. Sharpe, A.H.; Wherry, E.J.; Ahmed, R.; Freeman, G.J. The function of programmed cell death 1 and its ligands in regulating autoimmunity and infection. *Nat. Immunol.* **2007**, *8*, 239–245. [CrossRef]
24. Brahmer, J.R.; Tykodi, S.S.; Chow, L.Q.; Hwu, W.J.; Topalian, S.L.; Hwu, P.; Drake, C.G.; Camacho, L.H.; Kauh, J.; Odunsi, K.; et al. Safety and activity of anti-PD-L1 antibody in patients with advanced cancer. *N. Engl. J. Med.* **2012**, *366*, 2455–2465. [CrossRef]
25. Topalian, S.L.; Hodi, F.S.; Brahmer, J.R.; Gettinger, S.N.; Smith, D.C.; McDermott, D.F.; Powderly, J.D.; Carvajal, R.D.; Sosman, J.A.; Atkins, M.B.; et al. Safety, activity, and immune correlates of anti-PD-1 antibody in cancer. *N. Engl. J. Med.* **2012**, *366*, 2443–2454. [CrossRef]
26. Bertucci, F.; Finetti, P.; Birnbaum, D.; Mamessier, E. The PD1/PDL1 axis, a promising therapeutic target in aggressive breast cancers. *Oncoimmunology* **2016**, *5*, e1085148. [CrossRef] [PubMed]
27. McGee, H.S.; Yagita, H.; Shao, Z.; Agrawal, D.K. Programmed Death-1 antibody blocks therapeutic effects of T-regulatory cells in cockroach antigen-induced allergic asthma. *Am. J. Respir. Cell Mol. Biol.* **2010**, *43*, 432–442. [CrossRef] [PubMed]
28. Han, L.; Liu, F.; Li, R.; Li, Z.; Chen, X.; Zhou, Z.; Zhang, X.; Hu, T.; Zhang, Y.; Young, K.; et al. Role of programmed death ligands in effective T-cell interactions in extranodal natural killer/T-cell lymphoma. *Oncol. Lett.* **2014**, *8*, 1461–1469. [CrossRef] [PubMed]
29. Własiuk, P.; Putowski, M.; Giannopoulos, K. PD1/PD1L pathway, HLA-G and T regulatory cells as new markers of immunosuppression in cancers. *Postepy Hig. Med. Dosw.* **2016**, *70*, 1044–1058. [CrossRef] [PubMed]

30. Brockwell, N.K.; Owen, K.L.; Zanker, D.; Spurling, A.; Rautela, J.; Duivenvoorden, H.M.; Baschuk, N.; Caramia, F.; Loi, S.; Darcy, P.K.; et al. Neoadjuvant Interferons: Critical for Effective PD-1-Based Immunotherapy in TNBC. *Cancer Immunol. Res.* **2017**, *5*, 871–884. [CrossRef]
31. Mittendorf, E.A.; Philips, A.V.; Meric-Bernstam, F.; Qiao, N.; Wu, Y.; Harrington, S.; Su, X.; Wang, Y.; Gonzalez-Angulo, A.M.; Akcakanat, A.; et al. PD-L1 expression in triple-negative breast cancer. *Cancer Immunol. Res.* **2014**, *2*, 361–370. [CrossRef] [PubMed]
32. Adams, S.; Schmid, P.; Rugo, H.S.; Winer, E.P.; Loirat, D.; Awada, A.; Cescon, D.W.; Iwata, H.; Campone, M.; Nanda, R.; et al. Pembrolizumab monotherapy for previously treated metastatic triple-negative breast cancer: Cohort A of the phase II KEYNOTE-086 study. *Ann. Oncol.* **2019**, *30*, 397–404. [CrossRef] [PubMed]
33. Schmid, P.; Adams, S.; Rugo, H.S.; Schneeweiss, A.; Barrios, C.H.; Iwata, H.; Diéras, V.; Hegg, R.; Im, S.A.; Shaw, W.G.; et al. Atezolizumab and Nab-Paclitaxel in Advanced Triple-Negative Breast Cancer. *N. Engl. J. Med.* **2018**, *379*, 2108–2121. [CrossRef] [PubMed]
34. Baptista, M.Z.; Sarian, L.O.; Derchain, S.F.; Pinto, G.A.; Vassallo, J. Prognostic significance of PD-L1 and PD-L2 in breast cancer. *Hum. Pathol.* **2016**, *47*, 78–84. [CrossRef]
35. Muenst, S.; Schaerli, A.R.; Gao, F.; Däster, S.; Trella, E.; Droeser, R.A.; Muraro, M.G.; Zajac, P.; Zanetti, R.; Gillanders, W.E.; et al. Expression of programmed death ligand 1 (PD-L1) is associated with poor prognosis in human breast cancer. *Breast Cancer Res. Treat.* **2014**, *146*, 15–24. [CrossRef] [PubMed]
36. Li, M.; Li, A.; Zhou, S.; Xu, Y.; Xiao, Y.; Bi, R.; Yang, W. Heterogeneity of PD-L1 expression in primary tumors and paired lymph node metastases of triple negative breast cancer. *BMC Cancer* **2018**, *18*, 4. [CrossRef] [PubMed]
37. Sun, W.Y.; Lee, Y.K.; Koo, J.S. Expression of PD-L1 in triple-negative breast cancer based on different immunohistochemical antibodies. *J. Transl. Med.* **2016**, *14*, 173. [CrossRef]
38. Zhu, X.; Zhang, Q.; Wang, D.; Liu, C.; Han, B.; Yang, J.M. Expression of PD-L1 Attenuates the Positive Impacts of High-level Tumor-infiltrating Lymphocytes on Prognosis of Triple-negative Breast Cancer. *Cancer Biol. Ther.* **2019**, *20*, 1105–1112. [CrossRef] [PubMed]
39. Choi, S.H.; Chang, J.S.; Koo, J.S.; Park, J.W.; Sohn, J.H.; Keum, K.C.; Suh, C.O.; Kim, Y.B. Differential Prognostic Impact of Strong PD-L1 Expression and 18F-FDG Uptake in Triple-negative Breast Cancer. *Am. J. Clin. Oncol.* **2018**, *41*, 1049–1057. [CrossRef]
40. AiErken, N.; Shi, H.J.; Zhou, Y.; Shao, N.; Zhang, J.; Shi, Y.; Yuan, Z.Y.; Lin, Y. High PD-L1 Expression Is Closely Associated with Tumor-Infiltrating Lymphocytes and Leads to Good Clinical Outcomes in Chinese Triple Negative Breast Cancer Patients. *Int. J. Biol. Sci.* **2017**, *13*, 1172–1179. [CrossRef]
41. Botti, G.; Collina, F.; Scognamiglio, G.; Rao, F.; Peluso, V.; De Cecio, R.; Piezzo, M.; Landi, G.; De Laurentiis, M.; Cantile, M.; et al. Programmed Death Ligand 1 (PD-L1) Tumor Expression Is Associated with a Better Prognosis and Diabetic Disease in Triple Negative Breast Cancer Patients. *Int. J. Mol. Sci.* **2017**, *18*, 459. [CrossRef]
42. Schmid, P.; Rugo, H.S.; Adams, S.; Schneeweiss, A.; Barrios, C.H.; Iwata, H.; Diéras, V.; Henschel, V.; Molinero, L.; Chui, S.Y.; et al. Atezolizumab plus nab-paclitaxel as first-line treatment for unresectable, locally advanced or metastatic triple-negative breast cancer (IMpassion130): Updated efficacy results from a randomised, double-blind, placebo-controlled, phase 3 trial. *Lancet Oncol.* **2020**, *21*, 44–59. [CrossRef]
43. Nanda, R.; Chow, L.Q.; Dees, E.C.; Berger, R.; Gupta, S.; Geva, R.; Pusztai, L.; Pathiraja, K.; Aktan, G.; Cheng, J.D.; et al. Pembrolizumab in Patients With Advanced Triple-Negative Breast Cancer: Phase Ib KEYNOTE-012 Study. *J. Clin. Oncol.* **2016**, *34*, 2460–2467. [CrossRef] [PubMed]
44. Chen, L.; Gibbons, D.L.; Goswami, S.; Cortez, M.A.; Ahn, Y.H.; Byers, L.A.; Zhang, X.; Yi, X.; Dwyer, D.; Lin, W.; et al. Metastasis is regulated via microRNA-200/ZEB1 axis control of tumour cell PD-L1 expression and intratumoral immunosuppression. *Nat. Commun.* **2014**, *5*, 5241. [CrossRef] [PubMed]
45. Castagnoli, L.; Cancila, V.; Cordoba-Romero, S.L.; Faraci, S.; Talarico, G.; Belmonte, B.; Iorio, M.V.; Milani, M.; Volpari, T.; Chiodoni, C.; et al. WNT signaling modulates PD-L1 expression in the stem cell compartment of triple-negative breast cancer. *Oncogene* **2019**, *38*, 4047–4060. [CrossRef]
46. Maeda, T.; Hiraki, M.; Jin, C.; Rajabi, H.; Tagde, A.; Alam, M.; Bouillez, A.; Hu, X.; Suzuki, Y.; Miyo, M.; et al. MUC1-C Induces PD-L1 and Immune Evasion in Triple-Negative Breast Cancer. *Cancer Res.* **2018**, *78*, 205–215. [CrossRef]
47. Voorwerk, L.; Slagter, M.; Horlings, H.M.; Sikorska, K.; van de Vijver, K.K.; de Maaker, M.; Nederlof, I.; Kiuin, R.J.C.; Warren, S.; Ong, S.; et al. Immune induction strategies in metastatic triple-negative breast cancer to enhance the sensitivity to PD-1 blockade: The TONIC trial. *Nat. Med.* **2019**, *25*, 920–928. [CrossRef]
48. Plitas, G.; Konopacki, C.; Wu, K.; Bos, P.D.; Morrow, M.; Putintseva, E.V.; Chudakov, D.M.; Rudensky, A.Y. Regulatory T Cells Exhibit Distinct Features in Human Breast Cancer. *Immunity* **2016**, *45*, 1122–1134. [CrossRef]
49. Hashemi, V.; Maleki, L.A.; Esmaily, M.; Masjedi, A.; Ghalamfarsa, G.; Namdar, A.; Yousefi, M.; Yousefi, B.; Jadidi-Niaragh, F. Regulatory T cells in breast cancer as a potent anti-cancer therapeutic target. *Int. Immunopharmacol.* **2020**, *78*, 106087. [CrossRef]
50. Taylor, N.A.; Vick, S.C.; Iglesia, M.D.; Brickey, W.J.; Midkiff, B.R.; McKinnon, K.P.; Reisdorf, S.; Anders, C.K.; Carey, L.A.; Parker, J.S.; et al. Treg depletion potentiates checkpoint inhibition in claudin-low breast cancer. *J. Clin. Investig.* **2017**, *127*, 3472–3483. [CrossRef] [PubMed]
51. Jamiyan, T.; Kuroda, H.; Yamaguchi, R.; Nakazato, Y.; Noda, S.; Onozaki, M.; Abe, A.; Hayashi, M. Prognostic impact of a tumor-infiltrating lymphocyte subtype in triple negative cancer of the breast. *Breast Cancer* **2020**, *27*, 880–892. [CrossRef]
52. Zhang, L.; Wang, X.I.; Ding, J.; Sun, Q.; Zhang, S. The predictive and prognostic value of Foxp3+/CD25+ regulatory T cells and PD-L1 expression in triple negative breast cancer. *Ann. Diagn. Pathol.* **2019**, *40*, 143–151. [CrossRef]

53. Yeong, J.; Thike, A.A.; Lim, J.C.; Lee, B.; Li, H.; Wong, S.C.; Hue, S.S.; Tan, P.H.; Iqbal, J. Higher densities of Foxp3+ regulatory T cells are associated with better prognosis in triple-negative breast cancer. *Breast Cancer Res. Treat.* **2017**, *163*, 21–35. [CrossRef]
54. Tsiatas, M.; Kalogeras, K.T.; Manousou, K.; Wirtz, R.M.; Gogas, H.; Veltrup, E.; Zagouri, F.; Lazaridis, G.; Koutras, A.; Christodoulou, C.; et al. Evaluation of the prognostic value of CD3, CD8, and FOXP3 mRNA expression in early-stage breast cancer patients treated with anthracycline-based adjuvant chemotherapy. *Cancer Med.* **2018**, *7*, 5066–5082. [CrossRef] [PubMed]
55. Ge, Y.; Böhm, H.H.; Rathinasamy, A.; Xydia, M.; Hu, X.; Pincha, M.; Umansky, L.; Breyer, C.; Hillier, M.; Bonertz, A.; et al. Tumor-Specific Regulatory T Cells from the Bone Marrow Orchestrate Antitumor Immunity in Breast Cancer. *Cancer Immunol. Res.* **2019**, *7*, 1998–2012. [CrossRef] [PubMed]
56. Ortiz-Martínez, F.; Gutiérrez-Aviñó, F.J.; Sanmartín, E.; Pomares-Navarro, E.; Villalba-Riquelme, C.; García-Martínez, A.; Lerma, E.; Peiró, G. Association of Notch pathway down-regulation with Triple Negative/Basal-like breast carcinomas and high tumor-infiltrating FOXP3+ Tregs. *Exp. Mol. Pathol.* **2016**, *100*, 460–468. [CrossRef] [PubMed]
57. Martinez, L.M.; Robila, V.; Clark, N.M.; Du, W.; Idowu, M.O.; Rutkowski, M.R.; Bos, P.D. Regulatory T Cells Control the Switch From in situ to Invasive Breast Cancer. *Front. Immunol.* **2019**, *10*, 1942. [CrossRef] [PubMed]
58. Watanabe, M.A.; Oda, J.M.; Amarante, M.K.; Cesar, V.J. Regulatory T cells and breast cancer: Implications for immunopathogenesis. *Cancer Metastasis Rev.* **2010**, *29*, 569–579. [CrossRef] [PubMed]
59. Hao, S.; Chen, X.; Wang, F.; Shao, Q.; Liu, J.; Zhao, H.; Yuan, C.; Ren, H.; Mao, H. Breast cancer cell-derived IL-35 promotes tumor progression via induction of IL-35-producing induced regulatory T cells. *Carcinogenesis* **2018**, *39*, 1488–1496. [CrossRef]
60. Noy, R.; Pollard, J.W. Tumor-associated macrophages: From mechanisms to therapy. *Immunity* **2014**, *41*, 49–61. [CrossRef] [PubMed]
61. Zhou, J.; Wang, X.H.; Zhao, Y.X.; Chen, C.; Xu, X.Y.; Sun, Q.; Wu, H.Y.; Chen, M.; Sang, J.F.; Su, L.; et al. Cancer-Associated Fibroblasts Correlate with Tumor-Associated Macrophages Infiltration and Lymphatic Metastasis in Triple Negative Breast Cancer Patients. *J. Cancer* **2018**, *9*, 4635–4641. [CrossRef]
62. Zhang, W.J.; Wang, X.H.; Gao, S.T.; Chen, C.; Xu, X.Y.; Sun, Q.; Zhou, Z.H.; Wu, G.Z.; Yu, Q.; Xu, G.; et al. Tumor-associated macrophages correlate with phenomenon of epithelial-mesenchymal transition and contribute to poor prognosis in triple-negative breast cancer patients. *J. Surg. Res.* **2018**, *222*, 93–101. [CrossRef]
63. Yuan, Z.Y.; Luo, R.Z.; Peng, R.J.; Wang, S.S.; Xue, C. High infiltration of tumor-associated macrophages in triple-negative breast cancer is associated with a higher risk of distant metastasis. *Onco Targets Ther* **2014**, *7*, 1475–1480. [CrossRef]
64. Zhang, Q.; Le, K.; Xu, M.; Zhou, J.; Xiao, Y.; Yang, W.; Jiang, Y.; Xi, Z.; Huang, T. Combined MEK inhibition and tumor-associated macrophages depletion suppresses tumor growth in a triple-negative breast cancer mouse model. *Int. Immunopharmacol.* **2019**, *76*, 105864. [CrossRef]
65. Roux, C.; Jafari, S.M.; Shinde, R.; Duncan, G.; Cescon, D.W.; Silvester, J.; Chu, M.F.; Hodgson, K.; Berger, T.; Wakeham, A.; et al. Reactive oxygen species modulate macrophage immunosuppressive phenotype through the up-regulation of PD-L1. *Proc. Natl. Acad. Sci. USA* **2019**, *116*, 4326–4335. [CrossRef]
66. Jing, W.; Guo, X.; Wang, G.; Bi, Y.; Han, L.; Zhu, Q.; Qiu, C.; Tanaka, M.; Zhao, Y. Breast cancer cells promote CD169+ macrophage-associated immunosuppression through JAK2-mediated PD-L1 upregulation on macrophages. *Int. Immunopharmacol.* **2020**, *78*, 106012. [CrossRef] [PubMed]
67. Prasmickaite, L.; Tenstad, E.M.; Pettersen, S.; Jabeen, S.; Egeland, E.V.; Nord, S.; Pandya, A.; Haugen, M.H.; Kristensen, V.N.; Børresen-Dale, A.L.; et al. Basal-like breast cancer engages tumor-supportive macrophages via secreted factors induced by extracellular S100A4. *Mol. Oncol.* **2018**, *12*, 1540–1558. [CrossRef]
68. Lu, X.; Yang, R.; Zhang, L.; Xi, Y.; Zhao, J.; Wang, F.; Zhang, H.; Li, Z. Macrophage Colony-stimulating Factor Mediates the Recruitment of Macrophages in Triple negative Breast Cancer. *Int. J. Biol. Sci.* **2019**, *15*, 2859–2871. [CrossRef]
69. Donzelli, S.; Milano, E.; Pruszko, M.; Sacconi, A.; Masciarelli, S.; Iosue, I.; Melucci, E.; Gallo, E.; Terrenato, I.; Mottolese, M.; et al. Expression of ID4 protein in breast cancer cells induces reprogramming of tumour-associated macrophages. *Breast Cancer Res.* **2018**, *20*, 59. [CrossRef] [PubMed]
70. Donzelli, S.; Sacconi, A.; Turco, C.; Gallo, E.; Milano, E.; Iosue, I.; Blandino, G.; Fazi, F.; Fontemaggi, G. Paracrine Signaling from Breast Cancer Cells Causes Activation of ID4 Expression in Tumor-Associated Macrophages. *Cells* **2020**, *9*, 418. [CrossRef] [PubMed]
71. Sami, E.; Paul, B.T.; Koziol, J.A.; ElShamy, W.M. The Immunosuppressive Microenvironment in BRCA1-IRIS-Overexpressing TNBC Tumors Is Induced by Bidirectional Interaction with Tumor-Associated Macrophages. *Cancer Res.* **2020**, *80*, 1102–1117. [CrossRef]
72. Hamilton, M.J.; Banáth, J.P.; Lam, V.; Lepard, N.E.; Krystal, G.; Bennewith, K.L. Serum inhibits the immunosuppressive function of myeloid-derived suppressor cells isolated from 4T1 tumor-bearing mice. *Cancer Immunol. Immunother.* **2012**, *61*, 643–654. [CrossRef]
73. Li, F.; Zhao, Y.; Wei, L.; Li, S.; Liu, J. Tumor-infiltrating Treg, MDSC, and IDO expression associated with outcomes of neoadjuvant chemotherapy of breast cancer. *Cancer Biol. Ther.* **2018**, *19*, 695–705. [CrossRef] [PubMed]
74. Kumar, S.; Wilkes, D.W.; Samuel, N.; Blanco, M.A.; Nayak, A.; Alicea-Torres, K.; Gluck, C.; Sinha, S.; Gabrilovich, D.; Chakrabarti, R. ΔNp63-driven recruitment of myeloid-derived suppressor cells promotes metastasis in triple-negative breast cancer. *J. Clin. Investig.* **2018**, *128*, 5095–5109. [CrossRef] [PubMed]
75. Li, W.; Tanikawa, T.; Kryczek, I.; Xia, H.; Li, G.; Wu, K.; Wei, S.; Zhao, L.; Vatan, L.; Wen, B.; et al. Aerobic Glycolysis Controls Myeloid-Derived Suppressor Cells and Tumor Immunity via a Specific CEBPB Isoform in Triple-Negative Breast Cancer. *Cell Metab.* **2018**, *28*, 87–103.e6. [CrossRef] [PubMed]

76. Qian, X.; Zhang, Q.; Shao, N.; Shan, Z.; Cheang, T.; Zhang, Z.; Su, Q.; Wang, S.; Lin, Y. Respiratory hyperoxia reverses immunosuppression by regulating myeloid-derived suppressor cells and PD-L1 expression in a triple-negative breast cancer mouse model. *Am. J. Cancer Res.* **2019**, *9*, 529–545. [PubMed]
77. Dominguez, C.; McCampbell, K.K.; David, J.M.; Palena, C. Neutralization of IL-8 decreases tumor PMN-MDSCs and reduces mesenchymalization of claudin-low triple-negative breast cancer. *JCI Insight* **2017**, *2*, e94296. [CrossRef] [PubMed]
78. Gao, W.; Zhang, X.; Yang, W.; Dou, D.; Zhang, H.; Tang, Y.; Zhong, W.; Meng, J.; Bai, Y.; Liu, Y.; et al. Prim-O-glucosylcimifugin enhances the antitumour effect of PD-1 inhibition by targeting myeloid-derived suppressor cells. *J. Immunother. Cancer* **2019**, *7*, 231. [CrossRef]
79. Piranlioglu, R.; Lee, E.; Ouzounova, M.; Bollag, R.J.; Vinyard, A.H.; Arbab, A.S.; Marasco, D.; Guzel, M.; Cowell, J.K.; Thangaraju, M.; et al. Primary tumor-induced immunity eradicates disseminated tumor cells in syngeneic mouse model. *Nat. Commun.* **2019**, *10*, 1430. [CrossRef]
80. Wesolowski, R.; Duggan, M.C.; Stiff, A.; Markowitz, J.; Trikha, P.; Levine, K.M.; Schoenfield, L.; Abdel-Rasoul, M.; Layman, R.; Ramaswamy, B.; et al. Circulating myeloid-derived suppressor cells increase in patients undergoing neo-adjuvant chemotherapy for breast cancer. *Cancer Immunol. Immunother.* **2017**, *66*, 1437–1447. [CrossRef]
81. Ban, Y.; Mai, J.; Li, X.; Mitchell-Flack, M.; Zhang, T.; Zhang, L.; Chouchane, L.; Ferrari, M.; Shen, H.; Ma, X. Targeting Autocrine CCL5-CCR5 Axis Reprograms Immunosuppressive Myeloid Cells and Reinvigorates Antitumor Immunity. *Cancer Res.* **2017**, *77*, 2857–2868. [CrossRef] [PubMed]
82. Zhang, Y.; Lv, D.; Kim, H.J.; Kurt, R.A.; Bu, W.; Li, Y.; Ma, X. A novel role of hematopoietic CCL5 in promoting triple-negative mammary tumor progression by regulating generation of myeloid-derived suppressor cells. *Cell Res.* **2013**, *23*, 394–408. [CrossRef] [PubMed]
83. Levy, E.M.; Roberti, M.P.; Mordoh, J. Natural killer cells in human cancer: From biological functions to clinical applications. *J. Biomed. Biotechnol.* **2011**, *2011*, 676198. [CrossRef] [PubMed]
84. Tian, W.; Wang, L.; Yuan, L.; Duan, W.; Zhao, W.; Wang, S.; Zhang, Q. A prognostic risk model for patients with triple negative breast cancer based on stromal natural killer cells, tumor-associated macrophages and growth-arrest specific protein 6. *Cancer Sci.* **2016**, *107*, 882–889. [CrossRef] [PubMed]
85. Park, I.H.; Yang, H.N.; Lee, K.J.; Kim, T.S.; Lee, E.S.; Jung, S.Y.; Kwon, Y.; Kong, S.Y. Tumor-derived IL-18 induces PD-1 expression on immunosuppressive NK cells in triple-negative breast cancer. *Oncotarget* **2017**, *8*, 32722–32730. [CrossRef]
86. Foulds, G.A.; Vadakekolathu, J.; TMA, A.; Nagarajan, D.; Reeder, S.; Johnson, C.; Hood, S.; Moseley, P.M.; SYT, C.; Pockley, A.G.; et al. Immune-Phenotyping and Transcriptomic Profiling of Peripheral Blood Mononuclear Cells from Patients with Breast Cancer: Identification of a 3 Gene Signature Which Predicts Relapse of Triple Negative Breast Cancer. *Front. Immunol.* **2018**, *9*, 2028. [CrossRef] [PubMed]
87. Fregni, G.; Perier, A.; Avril, M.F.; Caignard, A. NK cells sense tumors, course of disease and treatments: Consequences for NK-based therapies. *Oncoimmunology* **2012**, *1*, 38–47. [CrossRef] [PubMed]
88. Choucair, K.; Duff, J.R.; Cassidy, C.S.; Albrethsen, M.T.; Kelso, J.D.; Lenhard, A.; Staats, H.; Patel, R.; Brunicardi, F.C.; Dworkin, L.; et al. Natural killer cells: A review of biology, therapeutic potential and challenges in treatment of solid tumors. *Future Oncol.* **2019**, *15*, 3053–3069. [CrossRef]
89. Habif, G.; Crinier, A.; André, P.; Vivier, E.; Narni-Mancinelli, E. Targeting natural killer cells in solid tumors. *Cell. Mol. Immunol.* **2019**, *16*, 415–422. [CrossRef]
90. Stojanovic, A.; Cerwenka, A. Natural killer cells and solid tumors. *J. Innate Immun.* **2011**, *3*, 355–364. [CrossRef]
91. Shenouda, M.M.; Gillgrass, A.; Nham, T.; Hogg, R.; Lee, A.J.; Chew, M.V.; Shafaei, M.; Aarts, C.; Lee, D.A.; Hassell, J.; et al. Ex vivo expanded natural killer cells from breast cancer patients and healthy donors are highly cytotoxic against breast cancer cell lines and patient-derived tumours. *Breast Cancer Res.* **2017**, *19*, 76. [CrossRef] [PubMed]
92. Nham, T.; Poznanski, S.M.; Fan, I.Y.; Vahedi, F.; Shenouda, M.M.; Lee, A.J.; Chew, M.V.; Hogg, R.T.; Lee, D.A.; Ashkar, A.A. Ex Vivo-expanded Natural Killer Cells Derived From Long-term Cryopreserved Cord Blood are Cytotoxic Against Primary Breast Cancer Cells. *J. Immunother.* **2018**, *41*, 64–72. [CrossRef] [PubMed]
93. Liu, X.; Ran, R.; Shao, B.; Rugo, H.S.; Yang, Y.; Hu, Z.; Wei, Z.; Wan, F.; Kong, W.; Song, G.; et al. Combined peripheral natural killer cell and circulating tumor cell enumeration enhance prognostic efficiency in patients with metastatic triple-negative breast cancer. *Chin. J. Cancer Res.* **2018**, *30*, 315–326. [CrossRef] [PubMed]
94. Hu, Z. Tissue factor as a new target for CAR-NK cell immunotherapy of triple-negative breast cancer. *Sci. Rep.* **2020**, *10*, 2815. [CrossRef] [PubMed]
95. Juliá, E.P.; Amante, A.; Pampena, M.B.; Mordoh, J.; Levy, E.M. Avelumab, an IgG1 anti-PD-L1 Immune Checkpoint Inhibitor, Triggers NK Cell-Mediated Cytotoxicity and Cytokine Production Against Triple Negative Breast Cancer Cells. *Front. Immunol.* **2018**, *9*, 2140. [CrossRef]
96. Roberti, M.P.; Juliá, E.P.; Rocca, Y.S.; Amat, M.; Bravo, A.I.; Loza, J.; Coló, F.; Loza, C.M.; Fabiano, V.; Maino, M.; et al. Overexpression of CD85j in TNBC patients inhibits Cetuximab-mediated NK-cell ADCC but can be restored with CD85j functional blockade. *Eur. J. Immunol.* **2015**, *45*, 1560–1569. [CrossRef]
97. Del, B.J.; Florès-Florès, R.; Josselin, E.; Goubard, A.; Ganier, L.; Castellano, R.; Chames, P.; Baty, D.; Kerfelec, B. A Bispecific Antibody-Based Approach for Targeting Mesothelin in Triple Negative Breast Cancer. *Front. Immunol.* **2019**, *10*, 1593.

98. Marrufo, A.M.; Mathew, S.O.; Chaudhary, P.; Malaer, J.D.; Vishwanatha, J.K.; Mathew, P.A. Blocking LLT1 (CLEC2D, OCIL)-NKRP1A (CD161) interaction enhances natural killer cell-mediated lysis of triple-negative breast cancer cells. *Am. J. Cancer Res.* **2018**, *8*, 1050–1063.
99. Vihervuori, H.; Autere, T.A.; Repo, H.; Kurki, S.; Kallio, L.; Lintunen, M.M.; Talvinen, K.; Kronqvist, P. Tumor-infiltrating lymphocytes and CD8+ T cells predict survival of triple-negative breast cancer. *J. Cancer Res. Clin. Oncol.* **2019**, *145*, 3105–3114. [CrossRef]
100. Ali, H.R.; Provenzano, E.; Dawson, S.J.; Blows, F.M.; Liu, B.; Shah, M.; Earl, H.M.; Poole, C.J.; Hiller, L.; Dunn, J.A.; et al. Association between CD8+ T-cell infiltration and breast cancer survival in 12,439 patients. *Ann. Oncol.* **2014**, *25*, 1536–1543. [CrossRef]
101. Miyashita, M.; Sasano, H.; Tamaki, K.; Hirakawa, H.; Takahashi, Y.; Nakagawa, S.; Watanabe, G.; Tada, H.; Suzuki, A.; Ohuchi, N.; et al. Prognostic significance of tumor-infiltrating CD8+ and FOXP3+ lymphocytes in residual tumors and alterations in these parameters after neoadjuvant chemotherapy in triple-negative breast cancer: A retrospective multicenter study. *Breast Cancer Res.* **2015**, *17*, 124. [CrossRef] [PubMed]
102. Papaioannou, E.; Sakellakis, M.; Melachrinou, M.; Tzoracoleftherakis, E.; Kalofonos, H.; Kourea, E. A Standardized Evaluation Method for FOXP3+ Tregs and CD8+ T-cells in Breast Carcinoma: Association With Breast Carcinoma Subtypes, Stage and Prognosis. *Anticancer Res.* **2019**, *39*, 1217–1232. [CrossRef] [PubMed]
103. Zheng, H.; Li, X.; Chen, C.; Chen, J.; Sun, J.; Sun, S.; Jin, L.; Li, J.; Sun, S.; Wu, X. Quantum dot-based immunofluorescent imaging and quantitative detection of TOP2A and prognostic value in triple-negative breast cancer. *Int. J. Nanomed.* **2016**, *11*, 5519–5529. [CrossRef]
104. Rao, N.; Qiu, J.; Wu, J.; Zeng, H.; Su, F.; Qiu, K.; Wu, J.; Yao, H. Significance of Tumor-Infiltrating Lymphocytes and the Expression of Topoisomerase IIα in the Prediction of the Clinical Outcome of Patients with Triple-Negative Breast Cancer after Taxane-Anthracycline-Based Neoadjuvant Chemotherapy. *Chemotherapy* **2017**, *62*, 246–255. [CrossRef]
105. Racioppi, L.; Nelson, E.R.; Huang, W.; Mukherjee, D.; Lawrence, S.A.; Lento, W.; Masci, A.M.; Jiao, Y.; Park, S.; York, B.; et al. CaMKK2 in myeloid cells is a key regulator of the immune-suppressive microenvironment in breast cancer. *Nat. Commun.* **2019**, *10*, 2450. [CrossRef]
106. Pantelidou, C.; Sonzogni, O.; De Oliveria Taveira, M.; Mehta, A.K.; Kothari, A.; Wang, D.; Visal, T.; Li, M.K.; Pinto, J.; Castrillon, J.A.; et al. PARP Inhibitor Efficacy Depends on CD8+ T-cell Recruitment via Intratumoral STING Pathway Activation in BRCA-Deficient Models of Triple-Negative Breast Cancer. *Cancer Discov.* **2019**, *9*, 722–737. [CrossRef]
107. Niavarani, S.R.; Lawson, C.; Boudaud, M.; Simard, C.; Tai, L.H. Oncolytic vesicular stomatitis virus-based cellular vaccine improves triple-negative breast cancer outcome by enhancing natural killer and CD8+ T-cell functionality. *J. Immunother. Cancer* **2020**, *8*, e000465. [CrossRef] [PubMed]
108. Li, X.; Gruosso, T.; Zuo, D.; Omeroglu, A.; Meterissian, S.; Guiot, M.C.; Salazar, A.; Park, M.; Levine, H. Infiltration of CD8+ T cells into tumor cell clusters in triple-negative breast cancer. *Proc. Natl. Acad. Sci. USA* **2019**, *116*, 3678–3687. [CrossRef]
109. Kalimutho, M.; Sinha, D.; Mittal, D.; Srihari, S.; Nanayakkara, D.; Shafique, S.; Raninga, P.; Nag, P.; Parsons, K.; Khanna, K.K. Blockade of PDGFRβ circumvents resistance to MEK-JAK inhibition via intratumoral CD8+ T-cells infiltration in triple-negative breast cancer. *J. Exp. Clin. Cancer Res.* **2019**, *38*, 85. [CrossRef]
110. Dong, M.B.; Wang, G.; Chow, R.D.; Ye, L.; Zhu, L.; Dai, X.; Park, J.J.; Kim, H.R.; Errami, Y.; Guzman, C.D.; et al. Systematic Immunotherapy Target Discovery Using Genome-Scale In Vivo CRISPR Screens in CD8 T Cells. *Cell* **2019**, *178*, 1189–1204.e23. [CrossRef] [PubMed]
111. Pyonteck, S.M.; Akkari, L.; Schuhmacher, A.J.; Bowman, R.L.; Sevenich, L.; Quail, D.F.; Olson, O.C.; Quick, M.L.; Huse, J.T.; Teijeiro, V.; et al. CSF-1R inhibition alters macrophage polarization and blocks glioma progression. *Nat. Med.* **2013**, *19*, 1264–1272. [CrossRef]
112. AHY, T.; Tu, W.; McCuaig, R.; Hardy, K.; Donovan, T.; Tsimbalyuk, S.; Forwood, J.K.; Rao, S. Lysine-Specific Histone Demethylase 1A Regulates Macrophage Polarization and Checkpoint Molecules in the Tumor Microenvironment of Triple-Negative Breast Cancer. *Front. Immunol.* **2019**, *10*, 1351.
113. Steenbrugge, J.; Breyne, K.; Demeyere, K.; De Wever, O.; Sanders, N.N.; Van Den Broeck, W.; Colpaert, C.; Vermeulen, P.; Van Laere, S.; Meyer, E. Anti-inflammatory signaling by mammary tumor cells mediates prometastatic macrophage polarization in an innovative intraductal mouse model for triple-negative breast cancer. *J. Exp. Clin. Cancer Res.* **2018**, *37*, 191. [CrossRef] [PubMed]
114. Araujo, J.M.; Gomez, A.C.; Aguilar, A.; Salgado, R.; Balko, J.M.; Bravo, L.; Doimi, F.; Bretel, D.; Morante, Z.; Flores, C.; et al. Effect of CCL5 expression in the recruitment of immune cells in triple negative breast cancer. *Sci. Rep.* **2018**, *8*, 4899. [CrossRef] [PubMed]
115. Liu, H.; Wang, S.H.; Chen, S.C.; Chen, C.Y.; Lin, T.M. Zoledronic acid blocks the interaction between breast cancer cells and regulatory T-cells. *BMC Cancer* **2019**, *19*, 176. [CrossRef] [PubMed]
116. Tiwari, P.; Blank, A.; Cui, C.; Schoenfelt, K.Q.; Zhou, G.; Xu, Y.; Khramtsova, G.; Olopade, F.; Shah, A.M.; Khan, S.A.; et al. Metabolically activated adipose tissue macrophages link obesity to triple-negative breast cancer. *J. Exp. Med.* **2019**, *216*, 1345–1358. [CrossRef] [PubMed]
117. TNT, U.; Lee, K.H.; Ahn, S.J.; Kim, K.W.; Min, J.J.; Hyun, H.; Yoon, M.S. Real-Time Tracking of Ex Vivo-Expanded Natural Killer Cells Toward Human Triple-Negative Breast Cancers. *Front. Immunol.* **2018**, *9*, 825.
118. Lehmann, B.D.; Bauer, J.A.; Chen, X.; Sanders, M.E.; Chakravarthy, A.B.; Shyr, Y.; Pietenpol, J.A. Identification of human triple-negative breast cancer subtypes and preclinical models for selection of targeted therapies. *J. Clin. Investig.* **2011**, *121*, 2750–2767. [CrossRef]
119. SMI, S.; Bertos, N.; Gruosso, T.; Gigoux, M.; Souleimanova, M.; Zhao, H.; Omeroglu, A.; Hallett, M.T.; Park, M. Identification of Interacting Stromal Axes in Triple-Negative Breast Cancer. *Cancer Res.* **2017**, *77*, 4673–4683.

120. Jézéquel, P.; Kerdraon, O.; Hondermarck, H.; Guérin-Charbonnel, C.; Lasla, H.; Gouraud, W.; Canon, J.L.; Gombos, A.; Dalenc, F.; Delaloge, S.; et al. Identification of three subtypes of triple-negative breast cancer with potential therapeutic implications. *Breast Cancer Res.* **2019**, *21*, 65. [CrossRef] [PubMed]
121. Jiang, Y.Z.; Ma, D.; Suo, C.; Shi, J.; Xue, M.; Hu, X.; Xiao, Y.; Yu, K.D.; Liu, Y.R.; Yu, Y.; et al. Genomic and Transcriptomic Landscape of Triple-Negative Breast Cancers: Subtypes and Treatment Strategies. *Cancer Cell* **2019**, *35*, 428–440.e5. [CrossRef] [PubMed]
122. Tseng, L.M.; Chiu, J.H.; Liu, C.Y.; Tsai, Y.F.; Wang, Y.L.; Yang, C.W.; Shyr, Y.M. A comparison of the molecular subtypes of triple-negative breast cancer among non-Asian and Taiwanese women. *Breast Cancer Res. Treat.* **2017**, *163*, 241–254. [CrossRef] [PubMed]
123. Quist, J.; Mirza, H.; MCU, C.; Telli, M.L.; O'Shaughnessy, J.A.; Lord, C.J.; ANJ, T.; Grigoriadis, A. A Four-gene Decision Tree Signature Classification of Triple-negative Breast Cancer: Implications for Targeted Therapeutics. *Mol. Cancer Ther.* **2019**, *18*, 204–212. [CrossRef] [PubMed]
124. He, Y.; Jiang, Z.; Chen, C.; Wang, X. Classification of triple-negative breast cancers based on Immunogenomic profiling. *J. Exp. Clin. Cancer Res.* **2018**, *37*, 327. [CrossRef]
125. Prado-Vázquez, G.; Gámez-Pozo, A.; Trilla-Fuertes, L.; Arevalillo, J.M.; Zapater-Moros, A.; Ferrer-Gómez, M.; Díaz-Almirón, M.; López-Vacas, R.; Navarro, H.; Maín, P.; et al. A novel approach to triple-negative breast cancer molecular classification reveals a luminal immune-positive subgroup with good prognoses. *Sci. Rep.* **2019**, *9*, 1538. [CrossRef]
126. DiNome, M.L.; JIJ, O.; Matsuba, C.; Manughian-Peter, A.O.; Ensenyat-Mendez, M.; Chang, S.C.; Jalas, J.R.; Salomon, M.P.; Marzese, D.M. Clinicopathological Features of Triple-Negative Breast Cancer Epigenetic Subtypes. *Ann. Surg. Oncol.* **2019**, *26*, 3344–3353. [CrossRef]
127. Zheng, S.; Zou, Y.; Xie, X.; Liang, J.Y.; Yang, A.; Yu, K.; Wang, J.; Tang, H.; Xie, X. Development and Validation of a Stromal Immune Phenotype Classifier for Predicting Immune Activity and Prognosis in Triple-Negative Breast Cancer. *Int. J. Cancer* **2020**, *147*, 542–553. [CrossRef]
128. Romero-Cordoba, S.; Meneghini, E.; Sant, M.; Iorio, M.V.; Sfondrini, L.; Paolini, B.; Agresti, R.; Tagliabue, E.; Bianchi, F. Decoding Immune Heterogeneity of Triple Negative Breast Cancer and Its Association with Systemic Inflammation. *Cancers* **2019**, *11*, 911. [CrossRef]
129. Zitvogel, L.; Kepp, O.; Kroemer, G. Immune parameters affecting the efficacy of chemotherapeutic regimens. *Nat. Rev. Clin. Oncol.* **2011**, *8*, 151–160. [CrossRef]
130. Kroemer, G.; Galluzzi, L.; Kepp, O.; Zitvogel, L. Immunogenic cell death in cancer therapy. *Annu. Rev. Immunol.* **2013**, *31*, 51–72. [CrossRef] [PubMed]
131. Voloshin, T.; Kaynan, N.; Davidi, S.; Porat, Y.; Shteingauz, A.; Schneiderman, R.S.; Zeevi, E.; Munster, M.; Blat, R.; Tempel, B.C.; et al. Tumor-treating fields (TTFields) induce immunogenic cell death resulting in enhanced antitumor efficacy when combined with anti-PD-1 therapy. *Cancer Immunol. Immunother.* **2020**, *69*, 1191–1204. [CrossRef] [PubMed]
132. Golden, E.B.; Frances, D.; Pellicciotta, I.; Demaria, S.; Helen, B.M.; Formenti, S.C. Radiation fosters dose-dependent and chemotherapy-induced immunogenic cell death. *Oncoimmunology* **2014**, *3*, e28518. [CrossRef]
133. Nakasone, E.S.; Askautrud, H.A.; Kees, T.; Park, J.H.; Plaks, V.; Ewald, A.J.; Fein, M.; Rasch, M.G.; Tan, Y.X.; Qiu, J.; et al. Imaging tumor-stroma interactions during chemotherapy reveals contributions of the microenvironment to resistance. *Cancer Cell* **2012**, *21*, 488–503. [CrossRef]
134. DeNardo, D.G.; Brennan, D.J.; Rexhepaj, E.; Ruffell, B.; Shiao, S.L.; Madden, S.F.; Gallagher, W.M.; Wadhwani, N.; Keil, S.D.; Junaid, S.A.; et al. Leukocyte complexity predicts breast cancer survival and functionally regulates response to chemotherapy. *Cancer Discov.* **2011**, *1*, 54–67. [CrossRef] [PubMed]
135. Kodumudi, K.N.; Woan, K.; Gilvary, D.L.; Sahakian, E.; Wei, S.; Djeu, J.Y. A novel chemoimmunomodulating property of docetaxel: Suppression of myeloid-derived suppressor cells in tumor bearers. *Clin. Cancer Res.* **2010**, *16*, 4583–4594. [CrossRef] [PubMed]
136. Sims, G.P.; Rowe, D.C.; Rietdijk, S.T.; Herbst, R.; Coyle, A.J. HMGB1 and RAGE in inflammation and cancer. *Annu. Rev. Immunol.* **2010**, *28*, 367–388. [CrossRef]
137. Bianchi, M.E.; Crippa, M.P.; Manfredi, A.A.; Mezzapelle, R.; Rovere, Q.P.; Venereau, E. High-mobility group box 1 protein orchestrates responses to tissue damage via inflammation, innate and adaptive immunity, and tissue repair. *Immunol. Rev.* **2017**, *280*, 74–82. [CrossRef] [PubMed]
138. Ladoire, S.; Enot, D.; Senovilla, L.; Ghiringhelli, F.; Poirier-Colame, V.; Chaba, K.; Semeraro, M.; Chaix, M.; Penault-Llorca, F.; Arnould, L.; et al. The presence of LC3B puncta and HMGB1 expression in malignant cells correlate with the immune infiltrate in breast cancer. *Autophagy* **2016**, *12*, 864–875. [CrossRef]
139. Ladoire, S.; Penault-Llorca, F.; Senovilla, L.; Dalban, C.; Enot, D.; Locher, C.; Prada, N.; Poirier-Colame, V.; Chaba, K.; Arnould, L.; et al. Combined evaluation of LC3B puncta and HMGB1 expression predicts residual risk of relapse after adjuvant chemotherapy in breast cancer. *Autophagy* **2015**, *11*, 1878–1890. [CrossRef]
140. Lee, H.J.; Kim, A.; Song, I.H.; Park, I.A.; Yu, J.H.; Ahn, J.H.; Gong, G. Cytoplasmic expression of high mobility group B1 (HMGB1) is associated with tumor-infiltrating lymphocytes (TILs) in breast cancer. *Pathol. Int.* **2016**, *66*, 202–209. [CrossRef]
141. Amornsupak, K.; Insawang, T.; Thuwajit, P.; O-Charoenrat, P.; Eccles, S.A.; Thuwajit, C. Cancer-associated fibroblasts induce high mobility group box 1 and contribute to resistance to doxorubicin in breast cancer cells. *BMC Cancer* **2014**, *14*, 955. [CrossRef] [PubMed]

142. Amornsupak, K.; Jamjuntra, P.; Warnnissorn, M.; O-Charoenrat, P.; Sa-Nguanraksa, D.; Thuwajit, P.; Eccles, S.A.; Thuwajit, C. High ASMA+ Fibroblasts and Low Cytoplasmic HMGB1+ Breast Cancer Cells Predict Poor Prognosis. *Clin. Breast Cancer* **2017**, *17*, 441–452.e2. [CrossRef]
143. Tanabe, Y.; Tsuda, H.; Yoshida, M.; Yunokawa, M.; Yonemori, K.; Shimizu, C.; Yamamoto, S.; Kinoshita, T.; Fujiwara, Y.; Tamura, K. Pathological features of triple-negative breast cancers that showed progressive disease during neoadjuvant chemotherapy. *Cancer Sci.* **2017**, *108*, 1520–1529. [CrossRef] [PubMed]
144. Huang, B.F.; Tzeng, H.E.; Chen, P.C.; Wang, C.Q.; Su, C.M.; Wang, Y.; Hu, G.N.; Zhao, Y.M.; Wang, Q.; Tang, C.H. HMGB1 genetic polymorphisms are biomarkers for the development and progression of breast cancer. *Int. J. Med. Sci.* **2018**, *15*, 580–586. [CrossRef]
145. Wang, L.; Kang, F.B.; Wang, J.; Yang, C.; He, D.W. Downregulation of miR-205 contributes to epithelial-mesenchymal transition and invasion in triple-negative breast cancer by targeting HMGB1-RAGE signaling pathway. *Anticancer Drugs* **2019**, *30*, 225–232. [CrossRef]
146. Su, Z.; Ni, P.; She, P.; Liu, Y.; Richard, S.A.; Xu, W.; Zhu, H.; Wang, J. Bio-HMGB1 from breast cancer contributes to M-MDSC differentiation from bone marrow progenitor cells and facilitates conversion of monocytes into MDSC-like cells. *Cancer Immunol. Immunother.* **2017**, *66*, 391–401. [CrossRef]
147. Park, S.J.; Kim, J.M.; Kim, J.; Hur, J.; Park, S.; Kim, K.; Shin, H.J.; Chwae, Y.J. Molecular mechanisms of biogenesis of apoptotic exosome-like vesicles and their roles as damage-associated molecular patterns. *Proc. Natl. Acad. Sci. USA* **2018**, *115*, E11721. [CrossRef]
148. Goh, C.Y.; Wyse, C.; Ho, M.; O'Beirne, E.; Howard, J.; Lindsay, S.; Kelly, P.; Higgins, M.; McCann, A. Exosomes in triple negative breast cancer: Garbage disposals or Trojan horses. *Cancer Lett.* **2020**, *473*, 90–97. [CrossRef] [PubMed]
149. Wang, T.; Gilkes, D.M.; Takano, N.; Xiang, L.; Luo, W.; Bishop, C.J.; Chaturvedi, P.; Green, J.J.; Semenza, G.L. Hypoxia-inducible factors and RAB22A mediate formation of microvesicles that stimulate breast cancer invasion and metastasis. *Proc. Natl. Acad. Sci. USA* **2014**, *111*, E3234–E3242. [CrossRef]
150. Stevic, I.; Müller, V.; Weber, K.; Fasching, P.A.; Karn, T.; Marmé, F.; Schem, C.; Stickeler, E.; Denkert, C.; van Mackelenbergh, M.; et al. Specific microRNA signatures in exosomes of triple-negative and HER2-positive breast cancer patients undergoing neoadjuvant therapy within the GeparSixto trial. *BMC Med.* **2018**, *16*, 179. [CrossRef]
151. PMM, O.; Alkhilaiwi, F.; Cavalli, I.J.; Malheiros, D.; de Souza Fonseca Ribeiro, E.M.; Cavalli, L.R. Extracellular vesicles from triple-negative breast cancer cells promote proliferation and drug resistance in non-tumorigenic breast cells. *Breast Cancer Res. Treat.* **2018**, *172*, 713–723.
152. Ni, Q.; Stevic, I.; Pan, C.; Müller, V.; Oliveira-Ferrer, L.; Pantel, K.; Schwarzenbach, H. Different signatures of miR-16, miR-30b and miR-93 in exosomes from breast cancer and DCIS patients. *Sci. Rep.* **2018**, *8*, 12974. [CrossRef]
153. Li, Y.; Liang, Y.; Sang, Y.; Song, X.; Zhang, H.; Liu, Y.; Jiang, L.; Yang, Q. MiR-770 suppresses the chemo-resistance and metastasis of triple negative breast cancer via direct targeting of STMN1. *Cell Death Dis.* **2018**, *9*, 14. [CrossRef] [PubMed]
154. Kavanagh, E.L.; Lindsay, S.; Halasz, M.; Gubbins, L.C.; Weiner-Gorzel, K.; MHZ, G.; McGoldrick, A.; Collins, E.; Henry, M.; Blanco-Fernández, A.; et al. Protein and chemotherapy profiling of extracellular vesicles harvested from therapeutic induced senescent triple negative breast cancer cells. *Oncogenesis* **2017**, *6*, e388. [CrossRef] [PubMed]
155. Gong, C.; Tian, J.; Wang, Z.; Gao, Y.; Wu, X.; Ding, X.; Qiang, L.; Li, G.; Han, Z.; Yuan, Y.; et al. Functional exosome-mediated co-delivery of doxorubicin and hydrophobically modified microRNA 159 for triple-negative breast cancer therapy. *J. Nanobiotechnology* **2019**, *17*, 93. [CrossRef] [PubMed]
156. Yu, M.; Gai, C.; Li, Z.; Ding, D.; Zheng, J.; Zhang, W.; Lv, S.; Li, W. Targeted exosome-encapsulated erastin induced ferroptosis in triple negative breast cancer cells. *Cancer Sci.* **2019**, *110*, 3173–3182. [CrossRef]
157. Tsuchida, J.; Nagahashi, M.; Takabe, K.; Wakai, T. Clinical Impact of Sphingosine-1-Phosphate in Breast Cancer. *Mediat. Inflamm.* **2017**, *2017*, 2076239. [CrossRef]
158. Kim, E.Y.; Choi, B.; Kim, J.E.; Park, S.O.; Kim, S.M.; Chang, E.J. Interleukin-22 Mediates the Chemotactic Migration of Breast Cancer Cells and Macrophage Infiltration of the Bone Microenvironment by Potentiating S1P/SIPR Signaling. *Cells* **2020**, *9*, 131. [CrossRef] [PubMed]
159. Sukocheva, O.A. Expansion of Sphingosine Kinase and Sphingosine-1-Phosphate Receptor Function in Normal and Cancer Cells: From Membrane Restructuring to Mediation of Estrogen Signaling and Stem Cell Programming. *Int. J. Mol. Sci.* **2018**, *19*, 420. [CrossRef]
160. Chakraborty, P.; Vaena, S.G.; Thyagarajan, K.; Chatterjee, S.; Al-Khami, A.; Selvam, S.P.; Nguyen, H.; Kang, I.; Wyatt, M.W.; Baliga, U.; et al. Pro-Survival Lipid Sphingosine-1-Phosphate Metabolically Programs T Cells to Limit Anti-tumor Activity. *Cell Rep.* **2019**, *28*, 1879–1893.e7. [CrossRef]
161. Weichand, B.; Popp, R.; Dziumbla, S.; Mora, J.; Strack, E.; Elwakeel, E.; Frank, A.C.; Scholich, K.; Pierre, S.; Syed, S.N.; et al. S1PR1 on tumor-associated macrophages promotes lymphangiogenesis and metastasis via NLRP3/IL-1β. *J. Exp. Med.* **2017**, *214*, 2695–2713. [CrossRef]
162. Liu, S.; Ni, C.; Zhang, D.; Sun, H.; Dong, X.; Che, N.; Liang, X.; Chen, C.; Liu, F.; Bai, J.; et al. S1PR1 regulates the switch of two angiogenic modes by VE-cadherin phosphorylation in breast cancer. *Cell Death Dis.* **2019**, *10*, 200. [CrossRef]
163. Rathinasamy, A.; Domschke, C.; Ge, Y.; Böhm, H.H.; Dettling, S.; Jansen, D.; Lasitschka, F.; Umansky, L.; Gräler, M.H.; Hartmann, J.; et al. Tumor specific regulatory T cells in the bone marrow of breast cancer patients selectively upregulate the emigration receptor S1P1. *Cancer Immunol. Immunother.* **2017**, *66*, 593–603. [CrossRef] [PubMed]

164. Xiao, S.; Yang, J. Preclinical study of the antitumor effect of sphingosine-1-phosphate receptor 1 antibody (S1PR1-antibody) against human breast cancer cells. *Investig. New. Drugs* **2019**, *37*, 57–64. [CrossRef] [PubMed]
165. Jung, M.; Ören, B.; Mora, J.; Mertens, C.; Dziumbla, S.; Popp, R.; Weigert, A.; Grossmann, N.; Fleming, I.; Brüne, B. Lipocalin 2 from macrophages stimulated by tumor cell-derived sphingosine 1-phosphate promotes lymphangiogenesis and tumor metastasis. *Sci. Signal.* **2016**, *9*, ra64. [CrossRef] [PubMed]
166. Acharya, S.; Yao, J.; Li, P.; Zhang, C.; Lowery, F.J.; Zhang, Q.; Guo, H.; Qu, J.; Yang, F.; Wistuba, I.I.; et al. Sphingosine Kinase 1 Signaling Promotes Metastasis of Triple-Negative Breast Cancer. *Cancer Res.* **2019**, *79*, 4211–4226. [CrossRef] [PubMed]
167. Maiti, A.; Takabe, K.; Hait, N.C. Metastatic triple-negative breast cancer is dependent on SphKs/S1P signaling for growth and survival. *Cell. Signal.* **2017**, *32*, 85–92. [CrossRef] [PubMed]
168. Nagahashi, M.; Yamada, A.; Katsuta, E.; Aoyagi, T.; Huang, W.C.; Terracina, K.P.; Hait, N.C.; Allegood, J.C.; Tsuchida, J.; Yuza, K.; et al. Targeting the SphK1/S1P/S1PR1 Axis That Links Obesity, Chronic Inflammation, and Breast Cancer Metastasis. *Cancer Res.* **2018**, *78*, 1713–1725. [CrossRef] [PubMed]
169. Lei, F.J.; Cheng, B.H.; Liao, P.Y.; Wang, H.C.; Chang, W.C.; Lai, H.C.; Yang, J.C.; Wu, Y.C.; Chu, L.C.; Ma, W.L. Survival benefit of sphingosin-1-phosphate and receptors expressions in breast cancer patients. *Cancer Med.* **2018**, *7*, 3743–3754. [CrossRef] [PubMed]
170. Wang, W.; Chapman, N.M.; Zhang, B.; Li, M.; Fan, M.; Laribee, R.N.; Zaidi, M.R.; Pfeffer, L.M.; Chi, H.; Wu, Z.H. Upregulation of PD-L1 via HMGB1-Activated IRF3 and NF-κB Contributes to UV Radiation-Induced Immune Suppression. *Cancer Res.* **2019**, *79*, 2909–2922. [CrossRef] [PubMed]
171. Zhang, X.; Shi, H.; Yuan, X.; Jiang, P.; Qian, H.; Xu, W. Tumor-derived exosomes induce N2 polarization of neutrophils to promote gastric cancer cell migration. *Mol. Cancer* **2018**, *17*, 146. [CrossRef] [PubMed]
172. Sheller-Miller, S.; Urrabaz-Garza, R.; Saade, G.; Menon, R. Damage-Associated molecular pattern markers HMGB1 and cell-Free fetal telomere fragments in oxidative-Stressed amnion epithelial cell-Derived exosomes. *J. Reprod. Immunol.* **2017**, *123*, 3–11. [CrossRef]
173. Song, S.; Min, H.; Niu, M.; Wang, L.; Wu, Y.; Zhang, B.; Chen, X.; Liang, Q.; Wen, Y.; Wang, Y.; et al. S1PR1 predicts patient survival and promotes chemotherapy drug resistance in gastric cancer cells through STAT3 constitutive activation. *EBioMedicine* **2018**, *37*, 168–176. [CrossRef]
174. Yang, S.; Yang, C.; Yu, F.; Ding, W.; Hu, Y.; Cheng, F.; Zhang, F.; Guan, B.; Wang, X.; Lu, L.; et al. Endoplasmic reticulum resident oxidase ERO1-Lalpha promotes hepatocellular carcinoma metastasis and angiogenesis through the S1PR1/STAT3/VEGF-A pathway. *Cell Death Dis.* **2018**, *9*, 1105. [CrossRef] [PubMed]
175. Lankadasari, M.B.; Aparna, J.S.; Mohammed, S.; James, S.; Aoki, K.; Binu, V.S.; Nair, S.; Harikumar, K.B. Targeting S1PR1/STAT3 loop abrogates desmoplasia and chemosensitizes pancreatic cancer to gemcitabine. *Theranostics* **2018**, *8*, 3824–3840. [CrossRef] [PubMed]

Review

How to Make Immunotherapy an Effective Therapeutic Choice for Uveal Melanoma

Mariarosaria Marseglia [1,†], Adriana Amaro [1,†], Nicola Solari [1], Rosaria Gangemi [1], Elena Croce [1,2], Enrica Teresa Tanda [1,2], Francesco Spagnolo [1], Gilberto Filaci [1,2], Ulrich Pfeffer [1,*] and Michela Croce [1,*]

1. IRCCS Ospedale Policlinico San Martino, 16132 Genoa, Italy; MariaRosaria.Marseglia@hsanmartino.it (M.M.); adriana.amaro@hsanmartino.it (A.A.); nicola.solari@hsanmartino.it (N.S.); rosaria.gangemi@hsanmartino.it (R.G.); croceelena91@gmail.com (E.C.); enrica.tanda@gmail.com (E.T.T.); francesco.spagnolo85@gmail.com (F.S.); gfilaci@unige.it (G.F.)
2. Department of Internal Medicine and Medical Specialties (DiMI), University of Genoa, 16132 Genoa, Italy
* Correspondence: ulrich.pfeffer@hsanmartino.it (U.P.); michela.croce@hsanmartino.it (M.C.); Tel.: +39-392 202 6881 (U.P.); +39 0105558376 (M.C.); Fax.: +39 0105558374 (M.C.)
† These authors contributed equally.

Simple Summary: Despite improvements in the early identification and successful control of primary uveal melanoma, 50% of patients will develop metastatic disease with only marginal improvements in survival. This review focuses on the tumor microenvironment and the cross-talk between tumor and immune cells in a tumor characterized by low mutational load, the induction of immune-suppressive cells, and the expression of alternative immune checkpoint molecules. The choice of combining different strategies of immunotherapy remains a feasible and promising option on selected patients.

Abstract: Uveal melanoma (UM), though a rare form of melanoma, is the most common intraocular tumor in adults. Conventional therapies of primary tumors lead to an excellent local control, but 50% of patients develop metastases, in most cases with lethal outcome. Somatic driver mutations that act on the MAP-kinase pathway have been identified, yet targeted therapies show little efficacy in the clinics. No drugs are currently available for the *G protein alpha subunits GNAQ* and *GNA11*, which are the most frequent driver mutations in UM. Drugs targeting the YAP–TAZ pathway that is also activated in UM, the tumor-suppressor gene *BRCA1 Associated Protein 1 (BAP1)* and the *Splicing Factor 3b Subunit 1 gene (SF3B1)* whose mutations are associated with metastatic risk, have not been developed yet. Immunotherapy is highly effective in cutaneous melanoma but yields only poor results in the treatment of UM: anti-PD-1 and anti-CTLA-4 blocking antibodies did not meet the expectations except for isolated cases. Here, we discuss how the improved knowledge of the tumor microenvironment and of the cross-talk between tumor and immune cells could help to reshape anti-tumor immune responses to overcome the intrinsic resistance to immune checkpoint blockers of UM. We critically review the dogma of low mutational load, the induction of immune-suppressive cells, and the expression of alternative immune checkpoint molecules. We argue that immunotherapy might still be an option for the treatment of UM.

Keywords: uveal; immunotherapy; *BAP1*; tumor microenvironment; anti-PD-1; anti-CTLA-4; TIL

1. Introduction

Uveal melanoma (UM) is the most common intraocular malignancy of adulthood. UM originates from melanocytes of the uvea, including the iris, ciliary body, and retinal choroid. Despite improvements in early identification and successful control of the primary tumor, approximately 20–30% of the patients develop metastatic disease within 5 years from diagnosis, while at 15 years, the percentage rises to 45%. UM metastatic sites are the liver, lung, soft tissue, and bone [1,2]. Most frequently, metastases involve the liver as the first or only target tissue, and untreated patients have a mean survival time of about

2 months that rises to close to 6 months upon treatment [1,3,4]. Distinct UM subtypes with different clinical outcomes and prognoses have been defined on the basis of various pathological parameters, with the contribution of different genetic abnormalities, through studies of gene expression profiles and The Cancer Genome Atlas (TCGA). Several driver mutations have been found, involving mainly *G protein alpha subunits GNAQ* and *GNA11* or, in a minor fraction of UM cases, the *Cysteinyl Leukotriene Receptor 2 (CYSLTR2)* [5], and the *Phospholipase C Beta 4 (PLCB4)* [6] genes. Mutations in *GNAQ* and *GNA11* are present in 75–95% of cases and occur early in the development of UM [2,7]. These mutations are mutually exclusive and lead to the constitutive activation of G alpha protein, which in turn leads to the activation of several downstream effectors, thus promoting cell growth and proliferation [8]. GNAQ and GNA11 activate the Phospholipase C/Protein Kinase C (PLC/PKC) pathway and several downstream signaling pathways, including the Rapidly Accelerated Fibrosarcoma/mitogen-activated protein kinase kinase/extracellular signal-regulated kinase (RAF/MEK/ERK), Phosphoinositide 3-kinase/AKT Serine/Threonine Kinase/Mechanistic Target Of Rapamycin Kinase (PI3K/AKT/MTOR), and Trio Rho Guanine Nucleotide Exchange Factor/Ras homologue family member/Rac family small GTPase 1/Yes associated protein 1 (Trio/Rho/Rac/YAP1) pathways [2]. Several molecules, such as CXCR4, c-MET, Hypoxia Inducible Factor 1 (HIF-1), and insulin-like-growth factor-1 (IGF-1) are involved in UM metastatic progression and thus considered as a target for new treatments [2]. Additional mutations in the calcium-signaling pathway, to which also *GNAQ* and *GNA11* belong, might also influence tumorigenesis [9].

The monosomy of chromosome 3 [10,11], loss of chromosome 3 heterozygosity [12], and inactivating mutations of the *BRCA1-associated protein 1 (BAP1)* oncosuppressor gene [13] are strongly associated with metastatic risk. On the contrary, somatic mutations in *Eukaryotic Translation Initiation Factor 1A X-Linked (EIF1AX)* and *Splicing Factor 3b subunit 1 (SF3B1)* genes prevalently occur in UM with disomy 3 [14,15]. According to data from whole-genome sequencing (WGS) and Sanger sequencing, *SF3B1*, *EIF1AX*, and *BAP1* mutations classify UM patients in different categories with different survival and metastatic risk. *EIF1AX* mutations are not associated with risk of metastasis and show, similar to tumors without *BAP1* and *SF3B1* mutations, prolonged survival. UM-bearing mutated *SF3B1* undergoes metastatic progression later, and tumors with mutated *BAP1* metastasize early and rapidly progress with poor survival rates [16]. *BAP1* is a tumor-suppressor gene located on chromosome 3; it encodes a deubiquitinating enzyme with tumor-suppressive activity [17,18]. Inactivating mutations of *BAP1* occur in nearly half of UM patients and approximately 84% of metastatic cases [13]. *BAP1* loss-of-function mutations correlate with a distinct DNA methylation profile [19]. Finally, germline *BAP1* mutations are associated with an early and increased incidence of UM [20] but also with an increased incidence of other malignancies [21]. Many secondary mutations were found by next-generation sequencing to occur in UM patients in the same G-protein-related pathways known as drivers, in particular in the calcium-signaling pathway [9]. These secondary driver mutations are likely to affect tumor development and progression.

Amplifications of the long arm of chromosome 8 confer an increased risk of metastasis in UM. Several genes such as *V-Myc Avian Myelocytomatosis Viral Oncogene Homolog (MYC)* and *Ankyrin Repeat and PH Domain 1 (ASAP1)*, located on the long arm of chromosome 8 have been proposed as mediators of the effects of 8q amplification [22]. Chromosome 6p amplifications exert a protective effect yet the molecular basis thereof has not been fully elucidated [22].

It is widely accepted that tumor mutational burden is an important biomarker to predict response to immune checkpoint blockers (ICB) in tumors. In UM, both primary tumor and metastases carry one of the lowest mutation burdens in adult solid tumors [23]. UM displays a mean mutation rate of 0.5 mutations per megabase (Mb) [6], as opposed to 49.2 in cutaneous melanoma (CM) [24]. The role of UV light has been proposed as the major cause for the differences in UM and CM mutational burden and a UV-associated mutational signature is expressed in CM [9,24,25]. Metastases from iris-UM, though rare, display a higher mutation load than the average of UM [24,26], and they are also connected to a UV signature [24]. The presence of germline mutations of *methyl-CpG-binding domain protein 4* (*MBD4*) was detected in a group of UM patients who experienced a disease stabilization and prolonged survival after ICB immunotherapy [27,28], thus suggesting a role for *MBD4* as a new predictor of response to immunotherapy in UM [29]. *MBD4* is thought to act as a tumor suppressor gene; it is located on chromosome 3, and mutations have recently been identified in approximately 2% of UM characterized by a high mutational burden and hypermutated tumors [27,29].

Treatment of primary UM (P-UM) consists in surgery or radiation. It has a low local recurrence rate, but almost 50% of the patients develop metastatic disease, prevalently to the liver [1]. At present, there are no effective therapies for metastatic UM (M-UM), and most patients survive less than 12 months after diagnosis of metastases [30,31]. Different therapeutic strategies, including targeted, immunotherapeutic, chemotherapeutic, and epigenetic, have been or are currently being investigated. Among different strategies pursued in clinical trials for UM, immunotherapy was the most promising, given the striking impact it had on CM patients' survival [32]. We refer to other recent reviews [33] for deeper insights into UM classification, epidemiology, genetic, and epigenetic [2,22,34], because this is beyond the purpose of this review.

In this paper, we review recent advances in innovative immune therapy options for UM in adjuvant and metastatic settings and develop perspectives for translating them in clinical practice. Special issues concerning an immune-suppressive tumor microenvironment (TME), poor mutational load and antigen expression, and signatures defining patients' responses will be addressed to define new immune therapeutic strategies for M-UM.

2. Immunobiology of Uveal Melanoma

The Melanoma Antigen Gene (MAGE) family proteins, tyrosinase, and gp100 are UM tumor-associated antigens (TAA) that are recognized by cells of the immune system [35]. Indeed, peripheral CD8+ cells from UM patients and tumor-infiltrating lymphocytes (TILs) can lysate UM cells in vitro [36,37]. Nevertheless, the immune privilege of the eye allows UM cells to escape the control of the immune system.

The most frequent site of UM metastases is the liver, but the mechanism that guides the liver tropism of UM remains elusive. The immunomodulatory nature of the liver is determined by its exposure to food antigens, allergens, and low levels of endotoxins, deriving from the gut. The liver microenvironment is composed of resident non-immune and immune cells, such as hepatocytes, liver sinusoidal endothelial cells (LSECs), Kupffer cells (KCs), T, NK, and NKT cells that strictly regulate the balance between tolerance and the defense against pathogens. UM cells that have escaped from the eye find further protection in the immune-modulatory microenvironment of the liver. Detailed mechanisms of immunosuppression in the eye and the liver will be described below.

2.1. Immunosuppressive Mechanisms in the Eye

Different mechanisms may contribute to immune suppression in UM, among which the site in which UM arises. The eye is a physiologically immune-privileged organ in order to protect it from destructive inflammation that may impair vision. This immune-privilege is maintained through different mechanisms, among which physical barriers such as the blood–retina barrier and the absence of efferent lymphatics [38,39].

Anterior chamber-associated immune deviation (ACAID), though difficult to be studied in humans, has been shown in different animal models, and it is responsible for the induction of complex immunoregulatory mechanisms and cells [33,40]. Characteristic of ACAID are the inhibition of Th1 differentiation and delayed-type hypersensitivity (DTH) [41].

A general immunosuppressive milieu in the eye avoids non-specific inflammatory reactions and immune responses. It is caused by the release of soluble factors (i.e., transforming growth factor-beta, TGF-β [42]), low MHC expression, the presence of neuropeptides, and expression of FAS ligand [43]. Primed T cells, activated in vitro in the presence of the aqueous humor, were reprogrammed to TGF-β producing regulatory T cells (Treg) and acquired immunosuppressive skills [42]. The aqueous humor also contains the pleiotropic cytokine Macrophage Migration Inhibitory Factor (MIF), which promotes immune privilege by inhibiting NK cell activity [44]. Finally, iris and ciliary body epithelial cells can prevent T cell activation and proliferation via direct cell-to-cell contact [45]. Specifically, in P-UM, soluble HLA class I (sHLA-I) has been detected in the anterior chamber aqueous humor and has been considered a prognostically unfavorable sign that may influence local immune responses. Indeed, sHLA-I was detected in monosomy 3 tumors, with gain of 8q and loss of BAP1 protein expression known to have a poor prognosis [46]. The immune-suppressive microenvironment of the eye is assumed to generate a niche in which UM can grow and proliferate without the pressure of both innate and adaptive immune cells until it breaks the blood–retina barrier and disseminates. Innate cells, especially NK cells, are believed to be able to prevent metastases or to kill tumor cells in the blood before they could reach the liver [47]. However, after leaving the eye, the ability of UM cells to express pro-oncogenic molecules such as indoleamine dioxygenase-1 (IDO-1, [48]), MIF [49], and PD-L1 [50] enhance their metastatic potential.

2.2. Immunosuppressive Mechanisms in the Liver

Considering that metastatic disease in UM patients may be diagnosed many years after the primary tumor, it has been proposed that UM cells that leave the eye and reach the liver remain stable for years until proliferation occurs. This characteristic has been called "UM cell dormancy" and implies that the disease was already disseminated at the time of diagnosis [51]. Dormant UM cells are quiescent cells blocked in the cell cycle that only occasionally undergo cell division, which is an adaptive mechanism used by cells in a hostile microenvironment. Dormancy consists in the regulation of cellular proliferation and includes autophagy, interaction with the extracellular matrix, hypoxia, impaired angiogenesis, inflammation, and immunity [51,52]. Liver UM metastases have been described, based on their growth pattern, as either infiltrative or nodular. The infiltrative pattern is characterized by UM cells lacking vascular endothelial growth factor (VEGF) expression, invading liver sinusoidal space, and creating pseudo-sinusoidal spaces for oxygen and nutrient supply. Differently, the nodular growth pattern arises in the peri-portal area, involves portal veins and, as the lesion becomes hypoxic, cells express Matrix Metallopeptidase 9 (MMP9) and VEGF, thus developing angiogenetic properties [53].

UM cells become resistant to NK cell-mediated cytolysis in the metastatic niche in the liver by producing TGF-β upregulating MHC-I molecules [54] and downregulating NK activating ligands for NKG2D [55]. Hepatic stellate cells are supposed to contribute to UM niche in the liver; they are recruited by UM cells and secrete pro-inflammatory factors and collagen [56].

2.3. Tumor-Infiltrating Lymphocytes

The presence of TILs is a marker of good prognosis for many cancers but not in UM where it is associated with a poor prognosis [57,58]. Why this is so is not fully understood, as there are contradictory reports on the immune cell subtypes populating liver metastases in UM [59–63]. It is of note that most studies that tried to characterize the immunosuppressive environment in UM metastases have been performed at the transcriptomic level on only very few immune cells. Robertson et al. [19] proposed a stratification based on CD8+ T-cell immune infiltrates and an altered transcriptional immune profile for P-UM bearing monosomy 3 and *BAP1* loss of function mutations. Using RNA-seq analysis, they showed an upregulation of CD8+ T cell-related genes in almost 30% of monosomic UM that was not observed in disomic cases. In addition, genes involved in interferon-γ (IFN-γ) signaling, T cell invasion, cytotoxicity, and immunosuppression were co-expressed with *CD8A*, as well as with *HLA* genes [19].

Chromosome 8q amplification is related to macrophage infiltration, and the loss of *BAP1* expression is associated with T cell infiltration in UM [64]. TILs do not seem to be cytotoxic CD8+ but mostly regulatory CD8+ T lymphocytes [65]. Moreover, *BAP1* loss correlated with the upregulation of several genes associated with a suppressive immune response, including *HLA-DRA*, *CD38*, and *CD74*, both in primary and metastatic tumors. Digital spatial profiling, a genomic analysis that maintains the spatial information of UM metastases, showed tumor-associated macrophages (TAMs) and TILs entrapped within peritumoral fibrotic areas expressing *IDO1*, *PD-L1*, and *β-catenin* (*CTNNB1*) [65]. Qin et al. [60] confirmed the more immunosuppressive TME in M-UM and found intra-tumoral rather than peripheral CD8+ infiltrates. However, a study considering 35 archival formalin-fixed, paraffin-embedded M-UM specimens described a tumor microenvironment in which M2-macrophages were the dominant subtype, CD4+ TILs were perivascular, and CD8+ lymphocytes were mainly peritumoral [59], suggesting that immune cells cannot invade the tumor to attack tumor cells. Recently, Coupland and coworkers classified UM hepatic metastases in four different groups: 'absent/cold' metastases with no TILs or TAMs in the tumor or at the tumor-normal liver interface, 'altered immunosuppressive' with a low scattered pattern of inflammatory cell infiltrate, 'altered excluded' where infiltrates of TILs or TAMs were low at the tumor center but high at the margin, and 'high/hot' where high infiltration of TILs or TAMs was present throughout the metastatic tissue [63]. The authors concluded that the predominant cell types present in M-UM and responsible for the immunosuppressed environment were M2-type TAMs and exhausted CD8+ TILs. Moreover, the absence of PD-L1 expression on UM tumors may explain the failure of anti-PD-1 monotherapy [60,63]. Indeed, several reports [26,59,62,65] and our unpublished observations find an elevated infiltration of CD8+ TIM-3+ and LAG-3+, but PD-1 negative cells suggesting that immune resistance in UM may occur via alternative immune checkpoints.

MART-1 and/or gp100 antigen-specific T cells were expanded in vitro from biopsy-derived TILs with IL-2. T cells displayed exhausted phenotype (PD-1+, CD39+, TIM-3+, TIGIT+, and LAG-3+) [26]. Similar results were obtained using single-cell (sc)RNA-sequencing by Durante et al. [62], who detected clonally expanded T cells and/or plasma cells in UM samples. Altogether, these data indicate that TILs may have mounted a response, despite the low tumor mutational burden.

TILs from a subset of a total of 13 UM patients have been identified and showed robust anti-tumor reactivity, similar to that frequently observed in TILs from CM patients. Interestingly, the number of TILs recovered from UM and CM were similar, but after two weeks of culture in the presence of IL-2, UM-derived TIL cultures were mainly CD4+T cells and produced IFN-γ in response to parental tumor cells [66]. In another setting, TILs from UM metastases from 5 patients were successfully expanded in vitro applying an agonistic anti-4-1BB and OKT3 antibodies (anti-CD3) with high dose IL-2 in a small device to produce immune cells for clinical use. The authors report that this method allows the proliferation of TILs in a short time frame, and TILs obtained after such expansion were mostly CD8+, not overly differentiated. The ability of these TILs to recognize and respond to autologous tumor cells was successfully pursued by the authors only in one case where TILs produced a discrete amount of IFN-γ [67].

The efficacy of in vitro expanded autologous TILs from UM metastasis in patients was addressed in a phase II clinical trial (ClinicalTrials.gov Identifier: NCT01814046) enrolling a total of 20 patients. Reinfusion of TILs after a non-myeloablative lymphodepleting conditioning regimen could induce objective tumor regression in 7/20 (35%) M-UM patients. Among the responders, one highly pre-treated patient demonstrated a durable complete regression of numerous hepatic metastases for two years (Table 1) [68]. Johansson et al. [69] found a direct correlation between the high infiltration of CD8+ T cells and macrophages with longer overall survival in patients before treatment with hyperthermic isolated hepatic perfusion (IHP). This is the only report indicating a positive correlation between the presence of immune cells and survival, although this may be related to the low numbers of metastatic biopsies studied.

Table 1. Immunotherapy in UM: published studies.

Study	Type of Study	Targeted Patients	No Patients (UM Patients)	ORR	Median OS	Median PFS	Rate 1-Year Surv	6 Months PFS	PR	CR	SD
Khoja [70]	Meta-analysis (2000–2016)	metastatic uveal melanoma	(912)	-	10.2	3.3	43%	27%	-	-	-
Chandran [68]	Phase II ClinicalTrials.gov Identifier: NCT01814046 (autologous TILs)	Metastatic Ocular Melanoma Metastatic Uveal Melanoma	(21)	NE	NE	NE	NE	NE	30%	5%	NE
Klemen [71]	Retrospective (Ipilimumab+Nivolumab)	metastatic melanoma	428 (30)	-	12.2	-	-	-	-	-	-
Bol [72]	Retrospective (Ipilimumab+Nivolumab)	metastatic UM	(126) Ipilimumab/Nivolumab $n = 19$	-	18.9	3.7	57.6%	3.7	21.1%	0	10.5%
Heppt [73]	Retrospective (Ipilimumab+Nivolumab)	metastatic or unresectable UM	(64) Ipi+nivo55	15.6	16.1	3	-	-	-	-	-
Kirchberg [74]	Real world (Ipilimumab+Nivolumab)	metastatic melanoma	33(9)	-	18.4	-	-	-	0	0	56%
Pfulats [75]	Phase II ClinicalTrials.gov Identifier: NCT02626962 (Ipilimumab+Nivolumab)	Metastatic uvela melanoma	(52)	-	12.7	3	51.9%	-	9.6%	1.9%	-
Pelster [76]	Phase II ClinicalTrials.gov Identifier: NCT01585194 (Ipilimumab+Nivolumab)	Metastatic uveal melanoma	(35)	18%	19.1	5.5	56%	-	15%	3%	33%
Middleton [77]	Phase I/II ClinicalTrials.gov Identifier:NCT01211262 (Tebentafusp)	Advanced melanoma	84 (19)	-	-	-	65%	-	16.6%	0	44.4%

Abbreviations: NE: not evaluated; ORR: overall response rate; PFS: progression free survival; surv: survival; OS: overall survival; PR: partial response; CR: complete response; SD: stable disease.

In summary, both P- and M-UM TILs display a phenotype mostly immunosuppressive or exhausted, and subsets of M-UM patients possess TILs that are antigen-specific and thus may potentially be responsive to immunotherapy. The use of antibodies/inhibitors of appropriate immune checkpoint expressed by M-UM may be the therapeutic option to be pursued, at least in a subset of UM patients.

2.4. Alternative Immune Checkpoint

The PD-1/PD-L1 immune checkpoint seems not to be as frequently upregulated in UM as in CM metastases; therefore, criticism on the strength of the rationale for this checkpoint blockade in UM has been raised [61]. Consistently, results from clinical trials with anti-PD-1 ICB are not so brilliant for M-UM patients [72]. This stimulated the search for new immune checkpoints, exhaustion markers, or immunosuppressive molecules that may become potential targets to be studied in clinical trials. The expression of the immunosuppressive molecule, IDO, and multiple immune checkpoint molecules, such as Vista, TIGIT, and LAG-3 on TILs in UM metastases has been shown [26,65]. TILs isolated from metastases and expanded in vitro, analyzed by flow cytometry, displayed in several cases tumor-reactive subsets of immune cells expressing the checkpoint receptors PD-1, TIM-3, LAG-3, and, to some extent, TIGIT [26]. The dominant exhaustion marker identified in UM was LAG-3 as analyzed by scRNA-seq and immunohistochemistry (IHC). This explains at least in part the failure of checkpoint blockade targeting CTLA-4 and PD-1. LAG-3 was found expressed mainly on CD8+ T cells but was also detected on some CD4+ T cells, FOXP3+ regulatory T cells, NK cells, and macrophages/monocytes [62]. Fusion protein and inhibitors of LAG-3 are in development or already tested in clinical trials either as a single agent or in association with anti-PD-L1, in different cancers, including UM (ClinicalTrials.gov Identifier: NCT02519322) (Table 2) [78].

Table 2. Ongoing clinical trials with adjuvant and not adjuvant therapies in UM.

ClinicalTrials.gov Identifier: NCT Number	Trial (Adjuvant)	Status	Phase	Targeted Patients	Actual Enrollment (Estimated Enrollment)	Principal Investigator	First Submitted Date	Last Update Posted Date
NCT02068586	A Randomized Phase II Study of Adjuvant Sunitinib or Valproic Acid in High-Risk Patients With Uveal Melanoma	Recruiting	Phase II	Ciliary Body and Choroid Melanoma Iris Melanoma Intraocular Melanoma	(150)	Takami Sato	19 February 2014	7 January 2021
NCT02223819	Phase II Trial of Adjuvant Crizotinib in High-Risk Uveal Melanoma Following Definitive Therapy	Active, not recruiting	Phase II	Uveal Melanoma	34 (30)	Richard Carvajal	20 August 2014	18 December 2019
NCT01983748	A non-commercial, multicenter, randomized, two-armed, open-label phase III study to evaluate the adjuvant vaccination with tumor RNA-loaded autologous dendritic cells versus observation of patients with resected monosomy 3 uveal melanoma	Recruiting	Phase III	Uveal Melanoma	(200)	Beatrice Schuler-Thurner	17 September 2013	6 January 2020
NCT01100528 [79]	Adjuvant Therapy for Patients With Primary Uveal Melanoma With Genetic Imbalance (Dacarbazine+IFNa-2B)	Completed	Phase II	Iris, Ciliary Body or Choroidal Melanoma	38(36)	Yogen Saunthararajah	7 April 2010	26 February 2019
NCT02519322 [78]	Neoadjuvant and Adjuvant Checkpoint Blockade (Ipi+Nivo+Relatlimab)	Recruiting	Phase II	Cutaneous Melanoma Mucosal Melanoma Ocular Melanoma	(53)	Rodabe N Amaria	4 August 2015	30 December 2020
NCT00254397	Study of the Modulatory Activity of an LHRH-Agonist (Leuprolide) on Melanoma Peptide Vaccines as Adjuvant Therapy in Melanoma Patients	Completed	Phase II	Melanoma	98	Patrick Hwu	14 November 2005	16 October 2019
NCT01989572 [80]	A Randomized, Placebo-Controlled Phase III Trial of Yeast Derived GM-CSF Versus Peptide Vaccination Versus GM-CSF Plus Peptide Vaccination Versus Placebo in Patients With "No Evidence of Disease" After Complete Surgical Resection of "Locally Advanced" and/or Stage IV Melanoma	Completed	Phase III	Ocular melanoma Cutaneous Melanoma Mucosal melanoma	815	David H Lawson	18 November 2013	7 July 2020

Table 2. Cont.

ClinicalTrials.gov Identifier: NCT number	Trial (other not adjuvant immunological therapies)	Status	Phase	Targeted Patients	Actual enrollment (Estimated enrollment)	Principal Investigator	First Submitted Date	Last Update Posted Date
NCT03070392	A Phase II Randomized, Open-label, Multi-center Study of the Safety and Efficacy of IMCgp100 Compared With Investigator Choice in HLA-A*0201 Positive Patients With Previously Untreated Advanced Uveal Melanoma	Active, not recruiting	Phase II	Uveal Melanoma	378 (327)	Mohammed Dar	14 February 2017	6 January 2021
NCT02570308	A Study of the Intra-Patient Escalation Dosing Regimen With IMCgp100 in Patients With Advanced Uveal Melanoma	Active, not recruiting	Phase I/Phase II	Uveal Melanoma	(150)	Not Provided	6 October 2015	6 January 2021
NCT03467516	A Phase II Study to Evaluate the Efficacy and Safety of Adoptive Transfer of Autologous Tumor-Infiltrating Lymphocytes in Patients With Metastatic Uveal Melanoma	Recruiting	Phase II	Uveal Neoplasms Melanoma, Uveal	(59)	Udai S Kammula	9 March 2018	18 February 2020
NCT00986661	A Phase I Study to Assess the Safety, Tolerability, and Pharmacokinetics of PV-10 Chemoablation of Cancer Metastatic to the Liver or Hepatocellular Carcinoma Not Amenable to Resection or Transplant	Recruiting	Phase I	Cancer Metastatic to the Liver Hepatocellular Carcinoma Metastatic Melanoma Metastatic Ocular Melanoma Metastatic Uveal Melanoma Metastatic Lung Cancer Metastatic Colon Cancer Metastatic Colorectal Cancer Metastatic Breast Cancer Metastatic Pancreatic Cancer	(78)	Eric Wachter	24 September 2009	5 March 2020
NCT01211262 [77]	A Phase I, Open-Label, Dose-Finding Study to Assess the Safety and Tolerability of IMCgp100, a Monoclonal T Cell Receptor Anti-CD3 scFv Fusion Protein in Patients With Advanced Malignant Melanoma	Completed	Phase I	Malignant Melanoma	84(50)	Namir Hassan	28 September 2010	8 July 2020
NCT04262466	Phase I/II Study of IMC-F106C in Advance PRAME-Positive Cancers	Recruiting	Phase I/Phase II	Select Advanced Solid Tumors	(170)	Shaad Abdullah, FACP	30 January 2020	16 February 2021
NCT02743611	A Phase I/II Dose-Finding Study to Evaluate the Safety, Feasibility, and Activity of BPX-701, a Controllable PRAME T-Cell Receptor Therapy, in HLA-A2+ Subjects With AML, Previously Treated MDS, or Metastatic Uveal Melanoma	Active, not recruiting	Phase I/Phase II	Acute Myeloid Leukemia Myelodysplastic Syndrome Uveal Melanoma	28 (36)	Bellicum Pharmaceuticals Senior Director	11 April 2016	27 April 2020

Table 2. Cont.

ClinicalTrials.gov Identifier: NCT number	Trial (other not adjuvant immunological therapies)	Status	Phase	Targeted Patients	Actual enrollment (Estimated enrollment)	Principal Investigator	First Submitted Date	Last Update Posted Date
NCT02697630 [81]	A Multicenter Phase II Open-Label Study to Evaluate Efficacy of Concomitant Use of Pembrolizumab and Entinostat in Adult Patients With Metastatic Uveal Melanoma	Active, not recruiting	Phase II	Metastatic Uveal Melanoma	(29)	Not Provided	22 February 2016	16 October 2019
NCT00338377 [82]	Lymphodepletion Plus Adoptive Cell Transfer With or Without Dendritic Cell Immunization in Patients With Metastatic Melanoma	Recruiting	Phase II	Melanoma	(189) 5 MU (primary site choroid)	Rodabe N. Amaria	10 February 2006	9 December 2020
NCT03635632	Phase I Study of Autologous T Lymphocytes Expressing GD2-specific Chimeric Antigen and Constitutively Active IL-7 Receptors for the Treatment of Patients With Relapsed or Refractory Neuroblastoma and Other GD2 Positive Solid Cancers(GAIL-N)	Recruiting	Phase I	Relapsed Neuroblastoma Refractory Neuroblastoma Relapsed Osteosarcoma Relapsed Ewing Sarcoma Relapsed Rhabdomyosarcoma Uveal Melanoma Phyllodes Breast Tumor	(94)	Bilal Omer	13 August 2018	9 December 2020
NCT03865212	Phase I Trial to Evaluate the Safety and Efficacy of Intratumoral and Intravenous Injection of Vesicular Stomatitis Virus Expressing Human Interferon Beta and Tyrosinase Related Protein 1 (VSV-IFNb-TYRP1) in Patients With Metastatic Ocular Melanoma and Previously Treated Patients With Unresectable Stage III/IV Cutaneous Melanoma	Recruiting	Phase I	Clinical Stage III Cutaneous Melanoma AJCC v8 Clinical Stage IV Cutaneous Melanoma AJCC v8 Metastatic Choroid Melanoma Metastatic Melanoma Metastatic Mucosal Melanoma Metastatic Uveal Melanoma Pathologic Stage III Cutaneous Melanoma AJCC v8 Pathologic Stage IIIA Cutaneous Melanoma AJCC v8 Pathologic Stage IIIB Cutaneous Melanoma AJCC v8 Pathologic Stage IIIC Cutaneous Melanoma AJCC v8 Pathologic Stage IIID Cutaneous Melanoma AJCC v8 Pathologic Stage IV Cutaneous Melanoma AJCC v8 Unresectable Melanoma	(72)	Roxana S Dronca	6 March 2019	18 November 2020

3. Immune Checkpoint Inhibitors: Retrospective, Real-World Studies, and Clinical Trials

There is no consensus on the standard treatment of UM, and the correct management of this disease remains a matter of discussion. To determine progression-free and overall survival benchmarks, Khoia et al. reported in 2019 [70] a meta-analysis of 912 M-UM patients from 29 trials published from 2000 to 2016. Among the selected trials, five studies used immunotherapy and only three of them used anti-CTLA-4. Considering the whole population, the median progression-free survival (PFS) was 3.3 months, the median overall survival (OS) was 10.2 months, and the 1-year OS rate was 43%. Liver-directed therapies appeared in this study as the best treatments.

UM is genetically and biologically different from CM [22,33] and is barely immunogenic due to its low number of mutations [6]. Surprisingly, a phase II trial (ClinicalTrials.gov Identifier: NCT01814046) with 21 M-UM patients treated with lymph-depleting chemotherapy (cyclophosphamide followed by fludarabine) and a single intravenous infusion of autologous TILs with high-dose IL-2, showed exciting results (Table 1) [68]. In this study, 7 (35%) patients demonstrated tumor regression, with 6 (30%) achieving a partial response (PR) and 1 achieving complete response (CR) (5%), justifying further investigations of other immunological approaches. A subsequent clinical trial with autologous TILs and IL-2 therapy in M-UM (ClinicalTrials.gov Identifier: NCT03467516) and another one in metastatic CM and UM (ClinicalTrials.gov Identifier: NCT00338377) are ongoing (Table 2).

An interesting approach is the use of dendritic cell (DC) vaccination in an adjuvant setting. The immunologic responses after adjuvant DC vaccination were studied in an open-label phase II clinical trial with high-risk UM. An increase in OS was observed in patients with a tumor antigen-specific immune response [83]. In addition, a multicenter, randomized, two-armed, open-label phase III study is currently ongoing to evaluate the adjuvant vaccination with tumor RNA-loaded autologous DC in patients with resected monosomy 3 UM (ClinicalTrials.gov Identifier: NCT01983748) (Table 2). A phase I trial is studying the side effects and best dose of a modified virus called Vesicular Stomatitis Virus, VSV-IFNbetaTYRP1 in patients with stage III-IV melanoma including M-UM (ClinicalTrials.gov Identifier: NCT03865212) (Table 2). The VSV has been modified to express two extra genes: *IFN-beta* and *TYRP1*. IFN-β may protect normal healthy cells from becoming infected with the virus and improve the antitumor efficacy due to its intrinsic antiproliferative effects and tyrosinase-related protein 1 (TYRP1) is a tumor-associated antigen expressed both in CM and UM.

Single ICB, anti-CTLA-4, or anti-PD-1 therapy gave only limited results in terms of efficacy in patients with M-UM with an overall response rate (ORR) that ranged from 0.5 to 6% [84]. Better results were expected from the combination of the two monoclonal antibodies. A real-world study [74] analyzed retrospectively 9 UM patients treated with low-dose anti-CTLA-4 (Ipilimumab, ipi) (1 mg/kg) and standard-dose anti-PD-1 (Pembrolizumab) (2 mg/kg). Median OS was 18.4 months with neither CR nor PR (0/9). No deaths for treatment-related adverse events occurred; however, 18% of patients had at least one grade 3 or 4 toxicity (Table 1).

A retrospective analysis by Klemen et al. [71] reported a single institutional experience using antibodies against CTLA-4, PD-1, and/or PD-L1 to treat 428 patients with metastatic melanoma histologically diagnosed as cutaneous, unknown, acral, mucosal, or uveal. For the 30 patients with M-UM, median OS was 12.2 months, and 5-year OS was 22%. Most of the longer survivors received both anti-CTLA-4 and anti-PD-1 or anti-PD-L1 either sequentially or in combination (Table 1). Clinical retrospective data of 126 patients diagnosed with M-UM in Denmark were analyzed before (pre-ICB, n = 32) and after (post-ICB, n = 94) the approval of first-line treatment with ICB [72]. The study shows a significant improvement of survival in patients post-ICB therapy: the combined ICB treatment (19 patients) achieved 18.9 months median OS and 57.6% of 1-year OS rate (Table 1). A multi-center retrospective study [73] analyzed 64 M-UM patients, 50 of which received combined checkpoint blockade as first-line systemic therapy. The median PFS was 3.0 months and the median OS was estimated to 16.1 months with an ORR of 15.6%. Severe treatment-related adverse events were experienced by 39.1% of patients (Table 1).

These retrospective studies showed better results than those obtained with Ipilimumab or anti-PD-1 (Nivolumab) monotherapy and established the basis for prospective clinical trials. At present, March 2021, there are 7 clinical trials involving combination immunotherapy listed by www.clinicaltrials.gov (Table 3).

A phase I pilot study (ClinicalTrials.gov Identifier: NCT03922880) plans to combine arginine depletion and ICB. Four phase I/II trials combine local liver therapy or immunoembolization with systemic administration of Ipilimumab and Nivolumab (Table 3). An open-label phase I basket study (ClinicalTrials.gov Identifier: NCT00986661, Table 2) is evaluating the safety and preliminary efficacy of intra-lesion PV-10 in patients with solid tumors of the liver including UM metastases. PV-10, a small molecule that accumulates in lysosomes inducing autolysis, can produce immunogenic cell death and therefore a T cell-mediated immune response against immunologically cold tumors, providing a rationale for the association with ICBs. Preliminary results were presented for 13 patients with stable disease (SD) in 62.5% and PR in 37.5% of patients [85]. Results from combination therapy with PV-10 and ICBs are awaited with interest. Complete results of the Spanish GEM-1402 study (ClinicalTrials.gov Identifier: NCT02626962) were recently published [75] (Tables 1 and 3). This phase II trial tested the efficacy of the combination of Nivolumab and Ipilimumab as first-line therapy in 52 patients with M-UM. Median OS was 12.7 months with a median PFS of 3 months. The outcome seems quite modest compared to benchmarks of UM responses. The authors claim that the short PFS may be related to the high levels of LDH, a serum marker of progression, at baseline. *GNAQ*, *GNA11*, and *SF3B1* gene mutational analysis and Multiplex Ligation Probe Amplification (MLPA) analysis to detect deletions and duplications in chromosomes 3 and 8 were performed in 25 patients (50% of total patients). Mutations and chromosomal aberrations did not appear to be related to ORR, although the number of patients analyzed was too small to obtain conclusive results. Treatment-related adverse events occurred in 49 of 52 patients with 1 death in a patient with thyroiditis and 1with Guillain–Barrè syndrome. Pelster et al. [76] (ClinicalTrials.gov Identifier: NCT01585194, PROSPER), reported on a phase II study of Nivolumab plus Ipilimumab an ORR of 18%, a median PFS of 5.5 months, and a median OS of 19.1 months in 33 patients, which is longer than the 6.8 to 9.6 months reported with monotherapy. Grade 3–4 treatment-related adverse events occurred in 40% of patients (Tables 1 and 3).

The influence of *BAP1* mutation or chromosome 3 monosomy has been considered only in a small fraction of patients. Patients at high risk of metastasis with monosomy 3 and/or *BAP1* mutation should be included in clinical trials of adjuvant therapy. Considering only clinical trials (www.clinicaltrials.gov) with updates starting in 2019 to March 2021, we found 7 clinical trials using adjuvant therapy in high-risk UM patients. One uses ICBs, and 4 other immunological approaches, whereas 2 exploit targeted therapies (Table 2). Interestingly, a randomized phase II study enrolling resectable metastatic melanoma including UM, uses Ipilimumab, Nivolumab, and Relatlimab, the latter blocking LAG-3 (ClinicalTrials.gov Identifier: NCT02519322, Table 2).

In summary, an increase in ORR was observed in combined ICB treatment compared to monotherapy although not comparable with the improvement obtained in CM.

Table 3. Combination immunotherapies in UM.

ClinicalTrials.gov Identifier: NCT Number	Trial	Status	Phase	Targeted Patients	Actual Enrollment (Estimated Enrollment)	Principal Investigator	First Submitted Date	Last Update Posted Date
NCT01585194 [76]	Phase II Study of Nivolumab in Combination With Ipilimumab for Uveal Melanoma	Active, not recruiting	Phase II	Metastatic Uveal Melanoma Stage IV Uveal Melanoma AJCC v7	67 (141)	Sapna Patel	23 April 2012	10 December 2020
NCT02626962 [75]	Phase II Multicenter, Non-Randomized, Open-Label Trial of Nivolumab in Combination With Ipilimumab in Subjects With Previously Untreated Metastatic Uveal Melanoma	Active, not recruiting	Phase II	Uveal Melanoma	48 (48)	Josep Maria Piulats	1 December 2015	19 October 2020
NCT03922880	Pilot Study Combining Arginine Depletion and Checkpoint Inhibition in Uveal Melanomas	Active, not recruiting	Phase I	Uveal Melanoma	9 (9)	Alexander Shoushtari	18 April 2019	11 January 2021
NCT02913417	A Feasibility Study of Sequential Hepatic Internal Radiation and Systemic Ipilimumab and Nivolumab in Patients With Uveal Melanoma Metastatic to Liver	Recruiting	Phase I/Phase II	Uveal Melanoma Hepatic Metastases	(26)	David R. Minor	21 September 2016	25 August 2020
NCT04463368	SCANDIUM II Trial—A Phase I Randomized Controlled Multicentre Trial of Isolated Hepatic Perfusion in Combination With Ipilimumab and Nivolumab in Patients With Uveal Melanoma Metastases	Not yet recruiting	Phase I	Uveal Melanoma Liver Metastases	(18)	Roger Olofsson Bagge	5 July 2020	1 September 2020
NCT04283890	Phase Ib/2 Study Combining Hepatic Percutaneous Perfusion With Ipilimumab Plus Nivolumab in Advanced Uveal Melanoma	Recruiting	Phase I/Phase II	Uveal Melanoma, Metastatic	(88)	Ellen W. Kapiteijn	21 February 2020	25 February 2020
NCT03472586	Ipilimumab and Nivolumab in Combination With Immunoembolization for the Treatment of Metastatic Uveal Melanoma	Recruiting	Phase II	Metastatic Uveal Melanoma	(35)	Marlana Orloff	14 March 2018	28 May 2020

A new approach is based on the bispecific soluble molecule Tebentafusp. This fusion molecule binds with high affinity the GP100 peptide presented by HLA-A*02:01 on tumor cells and, with the anti-CD3 effector domain, induces a polyclonal activation of naive T cells. Tebentafusp activates T cells independently of their natural TCR specificity. The phase I/II trial of Tebentafusp in metastatic melanoma (ClinicalTrials.gov Identifier: NCT01211262, Tables 1 and 2) enrolling previously treated cutaneous (n = 61) and M-UM patients (n = 18) recently reported a one-year OS rate of 65%, 16.6% PR, and 44.4% SD [77]. IFN-γ related markers (CXCL10, CXCL11, IL6, IL10, IL15, and IFN-γ) were measured in the serum at baseline and on treatment, and an increase was found in 11/18 UM patients analyzed. At present, March 2021, 2 additional clinical trials studying Tebentafusp are listed in www.clinicaltrials.gov (https://clinicaltrials.gov/ct2/results?term=Tebentafusp+and+uveal+melanoma&Search=Search): 1 phase II randomized, open-label, multicenter study in untreated, advanced UM (ClinicalTrials.gov Identifier: NCT03070392) and 1 phase I/II (ClinicalTrials.gov Identifier: NCT02570308) intra-patient escalation dosing in advanced UM (Table 2). Preliminary results were presented at the ESMO Immuno-Oncology Virtual Congress 2020. Following Tebentafusp, the ORR was 5% with only PRs. Stable disease was achieved by 45% of patients. The median duration of response was 8.7 months. With a median follow-up of 19.6 months, the median OS was 16.8 months. Patients (64%) developing rash within 7 days of Tebentafusp initiation demonstrated a superior median OS of 22.5 months compared to 10.3 months in patients with no rash, suggesting an immune-related effect. Further results need to clarify a real improvement in survival by bispecific molecules, although the U.S. Food and Drug Administration (FDA) has granted breakthrough therapy designation to Tebentafusp (IMCgp100 for HLA-A*02:01-positive patients) in UM [86].

The expression of Preferentially expressed Antigen in Melanoma (PRAME) correlates with high metastatic risk in UM [87] and is presently under investigation in several clinical trials as an immunotherapeutic target antigen of M-UM (ClinicalTrials.gov Identifier: NCT04262466 and NCT02743611, Table 2). In ClinicalTrials.gov Identifier: NCT04262466, a bispecific molecule consisting of a TCR targeting HLA-A*02:01 plus PRAME and anti-CD3 scFv will be used in association with anti-PD-L1 to treat PRAME positive patients. ClinicalTrials.gov Identifier: NCT02743611 exploits participants T cells that are modified to recognize and target PRAME on cancer cells.

The identification of an immunotherapy response signature would be of great advantage to spare potential non-responders from elevated toxicity. A recent attempt to identify molecular markers of immunotherapy resistance of metastatic CM was reported by Beck et al. using a clinical proteomic approach [88].

4. Immune Signatures

Several studies have characterized the immune infiltrate in metastatic UM, given the crucial prognostic role of TME in various types of metastatic cancers. Immune prognostic signatures have been proposed to identify those patients who could benefit from immunotherapy in an attempt to reduce the 5-year mortality rate. These signatures have been developed by digital cytometry working on retrospective, public datasets and require experimental validation to verify diagnostic reliability and clinical usefulness [89–92] (Table 4).

Table 4. Summary of the principal published immune signature.

Ref.	Signature	Aim of the Study
Li [91]	Immune-related gene signature based on two immune-related genes for predicting survival in UM.	Development of an immune-related prognostic and predictive signature to identify those patients who could benefit from immunotherapy. The signature is built on the TCGA-UM dataset and is significantly associated with tumor T stage and tumor basal diameter.
Wang [90]	Adaptive Immune Resistance Signature based on fifteen markers, to predict prognosis in UM.	Analysis of the immune and stromal infiltrate on gene expression data of the TCGA-UM and TCGA-CM datasets using different digital cytometry algorithms for significant prognostic marker selection. This signature could identify UM subgroups with a characteristic tumor microenvironment.
Zhang [89]	Immune cell-based prognosis signature to predict overall survival in UM. The signature is based on the contribution of CD8+, CD4+ T cells, monocytes, and Mast cells.	Tumor microenvironment landscape analysis by the CYBERSORT algorithm to classify the immune cell type profiles in the TCGA-UM patients. This signature highlights the impact of immune infiltrate components in the development of metastases.
Gong [92]	Immune and stromal prognostic signature based on published datasets. The signature is developed on a four-cell model (cytotoxic, Th1, Th2 cells, and myocytes).	Tumor microenvironment analysis by ESTIMATE algorithm for the identification of a four-cell model as a biomarker of overall survival in UM. This prognostic signature can stratify subgroups of patients with different classes of risk.

Patel's group recently proposed a study on a dataset of 47 P- and M-UM demonstrating, by IHC, that metastatic patients show significantly higher levels of immune infiltrate (CD3+, CD8+, FoxP3+, and CD68+ cells) compared to primary tumors [60]. They developed an IFN-γ signature using Nanostring technology between 2 responder and 4 non-responder patients to immunotherapy. Their data indicated that pre-treatment tumors of non-responders display a gene expression profile consistent with pro-inflammatory signaling, while responders have significantly elevated levels of *Suppressor Of Cytokine Signaling 1* (*SOCS1*) and *HLA* molecules. Two sets of genes that are differentially expressed between responders and non-responders were identified. Twelve genes were upregulated in responders at baseline before treatment and 13 showed significantly higher expression at baseline in non-responders. The authors identify, for the first time, a baseline tumor immune signature predicting response and resistance to immunotherapy in UM, that can be used to select those patients that are likely to respond to immunotherapy. A limit of this signature is the small number of patients ($n = 6$) analyzed. However, results from validation studies of this signature in larger cohorts of patients (GEM1402 and CA184-187) will provide more information on the resistance and response mechanisms of M-UM to immunotherapy, and prospective testing will establish clinical value. Figure 1 shows the application of this signature to the TCGA dataset of P-UM. The hierarchical clustering of this signature highlights three main clusters: high-risk with an immunotherapy responder profile (light blue), low-risk (pink), and high-risk non-responders (yellow). Among the high-risk UM, mostly metastatic cases with *BAP1* mutations, chromosome 3 monosomy, and chromosome 8q gain, 1 group (light blue) contains potential responders to immunotherapy.

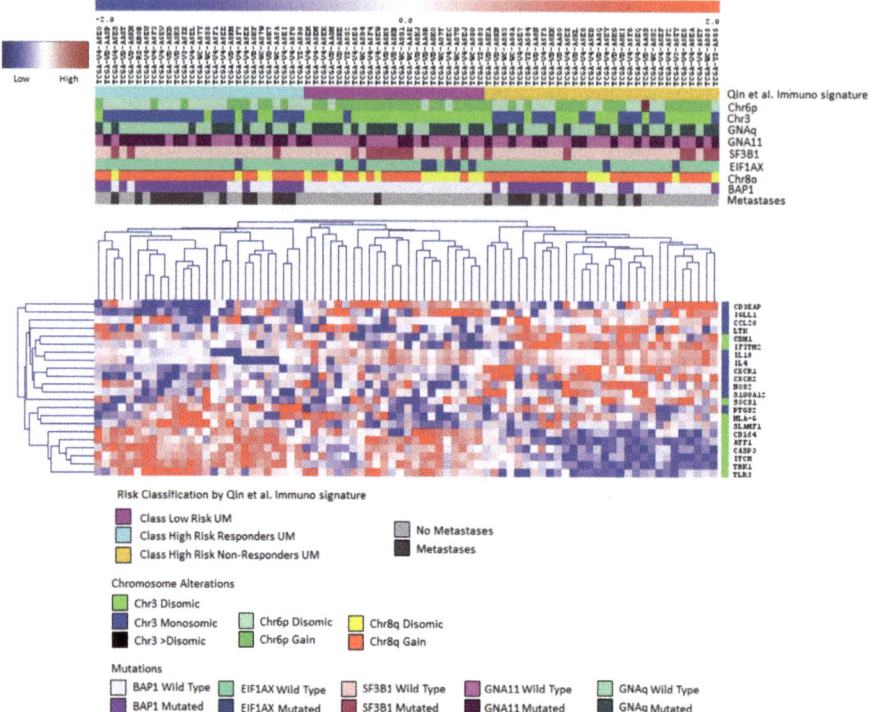

Figure 1. Application of the prognostic adaptive immune response signature developed by Qin et al. [60] to the TCGA-UM dataset. Euclidean hierarchical cluster Heatmap for 80 P-UM, highlighting mRNA expression levels of Qin et al. [60] immune signature genes. Responder genes are labeled in green, non-responder genes are labeled in blue. The expression values are reported by a color scale (blue = expression below the mean, red = expression above the mean, white = expression at the mean; the intensity is related to the distance from the mean). This signature shows three main clusters defined by differentially expressed profiles between responders versus non-responders genes.

5. Conclusions

Despite the considerable advancement in the diagnosis and classification of patients at low/high-risk of progression, UM still represents a challenge for oncologists. Indeed, still 50% of patients will develop metastatic disease with only marginal improvements in survival in decades. The origin from an immune-privileged site and the development of metastases in the liver, an immune-modulating organ, the low mutational burden, the few neoantigens, and the low expression of PD-L1 on tumor cells contribute to the poor response of UM to immunotherapy, compared to CM.

An increase in ORR was observed in patients receiving combined Ipilimumab and Nivolumab compared to monotherapy. The results achieved in UM are far from being comparable with the improvement obtained in CM, yet they are equal in terms of side effects. One of the reasons for this result is certainly the low expression of PD-1/PD-L1 in UM. Targeting LAG-3 that is expressed in UM at higher levels than PD-1 might yield better results. Immunotherapy is not only ICB treatment, and many different approaches are in development or already in clinical trials. Among these, Tebentafusp seems promising, since also the FDA has granted breakthrough therapy designation in UM. Yet, this treatment will be available for a small portion of patients because the drug is designed only for HLA-A*02:01-positive patients. This is a big issue, but it is strictly connected with immunotherapies that may require personalized drugs.

Single-cell omics studies and high-throughput data analysis are necessary and need to be improved to understand the mechanisms underlying the cross-talk between tumor and immune cells. This approach will provide new insights and identify new potentially actionable targets for immunotherapy. UM express few neo-antigens but high levels of TAA, such as MART1, GD2, Tyrosinase1, TRP1, gp100, and MAGE. Cell-based immunotherapies that are being developed exploit some of these TAA as targets. A phase I study using GD2-directed Chimeric Antigen Receptor T cells (CAR-T cells) is ongoing in patients of different cancers, including UM (ClinicalTrials.gov Identifier: NCT03635632, Table 2). Another antigen that has been successfully targeted by immunotherapy is PRAME either with bispecific TCR/anti-CD3 molecules (ClinicalTrials.gov Identifier: NCT04262466) or with autologous T cells engineered with PRAME-specific TCR (ClinicalTrials.gov Identifier: NCT02743611). Local liver chemotherapy and radiotherapy may release neoantigens and soluble mediators attracting cells from the immune system, into the tumor. The association of selective internal hepatic radiation (microspheres containing radioactive yttrium-90) with the combination of Ipilimumab and Nivolumab is exploited in an interventional open-label phase I/II clinical trial (ClinicalTrials.gov Identifier: NCT02913417, Table 2).

Recent observations highlight the expression of alternative immune-checkpoints: LAG-3 and TIM-3 should preferentially be targeted instead of PD-1/PD-L1, which are barely expressed by UM metastases. The clinical trial ClinicalTrials.gov Identifier: NCT02519322 that uses the association of anti-LAG-3 with Ipilimumab and Nivolumab and is, at present, enrolling patients will eventually show the advantage of LAG-3 targeting (Table 2).

Most studies exploiting new possibilities for ICB associations could be done in vitro in an autologous setting if lymphocytes and cells from the same patient were available. Syngeneic murine models are so far inappropriate, since many of them are obtained using melanoma cell lines to generate liver metastasis, thus resembling neither the biology nor the genetics of UM. Patient-Derived Xenografts (PDX), injected either subcutaneously or orthotopically, are also challenging to develop for M-UM, and they may be useful to test the tumor response to pharmacological or targeted therapy rather than to immunotherapy, since PDX cannot maintain immune cells alive. Humanized mice may be used to overcome this issue. Finally, the use of organoids, in vitro 3D culture systems, that keep the biological characteristics of the original tumor to simulate the in vivo tumor growth may be a useful method to study the effects of drugs before they come to the clinic.

Author Contributions: Conceptualization, M.C. and R.G.; Methodology, A.A., M.M., R.G., E.T.T., E.C., Formal Analysis, N.S., A.A., M.M.; Investigation, F.S., A.A., M.M.; Resources, U.P.; Data Curation, G.F.; Writing—Original Draft Preparation, A.A., M.C., R.G.; Writing—Review and Editing, M.C., R.G., U.P. All authors have read and agreed to the published version of the manuscript.

Funding: Support for this study was provided by the Italian Ministry of Health 5×1000 Funds 2013, Fondazione San Paolo 2016 (20067) and Associazione Italiana per la Ricerca sul Cancro (AIRC, IG 17103).

Acknowledgments: M.M. is a recipient of AIRC 5×1000 fellowship (ID 21073).

Conflicts of Interest: The authors have no conflict of interest to declare.

References

1. Diener-West, M.; Reynolds, S.M.; Agugliaro, D.J.; Caldwell, R.; Cumming, K.; Earle, J.D.; Hawkins, B.S.; Hayman, J.A.; Jaiyesimi, I.; Jampol, L.M.; et al. Development of metastatic disease after enrollment in the COMS trials for treatment of choroidal melanoma: Collaborative ocular melanoma study group report no. 26. *Arch. Ophthalmol.* **2005**, *123*, 1639–1643. [CrossRef] [PubMed]
2. Croce, M.; Ferrini, S.; Pfeffer, U.; Gangemi, R. Targeted therapy of uveal melanoma: Recent failures and new perspectives. *Cancers* **2019**, *11*, 846. [CrossRef] [PubMed]
3. Yang, J.; Manson, D.K.; Marr, B.P.; Carvajal, R.D. Treatment of uveal melanoma: Where are we now? *Ther. Adv. Med. Oncol.* **2018**, *10*. [CrossRef] [PubMed]
4. Rantala, E.S.; Hernberg, M.; Kivelä, T.T. Overall survival after treatment for metastatic uveal melanoma: A systematic review and meta-analysis. *Melanoma Res.* **2019**, *29*, 561–568. [CrossRef] [PubMed]

5. Moore, A.R.; Ceraudo, E.; Sher, J.J.; Guan, Y.; Shoushtari, A.N.; Chang, M.T.; Zhang, J.Q.; Walczak, E.G.; Kazmi, M.A.; Taylor, B.S.; et al. Recurrent activating mutations of G-protein-coupled receptor CYSLTR2 in uveal melanoma. *Nat. Genet.* **2016**, *48*, 675–680. [CrossRef] [PubMed]
6. Johansson, P.; Aoude, L.G.; Wadt, K.; Glasson, W.J.; Warrier, S.K.; Hewitt, A.W.; Kiilgaard, J.F.; Heegaard, S.; Isaacs, T.; Franchina, M.; et al. Deep sequencing of uveal melanoma identifies a recurrent mutation in PLCB4. *Oncotarget* **2016**, *7*, 4624–4631. [CrossRef] [PubMed]
7. Dono, M.; Angelini, G.; Cecconi, M.; Amaro, A.; Esposito, A.I.; Mirisola, V.; Maric, I.; Lanza, F.; Nasciuti, F.; Viaggi, S.; et al. Mutation frequencies of GNAQ, GNA11, BAP1, SF3B1, EIF1AX and TERT in Uveal melanoma: Detection of an activating mutation in the TERT gene promoter in a single case of uveal melanoma. *Br. J. Cancer* **2014**, *110*, 1058–1065. [CrossRef]
8. New, D.C.; Wong, Y.H. Molecular mechanisms mediating the G protein-coupled receptor regulation of cell cycle progression. *J. Mol. Signal.* **2007**, *2*, 2. [CrossRef]
9. Piaggio, F.; Tozzo, V.; Bernardi, C.; Croce, M.; Puzone, R.; Viaggi, S.; Patrone, S.; Barla, A.; Coviello, D.; Jager, M.J.; et al. Secondary somatic mutations in G-protein-related pathways and mutation signatures in uveal melanoma. *Cancers* **2019**, *11*, 1688. [CrossRef]
10. Horsman, D.E.; Sroka, H.; Rootman, J.; White, V.A. Monosomy 3 and isochromosome 8q in a uveal melanoma. *Cancer Genet. Cytogenet.* **1990**, *45*, 249–253. [CrossRef]
11. Prescher, G.; Bornfeld, N.; Hirche, H.; Horsthemke, B.; Jöckel, K.H.; Becher, R. Prognostic implications of monosomy 3 in uveal melanoma. *Lancet Lond. Eng.* **1996**, *347*, 1222–1225. [CrossRef]
12. Onken, M.D.; Worley, L.A.; Person, E.; Char, D.H.; Bowcock, A.M.; Harbour, J.W. Loss of heterozygosity of chromosome 3 detected with single nucleotide polymorphisms is superior to monosomy 3 for predicting metastasis in uveal melanoma. *Clin. Cancer Res. Off. J. Am. Assoc.* **2007**, *13*, 2923–2927. [CrossRef]
13. Harbour, J.W.; Onken, M.D.; Roberson, E.D.O.; Duan, S.; Cao, L.; Worley, L.A.; Council, M.L.; Matatall, K.A.; Helms, C.; Bowcock, A.M. Frequent mutation of BAP1 in metastasizing uveal melanomas. *Science* **2010**, *330*, 1410–1413. [CrossRef]
14. Harbour, J.W.; Roberson, E.D.O.; Anbunathan, H.; Onken, M.D.; Worley, L.A.; Bowcock, A.M. Recurrent mutations at codon 625 of the splicing factor SF3B1 in uveal melanoma. *Nat. Genet.* **2013**, *45*, 133–135. [CrossRef]
15. Martin, M.; Maßhöfer, L.; Temming, P.; Rahmann, S.; Metz, C.; Bornfeld, N.; van de Nes, J.; Klein-Hitpass, L.; Hinnebusch, A.G.; Horsthemke, B.; et al. Exome sequencing identifies recurrent somatic mutations in EIF1AX and SF3B1 in uveal melanoma with disomy 3. *Nat. Genet.* **2013**, *45*, 933–936. [CrossRef]
16. Yavuzyigitoglu, S.; Koopmans, A.E.; Verdijk, R.M.; Vaarwater, J.; Eussen, B.; van Bodegom, A.; Paridaens, D.; Kiliç, E.; de Klein, A. Rotterdam ocular melanoma study group uveal melanomas with SF3B1 mutations: A distinct subclass associated with late-onset metastases. *Ophthalmology* **2016**, *123*, 1118–1128. [CrossRef]
17. Ventii, K.H.; Devi, N.S.; Friedrich, K.L.; Chernova, T.A.; Tighiouart, M.; Van Meir, E.G.; Wilkinson, K.D. BRCA1-Associated protein-1 is a tumor suppressor that requires deubiquitinating activity and nuclear localization. *Cancer Res.* **2008**, *68*, 6953–6962. [CrossRef]
18. Jensen, D.E.; Proctor, M.; Marquis, S.T.; Gardner, H.P.; Ha, S.I.; Chodosh, L.A.; Ishov, A.M.; Tommerup, N.; Vissing, H.; Sekido, Y.; et al. BAP1: A novel ubiquitin hydrolase which binds to the BRCA1 RING finger and enhances BRCA1-mediated cell growth suppression. *Oncogene* **1998**, *16*, 1097–1112. [CrossRef]
19. Robertson, A.G.; Shih, J.; Yau, C.; Gibb, E.A.; Oba, J.; Mungall, K.L.; Hess, J.M.; Uzunangelov, V.; Walter, V.; Danilova, L.; et al. Integrative analysis identifies four molecular and clinical subsets in uveal melanoma. *Cancer Cell* **2017**, *32*, 204–220.e15. [CrossRef]
20. Gupta, M.P.; Lane, A.M.; DeAngelis, M.M.; Mayne, K.; Crabtree, M.; Gragoudas, E.S.; Kim, I.K. Clinical characteristics of uveal melanoma in patients with germline BAP1 mutations. *JAMA Ophthalmol.* **2015**, *133*, 881–887. [CrossRef]
21. Murali, R.; Wiesner, T.; Scolyer, R.A. Tumours associated with BAP1 mutations. *Pathology* **2013**, *45*, 116–126. [CrossRef] [PubMed]
22. Amaro, A.; Gangemi, R.; Piaggio, F.; Angelini, G.; Barisione, G.; Ferrini, S.; Pfeffer, U. The biology of uveal melanoma. *Cancer Metastasis Rev.* **2017**, *36*, 109–140. [CrossRef] [PubMed]
23. Alexandrov, L.B.; Nik-Zainal, S.; Wedge, D.C.; Aparicio, S.A.J.R.; Behjati, S.; Biankin, A.V.; Bignell, G.R.; Bolli, N.; Borg, A.; Børresen-Dale, A.-L.; et al. Signatures of mutational processes in human cancer. *Nature* **2013**, *500*, 415–421. [CrossRef] [PubMed]
24. Vergara, I.A.; Wilmott, J.S.; Long, G.V.; Scolyer, R.A. Genetic drivers of non-cutaneous melanomas: Challenges and opportunities in a heterogeneous landscape. *Exp. Dermatol.* **2021**. [CrossRef]
25. Saini, N.; Giacobone, C.K.; Klimczak, L.J.; Papas, B.N.; Burkholder, A.B.; Li, J.-L.; Fargo, D.C.; Bai, R.; Gerrish, K.; Innes, C.L.; et al. UV-exposure, endogenous DNA damage, and DNA replication errors shape the spectra of genome changes in human skin. *PLoS Genet.* **2021**, *17*, e1009302. [CrossRef]
26. Karlsson, J.; Nilsson, L.M.; Mitra, S.; Alsén, S.; Shelke, G.V.; Sah, V.R.; Forsberg, E.M.V.; Stierner, U.; All-Eriksson, C.; Einarsdottir, B.; et al. Molecular profiling of driver events in metastatic uveal melanoma. *Nat. Commun.* **2020**, *11*, 1894. [CrossRef]
27. Derrien, A.-C.; Rodrigues, M.; Eeckhoutte, A.; Dayot, S.; Houy, A.; Mobuchon, L.; Gardrat, S.; Lequin, D.; Ballet, S.; Pierron, G.; et al. Germline MBD4 mutations and predisposition to uveal melanoma. *J. Natl. Cancer Inst.* **2021**, *113*, 80–87. [CrossRef]
28. Rodrigues, M.; Mobuchon, L.; Houy, A.; Alsafadi, S.; Baulande, S.; Mariani, O.; Marande, B.; Ait Rais, K.; Van der Kooij, M.K.; Kapiteijn, E.; et al. Evolutionary routes in metastatic uveal melanomas depend on MBD4 alterations. *Clin. Cancer Res. Off. J. Am. Assoc.* **2019**, *25*, 5513–5524. [CrossRef]

29. Rodrigues, M.; Mobuchon, L.; Houy, A.; Fiévet, A.; Gardrat, S.; Barnhill, R.L.; Popova, T.; Servois, V.; Rampanou, A.; Mouton, A.; et al. Outlier response to anti-PD1 in uveal melanoma reveals germline MBD4 mutations in hypermutated tumors. *Nat. Commun.* **2018**, *9*, 1866. [CrossRef]
30. Blum, E.S.; Yang, J.; Komatsubara, K.M.; Carvajal, R.D. Clinical management of uveal and conjunctival melanoma. *Oncol. Williston Park N* **2016**, *30*, 29–32, 34–43, 48.
31. Chattopadhyay, C.; Kim, D.W.; Gombos, D.S.; Oba, J.; Qin, Y.; Williams, M.D.; Esmaeli, B.; Grimm, E.A.; Wargo, J.A.; Woodman, S.E.; et al. Uveal melanoma: From diagnosis to treatment and the science in between. *Cancer* **2016**, *122*, 2299–2312. [CrossRef] [PubMed]
32. Hodi, F.S.; Chiarion-Sileni, V.; Gonzalez, R.; Grob, J.-J.; Rutkowski, P.; Cowey, C.L.; Lao, C.D.; Schadendorf, D.; Wagstaff, J.; Dummer, R.; et al. Nivolumab plus ipilimumab or nivolumab alone versus ipilimumab alone in advanced melanoma (CheckMate 067): 4-Year outcomes of a multicentre, randomised, phase 3 trial. *Lancet Oncol.* **2018**, *19*, 1480–1492. [CrossRef]
33. Jager, M.J.; Shields, C.L.; Cebulla, C.M.; Abdel-Rahman, M.H.; Grossniklaus, H.E.; Stern, M.-H.; Carvajal, R.D.; Belfort, R.N.; Jia, R.; Shields, J.A.; et al. Uveal melanoma. *Nat. Rev. Dis. Primer* **2020**, *6*, 24. [CrossRef] [PubMed]
34. Fallico, M.; Raciti, G.; Longo, A.; Reibaldi, M.; Bonfiglio, V.; Russo, A.; Caltabiano, R.; Gattuso, G.; Falzone, L.; Avitabile, T. Current molecular and clinical insights into uveal melanoma (Review). *Int. J. Oncol.* **2021**, *58*, 10. [CrossRef]
35. Luyten, G.P.; van der Spek, C.W.; Brand, I.; Sintnicolaas, K.; de Waard-Siebinga, I.; Jager, M.J.; de Jong, P.T.; Schrier, P.I.; Luider, T.M. Expression of MAGE, Gp100 and tyrosinase genes in uveal melanoma cell lines. *Melanoma Res.* **1998**, *8*, 11–16. [CrossRef]
36. Kan-Mitchell, J.; Liggett, P.E.; Harel, W.; Steinman, L.; Nitta, T.; Oksenberg, J.R.; Posner, M.R.; Mitchell, M.S. Lymphocytes cytotoxic to uveal and skin melanoma cells from peripheral blood of ocular melanoma patients. *Cancer Immunol. Immunother.* **1991**, *33*, 333–340. [CrossRef]
37. Ksander, B.R.; Geer, D.C.; Chen, P.W.; Salgaller, M.L.; Rubsamen, P.; Murray, T.G. Uveal melanomas contain antigenically specific and non-specific infiltrating lymphocytes. *Curr. Eye Res.* **1998**, *17*, 165–173. [CrossRef]
38. McMenamin, P.G.; Saban, D.R.; Dando, S.J. Immune cells in the retina and choroid: Two different tissue environments that require different defenses and surveillance. *Prog. Retin. Eye Res.* **2019**, *70*, 85–98. [CrossRef]
39. Forrester, J.V.; Xu, H. Good news-bad news: The Yin and Yang of immune privilege in the eye. *Front. Immunol.* **2012**, *3*, 338. [CrossRef]
40. Vendomèle, J.; Khebizi, Q.; Fisson, S. Cellular and molecular mechanisms of anterior chamber-associated immune deviation (ACAID): What we have learned from knockout mice. *Front. Immunol.* **2017**, *8*, 1686. [CrossRef]
41. Niederkorn, J.Y. Immune escape mechanisms of intraocular tumors. *Prog. Retin. Eye Res.* **2009**, *28*, 329–347. [CrossRef]
42. Taylor, A.W.; Alard, P.; Yee, D.G.; Streilein, J.W. Aqueous humor induces transforming growth factor-beta (TGF-Beta)-producing regulatory T-cells. *Curr. Eye Res.* **1997**, *16*, 900–908. [CrossRef]
43. Ferguson, T.A.; Griffith, T.S. The role of fas ligand and TNF-related apoptosis-inducing ligand (TRAIL) in the ocular immune response. *Chem. Immunol. Allergy* **2007**, *92*, 140–154. [CrossRef]
44. Apte, R.S.; Sinha, D.; Mayhew, E.; Wistow, G.J.; Niederkorn, J.Y. Cutting edge: Role of macrophage migration inhibitory factor in inhibiting NK cell activity and preserving immune privilege. *J. Immunol.* **1998**, *160*, 5693–5696.
45. Yoshida, M.; Takeuchi, M.; Streilein, J.W. Participation of pigment epithelium of iris and ciliary body in ocular immune privilege. 1. Inhibition of T-cell activation in vitro by direct cell-to-cell contact. *Invest. Ophthalmol. Vis. Sci.* **2000**, *41*, 811–821.
46. Wierenga, A.P.A.; Gezgin, G.; van Beelen, E.; Eikmans, M.; Spruyt-Gerritse, M.; Brouwer, N.J.; Versluis, M.; Verdijk, R.M.; van Duinen, S.G.; Marinkovic, M.; et al. Soluble HLA in the aqueous humour of uveal melanoma is associated with unfavourable tumour characteristics. *Cancers* **2019**, *11*, 202. [CrossRef]
47. Javed, A.; Milhem, M. Role of Natural Killer Cells in Uveal Melanoma. *Cancers* **2020**, *12*, 694. [CrossRef]
48. Chen, P.W.; Mellon, J.K.; Mayhew, E.; Wang, S.; He, Y.G.; Hogan, N.; Niederkorn, J.Y. Uveal melanoma expression of indoleamine 2,3-deoxygenase: Establishment of an immune privileged environment by tryptophan depletion. *Exp. Eye Res.* **2007**, *85*, 617–625. [CrossRef]
49. Repp, A.C.; Mayhew, E.S.; Apte, S.; Niederkorn, J.Y. Human uveal melanoma cells produce macrophage migration-inhibitory factor to prevent lysis by NK cells. *J. Immunol.* **2000**, *165*, 710–715. [CrossRef]
50. Yang, W.; Li, H.; Chen, P.W.; Alizadeh, H.; He, Y.; Hogan, R.N.; Niederkorn, J.Y. PD-L1 Expression on human ocular cells and its possible role in regulating immune-mediated ocular inflammation. *Invest. Ophthalmol. Vis. Sci.* **2009**, *50*, 273–280. [CrossRef]
51. Blanco, P.L.; Lim, L.A.; Miyamoto, C.; Burnier, M.N. Uveal melanoma dormancy: An acceptable clinical endpoint? *Melanoma Res.* **2012**, *22*, 334–340. [CrossRef]
52. Vera-Ramirez, L.; Hunter, K.W. Tumor cell dormancy as an adaptive cell stress response mechanism. *F1000Research* **2017**, *6*. [CrossRef]
53. Grossniklaus, H.E.; Zhang, Q.; You, S.; McCarthy, C.; Heegaard, S.; Coupland, S.E. Metastatic ocular melanoma to the liver exhibits infiltrative and nodular growth patterns. *Hum. Pathol.* **2016**, *57*, 165–175. [CrossRef]
54. Jager, M.J.; Hurks, H.M.H.; Levitskaya, J.; Kiessling, R. HLA Expression in uveal melanoma: There is no rule without some exception. *Hum. Immunol.* **2002**, *63*, 444–451. [CrossRef]
55. Vetter, C.S.; Lieb, W.; Bröcker, E.-B.; Becker, J.C. Loss of nonclassical MHC molecules MIC-A/B expression during progression of uveal melanoma. *Br. J. Cancer* **2004**, *91*, 1495–1499. [CrossRef] [PubMed]
56. Piquet, L.; Dewit, L.; Schoonjans, N.; Millet, M.; Bérubé, J.; Gerges, P.R.A.; Bordeleau, F.; Landreville, S. Synergic interactions between hepatic stellate cells and uveal melanoma in metastatic growth. *Cancers* **2019**, *11*, 43. [CrossRef] [PubMed]

57. Whelchel, J.C.; Farah, S.E.; McLean, I.W.; Burnier, M.N. Immunohistochemistry of infiltrating lymphocytes in uveal malignant melanoma. *Invest. Ophthalmol. Vis. Sci.* **1993**, *34*, 2603–2606. [PubMed]
58. Bronkhorst, I.H.G.; Vu, T.H.K.; Jordanova, E.S.; Luyten, G.P.M.; van der Burg, S.H.; Jager, M.J. Different subsets of tumor-infiltrating lymphocytes correlate with macrophage influx and monosomy 3 in uveal melanoma. *Invest. Ophthalmol. Vis. Sci.* **2012**, *53*, 5370–5378. [CrossRef]
59. Krishna, Y.; McCarthy, C.; Kalirai, H.; Coupland, S.E. Inflammatory cell infiltrates in advanced metastatic uveal melanoma. *Hum. Pathol.* **2017**, *66*, 159–166. [CrossRef]
60. Qin, Y.; Bollin, K.; de Macedo, M.P.; Carapeto, F.; Kim, K.B.; Roszik, J.; Wani, K.M.; Reuben, A.; Reddy, S.T.; Williams, M.D.; et al. Immune profiling of uveal melanoma identifies a potential signature associated with response to immunotherapy. *J. Immunother. Cancer* **2020**, *8*. [CrossRef]
61. Qin, Y.; Petaccia de Macedo, M.; Reuben, A.; Forget, M.-A.; Haymaker, C.; Bernatchez, C.; Spencer, C.N.; Gopalakrishnan, V.; Reddy, S.; Cooper, Z.A.; et al. Parallel profiling of immune infiltrate subsets in uveal melanoma versus cutaneous melanoma unveils similarities and differences: A pilot study. *Oncoimmunology* **2017**, *6*, e1321187. [CrossRef] [PubMed]
62. Durante, M.A.; Rodriguez, D.A.; Kurtenbach, S.; Kuznetsov, J.N.; Sanchez, M.I.; Decatur, C.L.; Snyder, H.; Feun, L.G.; Livingstone, A.S.; Harbour, J.W. Single-cell analysis reveals new evolutionary complexity in uveal melanoma. *Nat. Commun.* **2020**, *11*, 496. [CrossRef] [PubMed]
63. Krishna, Y.; Acha-Sagredo, A.; Sabat-Pośpiech, D.; Kipling, N.; Clarke, K.; Figueiredo, C.R.; Kalirai, H.; Coupland, S.E. Transcriptome profiling reveals new insights into the immune microenvironment and upregulation of novel biomarkers in metastatic uveal melanoma. *Cancers* **2020**, *12*, 832. [CrossRef] [PubMed]
64. Gezgin, G.; Dogrusöz, M.; van Essen, T.H.; Kroes, W.G.M.; Luyten, G.P.M.; van der Velden, P.A.; Walter, V.; Verdijk, R.M.; van Hall, T.; van der Burg, S.H.; et al. Genetic evolution of uveal melanoma guides the development of an inflammatory microenvironment. *Cancer Immunol. Immunother.* **2017**, *66*, 903–912. [CrossRef]
65. Figueiredo, C.R.; Kalirai, H.; Sacco, J.J.; Azevedo, R.A.; Duckworth, A.; Slupsky, J.R.; Coulson, J.M.; Coupland, S.E. Loss of BAP1 expression is associated with an immunosuppressive microenvironment in uveal melanoma, with implications for immunotherapy development. *J. Pathol.* **2020**, *250*, 420–439. [CrossRef]
66. Rothermel, L.D.; Sabesan, A.C.; Stephens, D.J.; Chandran, S.S.; Paria, B.C.; Srivastava, A.K.; Somerville, R.; Wunderlich, J.R.; Lee, C.-C.R.; Xi, L.; et al. Identification of an immunogenic subset of metastatic uveal melanoma. *Clin. Cancer Res. Off. J. Am. Assoc.* **2016**, *22*, 2237–2249. [CrossRef]
67. Tavera, R.J.; Forget, M.-A.; Kim, Y.U.; Sakellariou-Thompson, D.; Creasy, C.A.; Bhatta, A.; Fulbright, O.J.; Ramachandran, R.; Thorsen, S.T.; Flores, E.; et al. Utilizing T-cell activation signals 1, 2, and 3 for tumor-infiltrating lymphocytes (TIL) expansion: The advantage over the sole use of interleukin-2 in cutaneous and uveal melanoma. *J. Immunother.* **2018**, *41*, 399–405. [CrossRef]
68. Chandran, S.S.; Somerville, R.P.T.; Yang, J.C.; Sherry, R.M.; Klebanoff, C.A.; Goff, S.L.; Wunderlich, J.R.; Danforth, D.N.; Zlott, D.; Paria, B.C.; et al. Treatment of metastatic uveal melanoma with adoptive transfer of tumour-infiltrating lymphocytes: A single-centre, two-stage, single-arm, phase 2 study. *Lancet Oncol.* **2017**, *18*, 792–802. [CrossRef]
69. Johansson, J.; Siarov, J.; Kiffin, R.; Mölne, J.; Mattsson, J.; Naredi, P.; Olofsson Bagge, R.; Martner, A.; Lindnér, P. Presence of tumor-infiltrating CD8+ T cells and macrophages correlates to longer overall survival in patients undergoing isolated hepatic perfusion for uveal melanoma liver metastasis. *Oncoimmunology* **2020**, *9*, 1854519. [CrossRef]
70. Khoja, L.; Atenafu, E.G.; Suciu, S.; Leyvraz, S.; Sato, T.; Marshall, E.; Keilholz, U.; Zimmer, L.; Patel, S.P.; Piperno-Neumann, S.; et al. Meta-analysis in metastatic uveal melanoma to determine progression free and overall survival benchmarks: An international rare cancers initiative (IRCI) ocular melanoma study. *Off. J. Eur. Soc. Med. Oncol.* **2019**, *30*, 1370–1380. [CrossRef]
71. Klemen, N.D.; Wang, M.; Rubinstein, J.C.; Olino, K.; Clune, J.; Ariyan, S.; Cha, C.; Weiss, S.A.; Kluger, H.M.; Sznol, M. Survival after checkpoint inhibitors for metastatic acral, mucosal and uveal melanoma. *J. Immunother. Cancer* **2020**, *8*. [CrossRef]
72. Bol, K.F.; Ellebaek, E.; Hoejberg, L.; Bagger, M.M.; Larsen, M.S.; Klausen, T.W.; Køhler, U.H.; Schmidt, H.; Bastholt, L.; Kiilgaard, J.F.; et al. Real-world impact of immune checkpoint inhibitors in metastatic uveal melanoma. *Cancers* **2019**, *11*, 489. [CrossRef]
73. Heppt, M.V.; Amaral, T.; Kähler, K.C.; Heinzerling, L.; Hassel, J.C.; Meissner, M.; Kreuzberg, N.; Loquai, C.; Reinhardt, L.; Utikal, J.; et al. Combined immune checkpoint blockade for metastatic uveal melanoma: A retrospective, multi-center study. *J. Immunother. Cancer* **2019**, *7*, 299. [CrossRef]
74. Kirchberger, M.C.; Moreira, A.; Erdmann, M.; Schuler, G.; Heinzerling, L. Real world experience in low-dose ipilimumab in combination with PD-1 blockade in advanced melanoma patients. *Oncotarget* **2018**, *9*, 28903–28909. [CrossRef]
75. Piulats, J.M.; Espinosa, E.; de la Cruz Merino, L.; Varela, M.; Alonso Carrión, L.; Martín-Algarra, S.; López Castro, R.; Curiel, T.; Rodríguez-Abreu, D.; Redrado, M.; et al. Nivolumab plus ipilimumab for treatment-naïve metastatic uveal melanoma: An open-label, multicenter, phase II trial by the spanish multidisciplinary melanoma group (GEM-1402). *J. Clin. Oncol. Off. J. Am. Soc.* **2021**, JCO2000550. [CrossRef]
76. Pelster, M.S.; Gruschkus, S.K.; Bassett, R.; Gombos, D.S.; Shephard, M.; Posada, L.; Glover, M.S.; Simien, R.; Diab, A.; Hwu, P.; et al. Nivolumab and ipilimumab in metastatic uveal melanoma: Results from a single-arm phase II study. *J. Clin. Oncol. Off. J. Am. Soc.* **2020**, JCO2000605. [CrossRef]
77. Middleton, M.R.; McAlpine, C.; Woodcock, V.K.; Corrie, P.; Infante, J.R.; Steven, N.M.; Evans, T.R.J.; Anthoney, A.; Shoushtari, A.N.; Hamid, O.; et al. Tebentafusp, A TCR/Anti-CD3 bispecific fusion protein targeting Gp100, potently activated antitumor immune responses in patients with metastatic melanoma. *Clin. Cancer Res. Off. J. Am. Assoc.* **2020**, *26*, 5869–5878. [CrossRef]

78. Amaria, R.N.; Reddy, S.M.; Tawbi, H.A.; Davies, M.A.; Ross, M.I.; Glitza, I.C.; Cormier, J.N.; Lewis, C.; Hwu, W.-J.; Hanna, E.; et al. Neoadjuvant immune checkpoint blockade in high-risk resectable melanoma. *Nat. Med.* **2018**, *24*, 1649–1654. [CrossRef] [PubMed]
79. Binkley, E.; Triozzi, P.L.; Rybicki, L.; Achberger, S.; Aldrich, W.; Singh, A. A Prospective trial of adjuvant therapy for high-risk uveal melanoma: Assessing 5-year survival outcomes. *Br. J. Ophthalmol.* **2020**, *104*, 524–528. [CrossRef]
80. Lawson, D.H.; Lee, S.; Zhao, F.; Tarhini, A.A.; Margolin, K.A.; Ernstoff, M.S.; Atkins, M.B.; Cohen, G.I.; Whiteside, T.L.; Butterfield, L.H.; et al. Randomized, placebo-controlled, phase III trial of yeast-derived granulocyte-macrophage colony-stimulating factor (GM-CSF) versus peptide vaccination versus GM-CSF plus peptide vaccination versus placebo in patients with no evidence of disease after complete surgical resection of locally advanced and/or stage IV melanoma: A trial of the Eastern Cooperative Oncology Group-American College of Radiology Imaging Network Cancer Research Group (E4697). *J. Clin. Oncol. Off. J. Am. Soc.* **2015**, *33*, 4066–4076. [CrossRef]
81. Jespersen, H.; Bagge, R.O.; Ullenhag, G.; Carneiro, A.; Helgadottir, H.; Ljuslinder, I.; Levin, M.; All-Eriksson, C.; Andersson, B.; Stierner, U.; et al. Phase II multicenter open label study of pembrolizumab and entinostat in adult patients with metastatic uveal melanoma (PEMDAC Study). *Ann. Oncol.* **2019**, *30*, v907. [CrossRef]
82. Joseph, R.W.; Peddareddigari, V.R.; Liu, P.; Miller, P.W.; Overwijk, W.W.; Bekele, N.B.; Ross, M.I.; Lee, J.E.; Gershenwald, J.E.; Lucci, A.; et al. Impact of clinical and pathologic features on tumor-infiltrating lymphocyte expansion from surgically excised melanoma metastases for adoptive T-cell therapy. *Clin. Cancer Res. Off. J. Am. Assoc.* **2011**, *17*, 4882–4891. [CrossRef] [PubMed]
83. Bol, K.F.; van den Bosch, T.; Schreibelt, G.; Mensink, H.W.; Keunen, J.E.E.; Kiliç, E.; Japing, W.J.; Geul, K.W.; Westdorp, H.; Boudewijns, S.; et al. Adjuvant dendritic cell vaccination in high-risk uveal melanoma. *Ophthalmology* **2016**, *123*, 2265–2267. [CrossRef] [PubMed]
84. Wessely, A.; Steeb, T.; Erdmann, M.; Heinzerling, L.; Vera, J.; Schlaak, M.; Berking, C.; Heppt, M.V. The role of immune checkpoint blockade in uveal melanoma. *Int. J. Mol. Sci.* **2020**, *21*, 879. [CrossRef]
85. Chua, V.; Mattei, J.; Han, A.; Johnston, L.; LiPira, K.; Selig, S.M.; Carvajal, R.D.; Aplin, A.E.; Patel, S.P. The latest on uveal melanoma research and clinical trials: Updates from the cure ocular melanoma (CURE OM) science meeting (2019). *Clin. Cancer Res. Off. J. Am. Assoc.* **2021**, *27*, 28–33. [CrossRef]
86. Tebentafusp Named FDA Breakthrough Therapy for Advanced Eye Cancer. Available online: https://immuno-oncologynews.com/2021/02/24/tebentafusp-wins-fda-breakthrough-therapy-designation-advanced-eye-cancer/ (accessed on 4 March 2021).
87. Field, M.G.; Durante, M.A.; Decatur, C.L.; Tarlan, B.; Oelschlager, K.M.; Stone, J.F.; Kuznetsov, J.; Bowcock, A.M.; Kurtenbach, S.; Harbour, J.W. Epigenetic reprogramming and aberrant expression of PRAME are associated with increased metastatic risk in class 1 and class 2 uveal melanomas. *Oncotarget* **2016**, *7*, 59209–59219. [CrossRef]
88. Beck, L.; Harel, M.; Yu, S.; Markovits, E.; Boursi, B.; Markel, G.; Geiger, T. Clinical proteomics of metastatic melanoma reveals profiles of organ specificity and treatment resistance. *Clin. Cancer Res. Off. J. Am. Assoc.* **2021**. [CrossRef]
89. Zhang, Z.; Ni, Y.; Chen, G.; Wei, Y.; Peng, M.; Zhang, S. Construction of immune-related risk signature for uveal melanoma. *Artif. Cells Nanomedicine Biotechnol.* **2020**, *48*, 912–919. [CrossRef]
90. Wang, Y.; Xu, Y.; Dai, X.; Lin, X.; Shan, Y.; Ye, J. The prognostic landscape of adaptive immune resistance signatures and infiltrating immune cells in the tumor microenvironment of uveal melanoma. *Exp. Eye Res.* **2020**, *196*, 108069. [CrossRef]
91. Li, Y.-Z.; Huang, Y.; Deng, X.-Y.; Tu, C.-S. Identification of an immune-related signature for the prognosis of uveal melanoma. *Int. J. Ophthalmol.* **2020**, *13*, 458–465. [CrossRef]
92. Gong, Q.; Wan, Q.; Li, A.; Yu, Y.; Ding, X.; Lin, L.; Qi, X.; Hu, L. Development and validation of an immune and stromal prognostic signature in uveal melanoma to guide clinical therapy. *Aging* **2020**, *12*, 20254–20267. [CrossRef]

Article

ACKR4 in Tumor Cells Regulates Dendritic Cell Migration to Tumor-Draining Lymph Nodes and T-Cell Priming

Dechen Wangmo [1], Prem K. Premsrirut [2], Ce Yuan [1], William S. Morris [1], Xianda Zhao [1,*] and Subbaya Subramanian [1,3,4,*]

1. Department of Surgery, University of Minnesota Medical School, Minneapolis, MN 55455, USA; wangm005@umn.edu (D.W.); yuanx236@umn.edu (C.Y.); morr0745@umn.edu (W.S.M.)
2. Mirimus Inc., Brooklyn, NY 11226, USA; prem@mirimus.com
3. Masonic Cancer Center, University of Minnesota, Minneapolis, MN 55455, USA
4. Center for Immunology, University of Minnesota, Minneapolis, MN 55455, USA
* Correspondence: zhaox714@umn.edu (X.Z.); subree@umn.edu (S.S.)

Simple Summary: Our study demonstrated that Atypical Chemokine Receptor 4 (ACKR4) was downregulated in human colorectal cancer (CRC) compared with normal colon tissues. Loss of ACKR4 in human CRC was associated with a weak anti-tumor immune response. Knockdown of ACKR4 in tumor cells impairs the dendritic cell migration from the tumor to the tumor-draining lymph nodes (TdLNs), causing inadequate tumor-specific T-cell expansion and insensitivity to immune checkpoint blockades. However, loss of ACKR4 in stromal cells does not significantly affect anti-tumor immunity. In human CRC, high expression of microRNA-552 was a mechanism leading to ACKR4 downregulation. Our study revealed a novel mechanism that leads to the poor immune response in a subset of CRC and will contribute to the framework for identifying new therapies against this deadly cancer.

Abstract: Colorectal cancer (CRC) is one of the most common malignancies in both morbidity and mortality. Immune checkpoint blockade (ICB) treatments have been successful in a portion of mismatch repair-deficient (dMMR) CRC patients but have failed in mismatch repair-proficient (pMMR) CRC patients. Atypical Chemokine Receptor 4 (ACKR4) is implicated in regulating dendritic cell (DC) migration. However, the roles of ACKR4 in CRC development and anti-tumor immunoregulation are not known. By analyzing human CRC tissues, transgenic animals, and genetically modified CRC cells lines, our study revealed an important function of ACKR4 in maintaining CRC immune response. Loss of ACKR4 in CRC is associated with poor immune infiltration in the tumor microenvironment. More importantly, loss of ACKR4 in CRC tumor cells, rather than stromal cells, restrains the DC migration and antigen presentation to the tumor-draining lymph nodes (TdLNs). Moreover, tumors with ACKR4 knockdown become less sensitive to immune checkpoint blockade. Finally, we identified that microRNA miR-552 negatively regulates ACKR4 expression in human CRC. Taken together, our studies identified a novel and crucial mechanism for the maintenance of the DC-mediated T-cell priming in the TdLNs. These new findings demonstrate a novel mechanism leading to immunosuppression and ICB treatment resistance in CRC.

Keywords: colorectal cancer; immune checkpoints; dendritic cells; Atypical Chemokine Receptor 4 (ACKR4); T-cell priming; immune checkpoint blockade

1. Introduction

Colorectal cancer (CRC) is the third most commonly diagnosed malignancy and the third leading cause of cancer-related deaths in the United States [1]. By 2030, the global CRC burden is expected to increase by 60% and surpass 2.2 million new cases and 1.1 million deaths [2]. The paradigm shift in cancer treatment brought by immunotherapy has been a major scientific and clinical breakthrough. Since the first immune checkpoint blockade

(ICB) therapy approval for melanoma, ICB is considered the standard of care for multiple types of cancer types, including the mismatch repair-deficient (dMMR)/microsatellite instability-high (MSI-H) CRC tumors [3]. However, not all dMMR/MSI-H CRC tumors are sensitive to ICB, and all of the mismatch repair-proficient (pMMR)/microsatellite instability-low (MSI-L)/microsatellite stability (MSS) CRC tumors are resistant to ICB [4]. Therefore, understanding the mechanisms of immunosuppression and immune therapy resistance is critical for designing novel treatments for CRC patients.

The immunogenicity of tumors is fundamental for ICB treatment. Low immunogenic tumors present a hallmark feature of sparse tumor T-cell infiltration. One of the key mechanisms involved in poor T-cell infiltration has been attributed to defects in the antigen presentation process, which significantly weakens the tumor-specific T-cell priming and precludes the T-cell mediated killing of cancer cells [5]. Dendritic cells (DCs) are the most potent antigen-presenting cells necessary to prime and activate tumor-antigen specific T-cells to induce an effective anti-tumor immune response [6–8]. Previous studies have shown that dysfunction of DCs caused defective antigen presentation and T-cell priming, leading to uncontrolled tumor development and ICB resistance in multiple cancers [9–11].

Successful antigen presentation by the DCs involves efficient migration of DCs from the tumor tissue to the regional lymph nodes. DC migration heavily depends on CCR7, a G-protein coupled receptor for two chemokines: CCL19 and CCL21 [12–14]. CCL21 has an extended positively charged C terminus that limits its interstitial diffusion, causing a stable gradient of CCL21 that directs the CCR7 expressing DCs from the tissue interstitium into lymphatic vessels [12,15]. On the other hand, both CCL19 and CCL21 are ligands for the atypical chemokine receptor 4 (ACKR4), a scavenging and decoy receptor that internalizes and mediates lysosomal degradation of CCL19/21 [15]. It is established that ACKR4 controls the bioavailability of CCL19/21, creating a CCL19/21 chemokine gradient that facilitates the directional migration of DCs from the non-lymphatic tissue to the draining lymph node [12–16]. However, the effects of ACKR4 in CRC progression and immunoregulation are largely unknown. Here, we examined the function of ACKR4 in CRC progression and anti-tumor immunity, emphasizing its role in the DC-mediated antigen presentation process and subsequent T-cell activation. Our study provides deeper insights into the immunoregulation in CRC and potentially leads to novel approaches for maximizing CRC response to ICB.

2. Material and Methods

2.1. Cell Lines and Organoids

Murine CRC cell lines MC38 and CT26 were used in the study. The source and detailed methods of cell culture are described in our previous publication [17].

2.2. Immunofluorescence and Histology

Human CRC tissues were fixed in 10% formalin before paraffin embedding. Sections of formalin-fixed paraffin-embedded (FFPE) tissues were deparaffinized with xylene and rehydrated with ethanol (twice in 100%, 90%, 80%, and 70%). The sections were heated in a boiling water bath with citric buffer for 12 min to retrieve antigens. Next, the sections were blocked by incubating for 30 min in 5% bovine serum albumin buffer. Tissues were incubated overnight at 4 °C with primary antibodies: anti-ACKR4 antibody (Novus, Centennial, CO, USA), anti-CD3 (Abcam, Cambridge, United Kingdom), and anti-CD11c (Abcam). The next day, the sections were washed and incubated with fluorescence-conjugated secondary antibodies (1:1000 dilution, ThermoFisher, Waltham, MA, USA) at room temperature for 1 h. After washing, the slides were mounted with ProLong Gold antifade mountant with DAPI and imaged. The researchers were blind to the ACKR4 expression level when evaluating the tumor immune infiltration. The information of primary antibodies is included in Table S1.

2.3. Western Blotting of ACKR4

Total protein of 40 µg was prepared from each sample and quantified by the Pierce™ BCA Protein Assay Kit (ThermoFisher). We ran the protein in sodium dodecyl sulfate–polyacrylamide (SDS) gel electrophoresis. The proteins from the gel were transferred to polyvinylidene difluoride membranes (ThermoFisher), blocked with 5% BSA, and incubated in primary antibodies overnight at 4 °C. The primary antibodies were anti-ACKR4 (Abcam) and anti-β-actin (Cell Signaling Technology, Danvers, MA, USA). The next day, the membranes were washed and incubated in peroxidase-linked anti-rabbit IgG and peroxidase-linked anti-mouse IgG for 1 h at room temperature. Pierce™ ECL Western blotting substrate (ThermoFisher) was used to image the membranes.

2.4. Cell Line Transfection and Transduction

We used the *ACKR4* shRNA expressing lentiviral vectors to knock down ACKR4 expression in the MC38 cell line. Briefly, 5 µg of DNA (2.5 µg of mixed shRNA expressing plasmids and 2.5 µg of pPACKH1-XL packaging vector) was mixed with 10 µL P3000™ reagent in 250 µL Opti-MEM medium. The pGIPZ vector was used as the backbone of *ACKR4* shRNA expression. Then the diluted DNA was added to 250 µL Lipofectamine™ 3000 Transfection Reagent and incubated for 15 min at room temperature. The mixture was added to 5×10^5 HEK293TN cells in one well of a 6-well plate. Another 500 µL Opti-MEM medium was added to make the final volume of 1000 µL. Then 24 h after the transfection, we changed the Opti-MEM medium to normal cell growth media and cultured the cells for another 24 h. Then the virus-containing media were collected and added to wild-type MC38 at different titrations. Empty shRNA vectors served as the negative control. Three days after the transduction, the transduced MC38 cells were subjected to antibiotic selection. After one week of antibiotic selection, we performed a Western blotting analysis of ACKR4 to validate the knockdown.

2.5. Dendritic Cell Isolation

A Dynabeads Untouched Mouse DC Enrichment Kit (ThermoFisher) was used, and the manufacturer's instructions were followed. Briefly, murine PBMCs were isolated from spleen, bilateral inguinal, brachial, and axillary lymph nodes by gradient centrifugation. The cells were incubated in antibody mix for 20 mins at 2 °C to 8 °C, washed, and then incubated with Depletion MyOne SA Dynabeads magnetic beads for 15 mins at 2 °C to 8 °C. The tube was placed on a magnet, and the untouched DCs in the supernatant were cultured in Iscove's Modified Dulbecco's Medium (IMDM) supplemented with 2000 IU/mL IL4, 2000 IU/mL granulocyte–macrophage colony-stimulating factor, and 2000 IU/mL tumor necrosis factor. All cytokines were purchased from R&D Systems (Minneapolis, MN, USA), Cat #: CDK008).

2.6. In Vivo DC Migration Assay

We resuspended 3×10^5 freshly enriched CD45.1$^+$ DCs in 50 µL PBS. We then injected them into multiple sites of MC38 subcutaneous tumors growing in C57BL/6 mouse with different ACKR4 expression (~500 mm^3, 3×10^5/tumor) by a syringe with a 30 G needle. Thirty-six hours after the injection, we sampled the tumor-draining lymph nodes (the unilateral inguinal and axillary lymph nodes). Then we isolated single cells from the tumor-draining lymph nodes (TdLNs) for detecting CD45.1$^+$ DCs by FACS analysis.

2.7. Flow Cytometry

Mouse tumor tissues were minced into small pieces ($2 \times 2 \times 2$ mm^3) and digested with collagenase IV (0.5 mg/mL) and deoxyribonuclease (50 units/mL) for 1 h at 37 °C. The digested tumor tissues and lymphatic tissues (TdLNs and spleens) were meshed and flushed through 70 µM and 40 µM strainers, respectively. Red blood cells were lysed by incubating the cells with red blood cell lysis buffer for 15 min and neutralizing with PBS. The cells were counted using a hemacytometer. Zombie Green fixable viability dye

(BioLegend, San Diego, CA, USA) was used to count live and dead cells. All the cells were stained with primary antibody cocktails for cell surface markers. For cytoplasmic staining, cells were treated with the Cyto-Fast Fix-Perm Buffer set (BioLegend). All samples were fixed after staining. The samples were immediately analyzed in a BD FACSCanto (BD Biosciences, Franklin Lakes, NJ, USA) cytometry to prevent signal deterioration. All the data were analyzed with the FlowJo (Version 10.7.2, BD Biosciences, Franklin Lakes, NJ, USA). The information of primary antibodies is included in Table S1.

2.8. In Vivo T-Cell Priming Assay

We cultured 3×10^5 freshly enriched DCs in 2 mL Dendritic Cell Base Media (R&D Systems) plus 10% FBS. A total of 40 µg of ovalbumin (OVA) peptides (257–264, AnaSpec) was supplied to the DC culture for a final concentration of 20 µg/mL. We also pulsed the DCs with lipopolysaccharide (LPS) (1 µg/mL) as a positive control of the DC maturation test. After 18 h of DC pulsing, we collected the DCs and injected them into multiple sites of MC38 subcutaneous tumors growing in C57BL/6 mouse with different ACKR4 expression (~300 mm^3, 3×10^5 million/tumor) by a syringe with a 30 G needle. Two weeks later, we collected the TdLNs for OVA-specific T-cell analysis.

Single cells were isolated from the TdLNs by mechanical tissue dissociation. Then, 3×10^5 single cells were resuspended in 100 µL PBS with 0.1 µL Zombie Green Fixable Viability dye and incubated at room temperature for 15 min. After washing, the cells were blocked with TruStain FcX™ PLUS (0.25 µg, Biolegend) and stained with Tetramer/BV421-H-2 Kb OVA (5 µL, MBL International, Woburn, MA, USA) for 40 min at room temperature. According to the manufacture's instruction and our preliminary experiment optimization, we used an anti-CD8 (clone KT15) antibody (MBL International) to minimize the false-positive rate of the tetramer staining. Lymphatic cells from naïve mice were used as a negative control.

2.9. Mouse Subcutaneous Models

The subcutaneous model was established by resuspending 5×10^5 MC38 cells in 100 µL Matrigel (BD Biosciences) and injecting the tumor cell suspension into the right flank of naïve C57BL/6 mice. Following injection, using an electronic caliper, tumor growth was monitored and measured 1–2 times a week. Tumor volume was calculated using the formula

$$(length*width^2)/2. \tag{1}$$

2.10. Mouse Treatment

Mice were treated with either IgG (5 mg/kg as an anti-4-1BB control, 10 mg/kg as an anti-PD-1 control, BioXcell, Lebanon, NH, USA), anti-4-1BB agonist (5 mg/kg, clone: 3H3, BioXcell), or anti-PD-1 (10 mg/kg, clone: RMP1-14, BioXcell) on day 10, 14, and 18. All treatments were given intraperitoneally (i.p.).

2.11. Quantitative PCR (qPCR) Analysis

The mirVana microRNA (miRNA) Isolation Kit (ThermoFisher Scientific) was used to extract total RNA from tumor cell lines and tissues. A total of 500 ng of total RNA was used for establishing the cDNA library with the miScript II RT Kit (Qiagen, Hilden, Germany). qRT-PCR was performed with the SYBR Green I Master kit (Roche Applied Science, Penzberg, Germany) in a LightCycler 480. The following forward primers were used: miR-552: GTTTAACCTTTTGCCTGTTGG and U6 snRNA: AAGGATGACACGCAAATTCG. The RT kit provides the universal reverse primer.

2.12. Enzyme-Linked Immunosorbent Assay (ELISA) for CCL21

CCL21 was quantified in tumor tissues and tumor-draining lymph nodes using an ELISA kit (Abcam). Briefly, tissue lysate samples were prepared by homogenizing tumor tissues and tumor-draining lymph nodes. We normalized the protein concentration between different samples before loading them to the experiment. The manufacturer's instructions were followed every step.

2.13. Statistical Analysis

We performed all statistical analyses and graphing using GraphPad Prism software (Version 8, San Diego, CA, USA). Data were displayed as means ± SEMs. For comparison of two groups' quantitative data, paired or unpaired Student's t-tests were used. For multiple group comparison, one-way analysis of variance (ANOVA) was used followed by Bonferroni correction. Kaplan–Meier curves and log-rank tests were used to compare survival outcomes between groups. We used the chi-square test to compare two variables in a contingency table to see if they were related. A two-tail p-value of less than 0.05 was considered statistically significant.

3. Results

3.1. ACKR4 Is Downregulated in CRC Compared with Normal Colon

To investigate the immunoregulatory role of ACKR4 in CRC, we first evaluated the ACKR4 expression in CRC and normal colon tissues. Analysis of the CRC dataset in the The Cancer Genome Atlas (TCGA) and another independent dataset reported by Vasaikar et al. [18] showed that ACKR4 expression was lower in CRC than in normal colon tissues (Figure 1A,B). Further stratification of the CRC cases based on the MSI/MSS statuses indicated that ACKR4 expression was lower in MSS/MSI-L tumors than the MSI-H tumors (Figure 1A,B). The immunofluorescence staining on sections of 68 human CRC and 17 normal colon tissues revealed that 88% of normal colon tissues and 78% of MSI-CRC tissues have abundant ACKR4 expression. In contrast, only 45% of MSS-CRC tissues have a similar ACKR4 level. These data confirmed the downregulation of ACKR4 in CRC tissues, especially in the MSS subtype (Figure 1C). Next, we evaluated the prognostic significance of ACKR4 in the TCGA cohort (Figure 1D). Although not statistically significant, patients with higher ACKR4 expression are more likely to have a longer median survival time than patients with lower ACKR4 expression (Figure 1D). To control the influence of MSS/MSI status on the survival benefit, we removed the MSI-H cases and performed a subgroup analysis with the MSS and MSI-L samples. Again, the ACKR4 high cases are more likely to have a better prognosis (Figure 1D). Finally, we determined the ACKR4 level in the mouse CRC cell lines, which are widely used in immunological studies. Notably, the mouse CRC cell line MC38 (MSI phenotype) had significantly higher ACKR4 expression than the CT26 cell line (MSS phenotype) (Figure 1E).

3.2. Knockdown of ACKR4 in Tumor Cells but Not the Host Tissues Accelerate Tumor Growth

Next, we sought to determine the impact of ACKR4 downregulation in CRC development. Using the vector-based short hairpin RNA (shRNA) interference technology, we knocked down ACKR4 expression in the MC38 cell line, which has relatively high endogenous ACKR4 expression ((Figure 1E and (Figure 2A). Knockdown of ACKR4 did not significantly influence the MC38 cell proliferation in vitro (Figure 2A). We then injected the MC38 cells subcutaneously into naïve C57BL/6 mice. Notably, the knockdown of ACKR4 in the tumor cells accelerated tumor growth in vivo (Figure 2B). To see whether the ACKR4 level in the host tissue also affects tumor development, we established a conditional ACKR4 knockdown mouse model (Figure 2C). We knocked down ACKR4 expression in the host mice by doxycycline treatment before MC38 tumor cell injection. However, the knockdown of ACKR4 in host tissue did not significantly alter the tumor development (Figure 2D). Our results indicated that ACKR4 of tumor cells is more competent in regulating tumor growth than the host ACKR4.

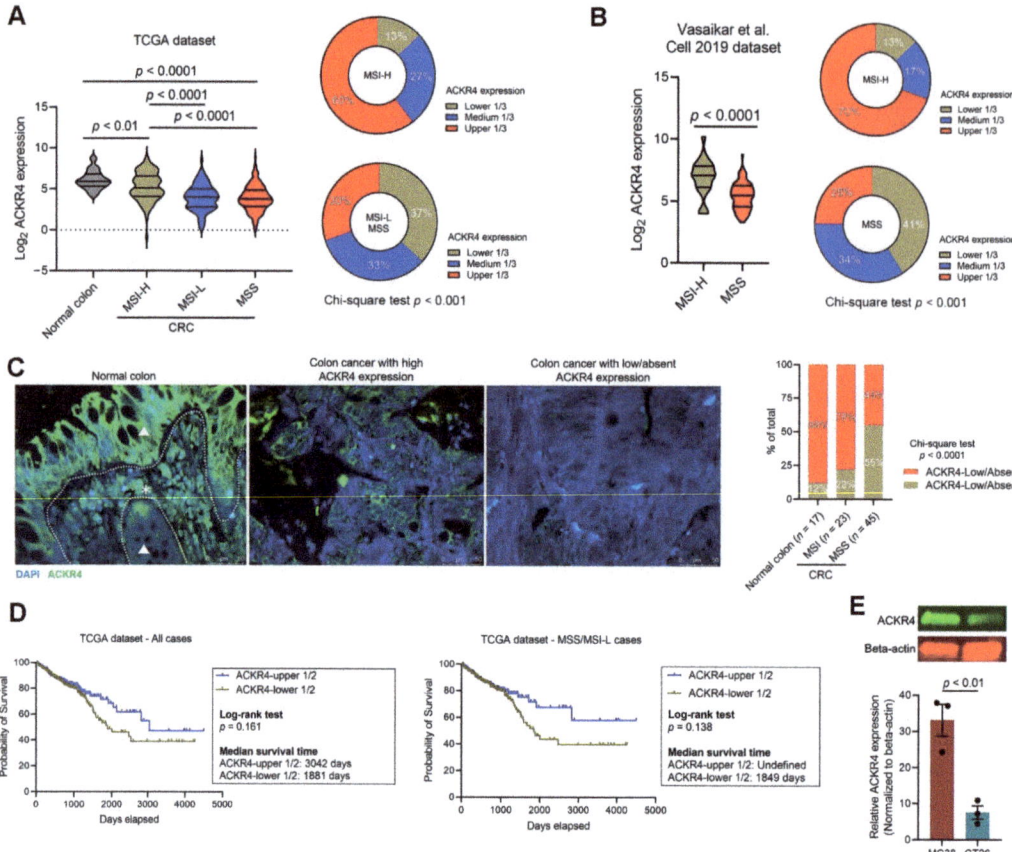

Figure 1. ACKR4 expression in human CRC tissue sample and cell lines. (**A**) *ACKR4* mRNA expression in the TCGA CRC dataset. The normal colon tissues had a higher ACKR4 expression level than the CRC tissues. The MSI-H subtype tumors had an elevated ACKR4 expression level compared to the MSI-L and MSS subtype tumors. (**B**) The *ACKR4* transcript levels in another independent Vasaikar et al. [18] dataset. (**C**) Immunofluorescence staining of ACKR4 in human normal colon tissues ($n = 17$) and CRC tumor tissues ($n = 23$ for MSI tumors and $n = 45$ for MSS tumors). The representative micrographs showed the low and high ACKR4 expression cases (The white dot line indicates the border of epithelium and stroma. The star indicates tumor stroma. The triangle indicates epithelium). (**D**) The overall survival curve of CRC patients with high or low ACKR4 expression (for the TCGA dataset, the median value of ACKR4 expression was used as the cut-off point). The comparisons were made in all CRC cases (left panel) or MSS and MSI-L cases (right panel; undefined means more than 50% of patients survive at the follow-up). (**E**) Western blotting analysis of ACKR4 expression in mouse CRC cell lines ($n = 3$). The ACKR4 expression level was normalized to the β-actin levels. (For more than two group statistical analyses, the uppermost *p*-value indicates the ANOVA-analysis, and other *p*-values indicate the posthoc analysis between two specific groups. TCGA: The Cancer Genome Atlas, MSS: Microsatellite stability, MSI-L: Microsatellite instability-low, MSI-H: Microsatellite instability-high, ACKR4: Atypical Chemokine Receptor 4, CRC: Colrectal cancer, ANOVA: Analysis of variance). Detailed information about the Western blotting can be found in Figure S3.

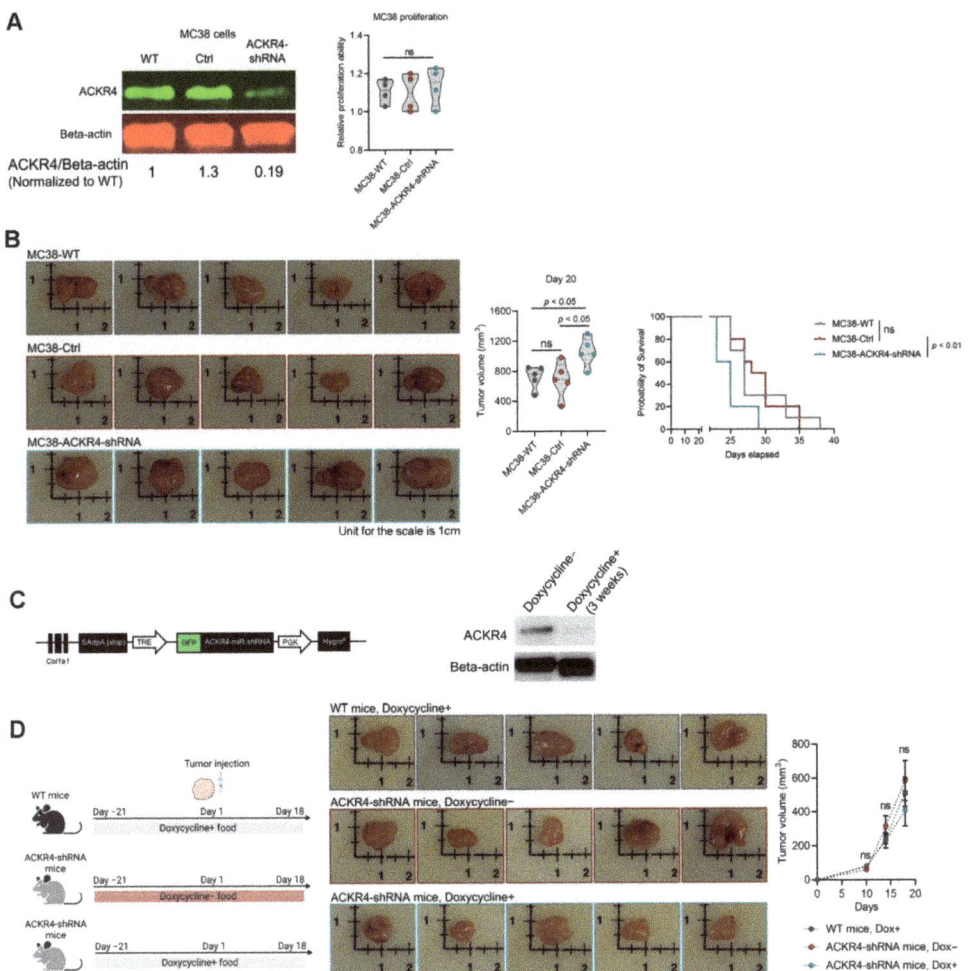

Figure 2. ACKR4 expression and tumor development. (**A**) Western blot analysis of ACKR4 knockdown in MC38 cell line (The grey area defines the data distribution. The dot lines in the violin plots indicate quartiles). (**B**) Knockdown of ACKR4 accelerated MC38 tumor growth in naïve C57BL/6J mice ($n = 5$ for tumor growth analysis and $n = 10$ for survival analysis). (**C**) The induction and confirmation of ACKR4 knockdown in transgenic mice. Doxycycline treatment for 3 weeks significantly reduced ACKR4 expression in the mouse skin and subcutaneous connective tissue. (**D**) Knockdown of ACKR4 in the host mice did not significantly affect MC38 tumor growth ($n = 5$). (For more than two group statistical analyses, the uppermost p-value indicates the ANOVA-analysis, and other p-values indicate the posthoc analysis between two specific groups. WT: Wild type, Ctrl: Control, shRNA: Short hairpin RNA, Dox: Doxycycline, Col1a1: Collagen, type I, alpha 1, GFP: Green fluorescent protein, Hygro^R: Hygromycin resistance, PGK: Phosphoglycerate kinase, TRE: Tetracycline response element, ns: No significance, ANOVA: Analysis of variance). Detailed information about the Western blotting can be found in Figures S4 and S5.

3.3. Loss of ACKR4 Reduces Tumor T-Cell Infiltration

To study whether the tumor growth caused by ACKR4 knockdown was associated with anti-tumor immunity, we analyzed the tumor immune infiltration in the TCGA CRC dataset by the CIBERSORT algorithm (Figures 3A,B and S1). Tumors with higher ACKR4 expression had elevated immune cell infiltration, including the total T-cells, CD8$^+$ T-cells,

CD4+ T-cells, regulatory T-cells (Treg), and total DCs, compared to tumors with lower ACKR4 expression (Figures 3B and S1A–C). Higher ACKR4 expression was also associated with more total NK cells, B-cells, and polarized macrophages in the tumor microenvironment (Figure S1D–H). Histological analysis on human CRC tissues confirmed that ACKR4 high-expressing tumors are associated with a higher number of tumor-infiltrating T-cells (Figure 3C). However, there was no difference in DC infiltration between the ACKR4-high and -low groups (Figure 3C). Next, we investigated the immune infiltration in ACKR4 knockdown tumor models (Figure S2A,B). Our results show that ACKR4 knockdown tumors have fewer CD4+ T-cells but a higher proportion of exhausted CD4+ T-cells in their tumor microenvironment than the control group (Figure 3D). However, the frequencies of tumor-infiltrating CD8+ T-cells and DCs are not influenced by ACKR4 expression (Figures 3D and S2C). The ACKR4 level in tumor cells also does not systemically change the frequency and function of immune cells in the tumor-draining lymph nodes (Figure S2D).

3.4. Loss of ACKR4 Impairs DC Migration to Tumor-Draining Lymph Nodes and Tumor-Specific T-Cell Expansion

Since ACKR4 regulates the CCL21 chemokine gradient [12], we hypothesized that loss of ACKR4 in tumor tissue would increase the CCL21 levels in the tumor microenvironment. An increase of CCL21 in the tumor tissue will potentially impede DC migration, mediated by the CCL21 chemokine gradient between the tumor tissue and the tumor-draining lymph nodes (TdLNs). To validate this hypothesis, we injected the CD45.1+ DCs into tumors with wild-type or knocked-down ACKR4 expression. We then analyzed the amount of CD45.1+ DCs in the TdLNs. Notably, DCs in the wild-type and control tumors are more likely to migrate to the TdLNs than the DCs in the ACKR4 knockdown tumors (Figure 4A). To observe whether the reduction of DC migration would cause the impaired tumor-specific T-cell priming in the TdLNs, we tested for the antigen-specific T-cells in the TdLNs. We first pulsed the DCs with the ovalbumin (OVA) antigen and then injected them into the tumors. We confirmed the DCs we used expressing DC maturation markers, CD80 and CD86 (Figure 4B). We analyzed the OVA-specific CD8+ T-cells in the TdLNs and found that AKCR4 knockdown in the tumor significantly reduced the DC mediated antigen-specific T-cell priming in the TdLNs (Figure 4B). We also confirmed the finding with the endogenous tumor antigen (Figure 4C). Finally, we determined that the CCL21 level in the ACKR4 knockdown tumor tissues was significantly higher than in the wild-type and control groups (Figure 4D).

3.5. Loss of ACKR4 Weakens Tumor Response to Immune Checkpoint Blockade

Because ACKR4 knockdown reduces the tumor infiltrating T-cells and DC mediated tumor-specific T-cell expansion in the TdLNs (Figures 3 and 4), we next evaluated whether ACKR4 knockdown affects the tumor response to immune checkpoint blockade. We treated the wild-type, control, and ACKR4 knockdown tumors with anti-PD-1 or anti-4-1BB antibodies. The ACKR4 knockdown tumors were less sensitive to anti-PD-1 or anti-4-1BB treatments than wild-type and control tumors (Figure 5). This result suggested that loss of ACKR4 could be implicated in the immune checkpoint blockade resistance in CRC.

3.6. MicroRNA miR-552 Downregulates ACKR4 in CRC

Our previous microRNA (miRNA) expression profiling analysis had shown that miR-552 is highly expressed in MSS-CRC, which does not respond to immune checkpoint blockade [19]. Further sequence match analysis showed that miR-552 potentially binds to human ACKR4 transcript and subsequently downregulates ACKR4 expression (Figure 6A). Our dual luciferase assay and flow cytometry analysis confirmed the effects of miR-552 on ACKR4 downregulation in human CRC cell lines (Figure 6A,B). Analysis of the TCGA-CRC dataset further confirmed the negative correlation between miR-552 and ACKR4 (Figure 6C).

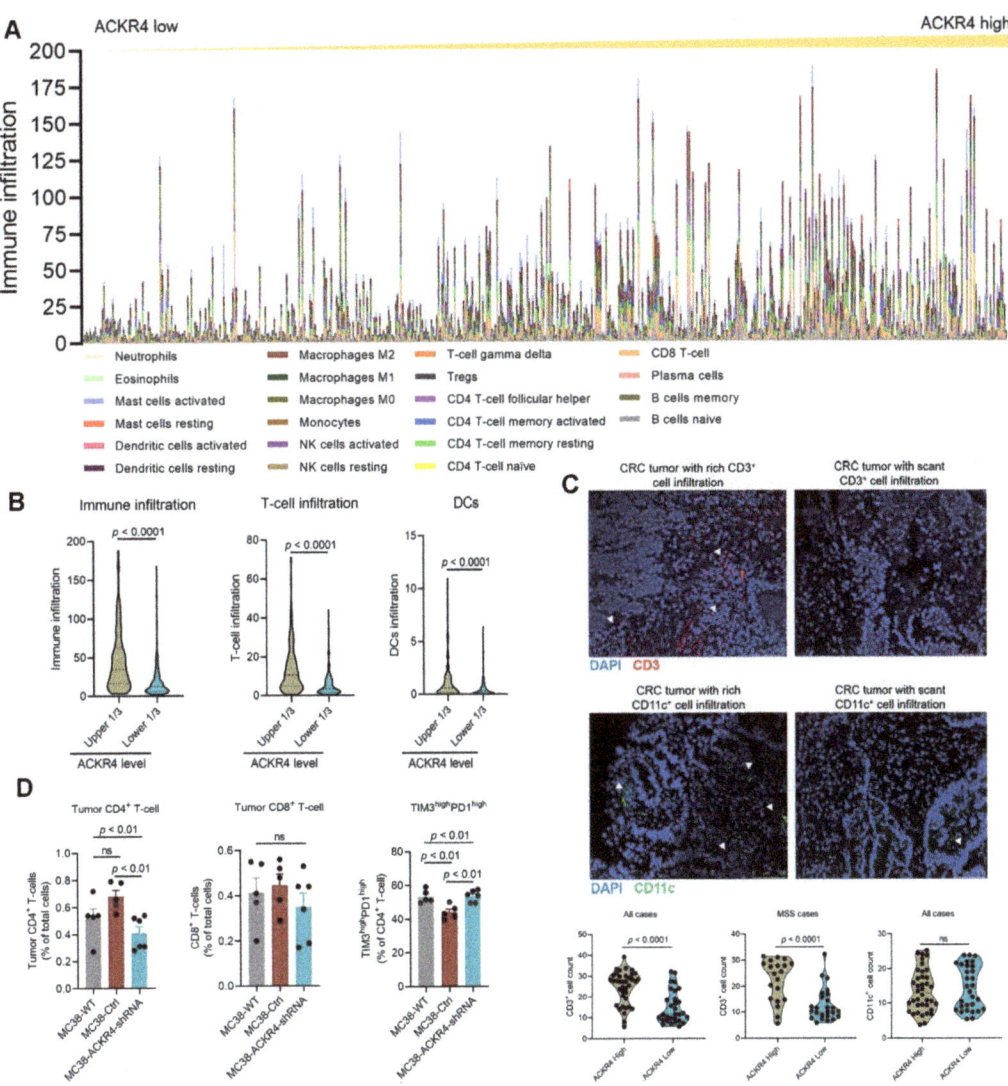

Figure 3. ACKR4 expression and tumor immune cell infiltration. (**A,B**) The immune profiles of CRC cases in the TCGA dataset generated by the CIBERSORT. Elevated ACKR4 expression is associated with higher total immune cells, T-cells, and DC infiltration (The dot lines in the violin plots indicate quartiles). (**C**) Immunofluorescence analysis of CD3 and CD11c on human CRC tissues. High ACKR4 expression was associated with high T-cell (CD3$^+$) but not dendritic cell (CD11c$^+$) infiltration (n = 68, the triangles indicate positive staining, the dot lines in the violin plots indicate quartiles). (**D**) FACS analysis on tumor-infiltrating T-cells on MC38 tumor models. ACKR4 knockdown MC38 tumors had fewer CD4$^+$ T-cells in their tumor microenvironment. The percentage of exhausted CD4$^+$ T-cells was higher in the ACKR4 knockdown tumors than in the controls (n = 5–6). (For more than two group statistical analyses, the uppermost p-value indicates the ANOVA-analysis, and other p-values indicate the posthoc analysis between two specific groups. DCs: Dendritic cells, CD8: Cluster of differentiation 8, CD4: Cluster of differentiation 4, CD3: Cluster of differentiation 3, CD11c: Cluster of differentiation 11c, WT: Wild type, Ctrl: Control, shRNA: Short hairpin RNA, TCGA: The Cancer Genome Atlas, CRC: Colrectal cancer, Treg: Regulatory T-cell, TIM3: T-cell immunoglobulin domain and mucin domain 3, PD1: Programmed cell death protein 1, MSS: Microsatellite stability, DAPI: 4′,6-diamidino-2-phenylindole, ANOVA: Analysis of variance, ns: No significance).

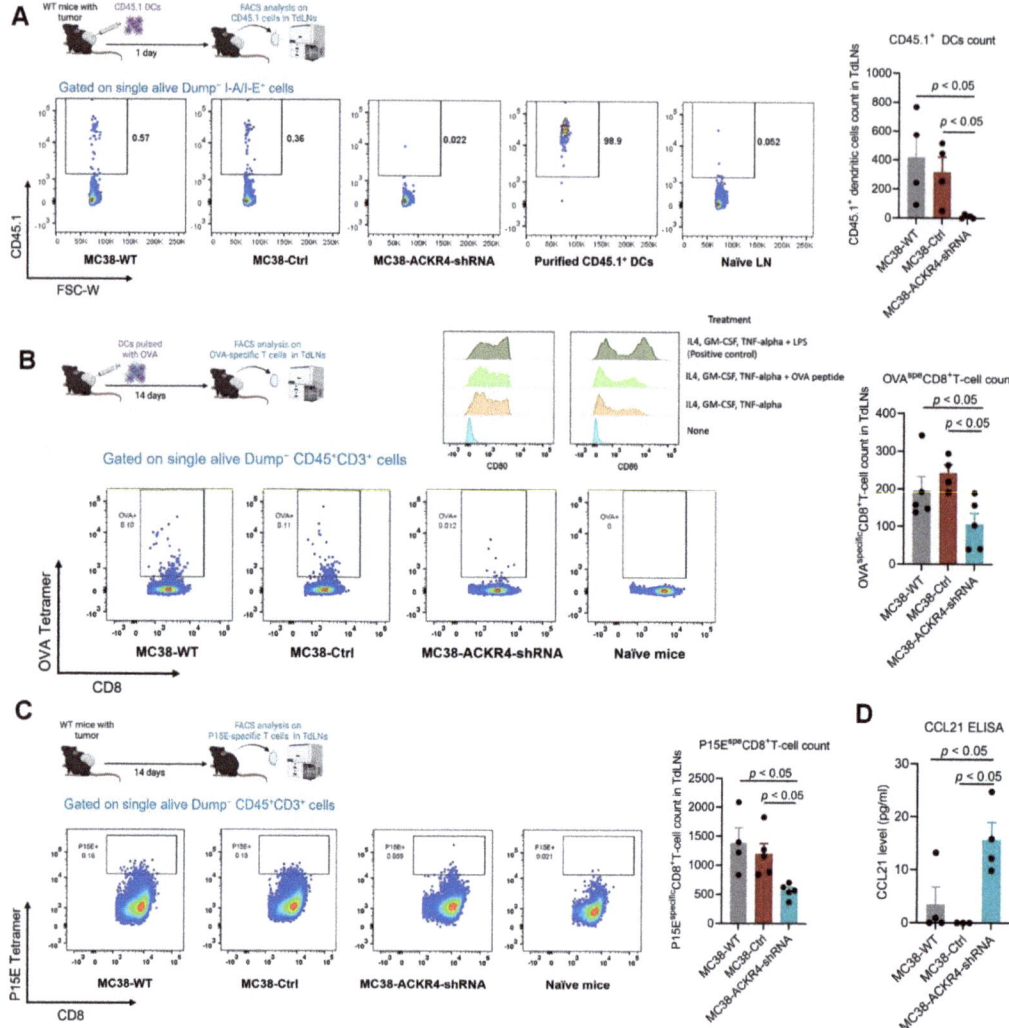

Figure 4. ACKR4 expression and DC migration and T-cell priming. (**A**) Enriched CD45.1+ DCs were injected into the MC38 tumor and analyzed in TdLNs 1 day post-injection. ACKR4 knockdown in MC38 tumor cells impaired DC migration from the tumor to the TdLNs ($n = 4$). (**B**) DCs loaded with OVA antigens were injected into the MC38 tumor microenvironment, and the OVA-specific CD8+ T-cells were analyzed in the TdLNs. ACKR4 knockdown in MC38 tumor cells impaired DC mediated T-cell priming ($n = 4$–5). The histogram shows CD80 and CD86 expression on DCs used in the study. (**C**) P15E (a tumor-associated antigen in MC38 cells)-specific CD8+ T-cell counts in TdLNs of MC38 tumors with various ACKR4 expression levels ($n = 4$–5). (**D**) CCL21 quantification in MC38 tumors with different ACKR4 expression levels ($n = 3$–4). (For more than two group statistical analyses, the uppermost p-value indicates the ANOVA-analysis, and other p-values indicate the posthoc analysis between two specific groups. DCs: Dendritic cells, OVA: Ovalbumin, CD8: Cluster of differentiation 8, CD3: Cluster of differentiation 3, CD80: Cluster of differentiation 80, CD86: Cluster of differentiation 86, CCL21: Chemokine (C-C motif) ligand 21, FACS: Fluorescence-activated cell sorting, TdLNs: Tumor-draining lymph nodes, P15E: Murine leukemia virus envelope protein P15E, FSC-W: Forward light scatter width, CD45.1: Cluster of differentiation 45.1, CD45: Cluster of differentiation 45, WT: Wild type, Ctrl: Control, shRNA: Short hairpin RNA, IL4: Interleukin 4, GM-CSF: Granulocyte-macrophage colony-stimulating factor, TNF: Tumor necrosis factor, LPS: Lipopolysaccharides, Spe: Specific, ANOVA: Analysis of variance).

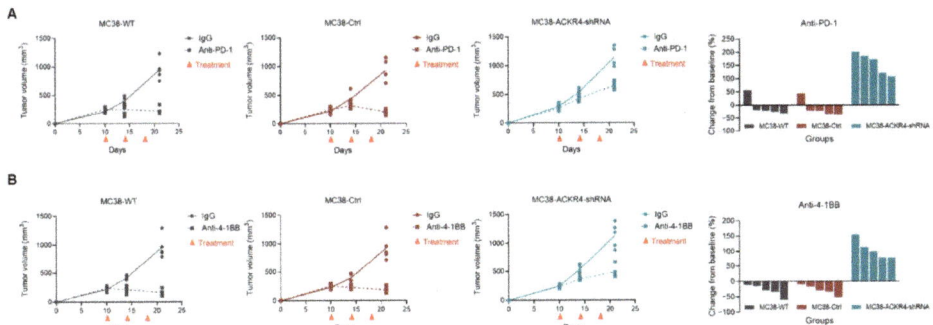

Figure 5. Immunotherapy response on MC38 tumors with different ACKR4 expression levels. (**A**) The mice were treated by anti-PD-1 on days 10, 14, and 18. The waterfall plot shows the individual tumor volume change post-treatment. The response of the ACKR4 knockdown group to anti-PD-1 treatment was worse than that of the other groups. (**B**) The anti-4-1BB agonist treatment showed similar results to the anti-PD-1 treatment. (WT: Wild type, Ctrl: Control, shRNA: Short hairpin RNA, PD-1: Programmed cell death protein 1, 4-1BB: CD137/Tumor necrosis factor receptor superfamily 9, IgG: Immunoglobulin G, ANOVA: Analysis of variance).

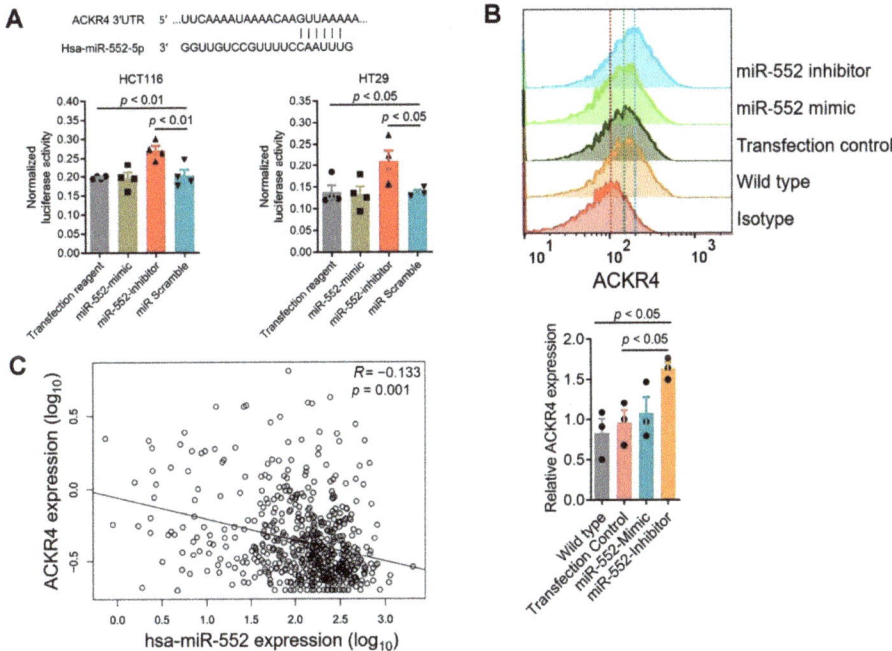

Figure 6. miR-552 downregulates ACKR4 expression in CRC tumors. (**A**) The sequence match between the miR-552 and the *ACKR4* 3′-untranslated region (UTR). Dual-luciferase assay confirmed that miR-552 binds to the 3′-UTR of *ACKR4* ($n = 4$). (**B**) miR-552 inhibitors enhanced ACKR4 expression in HCT116 cells ($n = 3$, the red vertical dot line indicates the isotype group's mean signal intensity, the green dot line indicates the transfection control group's mean signal intensity, and the blue vertical dot line indicates the miR-552 inhibitor group's mean signal intensity). (**C**) A negative correlation between *ACKR4* and miR-552 in the TCGA colorectal cancer dataset (The black line is the regression line). (For more than two group statistical analyses, the uppermost *p*-value indicates the ANOVA-analysis, and other *p*-values indicate the posthoc analysis between two specific groups. miR: MicroRNA, Hsa: Homo sapiens, TCGA: The Cancer Genome Atlas, ANOVA: Analysis of variance).

4. Discussion

Investigating the regulatory mechanism of tumor immunity is essential to alleviate drug resistance and improve the effect of immunotherapy [20,21]. As the key cell type in the process of antigen presentation, DCs and their function are closely associated with the intensity of tumor immunity [9–11]. The CCR7 expressed on DCs and the CCL19/21 gradient in the interstitial compartment largely regulates DC migration [12,13]. ACKR4 shapes the CCL19/21 gradient between the non-lymphatic and lymphatic tissues by scavenging both the soluble and immobilized CCL19/CCL21 [12,13]. In breast cancer, nasopharyngeal cancer, liver cancer, and cervical cancer, ACKR4 negatively regulates tumor growth and metastasis, implying a protective role in tumorigenesis [22–25]. However, the role of ACKR4 in tumor immunogenicity and overall anti-tumor immunity of CRC has not been determined.

Our study first evaluated the expression of ACKR4 in human normal colon and CRC tissues and revealed that ACKR4 was downregulated in CRC. This result corroborates a recent study showing that villous colon adenomas have less ACKR4 expression than the normal colon tissues [26]. Further analysis indicated that the MSI-H CRC had relatively higher expression of ACKR4 than the MSI-L/MSS CRC samples. These data showed the correlation between ACKR4 expression and CRC progression, providing the cornerstone for further studying the implications of ACKR4 in CRC pathobiology.

A key question is whether AKCR4 of tumor cells or ACKR4 of tumor-associated stromal cells affects tumor growth. Taking advantage of the inducible ACKR4 knockdown mice model, we were able to allow the mice to mature with intact ACKR4 expression and selectively downregulate the ACKR4 expression in the host right before and during wild-type MC38 tumor development. In another model, we knocked down ACKR4 in MC38 cells, which have a relatively high endogenous ACKR4 expression, and injected those cells into wild-type mice. Notably, ACKR4 knockdown in MC38 cells significantly accelerated tumor growth. However, ACKR4 expression in the stromal cells did not affect tumor growth. These results highlighted the distinct functions of ACKR4 in tumor cell and stromal cell compartments. Our data are distinctive from the previous study showing that ACKR4 knockout mice delayed E0771 mammary tumor growth [27]. These differences may be attributed to the different tumor cell lines tested. Although there are still controversies, permanent germline ACKR4 knockout may cause abnormalities in immune organ development [28–30]. This might be another reason why our results from inducible ACKR4 knockdown mice are different from embryonic ACKR4 knockout mice.

DCs have been identified as the most potent antigen-presenting cells in tumor antigen presentation and T-cell priming [6,9–11]. ACKR4, a decoy receptor that binds and degrades CCR7 ligands CCL19/CCL21, regulates DC migration from skin to the regional lymph nodes [12,13]. However, whether similar effects exist in tumor conditions remains unknown. Our work demonstrated that in the case of ACKR4 knockdown, tumor-infiltrating DCs are less likely to migrate towards TdLNs, causing a weak tumor-specific T-cell expansion in TdLNs. Consequently, the intensity of anti-tumor immunity and response to ICB was significantly restricted by ACKR4 downregulation. These data support our previous work showing that the immune response that occurs in TdLNs is extremely critical for initiating anti-tumor immunity [31]. In addition, our study also indicates that miR-552 negatively regulates ACKR4, and blocking the function of miR-552 increases ACKR4 expression in human CRC cell lines. Those results provided a potential target to rescue the ACKR4 expression in tumors.

Although our work has efficiently demonstrated the ACKR4 function in anti-tumor immunity, a few limitations remain. First, we did not investigate whether the ACKR4 function is dependent on the CCR7. However, it is the next step to determine if the ACKR4-mediated immunoregulation relies entirely on the CCR7 signaling or other pathways. Moreover, our work is restricted to the MC38 cell line in wild-type and ACKR4 knockdown mice. Due to technical difficulties, we could not overexpress ACKR4 in another widely used CRC cell line, CT26, which has a low ACKR4 expression. Further work with additional preclinical models are needed to confirm the conserved mechanism of ACKR4 mediated immunoregulation.

5. Conclusions

In conclusion, our work indicated that loss of ACKR4 in CRC is associated with poor anti-tumor immune infiltration. Mechanistically, the knockdown of ACKR4 in tumor cells restricts DC migration from tumor tissue to the tumor draining lymph nodes, thus impairing the tumor-specific T-cell priming and response to ICB. These data, collectively, describe a novel immunosuppressive mechanism and increase our understanding of how intrinsic tumor factors affect DC-mediated immune response in CRC.

Supplementary Materials: The following are available online at https://www.mdpi.com/article/10.3390/cancers13195021/s1, Figure S1: ACKR4 expression and tumor immune infiltration in TCGA CRC dataset, Figure S2: ACKR4 expression and tumor immune infiltration, Figure S3: Full Western blot images for Figure 1E, Figure S4: Full Western blot images for Figure 2A, Figure S5: Full Western blot images for Figure 2C, Table S1: Key resources.

Author Contributions: Conceptualization, D.W., X.Z. and S.S.; Data curation, D.W. and X.Z.; Formal analysis, D.W., C.Y. and X.Z.; Funding acquisition, X.Z. and S.S.; Investigation, D.W., W.S.M. and X.Z.; Methodology, P.K.P. and X.Z.; Project administration, S.S.; Resources, P.K.P. and S.S.; Software, C.Y.; Visualization, X.Z.; Writing—original draft, D.W. and X.Z.; Writing—review and editing, C.Y., W.S.M. and S.S. All authors have read and agreed to the published version of the manuscript.

Funding: This study was supported by the Minnesota Colorectal Cancer Research Foundation, Mezin Koats Colorectal Cancer Funds, Department of Surgery, University of Minnesota research funds, and the National Institutes of Health grant R03CA219129. Dechen Wangmo was supported by the National Institutes of Health's National Center for Advancing Translational Sciences, grants TL1R002493 and UL1TR002494. The content is solely the responsibility of the authors and does not necessarily represent the official views of the National Institutes of Health's National Center for Advancing Translational Sciences.

Institutional Review Board Statement: The study was conducted according to the guidelines of the Declaration of Helsinki and approved by the Institutional Review Board of the University of Minnesota (IRB ID: 1611M00002, latest approved on 8 July 2021).

Informed Consent Statement: Informed consent was obtained from all subjects involved in the study.

Data Availability Statement: Data available on request from the corresponding author due to privacy.

Conflicts of Interest: The authors declare no conflict of interest.

References

1. Siegel, R.L.; Miller, K.D.; Fuchs, H.E.; Jemal, A. Cancer Statistics, 2021. *CA A Cancer J. Clin.* **2021**, *71*, 7–33. [CrossRef] [PubMed]
2. Siegel, R.L.; Torre, L.A.; Soerjomataram, I.; Hayes, R.B.; Bray, F.; Weber, T.K.; Jemal, A. Global patterns and trends in colorectal cancer incidence in young adults. *Gut* **2019**, *68*, 2179–2185. [CrossRef] [PubMed]
3. Stein, A.; Moehler, M.; Trojan, J.; Goekkurt, E.; Vogel, A. Immuno-oncology in GI tumours: Clinical evidence and emerging trials of PD-1/PD-L1 antagonists. *Crit. Rev. Oncol./Hematol.* **2018**, *130*, 13–26. [CrossRef] [PubMed]
4. Lumish, M.A.; Cercek, A. Immunotherapy for the treatment of colorectal cancer. *J. Surg. Oncol.* **2021**, *123*, 760–774. [CrossRef]
5. Wang, S.; He, Z.; Wang, X.; Li, H.; Liu, X.S. Antigen presentation and tumor immunogenicity in cancer immunotherapy response prediction. *eLife* **2019**, *8*, e49020. [CrossRef] [PubMed]
6. Wang, Y.; Xiang, Y.; Xin, V.W.; Wang, X.W.; Peng, X.C.; Liu, X.Q.; Wang, D.; Li, N.; Cheng, J.T.; Lyv, Y.N.; et al. Dendritic cell biology and its role in tumor immunotherapy. *J. Hematol. Oncol.* **2020**, *13*, 107. [CrossRef]
7. Wculek, S.K.; Cueto, F.J.; Mujal, A.M.; Melero, I.; Krummel, M.F.; Sancho, D. Dendritic cells in cancer immunology and immunotherapy. *Nat. Rev. Immunol.* **2020**, *20*, 7–24. [CrossRef]
8. Mayoux, M.; Roller, A.; Pulko, V.; Sammicheli, S.; Chen, S.; Sum, E.; Jost, C.; Fransen, M.F.; Buser, R.B.; Kowanetz, M.; et al. Dendritic cells dictate responses to PD-L1 blockade cancer immunotherapy. *Sci. Transl. Med.* **2020**, *12*, eaav7431. [CrossRef] [PubMed]
9. Roberts, E.W.; Broz, M.L.; Binnewies, M.; Headley, M.B.; Nelson, A.E.; Wolf, D.M.; Kaisho, T.; Bogunovic, D.; Bhardwaj, N.; Krummel, M.F. Critical Role for CD103(+)/CD141(+) Dendritic Cells Bearing CCR7 for Tumor Antigen Trafficking and Priming of T-cell Immunity in Melanoma. *Cancer Cell* **2016**, *30*, 324–336. [CrossRef]

10. Salmon, H.; Idoyaga, J.; Rahman, A.; Leboeuf, M.; Remark, R.; Jordan, S.; Casanova-Acebes, M.; Khudoynazarova, M.; Agudo, J.; Tung, N.; et al. Expansion and Activation of CD103(+) Dendritic Cell Progenitors at the Tumor Site Enhances Tumor Responses to Therapeutic PD-L1 and BRAF Inhibition. *Immunity* **2016**, *44*, 924–938. [CrossRef]
11. Binnewies, M.; Mujal, A.M.; Pollack, J.L.; Combes, A.J.; Hardison, E.A.; Barry, K.C.; Tsui, J.; Ruhland, M.K.; Kersten, K.; Abushawish, M.A.; et al. Unleashing Type-2 Dendritic Cells to Drive Protective Antitumor CD4(+) T-cell Immunity. *Cell* **2019**, *177*, 556–571.e516. [CrossRef]
12. Bastow, C.R.; Bunting, M.D.; Kara, E.E.; McKenzie, D.R.; Caon, A.; Devi, S.; Tolley, L.; Mueller, S.N.; Frazer, I.H.; Harvey, N.; et al. Scavenging of soluble and immobilized CCL21 by ACKR4 regulates peripheral dendritic cell emigration. *Proc. Natl. Acad. Sci. USA* **2021**, *118*, e2025763118. [CrossRef]
13. Bryce, S.A.; Wilson, R.A.; Tiplady, E.M.; Asquith, D.L.; Bromley, S.K.; Luster, A.D.; Graham, G.J.; Nibbs, R.J. ACKR4 on Stromal Cells Scavenges CCL19 To Enable CCR7-Dependent Trafficking of APCs from Inflamed Skin to Lymph Nodes. *J. Immunol.* **2016**, *196*, 3341–3353. [CrossRef]
14. Hjorto, G.M.; Larsen, O.; Steen, A.; Daugvilaite, V.; Berg, C.; Fares, S.; Hansen, M.; Ali, S.; Rosenkilde, M.M. Differential CCR7 Targeting in Dendritic Cells by Three Naturally Occurring CC-Chemokines. *Front. Immunol.* **2016**, *7*, 568. [CrossRef]
15. Nibbs, R.J.; Graham, G.J. Immune regulation by atypical chemokine receptors. *Nat. Rev. Immunol.* **2013**, *13*, 815–829. [CrossRef] [PubMed]
16. Matti, C.; Salnikov, A.; Artinger, M.; D'Agostino, G.; Kindinger, I.; Uguccioni, M.; Thelen, M.; Legler, D.F. ACKR4 Recruits GRK3 Prior to beta-Arrestins but Can Scavenge Chemokines in the Absence of beta-Arrestins. *Front. Immunol.* **2020**, *11*, 720. [CrossRef] [PubMed]
17. Zhao, X.; Yuan, C.; Wangmo, D.; Subramanian, S. Tumor-Secreted Extracellular Vesicles Regulate T-Cell Costimulation and Can Be Manipulated To Induce Tumor-Specific T-Cell Responses. *Gastroenterology* **2021**, *161*, 560–574.e11. [CrossRef] [PubMed]
18. Vasaikar, S.; Huang, C.; Wang, X.; Petyuk, V.A.; Savage, S.R.; Wen, B.; Dou, Y.; Zhang, Y.; Shi, Z.; Arshad, O.A.; et al. Proteogenomic Analysis of Human Colon Cancer Reveals New Therapeutic Opportunities. *Cell* **2019**, *177*, 1035–1049.e1019. [CrossRef]
19. Sarver, A.L.; French, A.J.; Borralho, P.M.; Thayanithy, V.; Oberg, A.L.; Silverstein, K.A.; Morlan, B.W.; Riska, S.M.; Boardman, L.A.; Cunningham, J.M.; et al. Human colon cancer profiles show differential microRNA expression depending on mismatch repair status and are characteristic of undifferentiated proliferative states. *BMC Cancer* **2009**, *9*, 401. [CrossRef] [PubMed]
20. Zhao, X.; Subramanian, S. Intrinsic Resistance of Solid Tumors to Immune Checkpoint Blockade Therapy. *Cancer Res.* **2017**, *77*, 817–822. [CrossRef]
21. Perez-Ruiz, E.; Melero, I.; Kopecka, J.; Sarmento-Ribeiro, A.B.; Garcia-Aranda, M.; De Las Rivas, J. Cancer immunotherapy resistance based on immune checkpoints inhibitors: Targets, biomarkers, and remedies. *Drug Resist. Updates: Rev. Comment. Antimicrob. Anticancer Chemother.* **2020**, *53*, 100718. [CrossRef]
22. Hou, T.; Liang, D.; Xu, L.; Huang, X.; Huang, Y.; Zhang, Y. Atypical chemokine receptors predict lymph node metastasis and prognosis in patients with cervical squamous cell cancer. *Gynecol. Oncol.* **2013**, *130*, 181–187. [CrossRef]
23. Feng, L.Y.; Ou, Z.L.; Wu, F.Y.; Shen, Z.Z.; Shao, Z.M. Involvement of a novel chemokine decoy receptor CCX-CKR in breast cancer growth, metastasis and patient survival. *Clin. Cancer Res. Off. J. Am. Assoc. Cancer Res.* **2009**, *15*, 2962–2970. [CrossRef]
24. Ju, Y.; Sun, C.; Wang, X. Loss of atypical chemokine receptor 4 facilitates C-C motif chemokine ligand 21-mediated tumor growth and invasion in nasopharyngeal carcinoma. *Exp. Ther. Med.* **2019**, *17*, 613–620. [CrossRef] [PubMed]
25. Shi, J.Y.; Yang, L.X.; Wang, Z.C.; Wang, L.Y.; Zhou, J.; Wang, X.Y.; Shi, G.M.; Ding, Z.B.; Ke, A.W.; Dai, Z.; et al. CC chemokine receptor-like 1 functions as a tumour suppressor by impairing CCR7-related chemotaxis in hepatocellular carcinoma. *J. Pathol.* **2015**, *235*, 546–558. [CrossRef]
26. Lewandowska, P.; Wierzbicki, J.; Zawadzki, M.; Agrawal, A.; Krzystek-Korpacka, M. Biphasic Expression of Atypical Chemokine Receptor (ACKR) 2 and ACKR4 in Colorectal Neoplasms in Association with Histopathological Findings. *Biomolecules* **2020**, *11*, 8. [CrossRef] [PubMed]
27. Whyte, C.E.; Osman, M.; Kara, E.E.; Abbott, C.; Foeng, J.; McKenzie, D.R.; Fenix, K.A.; Harata-Lee, Y.; Foyle, K.L.; Boyle, S.T.; et al. ACKR4 restrains antitumor immunity by regulating CCL21. *J. Exp. Med.* **2020**, *217*, e20190634. [CrossRef]
28. Lucas, B.; White, A.J.; Ulvmar, M.H.; Nibbs, R.J.; Sitnik, K.M.; Agace, W.W.; Jenkinson, W.E.; Anderson, G.; Rot, A. CCRL1/ACKR4 is expressed in key thymic microenvironments but is dispensable for T lymphopoiesis at steady state in adult mice. *Eur. J. Immunol.* **2015**, *45*, 574–583. [CrossRef] [PubMed]
29. Eckert, N.; Werth, K.; Willenzon, S.; Tan, L.; Forster, R. B cell hyperactivation in an ACKR4-deficient mouse strain is not caused by lack of ACKR4 expression. *J. Leukoc. Biol.* **2020**, *107*, 1155–1166. [CrossRef] [PubMed]
30. Werth, K.; Hub, E.; Gutjahr, J.C.; Bosjnak, B.; Zheng, X.; Bubke, A.; Russo, S.; Rot, A.; Forster, R. Expression of ACKR4 demarcates the "peri-marginal sinus," a specialized vascular compartment of the splenic red pulp. *Cell Rep.* **2021**, *36*, 109346. [CrossRef]
31. Zhao, X.; Kassaye, B.; Wangmo, D.; Lou, E.; Subramanian, S. Chemotherapy but Not the Tumor Draining Lymph Nodes Determine the Immunotherapy Response in Secondary Tumors. *iScience* **2020**, *23*, 101056. [CrossRef] [PubMed]

Article

Antigen Presenting Cells from Tumor and Colon of Colorectal Cancer Patients Are Distinct in Activation and Functional Status, but Comparably Responsive to Activated T Cells

Frank Liang [1,*], Azar Rezapour [1], Louis Szeponik [1], Samuel Alsén [1], Yvonne Wettergren [2], Elinor Bexe Lindskog [2], Marianne Quiding-Järbrink [1] and Ulf Yrlid [1,*]

1 Department of Microbiology and Immunology, Institute of Biomedicine, Sahlgrenska Academy, University of Gothenburg, 405 30 Gothenburg, Sweden; azar.rezapour@gu.se (A.R.); louis.szeponik@gu.se (L.S.); samuel.alsen@gu.se (S.A.); marianne.quiding-jarbrink@microbio.gu.se (M.Q.-J.)
2 Department of Surgery, Institute of Clinical Sciences, Sahlgrenska University Hospital, University of Gothenburg, 413 45 Gothenburg, Sweden; yvonne.wettergren@gu.se (Y.W.); elinor.bexe-lindskog@surgery.gu.se (E.B.L.)
* Correspondence: frank.liang@gu.se (F.L.); ulf.yrlid@gu.se (U.Y.)

Simple Summary: Colorectal cancer (CRC) remains the third most common cancer. Associations between intratumoral T cells, also known as tumor infiltrating lymphocytes (TILs), and the CRC patients' responses to treatment have been described. Traditionally, TILs and antigen presenting cells (APCs) are studied separately on preserved CRC biopsies, disregarding the adjacent colonic tissue that would also be exposed to the administrated chemotherapy or radiotherapy. Thus, combined data sets on the subset composite and functional capacity of APCs and T cells within the same tumor, as well as colonic tissue, remain infrequent. Our phenotypic and functional comparison of T cell and APC subsets in tumor vs. colon from patients with CRC may give further insights into their propensity to maintain CRC treatment-induced immune responses locally in tumor and off-target colonic tissue.

Abstract: Although mouse models of CRC treatments have demonstrated robust immune activation, it remains unclear to what extent CRC patients' APCs and TILs interact to fuel or quench treatment-induced immune responses. Our ex vivo characterization of tumor and adjacent colon cell suspensions suggest that contrasting environments in these tissues promoted inversed expression of T cell co-stimulatory CD80, and co-inhibitory programmed death (PD)-ligand1 (PD-L1) on intratumoral vs. colonic APCs. While putative tumor-specific CD103+CD39+CD8+ TILs expressed lower CD69 (early activation marker) and higher PD-1 (extended activation/exhaustion marker) than colonic counterparts, the latter had instead higher CD69 and lower PD-1 levels. Functional comparisons showed that intratumoral APCs were inferior to colonic APCs regarding protein uptake and upregulation of CD80 and PD-L1 after protein degradation. Our attempt to model CRC treatment-induced T cell activation in vitro showed less interferon (IFN)-γ production by TILs than colonic T cells. In this model, we also measured APCs' CD80 and PD-L1 expression in response to activated co-residing T cells. These markers were comparable in the two tissues, despite higher IFN-γ exposure for colonic APCs. Thus, APCs within distinct intratumoral and colonic milieus showed different activation and functional status, but were similarly responsive to signals from induced T cell activation.

Keywords: colorectal cancer; tumor microenvironment; antigen presenting cells; T cells

1. Introduction

Cancer-related mortality remains high in patients with CRC [1,2]. It is widely accepted that TILs have prognostic value in CRC [3–5], and this has prompted the phenotypic definition of tumor-specific TILs. Previous studies have proposed the co-expression of CD103 and CD39 as markers for tumor-specificity on CD8+ TILs [6,7]. These memory T cells infiltrating the tumor likely interact with intratumoral APCs comprised of macrophages (MPs) and dendritic cells (DCs). The MPs are commonly divided into M1 (anti-tumor) and M2 (pro-tumor) subsets, although such distinction seems oversimplified due to the heterogeneity and plasticity of MPs [8,9]. Pro- or anti-tumor dichotomy has also been applied to DCs, since conventional DC 1(CDC1) may cross-present tumor antigens to CD8 T cells with tumor-killing potential, and CDC2 stimulate CD4 T cells that could promote or impede anti-tumor responses [10,11]. During APC-T cell interaction, MPs and DCs have the capacity to express cytokines or cell membrane-associated ligands that activate or inhibit effector functions of the interacting T cell. However, the role of DCs and MPs in CRC prognosis remains inconclusive [12–15].

Emerging reports associate TILs with clinical response to CRC treatments since treatment-induced immune activation indicated by the increase in TIL numbers was coupled with improved survival [16–19]. To date, immune checkpoint inhibition (ICI) with antibodies targeting PD-1 is approved for a subset of CRC patients. Although anti-PD-1 antibodies directly target T cells, models of ICI have demonstrated the prerequisite of specific DCs for tumor rejection [11,20,21]. Interestingly, the absence of co-stimulation is required for complete PD-1-mediated T cell inhibition [22]. Thus, to block inhibition conveyed by PD-1, mice treated with anti-PD-1 antibodies required specific T cell co-stimulation for efficient tumor elimination [23]. Since activated APCs highly express co-stimulatory molecules, they likely facilitate the unleashing of T cells in PD-1-targeting ICI. Further, mechanistic mouse models have also shown that radiotherapy indirectly augments T cell immunity, as initial APC activation by danger-associated molecules released from irradiated tumors were essential for tumor rejection [24,25]. Chemotherapy has also been reported to induce extracellular danger-associated molecules that activate APCs [26]. Altogether, potent and durable CRC treatment-induced intratumoral immune responses might depend on APC-T cell interactions and their mutual regulation.

Chemoradiation or ICI induce systemic immune responses [27,28], which involve off-target sites such as the tumor-adjacent colon. Here, we phenotypically and functionally characterized the APCs and T cells co-residing in the same tumor from CRC patients, as well as autologous colon tissue to ultimately address the responses of APCs exposed to induced activation of neighboring T cells.

First, we determined the ex vivo composition and activation status of T cell and APC subsets. We analyzed the expression of co-stimulatory CD80, and co-inhibitory PD-L1 on CD11c+CD64+CD14+ MPs and CD11c+CD64- DCs, plus subsets of the latter (i.e., CD141+ CDC1 and CD1c+ CDC2). We also evaluated the activation markers, CD69 and PD-1 on T cell subsets defined by CD103 and CD39 expression. Our concurrent assessment of surface activation markers enabled correlation analyses of APC and T cell subsets in the autologous tumor and colon to address potential relationships in their ex vivo activation status. Secondly, protein processing of APCs and T cell cytokine responses were determined in both tumor and colonic tissues. Lastly, we attempted to model general T cell activation potentially induced by direct or indirect T cell-targeting treatments via in vitro polyclonal T cell stimulation. We used this model to determine the regulation of CD80 and PD-L1 expression on tissue APCs by activated T cells.

Overall, our data obtained from a total of fifty-five CRC patients could provide further insight on whether intratumoral and colonic APCs exposed to activated T cells promote or limit the CRC treatment-induced immunological landscape.

2. Results
2.1. APCs in Both Tumor and Adjacent Colon Are Predominantly CD64+CD14+ MPs

Intratumoral APCs are instrumental in stimulation of anti-tumor responses of TILs [10,29]. Hence, we initiated our APC analyses by immunofluorescent staining of histological sections from the first of three patient cohorts included in the current study. The CD11c+

APCs were mainly found in the EpCAM- tumor stroma alongside CD8+ TILs (Figure 1A). Within the CD11c+ APCs, MPs were distinguished from DCs by their co-expression of CD64 and CD163 (Figure 1B), as these markers were absent on DCs (Figure 1C). In general, CD64 and CD163 plus an array of other markers are used for defining anti-tumor and pro-tumor MPs, respectively. However, expression of CD64 or CD163 is not always mutually exclusive since CD64+CD163+ MPs have been reported in CRC patients [9,30], which likely reflect the plasticity or heterogeneity of MPs.

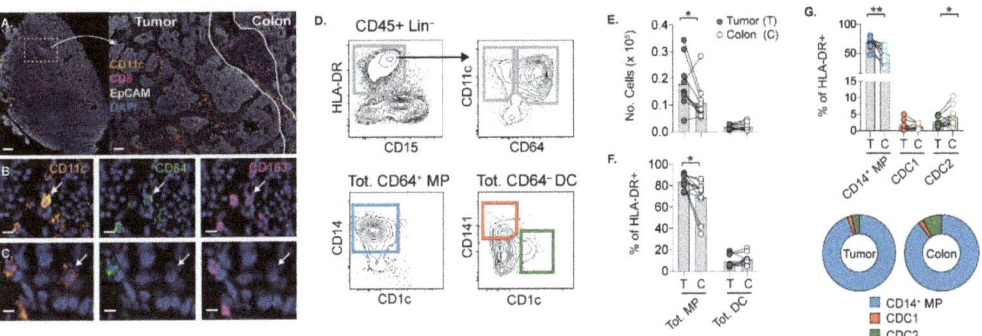

Figure 1. Characterization of APCs in tumor and adjacent colon from patients with CRC. (**A**) Tumor section showing CD11c+ APCs, CD8+ TILs and EpCAM+ epithelial cells. Arrows denote CD11c+CD64+CD163+ MPs (**B**) and CD11c+ CD64-CD163- DCs (**C**). Cell nuclei are DAPI+. Scale bars: (**A**) 500 μm (left) and 100 μm (right), (**B**) 10 μm and (**C**) 5 μm. (**D**) Gating strategy on tumor suspension showing CD45+ lineage (CD3/CD19/CD56)- HLA–DR+CD15-CD11c+ APCs divided into total CD64+ MPs containing CD14+ MPs (lower left panel), and total CD64- DCs comprising CD141+ CDC1 and CD1c+ CDC2 (lower right panel). Total MPs and DCs in tumor (T) and colon (C) by numbers per 1 million cells (**E**), or percent of HLA–DR+ APCs (**F**). (**G**) Percent and proportion of APC subsets within HLA–DR+ APCs. Bars show the mean. Connecting lines indicate autologous samples. (* $p < 0.05$, ** $p < 0.01$, Wilcoxon test).

In the second cohort, frequencies and phenotypic activation of APCs were determined in cell suspensions of tumor, adjacent colon and peripheral blood mononuclear cells (PBMCs). In line with our in situ analyses, the lineage- HLA-DR+ CD11c+ APCs among total CD45+ leukocytes in tumor suspension comprised of CD64+ MPs and CD64- DCs, which, hereafter, are referred to as MPs and DCs, respectively (Figure 1D). The majority of MPs co-expressed CD14, and DCs contained the CD141+ CDC1 and CD1c+ CDC2 subsets. Relative to colon, MPs were significantly more prominent in the tumor (Figure 1E,F), as previously described [9,31]. A similar elevation in frequency or number of intratumoral DCs was not observed. At the subset level, CD14+ MPs remained elevated in the tumor, and the percentage of CDC2 was significantly decreased in the tumor, along with a slight increase in CDC1 (Figure 1G). Both tumor and colon had a large proportion of CD14+ MPs within the HLA-DR+ APCs, and CDC2 were proportionally higher than CDC1 in colon (Figure 1G). Collectively, these analyses show alterations in the composition of APCs subsets in tumors compared to the adjacent colon.

2.2. The Level of MPs Relative to TILs in MSS Tumors Exceeds That of MSI-H Tumors

Tumors with high microsatellite instability (MSI-H) generally display immune activation represented by, e.g., increased T cell infiltration [32,33]. We found no apparent correlation between the frequencies of MPs vs. DCs among the leukocytes in microsatellite stable (MSS) and MSI-H tumors (Supplementary Figure S1A). There were no correlations between tumor stage and the frequency of TILs and APCs (Supplementary Figure S1B).

As infiltration of immune cells may indicate vigorous immune responses, we assessed whether frequencies of total CD3+ TILs associated with levels of MPs or DCs. We observed an inversed correlation between frequencies of TILs and MPs (Supplementary Figure S1C).

However, percentages of DCs and TILs in MSI-H tumors displayed a trend towards a positive association, which was not statistically significant. Regarding the APC to TIL ratio, a significantly higher MP to TIL ratio was found in MSS tumors, which also surpassed the DC to TIL ratio in the same tumors (Supplementary Figure S1D). The ratios of the combined levels of MPs and DCs relative to TILs were also higher in MSS tumors, which could be due to, e.g., lower infiltration, survival and/or proliferation of T cells in these tumors.

So far, T cell infiltration was assessed by the pan-T cell marker CD3 in cell suspensions for a general overview on APC and T cell frequencies. We revisited the cryopreserved tumors in the previous cohort to address the intratumoral location and levels of CD4 vs. CD8 TILs. The mean in situ numbers of CD3+ CD4 or CD8 TILs in MSS tumor centers tended to be lower than in MSI-H tumors (Supplementary Figure S1E), which may explain the higher APC to TIL ratio in MSS tumor suspensions from the second cohort. Combined numbers of CD4 and CD8 TILs in the tumor center or stroma were rather similar, regardless of MSI status. Despite the presence of more TILs in the center of some of the MSI-H tumors, increased mutational burden in our patient cohort was not related to increased levels of TILs or APCs.

2.3. MPs in Tumor and Colon Inversely Express Co-Stimulatory CD80 and Co-Inhibitory PD-L1

The abundance of TILs in the tumor stroma (Supplementary Figure S1E), where the majority of APCs also reside (Figure 1A–C), would enable frequent APC-T cell interactions. We, therefore, addressed the expression level of co-stimulatory CD80 and co-inhibitory PD-L1 by APCs, since these ligands have opposed effects on T cell activation (Figure 2A). As the microenvironment impacts APC-T cell interactions, we also included analyses of the adjacent colon and autologous PBMCs.

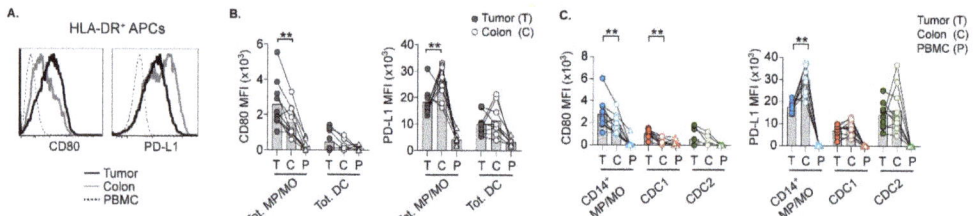

Figure 2. Expression of CD80 and PD-L1 on intratumoral and colonic APCs. (**A**) Histograms representing staining of CD80 or PD-L1 on total APCs in Tumor (T), Colon (C), and PBMCs (P). Mean fluorescence intensity (MFI) of CD80 and PD-L1 on total MPs or monocytes (MOs) and DCs (**B**), or on CD14+MPs/MOs, CDC1 and CDC2 (**C**). Bars show the mean. Connecting lines indicate autologous samples. (** $p < 0.01$, Wilcoxon test).

While intratumoral MPs had higher CD80 levels than colonic MPs, the latter also expressed higher PD-L1 (Figure 2B). This inversed CD80 and PD-L1 expression was not equally apparent on the DCs. As CD14+ MPs were the predominant MP subset (Figure 1D), their inversed pattern of CD80 and PD-L1 expression in tumor and colon remained unchanged (Figure 2C). The expression of CD80, but not PD-L1, was significantly upregulated on intratumoral CDC1. However, CDC2 in both tumor and colon had diverse expression of these markers. In contrast to tissue APCs, monocytes and DCs in PBMCs hardly expressed CD80 or PD-L1. The expression pattern of CD80 and PD-L1 in tumor and colon suggests that APCs within these discrete tissue environments have discrepancies in their capacity to promote and limit T cell activation.

2.4. Tissue-Resident Memory T Cells in Tumor and Colon Express Opposed PD-1 and CD69 Levels

Additional regulation of tissue APCs' CD80 and PD-L1 expression could stem from effector molecules from neighboring memory T cells acting on APCs. In this regard, the previously reported CD8+CD39+CD103+ tumor-specific TILs [6,7] could potentially be instrumental in regulation of these APC markers during tumor antigen-specific restimulation. However, CD39 or CD103 expression are also assessed in other aspects. For example, CD39 identifies regulatory T cells (Tregs) [34], and CD103 indicates tissue retention on tissue-resident memory T cells (TRMs) [35]. To this end, whether the frequency of CD4 or CD8 T cell subsets solely defined by CD103 and CD39 expression differ at several autologous sites with distinct milieus is largely unexplored. Thus, we assessed these T cell subsets in the second patient cohort (Figure 3A), from which we had access to matched tumor, colon and PBMC samples.

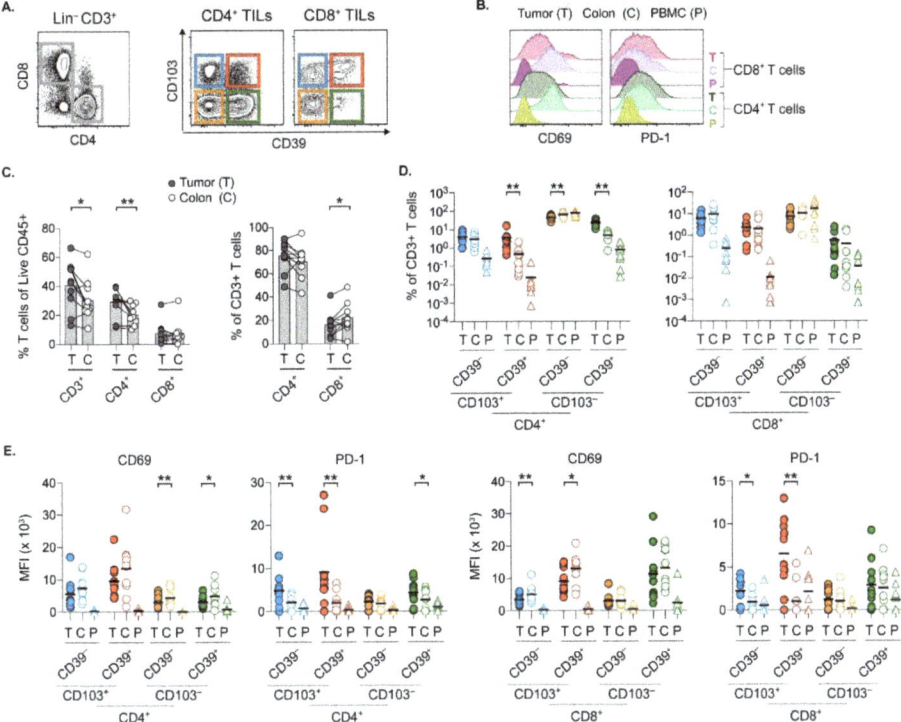

Figure 3. Phenotypic assessment of TILs and colonic T cells. (**A**) Gating strategy on tumor suspension where (CD11c/CD15/CD19)- CD3+ TILs are separated into CD4+ vs. CD8+ T cells and from which, four T cell subsets are subsequently defined by CD103 and CD39 expression. (**B**) Histogram overlays of CD69 and PD-1 staining on CD4 vs. CD8 T cells at specified sites. (**C**) Percent of total CD3+ T cells and CD4 or CD8 T cells within indicated population. (**D**) Percent of T cell subsets defined by CD103 and CD39 within CD3+ T cells. (**E**) MFI of CD69 and PD-1 staining on indicated T cell subsets and sites. Bars and bold horizontal line show the mean. Connecting lines indicate autologous samples. (* $p < 0.05$, ** $p < 0.01$, Wilcoxon test).

Initial T cell characterization showed higher levels of CD69 and PD-1 on CD4 and CD8 T cells in tissues, compared to PBMCs (Figure 3B). The early T cell activation marker CD69, also identifies TRMs at e.g. mucosal sites and in certain non-CRC tumors [35,36]. Thus, the wide range of CD69 levels on TILs, and more uniform CD69 expression on colonic T cells could represent varying degrees of tissue retention and/or T cell activation. Further, PD-1 upregulation by T cells denotes extended activation with or without exhaustion. Relative to the colon, frequencies of CD3+ TILs and CD4+ TILs within leukocytes were increased in the tumor (Figure 3C). Conversely, the percentage of CD8+ T cells among total T cells were slightly lower in the tumor compared to adjacent colonic tissue.

The frequency of CD4+CD39+ TILs, which likely contain the Tregs were significantly higher in the tumor regardless of their CD103 expression (Figure 3D). Both tumor and colon had similar levels of CD103+CD39− CD4 or CD8 TRMs. As expected, circulating T cells in PBMCs had low CD103 expression. The CD103− T cells in tissues might represent T cells that have yet to establish tissue residency and/or those about to exit the tissue. It is unclear if CD103 and CD39 on colonic CD8 T cells correspond to specificity for colonic antigens as co-expression of these markers on CD8 TILs suggest tumor specificity [6,7]. Relative to the colon, the frequency of potentially tumor-specific CD8+CD103+CD39+ TILs was not increased (Figure 3D). However, these TILs had significantly higher PD-1 and lower CD69 expression compared to their counterparts in colon (Figure 3E). In fact, the pattern of elevated PD-1 and decreased CD69 was observed on CD103+ CD4 or CD8 TILs, linking tumor residency to specific degree of TIL activation status.

2.5. Association of Activation Status between Particular APCs and T cells Are Tissue-Specific

Our simultaneous characterization of multiple co-residing APC and T cell subsets encompassed the relatively rare opportunity to address whether the activation status of these cell subsets could be correlated (Figure 4A–D and Supplementary Figure S2A–D). In this regard, the opposed expression of PD-1 and CD69 on specific T cells (Figure 3E), as well as CD80 and PD-L1 on APCs (Figure 2A) in tumor vs. colon tissue, suggest discrete activation patterns in these tissues.

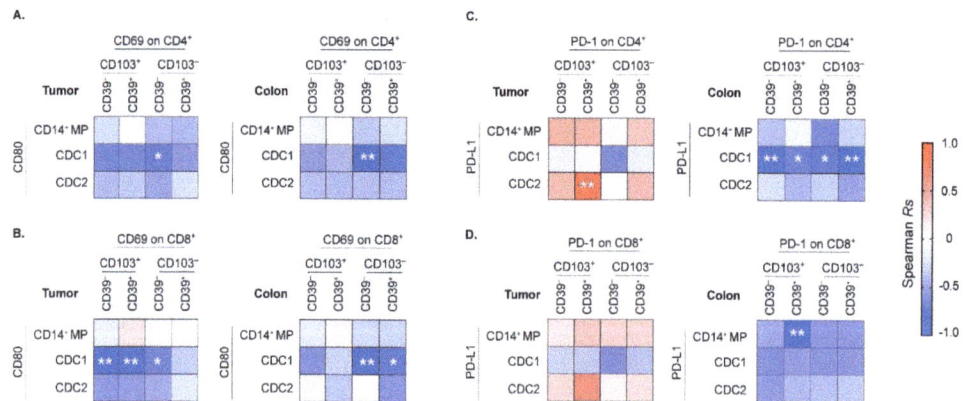

Figure 4. Correlation analyses of ex vivo surface activation markers of APC and T cell subsets. Heat maps from Spearman correlation between CD80 MFI of APC subsets vs. CD69 MFI of CD4 (**A**), or CD8 (**B**) T cell subsets in tumor and adjacent colon. Similar analyses on PD-L1 MFI of APC subsets vs. PD-1 MFI on CD4+ (**C**), or CD8+ (**D**) T cell subsets. (* $p < 0.05$, ** $p < 0.01$). Spearman R_s values and p-values are specified in Supplementary Figure S2A–D.

Correlation analyses showed that associations between CD80 on APC subsets vs. CD69 on T cell subsets defined by CD103 and CD39 were generally negative in the tumor and colon (Figure 4A,B). In contrast, PD-L1 on intratumoral CD14+ MPs tended to be positively linked to PD-1 expression on CD4 and CD8 TIL subsets (Figure 4C,D), but an opposite trend was observed in the colon. This may suggest that MPs in the tumor support PD-1-mediated T cell inhibition, but colonic MPs prefer to reduce signals conveyed by PD-1.

Higher PD-L1 levels on CDC1 in both tumor and colon were linked to lower PD-1 expression on T cell subsets (Figure 4C,D). In contrast, PD-L1 on intratumoral CDC2 was especially associated with increased PD-1 expression on CD4+CD103+CD39+ TILs containing the Tregs (Figure 4C). Furthermore, elevated CD80 levels on intratumoral CDC1 were highly associated with lower CD69 expression on CD103+ CD8 TILs (Figure 4B), but conversely, the association was more pronounced with CD103- CD8 T cells in the colon. This may indicate that DCs are interacting with different T cell subsets in these two tissues. For example, activated CDC1 preferentially promote CD69 downregulation on CD8+ TRMs in the tumor, whereas in the colon, the CDC1 mediate this effect on CD8+ former TRMs or non-TRMs.

2.6. APCs in Tumor and Colon Display Opposed Co-Stimulatory Capacity upon Protein Digestion

Next, we assessed the functional capacity of APCs from the distinct environments of the colon and tumor. The expression of co-stimulatory and co-inhibitory markers on APC subsets from these tissues likely impact T cell stimulation during antigen presentation. Thus, we determined the APCs' capacity to take up and degrade the protein antigen ovalbumin (OVA), and their subsequent expression of activation or inhibitory markers. In the presence of low concentration of OVA that only becomes fluorescent upon degradation (DQ-OVA), there were similar frequencies of DQ-OVA+ APCs in cell suspensions of both colon and tumor (Figure 5A). However, when evaluating both OVA uptake and degradation using low amounts of AF647-labeled OVA, the colonic APCs were more efficient (Figure 5B).

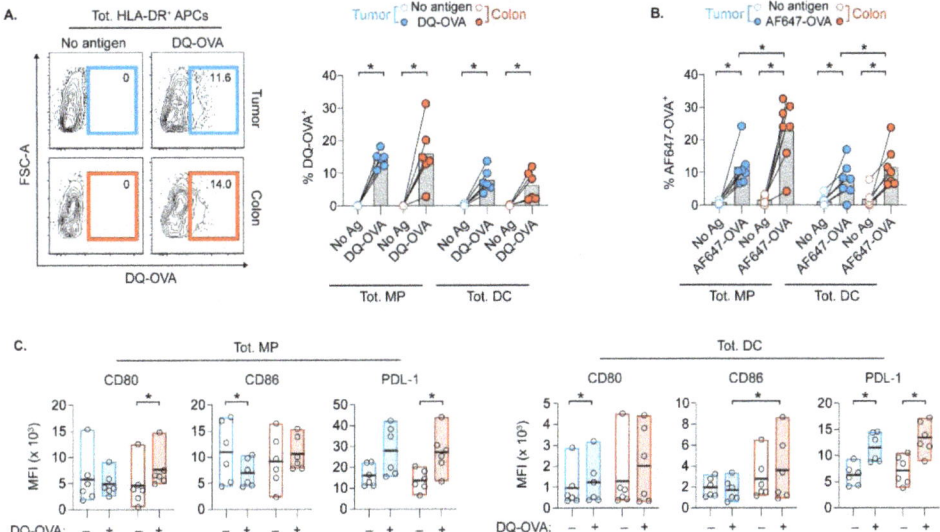

Figure 5. Protein uptake and degradation capacity of APCs from tumor and colon. (**A**) Flow cytometry plots on HLA-DR+ APCs with values indicating percent of APCs that degraded DQ-OVA protein (i.e., DQ-OVA+), and compiled data on DQ-OVA+ MPs and DCs. (**B**) Percent of MPs and DCs that ingested and/or degraded AF647-conjugated OVA. (**C**) Comparison of CD80, CD86 and PD-L1 MFIs on MPs and DCs in DQ-OVA–cultures that degraded DQ-OVA (DQ-OVA+), or not (DQ-OVA–). Bars and bold line in min-max bars show the mean. Connecting lines show control and treated cell suspension from the same patients. (* $p < 0.05$, Wilcoxon test).

We also assessed whether protein degradation would subsequently enhance the APCs' capacity to stimulate T cells. Comparison of DQ-OVA+ and DQ-OVA- APCs within the same cultures receiving DQ-OVA allows analyses of co-stimulatory CD80, CD86 and co-inhibitory PD-L1 on APCs that degraded OVA vs. those that were less efficient. Relative to internal DQ-OVA- control, there was a significant downregulation of CD86 on DQ-OVA+ MPs in the tumor, while upregulation of CD80 was observed on DQ-OVA+ colonic MPs (Figure 5C). Colonic DQ-OVA+ DCs upregulated CD86 at significantly higher levels than DQ-OVA+ DCs from the tumors. Interestingly, APCs in both tissues upregulated PD-L1 after OVA degradation. In conclusion, OVA degradation efficiency of APCs was generally not impacted by their tissue-specific stimulatory and inhibitory states. However, APCs from tumor vs. colon had distinct levels of co-stimulatory markers following OVA digestion.

2.7. APCs Co-Express CD80 and PD-L1 in Presence of Activated T Cells In Vitro

APCs providing T cell stimulation was vital for tumor rejection in mouse models of CRC treatments [23–25]. We, therefore, determined the expression of CD80 and PD-L1 on APCs in our attempts to model treatment-induced T cell activation in vitro. As ICI and chemoradiation directly or indirectly target T cells regardless of their specificity, we simulated global T cell activation using microbeads loaded with anti-CD2/CD3/CD28 antibodies (referred to as microbeads) in the third CRC patient cohort (Figure 6A–D). Since off-target tissues are also exposed to CRC treatments, we included microbead-stimulated colonic suspensions to our analyses.

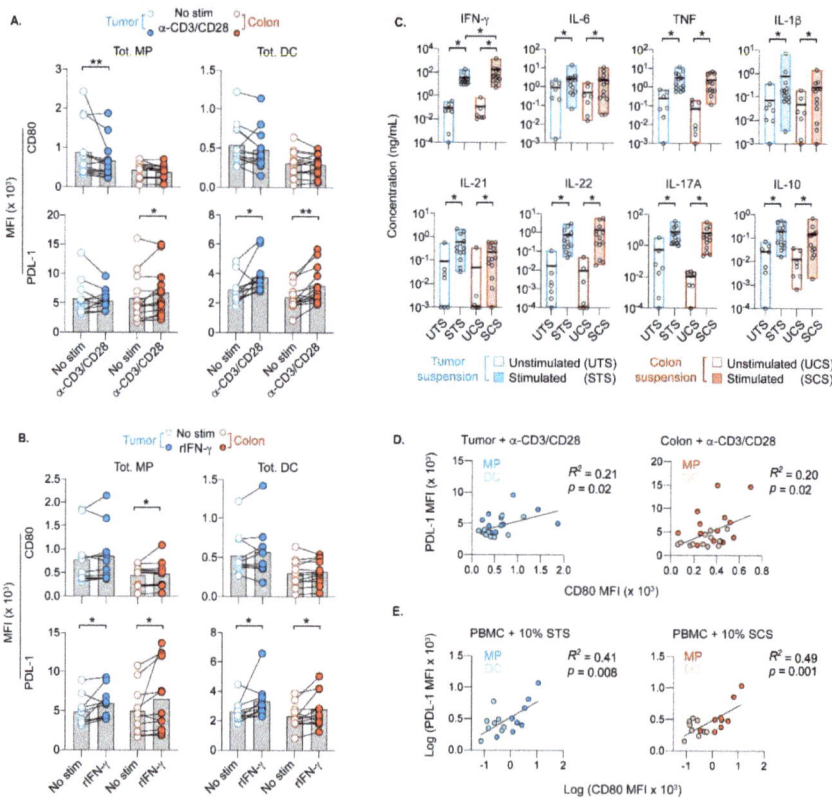

Figure 6. CD80 and PD–L1 expression on APCs in the presence of induced T cell activation. CD80 and PD–L1 MFI on MPs and DCs in tumor or colon suspensions after 6 h culture in media alone, or with microbeads loaded with anti-CD2/CD3/CD28 antibodies (referred to as microbeads) (**A**), or recombinant IFN-γ (**B**). (**C**) Cytokine levels after 6 h of culture. UTS and STS denote supernatants from unstimulated and microbeads-stimulated tumor suspensions, respectively. Likewise, supernatants from colon suspensions are referred to as UCS and SCS. Linear regression analyses of CD80 vs. PD–L1 MFI of APCs in microbeads-stimulated tissue suspensions (**D**), or in PBMCs exposed to 10% STS or SCS (**E**). Bars and bold line in min–max bars show the mean. Connecting lines show control and treated cell suspensions from the same patients (* $p < 0.05$, ** $p < 0.01$, Wilcoxon test).

In the midst of T cells activated by microbeads, downregulation of CD80 was observed only on intratumoral MPs, while DCs in tumor and colon retained their CD80 levels (Figure 6A). The colonic MPs and DCs that retained CD80 expression, instead upregulated PD-L1. Activated T cells regulate other cells by, e.g., IFN-γ [37]. Addition of recombinant IFN-γ (rIFN-γ) to the cell suspensions, upregulated CD80 on colonic MPs, but did not reduce CD80 levels on intratumoral MPs (Figure 6B). The CD80 expression on DCs was also not significantly altered by rIFN-γ. In line with previous murine studies [38], rIFN-γ upregulated PD-L1 on APCs (Figure 6B).

Relative to the unstimulated control, pro-inflammatory and inhibitory cytokines were increased in the supernatants of microbeads cultures (Figure 6C). Increased levels of IL-1β, IL-6 and tumor necrosis factor (TNF) are indicative of APC activation and confirm the ability of APCs to respond to activated T cells. Although microbeads-stimulated colon suspensions had higher IFN-γ levels (Figure 6C), the colonic APCs in these suspensions were not more efficient in PD-L1 upregulation than intratumoral APCs (Figure 6A). There was a positive correlation of CD80 and PD-L1 on APCs in tumor and colon (Figure 6D), which suggests a negative feedback loop where CD80+ APCs that responded to activated T cells upregulate PD-L1 to reduce stimulatory cues from T cells. Substituting 10% of the culture media with supernatants from microbeads-stimulated cultures resulted in a similar correlated expression of CD80 and PD-L1 on autologous APCs in PBMC cultures (Figure 6E).

Overall, induced T cell activation in cell suspensions from colon or tumor did not result in distinct patterns of CD80 and PD-L1 expression on co-residing APCs. Thus, in the scenario of CRC treatment-induced T cell activation, APCs in target and off-target tissues are likely comparably responsive to signals from activated T cells.

3. Discussion

Local immune responses in the distinct environment of tumor and adjacent colon, which stem from the interactions between infiltrating and/or tissue-residing T cells and APCs are largely unknown. Here, we characterized multiple intratumoral and colonic APC and T cell subsets from non-treated CRC patients, with regards to their ex vivo frequencies, phenotypic activation, and functional status. In addition, we attempted modelling of T cell activation after parenterally administered CRC treatments, capable of direct or indirect targeting of T cells irrespective of their specificity, or anatomical location. Tissue-specific expression pattern of T cell co-stimulatory CD80 and T cell co-inhibitory PD-L1 were observed, especially on the MPs. Higher PD-L1 and lower CD80 expression on colonic APCs may imply greater propensity towards inhibitory responses compared to intratumoral APCs with opposed levels of these markers. It is possible that commensal microbes specifically in the colon, render colonic MPs more resistant to induce T cell activation. On the other hand, the abundance of intratumoral CD80+ APCs might represent a persistent tumor infiltration of circulating APCs that upregulate CD80 as they enter the non-sterile tumor environment.

MSI-H tumors are considered more immunologically active than MSS tumors [32,33], which would impact infiltration of APCs. Correlation analyses implied DC infiltration was somewhat linked to homing of T cells to MSI-H tumors but, overall, the association between APC and T cell frequencies displayed a negative trend regardless of MSI status. Further, we did not observe significant differences in TIL numbers at the center or stroma of MSI-H vs. MSS tumors, which likely reflect the limited number of patients in the initial cohort for in situ analyses. These in situ observations and the predominance of MSS tumors in subsequent cohorts did not prompt further comparisons of TIL responses based on MSI status. Instead, we detailed the T cells infiltrating the tumor and adjacent colon by their expression of CD103 and CD39. These markers have been postulated to identify rare tumor-specific CD103+CD39+CD8+ TILs in solid tumors, including CRC [6,7].

Although the frequency of CD8 TILs co-expressing CD103 and CD39 was not increased relative to adjacent colon, these CD8 TILs had also significantly higher PD-1 and lower

CD69 expression. Of note, CD4 and CD8 TILs with elevated PD-1 and decreased CD69 expression also co-expressed CD103, commonly used in the identification of TRMs. As high CD103+CD8+ TIL levels are linked with improved CRC patient survival [39], our analyses suggest that colonic and intratumoral CD103+CD8+ T cells with prominent PD-1 expression would be frequently targeted by anti-PD-1 ICI. Further, the relationship between the ex vivo activation status of APCs and T cells was observed on specific cell subsets and tissue. For example, activated CDC1 in tumor, but not colon, seem to confer CD69 downregulation on co-residing CD8+CD103+ TRMs. As CD69 identifies recently activated T cells but also tissue residency, it remains unclear if CD69 downregulation on CD103+ TRMs represent initial steps towards exiting the tissue, or potential sign of stimulation beyond the time frame for early activation.

Having completed the overview of the ex vivo phenotype and activation status of APC and TIL subsets, we proceeded to determine their predisposition to perform some of the essential immune functions. As protein processing by APCs precedes antigen presentation and stimulation of cognate T cells, we assessed uptake and degradation of protein by MPs and DCs. We have used protein labeled with AF647 dye, which remains detectable in intracellular acidic compartments containing degraded proteins. The higher frequencies of OVA-AF647+ colonic MPs and DCs would represent more efficient uptake, as well as degradation of the scarce amount of protein compared to corresponding subsets in the tumor. Further, evaluation of protein degradation alone via DQ-OVA showed that both colonic and intratumoral APCs were equally efficient. However, while colonic APCs that degraded protein significantly upregulated both CD80 and PD-L1, the intratumoral counterpart was more prone to only upregulate PD-L1. Whether upregulation of both CD80 and PD-L1 on colonic APCs during protein degradation would simultaneously convey contradicting signals to T cells during antigen presentation is unclear. As the intestine contains large numbers of memory T cells, it might be reasonable that CD80 and PD-L1 co-expression inhibit unjustified activation of tissue-residing T cells while the APCs process antigens. Relative to the tumor of CRC patients, higher PD-L1 expression on histological sections of the adjacent colon has also been reported earlier [31].

Treatment-induced modulations of immune cells likely shape the potency of anti-tumor immunity, severity of immune-related adverse events and overall survival of patients with CRC. Although, mouse models of CRC treatments have demonstrated the central role of APCs for induction and maintenance of T cell-mediated tumor rejection [24,25], the precise effects chemoradiation or ICI exert on the functional capacity of APCs and T cells located in target (tumor) and off-target (e.g., adjacent colon) tissues remain unclear. Therefore, we assessed the functional capacity of APCs subsequent to induced polyclonal T cell activation, since effector molecules from activated TILs would, in turn, uphold or limit APC functions during generation of anti-tumor immunity. Our attempts to simulate CRC treatment-induced global activation of T cells by stimulating T cells with microbeads, supported concurrent upregulation of CD80 and PD-L1 on APCs, regardless of whether they were from colon, tumor or blood. As IFN-γ induces PD-L1 on APCs [38], it is likely that CD80+ APCs would eventually also co-express PD-L1. Despite significantly higher IFN-γ concentration in stimulated colon suspensions, the colonic and intratumoral APCs were both efficient in upregulation of CD80 and PD-L1 in the presence of induced T cell activation.

To this end, the tissue cell suspensions used in the study also contain other cells that may augment or interfere the functional responses of APCs. Highly purified APC and TIL subsets would be ideal for our in vitro assays, but the cell yields from available tissues were insufficient for in vitro assessments with purified cells. To simultaneously address whether tumor cells, stromal cells, granulocytes, myeloid-derived suppressor cells (MDSCs) or other lymphocytes, influence APC's responses to induced T cell activation would require systems biological assessment of, e.g., the transcriptome of tumor vs. colonic tissues. Further, the current study is based on untreated CRC patients and whether intratumoral and colonic APCs respond differently to activated T cells after exposure to

ICI or chemoradiation remains undetermined. In most cases, colonic or rectal resection is not considered for recently treated patients according to clinical guidelines, which greatly restricts the availability of these patient materials for study purposes. However, a few punch biopsies collected during routine post-treatment examination may suffice for a limited number of in vitro assays with bulk cell suspension.

4. Materials and Methods

4.1. CRC Patients

The patients were divided into three cohorts (n = 28, 10, and 17, respectively) based on the experimental procedure applied to collected samples. Colonic or rectal resection was performed at the Sahlgrenska University Hospital and additional patient data are provided in Supplementary Table S1. The patients had not been treated with immune-modulatory drugs or (chemo)radiation prior surgery. After collection of tumor tissue from the resectate, macroscopically tumor-free tissues were collected approximately 10 cm away from the tumor. Biopsies from tumor and colon were snapfrozen in liquid nitrogen for MSI analysis, or embedded in OCT (Histolabs, Gothenburg, Sweden) before sequential freezing in isopentane and liquid nitrogen for immunohistochemistry. Remaining tissues were transported in PBS on ice for generation of single-cell suspensions within one hour. Venous blood was collected into heparin tubes (BD, San Jose, CA, USA) during surgery. Tumor stages were specified in the pathology report on tumor invasion (T1-T4), lymph node involvement (N0-N2) and distant metastases (M0-M1). An overall stage (I-IV), based on TNM status, was subsequently determined. Microsatellite instability was analyzed by MSI Analysis System, version 1.2. (Promega, Nacka, Sweden) via fluorescently-labeled primers for co-amplification of mononucleotide repeat markers (BAT-25, BAT-26, NR-21, NR-24 and MONO-27) and pentanucleotide repeat markers (Penta C and Penta D). MSI of tumors were defined by peak alterations in the marker electropherogram relative to tumor-free control tissue. While instability in >1 mononucleotide repeat markers indicate MSI-H tumors, instability in 1 of these markers is considered MSI-L. Absence of MSI represents MSS tumors.

4.2. Immunohistochemistry

OCT embedded tissues stored in −80 °C were placed inside −20 °C cryostat for 20 min prior sectioning. The 7 μm thick sections were fixed with 2% paraformaldehyde for 10 min, washed with PBS for 5 min, permeabilized with 0.1% Triton-X 100 (Sigma, St. Louis, MO, USA) and washed once more. Endogenous avidin and biotin were sequentially blocked for 10 min, respectively using Avidin/Biotin blocking kit (Biocare Medical, Pacheco, CA, USA) with PBS wash between and after these blocking steps. Sections were then incubated for 1 h at room temperature with a mixture of monoclonal antibodies diluted in PBS with 0.1% BSA (Sigma); anti- CD8 (clone RPA-T8), CD64 (clone 10.1), CD163 (clone RM3/1), EpCAM (clone G8.8) (Biolegend, San Diego, CA, USA), CD3 (clone UCHT1, BD) and CD11c (clone EP1347Y, Abcam, Cambridge, UK). All antibodies, except rabbit-anti CD11c, are fluorescence-conjugated and raised in mice. Thus, stained sections were incubated for 40 min with fluorescence-conjugated donkey anti-rabbit IgG (Jackson ImmunoResearch, West Grove, PA, USA). Finally, sections were mounted with Prolong Diamond Antifade with DAPI (Invitrogen, Thermo Fisher, Carlsbad, CA, USA). Sections were scanned with Metafer Slide Scanning Platform (Metasystems, Heidelberg Germany) with Axio Imager.Z2 Microscope, 20×/0.8/air objective (Zeiss, Oberkochen, Germany) and SpectraSplit filter set (Kromnigon, Gothenburg, Sweden). TILs and APCs were quantified within 1 mm² region by Fiji/ImageJ software with Cell Counter plugin. For simplification, the sole MSI-L tumor was allocated to the MSS tumor group for immunohistochemistry.

4.3. Generation of Single Cell Suspensions

Cell suspensions of tumor and adjacent control tissue were generated as described [37]. Briefly, 3 × 3 mm tissue pieces were washed in HBSS without Mg^{++}/Ca^{++} (Gibco, Thermo

Fisher Carlsbad, CA, USA) supplemented with 2% FBS (Gibco), 1% Hepes buffer and 2 mM EDTA (Lonza, Basel, Switzerland). Tissues were digested with 70 µg/mL Liberase TM (Roche, Basel, Switzerland) and 20 µg/mL DNase I (Sigma) in complete media consisting of RPMI 1640 with 10% FCS, 1% Penicillin/Streptomycin, 1% Hepes buffer and 0.1% Gentamycin (Gibco). Cell suspensions were filtered, washed and resuspended to 1×10^6 cells/mL. PBMCs were isolated via Ficoll gradient (GE Healthcare, Uppsala, Sweden).

4.4. OVA Uptake and Degradation

For evaluation of uptake and degradation, tumor and colon tissue suspensions were cultured overnight in presence of 0.4 µg AlexaFluor 647 (AF647)-conjugated OVA (Invitrogen). To solely assess degradation, 0.6 µg DQ-OVA (Invitrogen) was used instead. Efficiency of uptake and degradation of OVA were analyzed with flow cytometry.

4.5. Stimulation of APCs by Activated T Cells

Single cell suspensions from tumor and colon tissues were resuspended in complete medium to 1 million cells/mL in 5-mL polystyrene tubes (Corning, Amsterdam, Netherlands). Stimulation of tissue APCs by polyclonally activated co-residing T cells was performed by addition of 2.5 µL microbeads loaded with anti-CD2, CD3 and CD28 antibodies (Miltenyi Biotec, Auburn, CA, USA), which represent 1 bead: 4 cell ratio. In parallel, cell suspensions were stimulated with 0.2 µg/mL recombinant IFN-γ (Biolegend, San Diego, CA, USA). After 6 h of culture, supernatants from unstimulated and microbeads-stimulated cell suspensions were collected from the top layer of the culture medium to avoid contamination of sedimented cells or microbeads. Moreover, autologous PBMCs were also cultured for 12 h in 90% complete media and 10% supernatant from microbeads-stimulated tissue suspensions. Remaining supernatants were stored at −20 °C until use. Activation of APCs was analyzed by flow cytometry.

4.6. Supernatant Cytokine Analyses

Cytokines (IFN-γ, IL-1β, IL-6, IL-10, IL-17A, IL-21, IL-22 and TNF) in supernatants from microbeads-stimulated tumor and control tissue suspensions were analyzed by Meso Scale discovery (MSD) multiplex platform using the U-PLEX TH17 Combo2 kit (Meso Scale Diagnostics, Rockville, MD, USA), according to manufacturer's protocol.

4.7. Flow Cytometry

Cell suspensions were stained, as described [37,40]. Briefly, cell viability in 1 mL cell suspension containing 1 million cells was determined by Zombie Red Fixable viability dye (Biolegend), followed by 20 min incubation with fluorescence-conjugated antibodies (Supplementary Table S2). Stained cells were fixed with 2% paraformaldehyde. AccuCount beads (Spherotech, Lake Forest, IL, USA) were used for cell enumeration, according to manufacturer's protocol. Samples were acquired using BD LSRFortessa flow cytometer and analyzed with FlowJo v.9.9.6 (Tree Star, Ashland, OR, USA).

4.8. Statistical Analysis

Paired and unpaired comparisons were made using Wilcoxon signed-rank test and Mann–Whitney test, respectively. Correlation matrixes and bubble plots were based on non-parametric Spearman correlation analyses. Where indicated, linear regression was applied instead. All analyses were performed using GraphPad Prism software, v.9.0.2. (San Diego, CA, USA) and considered significant at $p < 0.05$.

5. Conclusions

In conclusion, our extensive characterization of APCs and T cells from the contrasting environments of tumor and colon of patients with CRC enabled determination of tissue-specific activation status and functional capacity of specific APC and T cells. The distinct milieus of tumor and colon are likely reflected on the opposed expression of CD80 and PD-L1 on MPs, as well as PD-1 and CD69 on CD8+CD103+CD39+ T cells. The colonic APCs' superior protein uptake and subsequent upregulation of CD80 and PD-L1 compared to tumor APCs, plus the enhanced IFN-γ responses of colonic T cells might indicate different functional propensity fostered in tumor vs. colon. Overall, colonic or intratumoral APCs displayed similar capacity to respond to activated T cell-mediated regulation of their co-stimulatory CD80 and co-inhibitory PD-L1 expression. The presented data could provide insights into tissue APCs' post-treatment induction and/or maintenance of local anti-tumor responses or immune-related adverse events affecting the normal colon.

Supplementary Materials: The following are available online at https://www.mdpi.com/article/10.3390/cancers13205247/s1, Figure S1: Associations of APCs and TILs according to tumor stage and MSI status. Figure S2: Spearman correlation coefficients and p-values from association analyses of APC and T cell subsets. Table S1: Characteristics of colorectal cancer patients, Table S2: Fluorescence-conjugated antibodies for flow cytometry.

Author Contributions: Conceptualization, F.L. and U.Y.; methodology, F.L., L.S., S.A. and Y.W.; validation, F.L., A.R. and L.S.; formal analysis, F.L., A.R. and L.S.; investigation, F.L., A.R., L.S., S.A. and Y.W.; resources, E.B.L., M.Q.-J. and U.Y.; data curation, F.L., A.R., L.S. and Y.W.; writing—original draft preparation, F.L.; writing—review and editing, F.L., A.R., M.Q.-J., E.B.L. and U.Y.; visualization, F.L., A.R. and U.Y.; supervision, F.L. and U.Y.; project administration, F.L. and U.Y. All authors have read and agreed to the published version of the manuscript.

Funding: This research was funded by: Swedish Cancer Society, 2018/724 and 19 0333 Pj; Swedish State under the agreement between Swedish government and the county councils—the ALF agreement, ALFGBG-723231 and ALFGBG-716581; Swedish Research Council, 2017-01103. F.L. is recipient of post-doctoral salary grant from the Swedish Society of Medical Research (P17-0024) Gunvor and Josef Anér's Foundation, Mary von Sydow's Foundation, Assar Gabrielsson's Foundation plus Anna-Lisa and Bror Björnsson's Foundation.

Institutional Review Board Statement: The study was conducted according to the guidelines of the Declaration of Helsinki, and approved by the Regional Board of Ethics in Gothenburg, Sweden (approval nr. 118-15 and 637-18 for study of material from colonic and rectal surgery, respectively).

Informed Consent Statement: Informed consent was obtained from all patients in the study.

Data Availability Statement: Data is contained within the article and supplementary material.

Acknowledgments: We thank the following colleagues at the Department of Surgery; Ann-Louise Helminen, Hillevi Björkqvist and Zunash Malik for initial processing of the resectates. Jaqueline Flach for assistance during MSI analyses and Andrew Boucher for language editing. We also thank Eva Angenete and Peter Falk for coordinating the sampling of patient material.

Conflicts of Interest: The authors declare no conflict of interest.

References

1. Arnold, M.; Sierra, M.S.; Laversanne, M.; Soerjomataram, I.; Jemal, A.; Bray, F. Global patterns and trends in colorectal cancer incidence and mortality. *Gut* **2017**, *66*, 683–691. [CrossRef]
2. Bray, F.; Ferlay, J.; Soerjomataram, I.; Siegel, R.L.; Torre, L.A.; Jemal, A. Global cancer statistics 2018: GLOBOCAN estimates of incidence and mortality worldwide for 36 cancers in 185 countries. *CA Cancer J. Clin.* **2018**, *68*, 394–424. [CrossRef] [PubMed]
3. Galon, J.; Costes, A.; Sanchez-Cabo, F.; Kirilovsky, A.; Mlecnik, B.; Lagorce-Pages, C.; Tosolini, M.; Camus, M.; Berger, A.; Wind, P.; et al. Type, density, and location of immune cells within human colorectal tumors predict clinical outcome. *Science* **2006**, *313*, 1960–1964. [CrossRef] [PubMed]
4. Pages, F.; Kirilovsky, A.; Mlecnik, B.; Asslaber, M.; Tosolini, M.; Bindea, G.; Lagorce, C.; Wind, P.; Marliot, F.; Bruneval, P.; et al. In situ cytotoxic and memory T cells predict outcome in patients with early-stage colorectal cancer. *J. Clin. Oncol.* **2009**, *27*, 5944–5951. [CrossRef]
5. Li, X.; Ling, A.; Kellgren, T.G.; Lundholm, M.; Lofgren-Burstrom, A.; Zingmark, C.; Rutegard, M.; Ljuslinder, I.; Palmqvist, R.; Edin, S. A Detailed Flow Cytometric Analysis of Immune Activity Profiles in Molecular Subtypes of Colorectal Cancer. *Cancers* **2020**, *12*, 3440. [CrossRef] [PubMed]

6. Duhen, T.; Duhen, R.; Montler, R.; Moses, J.; Moudgil, T.; de Miranda, N.F.; Goodall, C.P.; Blair, T.C.; Fox, B.A.; McDermott, J.E.; et al. Co-expression of CD39 and CD103 identifies tumor-reactive CD8 T cells in human solid tumors. *Nat. Commun.* **2018**, *9*, 2724. [CrossRef]
7. Simoni, Y.; Becht, E.; Fehlings, M.; Loh, C.Y.; Koo, S.L.; Teng, K.W.W.; Yeong, J.P.S.; Nahar, R.; Zhang, T.; Kared, H.; et al. Bystander CD8(+) T cells are abundant and phenotypically distinct in human tumour infiltrates. *Nature* **2018**, *557*, 575–579. [CrossRef]
8. Jayasingam, S.D.; Citartan, M.; Thang, T.H.; Mat Zin, A.A.; Ang, K.C.; Ch'ng, E.S. Evaluating the Polarization of Tumor-Associated Macrophages Into M1 and M2 Phenotypes in Human Cancer Tissue: Technicalities and Challenges in Routine Clinical Practice. *Front. Oncol.* **2019**, *9*, 1512. [CrossRef]
9. Norton, S.E.; Dunn, E.T.; McCall, J.L.; Munro, F.; Kemp, R.A. Gut macrophage phenotype is dependent on the tumor microenvironment in colorectal cancer. *Clin. Transl. Immunol.* **2016**, *5*, e76. [CrossRef]
10. Wculek, S.K.; Cueto, F.J.; Mujal, A.M.; Melero, I.; Krummel, M.F.; Sancho, D. Dendritic cells in cancer immunology and immunotherapy. *Nat. Rev. Immunol.* **2020**, *20*, 7–24. [CrossRef]
11. Sanchez-Paulete, A.R.; Cueto, F.J.; Martinez-Lopez, M.; Labiano, S.; Morales-Kastresana, A.; Rodriguez-Ruiz, M.E.; Jure-Kunkel, M.; Azpilikueta, A.; Aznar, M.A.; Quetglas, J.I.; et al. Cancer Immunotherapy with Immunomodulatory Anti-CD137 and Anti-PD-1 Monoclonal Antibodies Requires BATF3-Dependent Dendritic Cells. *Cancer Discov.* **2016**, *6*, 71–79. [CrossRef]
12. Forssell, J.; Oberg, A.; Henriksson, M.L.; Stenling, R.; Jung, A.; Palmqvist, R. High macrophage infiltration along the tumor front correlates with improved survival in colon cancer. *Clin. Cancer Res.* **2007**, *13*, 1472–1479. [CrossRef]
13. Gulubova, M.V.; Ananiev, J.R.; Vlaykova, T.I.; Yovchev, Y.; Tsoneva, V.; Manolova, I.M. Role of dendritic cells in progression and clinical outcome of colon cancer. *Int. J. Colorectal Dis.* **2012**, *27*, 159–169. [CrossRef]
14. Pinto, M.L.; Rios, E.; Duraes, C.; Ribeiro, R.; Machado, J.C.; Mantovani, A.; Barbosa, M.A.; Carneiro, F.; Oliveira, M.J. The Two Faces of Tumor-Associated Macrophages and Their Clinical Significance in Colorectal Cancer. *Front. Immunol.* **2019**, *10*, 1875. [CrossRef]
15. Li, S.; Xu, F.; Zhang, J.; Wang, L.; Zheng, Y.; Wu, X.; Wang, J.; Huang, Q.; Lai, M. Tumor-associated macrophages remodeling EMT and predicting survival in colorectal carcinoma. *Oncoimmunology* **2018**, *7*, e1380765. [CrossRef] [PubMed]
16. Lim, Y.J.; Koh, J.; Kim, S.; Jeon, S.R.; Chie, E.K.; Kim, K.; Kang, G.H.; Han, S.W.; Kim, T.Y.; Jeong, S.Y.; et al. Chemoradiation-Induced Alteration of Programmed Death-Ligand 1 and CD8(+) Tumor-Infiltrating Lymphocytes Identified Patients With Poor Prognosis in Rectal Cancer: A Matched Comparison Analysis. *Int. J. Radiat. Oncol. Biol. Phys.* **2017**, *99*, 1216–1224. [CrossRef]
17. Morris, M.; Platell, C.; Iacopetta, B. Tumor-infiltrating lymphocytes and perforation in colon cancer predict positive response to 5-fluorouracil chemotherapy. *Clin. Cancer Res.* **2008**, *14*, 1413–1417. [CrossRef] [PubMed]
18. Shibutani, M.; Maeda, K.; Nagahara, H.; Fukuoka, T.; Iseki, Y.; Matsutani, S.; Kashiwagi, S.; Tanaka, H.; Hirakawa, K.; Ohira, M. Tumor-infiltrating Lymphocytes Predict the Chemotherapeutic Outcomes in Patients with Stage IV Colorectal Cancer. *Vivo* **2018**, *32*, 151–158. [CrossRef]
19. Teng, F.; Meng, X.; Kong, L.; Mu, D.; Zhu, H.; Liu, S.; Zhang, J.; Yu, J. Tumor-infiltrating lymphocytes, forkhead box P3, programmed death ligand-1, and cytotoxic T lymphocyte-associated antigen-4 expressions before and after neoadjuvant chemoradiation in rectal cancer. *Transl. Res.* **2015**, *166*, 721–732.e1. [CrossRef] [PubMed]
20. Salmon, H.; Idoyaga, J.; Rahman, A.; Leboeuf, M.; Remark, R.; Jordan, S.; Casanova-Acebes, M.; Khudoynazarova, M.; Agudo, J.; Tung, N.; et al. Expansion and Activation of CD103(+) Dendritic Cell Progenitors at the Tumor Site Enhances Tumor Responses to Therapeutic PD-L1 and BRAF Inhibition. *Immunity* **2016**, *44*, 924–938. [CrossRef]
21. Alloatti, A.; Rookhuizen, D.C.; Joannas, L.; Carpier, J.M.; Iborra, S.; Magalhaes, J.G.; Yatim, N.; Kozik, P.; Sancho, D.; Albert, M.L.; et al. Critical role for Sec22b-dependent antigen cross-presentation in antitumor immunity. *J. Exp. Med.* **2017**, *214*, 2231–2241. [CrossRef] [PubMed]
22. Hui, E.; Cheung, J.; Zhu, J.; Su, X.; Taylor, M.J.; Wallweber, H.A.; Sasmal, D.K.; Huang, J.; Kim, J.M.; Mellman, I.; et al. T cell costimulatory receptor CD28 is a primary target for PD-1-mediated inhibition. *Science* **2017**, *355*, 1428–1433. [CrossRef] [PubMed]
23. Kamphorst, A.O.; Wieland, A.; Nasti, T.; Yang, S.; Zhang, R.; Barber, D.L.; Konieczny, B.T.; Daugherty, C.Z.; Koenig, L.; Yu, K.; et al. Rescue of exhausted CD8 T cells by PD-1-targeted therapies is CD28-dependent. *Science* **2017**, *355*, 1423–1427. [CrossRef] [PubMed]
24. Deng, L.; Liang, H.; Xu, M.; Yang, X.; Burnette, B.; Arina, A.; Li, X.D.; Mauceri, H.; Beckett, M.; Darga, T.; et al. STING-Dependent Cytosolic DNA Sensing Promotes Radiation-Induced Type I Interferon-Dependent Antitumor Immunity in Immunogenic Tumors. *Immunity* **2014**, *41*, 843–852. [CrossRef] [PubMed]
25. Han, C.; Godfrey, V.; Liu, Z.; Han, Y.; Liu, L.; Peng, H.; Weichselbaum, R.R.; Zaki, H.; Fu, Y.X. The AIM2 and NLRP3 inflammasomes trigger IL-1-mediated antitumor effects during radiation. *Sci. Immunol.* **2021**, *6*, eabc6998. [CrossRef]
26. Fang, H.; Ang, B.; Xu, X.; Huang, X.; Wu, Y.; Sun, Y.; Wang, W.; Li, N.; Cao, X.; Wan, T. TLR4 is essential for dendritic cell activation and anti-tumor T-cell response enhancement by DAMPs released from chemically stressed cancer cells. *Cell Mol. Immunol.* **2014**, *11*, 150–159. [CrossRef] [PubMed]
27. Kalanxhi, E.; Meltzer, S.; Schou, J.V.; Larsen, F.O.; Dueland, S.; Flatmark, K.; Jensen, B.V.; Hole, K.H.; Seierstad, T.; Redalen, K.R.; et al. Systemic immune response induced by oxaliplatin-based neoadjuvant therapy favours survival without metastatic progression in high-risk rectal cancer. *Br. J. Cancer* **2018**, *118*, 1322–1328. [CrossRef]

28. Cloughesy, T.F.; Mochizuki, A.Y.; Orpilla, J.R.; Hugo, W.; Lee, A.H.; Davidson, T.B.; Wang, A.C.; Ellingson, B.M.; Rytlewski, J.A.; Sanders, C.M.; et al. Neoadjuvant anti-PD-1 immunotherapy promotes a survival benefit with intratumoral and systemic immune responses in recurrent glioblastoma. *Nat. Med.* **2019**, *25*, 477–486. [CrossRef]
29. Pan, Y.; Yu, Y.; Wang, X.; Zhang, T. Tumor-Associated Macrophages in Tumor Immunity. *Front. Immunol.* **2020**, *11*, 583084. [CrossRef]
30. Cantero-Cid, R.; Casas-Martin, J.; Hernandez-Jimenez, E.; Cubillos-Zapata, C.; Varela-Serrano, A.; Avendano-Ortiz, J.; Casarrubios, M.; Montalban-Hernandez, K.; Villacanas-Gil, I.; Guerra-Pastrian, L.; et al. PD-L1/PD-1 crosstalk in colorectal cancer: Are we targeting the right cells? *BMC Cancer* **2018**, *18*, 945. [CrossRef]
31. Strasser, K.; Birnleitner, H.; Beer, A.; Pils, D.; Gerner, M.C.; Schmetterer, K.G.; Bachleitner-Hofmann, T.; Stift, A.; Bergmann, M.; Oehler, R. Immunological differences between colorectal cancer and normal mucosa uncover a prognostically relevant immune cell profile. *Oncoimmunology* **2019**, *8*, e1537693. [CrossRef] [PubMed]
32. Llosa, N.J.; Cruise, M.; Tam, A.; Wicks, E.C.; Hechenbleikner, E.M.; Taube, J.M.; Blosser, R.L.; Fan, H.; Wang, H.; Luber, B.S.; et al. The vigorous immune microenvironment of microsatellite instable colon cancer is balanced by multiple counter-inhibitory checkpoints. *Cancer Discov.* **2015**, *5*, 43–51. [CrossRef] [PubMed]
33. Guinney, J.; Dienstmann, R.; Wang, X.; de Reynies, A.; Schlicker, A.; Soneson, C.; Marisa, L.; Roepman, P.; Nyamundanda, G.; Angelino, P.; et al. The consensus molecular subtypes of colorectal cancer. *Nat. Med.* **2015**, *21*, 1350–1356. [CrossRef]
34. Ahlmanner, F.; Sundstrom, P.; Akeus, P.; Eklof, J.; Borjesson, L.; Gustavsson, B.; Lindskog, E.B.; Raghavan, S.; Quiding-Jarbrink, M. CD39(+) regulatory T cells accumulate in colon adenocarcinomas and display markers of increased suppressive function. *Oncotarget* **2018**, *9*, 36993–37007. [CrossRef]
35. Kumar, B.V.; Ma, W.; Miron, M.; Granot, T.; Guyer, R.S.; Carpenter, D.J.; Senda, T.; Sun, X.; Ho, S.H.; Lerner, H.; et al. Human Tissue-Resident Memory T Cells Are Defined by Core Transcriptional and Functional Signatures in Lymphoid and Mucosal Sites. *Cell Rep.* **2017**, *20*, 2921–2934. [CrossRef]
36. Okla, K.; Farber, D.L.; Zou, W. Tissue-resident memory T cells in tumor immunity and immunotherapy. *J. Exp. Med.* **2021**, *218*. [CrossRef]
37. Liang, F.; Rezapour, A.; Falk, P.; Angenete, E.; Yrlid, U. Cryopreservation of Whole Tumor Biopsies from Rectal Cancer Patients Enable Phenotypic and In Vitro Functional Evaluation of Tumor-Infiltrating T Cells. *Cancers* **2021**, *13*, 2428. [CrossRef]
38. Peng, Q.; Qiu, X.; Zhang, Z.; Zhang, S.; Zhang, Y.; Liang, Y.; Guo, J.; Peng, H.; Chen, M.; Fu, Y.X.; et al. PD-L1 on dendritic cells attenuates T cell activation and regulates response to immune checkpoint blockade. *Nat. Commun.* **2020**, *11*, 4835. [CrossRef]
39. Hu, W.; Sun, R.; Chen, L.; Zheng, X.; Jiang, J. Prognostic significance of resident CD103(+)CD8(+)T cells in human colorectal cancer tissues. *Acta Histochem* **2019**, *121*, 657–663. [CrossRef] [PubMed]
40. Hofving, T.; Liang, F.; Karlsson, J.; Yrlid, U.; Nilsson, J.A.; Nilsson, O.; Nilsson, L.M. The Microenvironment of Small Intestinal Neuroendocrine Tumours Contains Lymphocytes Capable of Recognition and Activation after Expansion. *Cancers* **2021**, *13*, 4305. [CrossRef] [PubMed]

Article

Qualitative Analysis of Tumor-Infiltrating Lymphocytes across Human Tumor Types Reveals a Higher Proportion of Bystander CD8⁺ T Cells in Non-Melanoma Cancers Compared to Melanoma

Aishwarya Gokuldass [1,†], Arianna Draghi [1,†], Krisztian Papp [2], Troels Holz Borch [1], Morten Nielsen [1], Marie Christine Wulff Westergaard [1], Rikke Andersen [1], Aimilia Schina [1], Kalijn Fredrike Bol [1], Christopher Aled Chamberlain [1], Mario Presti [1], Özcan Met [1,3], Katja Harbst [4,5], Martin Lauss [4,5], Samuele Soraggi [6], Istvan Csabai [2], Zoltán Szállási [7], Göran Jönsson [4,5], Inge Marie Svane [1] and Marco Donia [1,*]

1. National Center for Cancer Immune Therapy (CCIT-DK), Department of Oncology, Copenhagen University Hospital, 2730 Herlev, Denmark; aishwarya.gokuldass@regionh.dk (A.G.); arianna.draghi.01@regionh.dk (A.D.); troels.holz.borch@regionh.dk (T.H.B.); morten.nielsen.03@regionh.dk (M.N.); marie.christine.wulff.westergaard@regionh.dk (M.C.W.W.); Rikke.Andersen.02@regionh.dk (R.A.); aimilia.schina@regionh.dk (A.S.); kalijn.fredrike.bol@regionh.dk (K.F.B.); christopher.aled.chamberlain@regionh.dk (C.A.C.); mario.presti@regionh.dk (M.P.); Ozcan.Met@regionh.dk (Ö.M.); Inge.Marie.Svane@regionh.dk (I.M.S.)
2. Department of Physics of Complex Systems, ELTE Eötvös Loránd University, H-1117 Budapest, Hungary; pkrisz5@elte.hu (K.P.); csabai@complex.elte.hu (I.C.)
3. Department of Immunology and Microbiology, Faculty of Health and Medical Sciences, University of Copenhagen, 2200 Copenhagen, Denmark
4. Department of Clinical Sciences Lund, Division of Oncology and Pathology, Faculty of Medicine, Lund University, 221 00 Lund, Sweden; katja.harbst@med.lu.se (K.H.); martin.lauss@med.lu.se (M.L.); goran_b.jonsson@med.lu.se (G.J.)
5. Lund University Cancer Centre, Lund University, 221 00 Lund, Sweden
6. Bioinformatics Research Center, Aarhus University, 8000 Aarhus, Denmark; samuele@birc.au.dk
7. Danish Cancer Society Research Center, 2100 Copenhagen, Denmark; Zoltan.Szallasi@childrens.harvard.edu
* Correspondence: marco.donia@regionh.dk; Tel.: +45-38689339; Fax: +45-38683457
† These authors contributed equally to this paper.

Received: 7 October 2020; Accepted: 8 November 2020; Published: 12 November 2020

Simple Summary: Human tumors are often infiltrated by T cells; however, it remains unclear what proportion of T cells infiltrating tumors are bystander and non-tumor specific. We have investigated qualitative characteristics of these tumor-infiltrating lymphocytes (TILs) based on their gene-expression in the tumor-microenvironment or on their response to autologous tumor cells in vitro. Despite a considerable inter-sample variability, we found the overall proportion of bystander (non-tumor reactive) TILs to be remarkably high. Importantly, we observed a higher proportion of bystander TILs in non-melanoma tumors, compared to melanoma. This study suggests that immunotherapeutic strategies, especially when applied to non-melanoma tumors, should be tailored to reinvigorate the small proportion of tumor-reactive T cells infiltrating the tumor-microenvironment.

Abstract: *Background:* Human intratumoral T cell infiltrates can be defined by quantitative or qualitative features, such as their ability to recognize autologous tumor antigens. In this study, we reproduced the tumor-T cell interactions of individual patients to determine and compared the qualitative characteristics of intratumoral T cell infiltrates across multiple tumor types. *Methods:* We employed 187 pairs of unselected tumor-infiltrating lymphocytes (TILs) and autologous tumor cells from patients with melanoma, renal-, ovarian-cancer or sarcoma, and single-cell RNA sequencing data from a pooled cohort of 93 patients with melanoma or epithelial cancers. Measures of TIL quality including the proportion of tumor-reactive CD8$^+$ and CD4$^+$ TILs, and TIL response polyfunctionality were determined. *Results:* Tumor-specific CD8$^+$ and CD4$^+$ TIL responses were detected in over half of the patients in vitro, and greater CD8$^+$ TIL responses were observed in melanoma, regardless of previous anti-PD-1 treatment, compared to renal cancer, ovarian cancer and sarcoma. The proportion of tumor-reactive CD4$^+$ TILs was on average lower and the differences less pronounced across tumor types. Overall, the proportion of tumor-reactive TILs in vitro was remarkably low, implying a high fraction of TILs to be bystanders, and highly variable within the same tumor type. In situ analyses, based on eight single-cell RNA-sequencing datasets encompassing melanoma and five epithelial cancers types, corroborated the results obtained in vitro. Strikingly, no strong correlation between the proportion of CD8$^+$ and CD4$^+$ tumor-reactive TILs was detected, suggesting the accumulation of these responses in the tumor microenvironment to follow non-overlapping biological pathways. Additionally, no strong correlation between TIL responses and tumor mutational burden (TMB) in melanoma was observed, indicating that TMB was not a major driving force of response. No substantial differences in polyfunctionality across tumor types were observed. *Conclusions:* These analyses shed light on the functional features defining the quality of TIL infiltrates in cancer. A significant proportion of TILs across tumor types, especially non-melanoma, are bystander T cells. These results highlight the need to develop strategies focused on the tumor-reactive TIL subpopulation.

Keywords: tumor-infiltrating lymphocytes; tumor microenvironment; immunotherapy

1. Introduction

The success of cancer immunotherapy relies on the activation of a potent effector T cell response. These antitumor immune responses mediate tumor regression via recognition of tumor antigens presented on the surface of tumor cells by major histocompatibility complex (MHC) molecules. The most successful immune responses are believed to target antigens deriving from somatic tumor mutations, or neo-antigens, recognized with exquisite specificity by the T cell receptors (TCRs) expressed by tumor-infiltrating lymphocytes (TILs) [1]. An abundance of somatic tumor mutations (high tumor mutational burden, or TMB) translates into a high number of neo-antigens [2]. Immunologically active or "hot" tumors [3], especially those bearing a high TMB, are particularly sensitive to treatments stimulating adaptive immunity, such as cancer immunotherapy with immune checkpoint inhibitors [4]. Recent studies have demonstrated that TMB does not correlate well with the intratumoral immune activity of a given cancer when measured with standard methods, suggesting these two parameters to be independent.

Hot tumors are commonly defined by unspecific immunologic measures quantifying intratumoral immune activity, such as broad immune response gene-signatures related to T cell infiltration or interferon gamma (IFNγ) activity [5–8] and PD-L1 expression [9]. However, there is currently no singular definition of hot tumors. Recent in-depth TCR characterization of TILs in distinct tumor types revealed that the majority of CD8$^+$ TILs do not have the potential to recognize and kill autologous tumor cells ([10,11]). A fraction of CD8$^+$ TILs can be bystanders and recognize non-tumor related viral antigens [11,12], whereas CD4$^+$ TILs can be both bystander [13] and/or forkhead Box P3 (FOXP3)$^+$ tumor-specific T cells [14], which may be endowed with regulatory T cell functions. Activating non-tumor specific or regulatory T cells with immunotherapy, even in cases of significant T cell infiltration, is unlikely to induce tumor regression. These data have highlighted the issue of correctly identifying bona fide hot tumors, which are characterized not only by high immune cell infiltration but also an active tumor-directed immune response. Hence, novel approaches to study both the quantity and the quality of the tumor immune infiltrate at the functional level are highly warranted. The quality of a tumor immune infiltrate can be measured by the ability of TILs to recognize autologous tumor antigens and carry out functions such as type 1 cytokine secretion [15], mobilization of cytotoxic granules [16], or upregulation of costimulatory molecules [17].

In this study, we addressed this issue by measuring qualitative features of the TIL infiltrates based on recognition of autologous tumor antigens. This led to the identification of the proportion of tumor-reactive TILs and their corresponding functional profiles. We studied the association of TIL quality parameters to immunological and genomic biomarkers across multiple tumor types representative of the broad TMB spectrum. These data can serve as a reference for any future study describing the level of functional antitumor reactivity of a given population of tumor-reactive T cells.

2. Results

2.1. Antitumor Reactivity In Vitro: Testing Modalities

A total of 187 TILs/tumor cell pairs were obtained over a decade from individual patients spanning four distinct tumor types and five clinical cohorts (Table S1). Due to the large time frame of sample acquisition and analysis, as well as technical differences between cohorts, preliminary analyses were carried out to check for potential confounding variables and to estimate the comparability of all cohorts. The success rates of tumor cell line (TCL) generation that we reported previously were distinct across tumor types (Metastatic Melanoma (MM) 61% (consisting of two sub groups named as MM PD-1 naïve, including samples deriving from patients who were not treated with anti-PD1 previously, and MM PD-1 resistant (MM PD-1 res), including samples deriving from patients who were treated with anti-PD1 and progressed), renal cell carcinoma (RCC) 77%, ovarian cancer (OC) 32%, and sarcoma (SAR) 50% [18–22]), hence, different proportions of samples across cohorts were tested with fresh tumor digests (FTDs) only, or with autologous TCLs pre-treated or not with recombinant human IFNγ (Tables S2 and S3). A pooled analysis from all pairs that were tested with both TCLs/TCLs + IFNγ and FTDs (n = 71; 22 MM, 19 RCC, 21 OC and 9 SAR) showed that tests against FTDs yielded a lower reactivity for CD8$^+$ (p = 0.018, Figure S1A) and a higher reactivity for CD4$^+$ TILs (p = 0.039, Figure S1B) compared to testing against TCLs/TCLs + IFNγ. These data could indicate additional tumor-antigen presentation by non-tumor cells (e.g., stromal- or antigen presenting-cells) to CD4$^+$ T cells via MHC class II in assays using FTDs. When testing for potential differences in reactivity among pairs tested separately with Young TILs (Y TILs) and Rapidly Expanded TILs (REP TILs) (Tables S2 and S3), no significant differences were observed in CD8$^+$ (n = 132; 59 MM, 28 RCC and 45 OC; p = 0.25, Figure S1C) or CD4$^+$ (n = 128; 55 MM, 28 RCC and 45 OC; p = 0.66, Figure S1D) T cell reactivity. These data indicate that massive TIL-expansion with the rapid expansion protocol (REP) does not significantly impair the proportion of tumor-reactive TILs.

Overall, the differences in testing with TCLs/TCLs + IFNγ versus FTDs were statistically significant yet minor when compared to the larger differences observed when comparing cohorts. In addition, it cannot be ruled out that the ability to establish a TCLs is associated with a specific pattern of reactivity. Hence, we report pooled analyses using TCLs/TCLs + IFNγ or FTDs (showing only the highest reactivity), as well as data where only TCLs/TCLs + IFNγ were used.

2.2. Antitumor Reactivity In Vitro across Clinical Cohorts

In pooled analyses with TCLs/TCLs + IFNγ and FTDs (n = 186, Figure 1A), the mean proportion of tumor-reactive CD8$^+$ TILs of melanoma samples, regardless of previous anti-PD-1 therapy, far surpassed other cohorts ($p < 0.001$). Except RCC being greater than OC ($p < 0.01$), the other cohorts were highly similar to each other. However, there was considerable inter-sample variability within each clinical cohort, and individual samples in each of the other cohorts exceeded the mean reactivity level of CD8$^+$ TILs in MM. Of note, CD8$^+$ tumor-reactive TILs (above the detection limit of 0.5%) were detected in around 50% or more samples within each clinical cohort (range 50% in OC to 91% in MM PD-1 naïve). Analysis of reactivity against only TCLs/TCLs + IFNγ confirmed this pattern, except differences in non-melanoma (non-MM) tumors were no longer present (n = 143, Figure S2A).

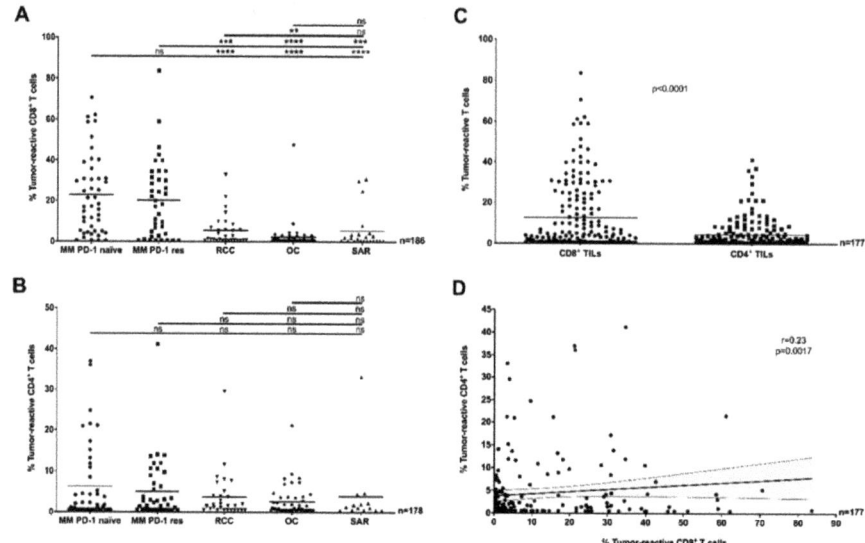

Figure 1. Antitumor-reactivity of tumor-infiltrating lymphocytes (TILs) across clinical cohorts and TIL sub-populations (in vitro, pooled data). (**A**) The proportion of tumor-reactive CD8$^+$ TILs was significantly greater in Metastatic Melanoma (MM) cohorts compared to other tumor types, with no difference related to previous exposure to anti-PD-1 therapy (Mann–Whitney U test, $p < 0.001$). (**B**) The proportion of tumor-reactive CD4$^+$ TILs was similar across all clinical cohorts (Mann–Whitney test, $p > 0.05$). (**C**) Comparing TIL subpopulations demonstrated that a greater proportion of CD8$^+$ TILs were tumor-reactive compared to CD4$^+$ TILs (Mann–Whitney test $p < 0.0001$, pooled clinical cohorts). (**D**) The proportion of tumor-reactive CD8$^+$ TILs was only weakly correlated (Spearman $r = 0.23$, $p = 0.0017$) to the proportion of tumor-reactive CD4$^+$ TILs in the same samples (pooled clinical cohorts). The solid line and dotted lines represent the best-fit regression line and 95% confidence interval, respectively. (**A–D**) In all panels, the recognition of TILs (Young TILs (Y TILs) and Rapidly Expanded TILs (REP TILs)) was tested against separate sets of autologous tumor cells (tumor cell lines (TCLs), TCLs + interferon gamma (IFNγ) or fresh tumor digests (FTDs)) and only the highest value reported. T cells were considered reactive if positive for at least one of TNF, IFNγ or CD107a, minus control. ** $p < 0.01$, *** $p < 0.001$, **** $p < 0.0001$, ns: no statistical significance.

The downstream effects of direct recognition of tumor-antigens by tumor-reactive CD4$^+$ TILs have not been well documented so far. Here, in pooled analyses with TCLs/TCLs + IFNγ and FTDs ($n = 177$), the proportion of tumor-reactive CD4$^+$ TILs was similar across all cohorts, and tumor-specific CD4$^+$ T cell responses were detected in over 50% of patients in each clinical cohort (range 58% in OC to 71% in RCC) (Figure 1B). A high inter-sample variability within individual cohorts was observed. When only responses to TCLs/TCLs + IFNγ were considered, responses were increased in MM PD-1 res samples compared to RCC ($p < 0.05$), in both melanoma cohorts (regardless of previous anti-PD-1 treatment) compared to OC ($p < 0.001$) and in SAR compared to OC ($p < 0.01$) ($n = 135$, Figure S2B). Differences between MM and non-MM tumors were considerably less pronounced compared to CD8$^+$ TIL responses.

In order to assess whether the differences in the proportion of tumor-reactive CD8$^+$ TILs observed in distinct tumor types were associated with other parameters of importance for T cell-mediated recognition of tumor-antigens, we re-analyzed data from The Cancer Genome Atlas (TCGA) and compared TCR richness [23] and antigen processing and presentation (APM) machinery activity across tumor types. Both parameters were not significantly increased in MM samples when compared to other tumor types (Figure S3A,B), hence neither of these parameters could explain the higher antitumor-reactivity of CD8$^+$ TILs observed in MM.

Interestingly, when comparing the proportion of tumor-reactive CD8$^+$ TILs to CD4$^+$ TILs, a significant difference in favor of CD8$^+$ TIL responses was observed when pooling all cohorts together ($p < 0.0001$, $n = 177$, Figure 1C). Segregation by cohort revealed that these differences were largely driven by MM (Figure S4), regardless of previous anti-PD-1 therapy. In addition, we found only a weak correlation linking the proportion of tumor-reactive CD8$^+$ and CD4$^+$ TILs in each sample ($r = 0.23$, $p < 0.0017$, $n = 177$, Figure 1D).

2.3. Antitumor Reactivity In Situ

In vitro studies with expanded TILs may not reflect the exact proportion of tumor-reactive T cells found in situ in the TME, as TIL expansion can result in culture-induced changes in clonal composition [24]. In order to further investigate the varying proportions of truly tumor-reactive TILs between distinct tumor types in situ, we re-analyzed single cell RNA-sequencing (scRNAseq) data from 101 tumor biopsies of 93 patients with MM [25–27] or non-MM epithelial [28–32] tumor types. After merging multiple datasets, we identified a total of 16,651 CD8$^+$ and 14,036 CD4$^+$ TILs and based on the gene expression of functional markers related to recent T cell activation (see Section 4.6) determined the proportion of tumor-reactive CD8$^+$ TILs (Figure 2A) or CD4$^+$ TILs (Figure 2B and Figure S5B) in MM versus non-MM epithelial samples, the proportion of tumor-reactive CD8$^+$ versus CD4$^+$ TILs (Figure 2C), and the correlation in the proportion of tumor-reactive CD8$^+$ and CD4$^+$ TILs in each sample (Figure 2D). These analyses largely reproduced the results obtained in vitro, shown in Figure 1, differing only in the numerically greater number of tumor-reactive TILs detected in situ, and in a marginally stronger ($r = 0.47$, Figure 2D) positive correlation of CD8$^+$ and CD4$^+$ TIL-responses.

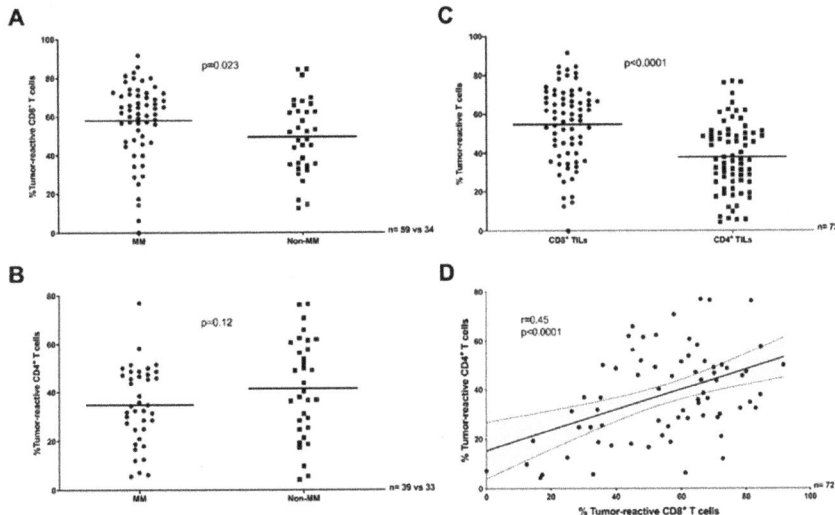

Figure 2. Antitumor-reactivity of TILs across clinical datasets and TIL sub-populations (scRNAseq in situ). (**A**) The proportion of tumor-reactive CD8+ TILs was greater in MM (PD-1 naïve plus PD-1 res) compared to non-MM (Mann–Whitney test, $p = 0.023$). (**B**) The proportion of tumor-reactive CD4+ TILs was comparable in MM (PD-1 naïve only) and non-MM (Unpaired t test, $p = 0.12$). (**C**) A greater proportion of CD8+ TILs were tumor-reactive compared to CD4+ TILs (Paired t test, $p < 0.0001$, MM PD-1 naïve plus non-MM). (**D**) The proportion of tumor-reactive CD8+ TILs was only moderately

correlated to the proportion of tumor-reactive CD4+ TILs in the same samples (Pearson $r = 0.45$, $p < 0.0001$, MM PD-1 naïve plus non-MM). The solid line and dotted lines represent the best-fit regression line and 95% confidence interval, respectively. (A–D) T cells were considered reactive if positive for the expression of least one of *TNF*, *IFNG*, or *TNFRSF9*.

2.4. Polyfunctionality of Responses In Vitro across Tumor Types

We recently demonstrated that tumor-reactive $CD8^+$ T cells derived from melanoma and renal cancer can be characterized by their functional patterns [20]. Therefore, we expanded this analysis by determining whether tumor-reactive TILs derived from multiple tumor types were endowed with distinct (poly)functional profiles.

The functional profile of $CD8^+$ and $CD4^+$ TILs isolated from the two cohorts of MM, anti-PD-1 naïve and MM PD-1 res, did not significantly differ ($p = 0.9$ and $p = 0.3$, Figure S6). Hence, additional analyses were performed comparing the four distinct tumor types regardless of previous anti-PD-1 therapy. Consistent with our previous data from a smaller but partially overlapping cohort [20], tumor-reactive $CD8^+$ T cells isolated from MM exhibited greater polyfunctionality compared to $CD8^+$ T cells isolated from RCC and, additionally, from SAR (Figure S7). However, these differences were not significant in all other tumor types comparisons (Figure S7). No major differences were observed within the $CD4^+$ tumor-reactive TILs, which appeared to be primarily characterized by tumor necrosis factor (TNF) production only, and a smaller population releasing both IFNγ and TNF (Figure S8). Overall, our analysis did not depict extensive differences in T cell-polyfunctional profiles across the tumor panel analyzed.

2.5. Tumor-Reactive TILs In Vitro and TMB in Melanoma

Although it is generally accepted that tumor types with a higher average TMB respond more frequently to immunotherapy [33,34], the value of TMB for predicting clinical outcome following immunotherapy within tumor types or subtypes is debated and several studies have failed to link TMB to immunological quantitative biomarkers in the TME [5–7,9]. On average, TMB is very high in melanoma, but the wide range observed highlights a high heterogeneity across patients [35]. Hence, we examined whether the high inter-sample variability in the proportion of tumor-reactive TILs across melanoma could be explained by the TMB of individual samples. Here, analysis of a smaller melanoma cohort ($n = 36$) did not show any obvious correlation between the proportion of tumor-reactive $CD8^+$ or $CD4^+$ TILs and TMB (Figure 3, data with TCL/TCL + IFNγ only are shown in Figure S9).

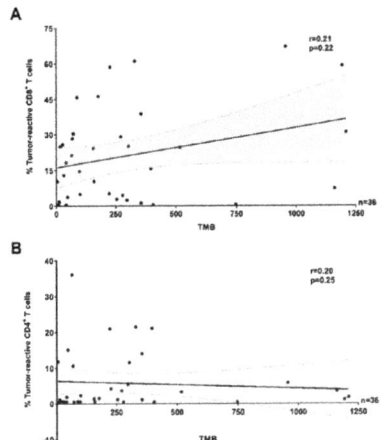

Figure 3. Antitumor-reactivity of TILs and tumor mutational burden in melanoma (pooled data). The proportion of tumor-reactive (**A**) $CD8^+$ and (**B**) $CD4^+$ TILs was not correlated (Spearman $r = 0.21$, $p = 0.22$ and Spearman $r = 0.20$, $p = 0.25$, respectively) to tumor mutational burden (TMB) (pooled MM clinical cohorts). The solid lines and dotted lines represent the best-fit regression line and 95% confidence interval, respectively. (**A,B**) In all panels, the recognition of TILs (Y TILs and REP TILs) was tested against separate sets of autologous tumor cells (TCLs, TCLs + IFNγ or FTDs) and only the highest value reported. T cells were considered reactive if positive for at least one of TNF, IFNγ, or CD107a, minus control.

3. Discussion

Here, we have presented a functional qualitative analysis of tumor-specific immune responses of TILs across multiple tumor types. Importantly, the proportion of tumor-specific T cells with functional capacity may define the quality of a TIL population, and although multiple parameters associated with an immunologically active TME have been positively associated with response to cancer immunotherapy [5,6,8,36] the predictive value of TIL quality is yet to be established in larger datasets. Early results have shown that, at least for adoptive cell therapy with unselected TILs, a parameter of TIL quality (i.e., defining the amount of tumor-reactive cells infused) may help identifying those patients with a higher likelihood of achieving tumor regression [37,38].

Recent studies have demonstrated that bystander T cells may represent the majority of infiltrating lymphocytes in cancer [10,11]. Here, we quantified the proportion of tumor-reactive T cells across multiple cancer types, and observed that although tumor-reactive $CD8^+$ and $CD4^+$ T cells could be identified in most samples across cohorts, the proportion was on average remarkably low (especially for non-MM tumors). The estimated proportion of tumor-reactive T cells in situ appeared on average higher when compared to the in vitro analyses, yet still failed to encompass the entire T cell repertoire for most patients and melanoma TILs were more reactive compared to other tumors. These results signify the need for effective T cell selection-strategies in clinical protocols of adoptive cell transfer, especially in non-melanoma tumors. Along this line, next-gen cellular therapy technologies based on T cell selection and selective expansion of tumor-reactive T cells may provide a solution for tumors with low natural immunogenicity. In addition, further studies establishing strategies for bona fide identification and stimulation of truly tumor-specific/tumor-reactive T cells in situ are highly warranted. These novel strategies should focus on avoiding stimulation of deleterious immune sub-populations, such as regulatory T cells, which may account for a fraction of TILs.

For decades, melanoma and RCC have been considered highly immunogenic tumors. This theory is largely supported by multiple reports of occasional spontaneous regression and durable, although minimal, responses to IL-2 [39]. Interestingly, in this study melanomas were on average infiltrated by a greater proportion of CD8$^+$ tumor-reactive TILs than other tumor types, yet melanoma PD-1 naïve and PD-1 resistant samples (collected after progression to anti-PD-1) were indistinguishable. This suggests that high infiltration with tumor-reactive CD8$^+$ TILs does not appear to guarantee a response to immune checkpoint inhibitors. Additionally, the increased infiltration of tumor-reactive CD8$^+$ TILs could not be explained by the differences that we observed in TCR richness or APM machinery activity when comparing melanoma to other tumor types.

Our analysis of polyfunctionality in vitro revealed differences in TIL-responses across tumors. Distinct functional profiles, albeit without dramatic differences, were detected for CD8$^+$ TILs, whereas profiling of CD4$^+$ TILs resulted in largely overlapping results across tumor types. At present, the impact of T cell response polyfunctionality on clinical parameters is largely unknown, but the absence of major differences across tumor types indicates that interventions to improve the (low) proportion of tumor-reactive T cell responses may represent a more urgent issue.

Most current data support a model where tumors with high TMB are endowed with a greater number of potential T cell targets (neo-antigens) [2] and are therefore more easily recognized by the immune system. However, TMB does not appear to be well correlated to quantitative biomarkers such as immune infiltration or immune activity in situ [5,6,9]. This is somewhat paradoxical, as both TMB and gene-signatures identifying tumors with high immune infiltration/immune activity in situ can identify patients with a higher likelihood to respond to immunotherapy with checkpoint inhibitors [5,6]. In this study, we expand on these associations by correlating TMB and T cell infiltrate quality (measured as the proportion of tumor-reactive T cells amongst all TILs). Here, although melanoma (the tumor with the highest average TMB) presented with the greatest CD8$^+$ TIL response, within melanoma a high TMB did not appear to represent a major driver for the accumulation of either tumor-reactive CD8$^+$ TILs or CD4$^+$ TILs. Hence, the biological pathways leading to high infiltration by tumor-reactive CD8$^+$ TILs in melanomas appears to be independent of TMB. These data confirm that TMB does not strongly correlate to any known immunological parameters across samples, and therefore still functions as a largely immune-independent biomarker. Of note, the major driving forces governing the accumulation of CD8$^+$ or CD4$^+$ tumor-reactive TILs in the TME are yet to be identified. In our study, we could not find a strong correlation linking the proportion of tumor-reactive CD8$^+$ and CD4$^+$ TILs in each sample. These data suggest that the accumulation of CD8$^+$ or CD4$^+$ TIL responses in the TME may follow non-overlapping biological pathways that may not be simultaneously present in an individual tumor or influenced by each other.

This study has some caveats, primarily the low coverage of tumor types (only four in vitro), differences between metastatic or primary tumor sites in distinct tumor types (similar to the composition of samples contained in The Cancer Genome Atlas for melanoma, RCC and SAR), limited number of samples for each of the non-MM tumors in situ and the heterogeneity of in vitro testing due to the extended timeframe (a decade) of sample acquisition and analysis. It was not possible to further characterize regulatory T cells among CD4$^+$ TILs, as the TILs used in our in vitro experiments were stimulated with IL-2, hence the expression of FOXP3 was induced in conventional T cells [40]. In addition, although antitumor reactivity testing with autologous cell lines in vitro still represents a gold-standard, the success rate of cell line establishment is variable across tumor types, thereby potentially resulting in a degree of sample selection bias. However, all samples used for in vitro analyses were acquired and analyzed at the same center, guaranteeing a reliable level of consistency. We observed an average higher apparent proportion of tumor-reactive T cells observed in situ compared to in vitro. Due to technical constraints, we used distinct markers to determine the proportion of tumor-reactive T cells in vitro and in situ; in addition, in vitro culturing may influence the clonal composition of TIL preparations, with depletion of tumor-reactive TIL clones because of poor proliferative capacity of dysfunctional cells. These factors may partially (but probably not fully) explain the higher apparent proportion of tumor-reactive T cells observed in situ.

4. Materials and Methods

4.1. Patients and Samples

Fresh tumor specimens were obtained via surgical resection or needle biopsy from patients with solid tumors over a ten-year period at the National Center for Cancer Immune Therapy, Copenhagen University Hospital, Herlev, Denmark. Samples were obtained via biopsy collection for enrolment in 11 clinical trials conducted between 2009 and 2020. Written informed consent was provided by all patients prior to obtaining any samples. All trials (NCT00937625, NCT02379195, NCT02354690, H-18055660, NCT02926053, NCT02482090, NCT03287674, NCT03296137, H-4-2012-118, H-15007073, H-2-2014-055) were approved by the relevant Ethics Committee and conducted in accordance with the Declaration of Helsinki and Good Clinical Practice. The clinical cohorts used in this study partially overlap with those that we previously published in other studies [18–22,41–43]

Two tumor types (MM and RCC) were selected for their known high immunogenicity and reported sensitivity to immunotherapy; two other tumor types (OC and SAR) were selected for their known relative resistance to checkpoint immunotherapy. All MM and OC originated from metastases, whereas SAR and the majority of RCC originated from primaries. To account for the potential biological differences of tumor samples recovered after progression to anti-PD-1 therapy, MM samples were sub-grouped as anti-PD-1 naïve (MM PD-1 naïve, not previously treated with ant-PD-1 regardless of response to any immunotherapy given after tumor collection) or anti-PD-1 resistant (MM PD-1 res). Overall, these four tumor types are representative of tumors with high (MM), intermediate (RCC and OC), and low (SAR) TMB, according to Yarchoan et al. [9].

4.2. Establishment of TILs, TCLs, and FTDs

TIL cultures were established in vitro with a two-step process; the initial expansion to generate "Young" or "minimally-cultured" TILs and the REP to generate REP TILs, as previously described in detail [18–22,41]. Short-term autologous TCLs (<10 in vitro passages) were established as described elsewhere using fragments or transport media following scalpel dissection from the same tumor lesion from which the TILs were generated [18–22,41]. All cell lines were generated internally and primarily authenticated via morphology (light microscopy) and in vitro patterns of growth. When in doubt, expression of lineage antigens by PCR or cytospin followed by morphologic evaluation (according to standard cytologic criteria of malignancy [44]) and immunohistochemistry staining of formalin-fixed, paraffin-embedded tissue was carried out. Mycoplasma testing was not performed for all samples. FTDs were obtained from tumor fragments via overnight digestion followed by immediate cryopreservation, as previously described in detail [19,20,22].

4.3. Assessment of TIL Reactivity Against TCLs or FTDs In Vitro

The level of bulk antitumor reactivity of Y TILs or REP TILs was tested in vitro separately against autologous TCLs, autologous TCLs pre-treated with IFNγ (Peprotech, Stockholm, Sweden; to upregulate tumor antigen processing machinery and presentation, as described elsewhere [45] or autologous FTDs. This was achieved by co-culturing effector (TILs) and target (TCLs, TCLs + IFNγ or FTDs) cells, followed by flow-cytometry analysis of three extensively described type 1 immune response activation markers; TNF, IFNγ and CD107a [18–20,22]. Antitumor reactivity was defined as the percentage of live $CD8^+$ or $CD4^+$ T cells staining positive for at least one of TNF, IFNγ and CD107a, minus control (TILs alone). To define the bulk antitumor reactivity in a given sample, only the highest value obtained from Y TILs or REP TILs tested against TCLs, TCLs + IFNγ or FTDs was reported. For the SAR cohort only, effector-target pairs (only TILs vs TCLs) were pre-tested with co-culture followed by IFNγ ELISPOT, as described previously [46,47]. Further testing with flow cytometry, as described in supplementary methods, was conducted only in those samples with suspected or confirmed ELISPOT reactivity. Therefore, the samples tested only in ELISPOT were not evaluable for $CD4^+$ reactivity, because we could not exclude that IFNγ pretreatment of tumors (TCLs + IFNγ) would have resulted in CD4⁺ T cell recognition and positive responses by ELISPOT [18]. SAR samples were not included in the analyses of Y TILs versus REP TILs responses due to the limited number of samples tested with both Y and REP TILs. For polyfunctional characterization of tumor-reactive cells, data were primarily analyzed in FlowJo V10 (BD). Analysis and presentation of distributions was performed using Pestle 2.0 (downloaded from https://niaid.github.io/spice/) and Simplified Presentation of Incredibly Complex Evaluations (SPICE) 6.0 (downloaded from https://niaid.github.io/spice/) according to manufacturer's instructions. Detailed information can be found in Supplementary Methods.

4.4. TCR Richness and Antigen Processing and Presentation Machinery (APM).

TCR richness and APM machinery were quantified in >1000 samples obtained from TCGA. A detailed description is provided in Supplementary Methods.

4.5. Processing of Single-Cell RNA-Sequencing Datasets

The literature was screened for single-cell RNA-sequencing datasets of tumor biopsies containing data on T cells from individual patients. Eight independent datasets containing single-cell RNA-sequencing data, from 101 tumor biopsies (93 patients) and covering six tumor types (2 breast [28], 14 non-small cell lung [29], 6 hepatocellular [30], 4 renal [31], 8 colorectal [32] cancer and 67 melanoma [25–27], were obtained from public repositories or requested directly from the authors and selected for inclusion in our study (Table S4). Only CD8⁺ and CD4⁺ T cells isolated from tumor tissues were utilized. Detailed information can be found in Supplementary Methods, Figure S5A and Figure S10.

4.6. Assessment of TIL Reactivity Against Tumor Cells in Situ

Antitumor reactivity within the CD8⁺ and CD4⁺ T cell compartment was defined as the expression of at least one of *TNF*, *IFNG*, and *TNFRSF9*. Here, although TNF and IFNγ were also used in vitro in all samples due to their high signal-to-noise ratio in activated TILs (barely detectable in TILs alone —significantly upregulated in a variable proportion of TILs recognizing autologous tumor cells), CD137 (encoded by the gene *TNFRSF9*) upregulation was not included in the in vitro analyses. Indeed, the samples used in the study were obtained over a decade, whereas we only recently began to use CD137 as additional tumor-specific (but function-agnostic) T cell activation marker, and confirmed that a proportion accounting for ~20% of the total CD8⁺ tumor-reactive repertoire may be identified by expression of CD137, but not TNF, IFNγ or CD107a [48]. As CD137 may be constitutively expressed on CD4⁺ regulatory T cells [49], we carried out additional analyses on intratumor CD4⁺ T cells using *TNF* and *IFNG* only (Figure S5). We determined that the expression of *LAMP1* (coding for CD107a or lysosome-associated membrane protein-1) could not be used as a T cell degranulation marker in the transcriptomic setting, as its function as part of pre-formed lytic granules in the T cell cytoplasm that are mobilized upon T cell activation requires constitutive mRNA expression, regardless of activation status [50]. In an additional study we could not detect significant upregulation of *LAMP1* on tumor-reactive T cells upon target-recognition (Draghi A et al., in preparation). Therefore, we did not expect *LAMP1* upregulation to be associated with T cell activation/degranulation. Other molecules, such as Granzyme-B (*GZMB*), were expressed in a significant proportion of resting T cells in vitro, and we could therefore not consider these molecules (and their relative mRNA) as bona fide markers of T cell activation.

4.7. Analysis of Tumor Mutational Burden

Thirty-six samples obtained from patients with melanoma enrolled in interventional clinical trials at the host institution had DNA sequencing data available. Whole-exome sequencing was carried out as previously described [51], and total TMB was calculated based on the somatic non-synonymous single-nucleotide variants detected.

4.8. Statistical Analyses

Statistical analyses were carried out using GraphPad Prism 8.4 or SPICE 6.0. Values below 0.5% derived from the subtraction of unstimulated samples from stimulated samples were converted to 0.5% for statistical purposes and generation of figures. All values were expressed as mean unless otherwise specified. The D'Agostino and Pearson normality test was performed to determine whether the data were normally distributed. Mann–Whitney or Wilcoxon-matched pairs tests were used to determine statistical significance in case of non-normally distributed data. Unpaired or paired T tests were used to determine statistical significance in case of normally distributed data. Correlations were expressed by Spearman and Pearson R value in case of non-normally and normally distributed data, respectively. Regarding statistical analyses on TCGA data, for pairwise comparisons of multiple groups, with non-equal variances and samples sizes, the non-parametric Games–Howell post-hoc test was used. Benjamini Hochberg method was selected as the adjustment method for *p*-values for multiple comparisons. Significance level for *p*-adjusted values was set to 0.05.

5. Conclusions

In conclusion, multiple studies have shown that exploiting the T cell-infiltrates of solid tumors has great therapeutic potential across tumor types [22,37,52–55]. However, a large proportion of tumor-infiltrating T cells, especially in TIL cultures obtained from non-MM tumors, are not tumor-relevant. These data indicate that future strategies employing immunotherapies based on T cell infusion or stimulation of T cells residing in the TME should be tailored to the tumor-reactive T cell subpopulation. Parameters such as TMB, which are related to the overall burden of neo-antigens, may not be relevant to address these issues.

Supplementary Materials: The following are available online at http://www.mdpi.com/2072-6694/12/11/3344/s1, Figure S1: Comparison of multiple sources of tumor targets and TILs (in vitro), Figure S2: Antitumor-reactivity of TILs across clinical cohorts (in vitro, TCL/TCL+IFNγ only), Figure S3: TCR richness and antigen processing and presentation machinery across four tumor types, Figure S4: Antitumor-reactivity of CD8+ and CD4+ TILs across tumor types (in vitro, pooled data), Figure S5: Proportion of tumor-reactive CD4+ TILs calculated using two markers, TNF and IFNG (scRNAseq in situ), Figure S6: Polyfunctional characterization of CD8+ tumor-reactive TILs in MM (in vitro, functions combination), Figure S7: Polyfunctional characterization of CD8+ tumor-reactive TILs across tumor types (in vitro, functions combination), Figure S8: Polyfunctional characterization of CD4+ tumor-reactive TILs across tumor types (in vitro, functions combination), Figure S9: Antitumor-reactivity of TILs and tumor mutational burden in melanoma (in vitro, TCL/TCL+IFNγ only), Figure S10: Proportion of tumor-reactive TILs in PD1-naïve and PD-1 res samples (scRNAseq in situ), Table S1: Overview of all samples used in the study (in vitro data), Table S2: Overview of CD8+ TILs analyses (in vitro data), Table S3: Overview of CD4+ TILs analyses (in vitro data), Table S4: Single-cell RNA-sequencing datasets accession numbers (in situ data).

Author Contributions: Conceptualization: A.G., A.D., M.D.; Supervision: Ö.M., I.C., Z.S., G.J., I.M.S., and M.D.; Formal analysis: A.G., A.D., T.H.B., M.N., M.C.W.W., R.A., K.F.B., C.A.C., M.P., K.H., M.L., and M.D.; Data curation: A.G., A.D., K.P., M.N., M.C.W.W., R.A., A.S., Ö.M., K.H., M.L., S.S., and M.D.; Statistical analysis: A.G., A.D., A.S., and M.D.; Data visualization: A.G., A.D., A.S., and M.D.; Writing—original draft: A.G., A.D., C.A.C., M.D.; Resources supply and funding acquisition: Ö.M., I.C., Z.S., G.J., I.M.S., and M.D. All authors critically revised and edited the manuscript and approved it for final submission. All authors have read and agreed to the published version of the manuscript.

Funding: The present study was funded by the Danish Cancer Society, grant R148-A9862 and R204-A12535; the Capital Region of Denmark Research Foundation, grant R146-A5693; the Lundbeck Foundation, grant R233-2016-3728; the Denmark Independent Research Fund, grant 8045-00067B; and The National Research, Development and Innovation Fund of Hungary, grant FIEK_16-1-2016-0005; Danish Cancer Research Fund, grant FID2038024.

Acknowledgments: The authors wish to thank all the patients who donated samples that were used in this study and the funding sources that generously supported this research. Morten Hansen and Michael Douglas Crowther are acknowledged for their technical assistance with the flow cytometry setup. Julie Westerlin Kjeldsen, Cathrine Lund Lorentzen, Sofie Kirial Mørk, Anders Kverneland, Henrik Lajer, Dorrit Hovgaard, Michael Mørk Petersen and Niels Junker are acknowledged for patient enrolment. Mads Hald Andersen is acknowledged for excellent scientific discussions regarding these works. Sandra Ullitz Færch, Betina Saxild and Susanne Wendt are acknowledged for technical support. The results shown here are partly based upon data generated by the TCGA Research Network: http://cancergenome.nih.gov/ and by other authors who deposited clinical datasets in public repositories.

Conflicts of Interest: Marco Donia has received honoraria for lectures from Roche and Novartis (past two years). Inge Marie Svane has received honoraria for consultancies and lectures from Novartis, Roche, Merck, and Bristol-Myers Squibb; a restricted research grant from Novartis; and financial support for attending symposia from Bristol-Myers Squibb, Merck, Novartis, Pfizer and Roche. All other authors declare that they have no conflict of interest.

References

1. Tran, E.; Robbins, P.F.; Rosenberg, S.A. "Final common pathway" of human cancer immunotherapy: Targeting random somatic mutations. *Nat. Immunol.* **2017**, *18*, 255–262. [CrossRef]
2. Efremova, M.; Finotello, F.; Rieder, D.; Trajanoski, Z. Neoantigens generated by individual mutations and their role in cancer immunity and immunotherapy. *Front. Immunol.* **2017**, *8*, 1–8. [CrossRef] [PubMed]
3. Galon, J.; Bruni, D. Approaches to treat immune hot, altered and cold tumours with combination immunotherapies. *Nat. Rev. Drug Discov.* **2019**, *18*, 197–218. [CrossRef] [PubMed]
4. Havel, J.J.; Chowell, D.; Chan, T.A. The evolving landscape of biomarkers for checkpoint inhibitor immunotherapy. *Nat. Rev. Cancer* **2019**, *19*, 133–150. [CrossRef] [PubMed]
5. Ott, P.A.; Bang, Y.-J.; Piha-Paul, S.A.; Razak, A.R.A.; Bennouna, J.; Soria, J.-C.; Rugo, H.S.; Cohen, R.B.; O'Neil, B.H.; Mehnert, J.M.; et al. T-Cell–Inflamed Gene-Expression Profile, Programmed Death Ligand 1 Expression, and Tumor Mutational Burden Predict Efficacy in Patients Treated With Pembrolizumab Across 20 Cancers: KEYNOTE-028. *J. Clin. Oncol.* **2019**, *37*, 318–327. [CrossRef]
6. Cristescu, R.; Mogg, R.; Ayers, M.; Albright, A.; Murphy, E.; Yearley, J.; Sher, X.; Liu, X.Q.; Lu, H.; Nebozhyn, M.; et al. Pan-tumor genomic biomarkers for PD-1 checkpoint blockade-based immunotherapy. *Science* **2018**, *362*, eaar3593. [CrossRef]
7. Spranger, S.; Luke, J.J.; Bao, R.; Zha, Y.; Hernandez, K.M.; Li, Y.; Gajewski, A.P.; Andrade, J.; Gajewski, T.F. Density of immunogenic antigens does not explain the presence or absence of the T-cell–inflamed tumor microenvironment in melanoma. *Proc. Natl. Acad. Sci. USA* **2016**, *113*, E7759–E7768. [CrossRef]
8. Ayers, M.; Ribas, A.; Mcclanahan, T.K.; Ayers, M.; Lunceford, J.; Nebozhyn, M.; Murphy, E.; Loboda, A.; Kaufman, D.R.; Albright, A.; et al. IFN- g-related mRNA profile predicts clinical response to PD-1 blockade Find the latest version: IFN-γ-related mRNA profile predicts clinical response to PD-1 blockade. *J. Clin. Invest.* **2017**, *127*, 2930–2940. [CrossRef]
9. Yarchoan, M.; Albacker, L.A.; Hopkins, A.C.; Montesion, M.; Murugesan, K.; Vithayathil, T.T.; Zaidi, N.; Azad, N.S.; Laheru, D.A.; Frampton, G.M.; et al. PD-L1 expression and tumor mutational burden are independent biomarkers in most cancers. *JCI Insight* **2019**, *4*. [CrossRef]
10. Scheper, W.; Kelderman, S.; Fanchi, L.F.; Linnemann, C.; Bendle, G.; de Rooij, M.A.J.; Hirt, C.; Mezzadra, R.; Slagter, M.; Dijkstra, K.; et al. Low and variable tumor-reactivity of the intratumoral TCR repertoire in human cancers. *Nat. Med.* **2019**. [CrossRef]
11. Simoni, Y.; Becht, E.; Fehlings, M.; Loh, C.Y.; Koo, S.L.; Teng, K.W.W.; Yeong, J.P.S.; Nahar, R.; Zhang, T.; Kared, H.; et al. Bystander CD8+T cells are abundant and phenotypically distinct in human tumour infiltrates. *Nature* **2018**, *557*, 575–579. [CrossRef] [PubMed]
12. Andersen, R.S.; Thrue, C.A.; Junker, N.; Lyngaa, R.; Donia, M.; Ellebæk, E.; Svane, I.M.; Schumacher, T.N.; Thor Straten, P.; Hadrup, S.R. Dissection of T-cell antigen specificity in human melanoma. *Cancer Res.* **2012**, *72*, 1642–1650. [CrossRef] [PubMed]
13. Simoni, Y.; Li, S.; Zhuang, S.; Heit, A.; Koo, S.-L.; Chow, I.-T.; Kwok, W.; Tan, I.B.; Tan, D.; Newell, E. Bystander CD4 + T cells infiltrate human tumors and are phenotypically distinct. *bioRxiv* **2020**. [CrossRef]
14. Ahmadzadeh, M.; Pasetto, A.; Jia, L.; Deniger, D.C.; Stevanović, S.; Robbins, P.F.; Rosenberg, S.A. Tumor-infiltrating human CD4 + regulatory T cells display a distinct TCR repertoire and exhibit tumor and neoantigen reactivity. *Sci. Immunol.* **2019**, *4*, eaao4310. [CrossRef]
15. Melief, J.; Wickström, S.; Kiessling, R.; Pico de Coaña, Y. Assessment of antitumor T-cell responses by flow cytometry after coculture of tumor cells with autologous tumor-infiltrating lymphocytes. In *Methods in Molecular Biology*; Humana Press Inc.: Totowa, NJ, USA, 2019; Volume 1913, pp. 133–140.
16. Rubio, V.; Stuge, T.B.; Singh, N.; Betts, M.R.; Weber, J.S.; Roederer, M.; Lee, P.P. Ex vivo identification, isolation and analysis of tumor-cytolytic T cells. *Nat. Med.* **2003**, *9*, 1377–1382. [CrossRef]

17. Ye, Q.; Song, D.-G.; Poussin, M.; Yamamoto, T.; Best, A.; Li, C.; Coukos, G.; Powell, D.J. CD137 Accurately Identifies and Enriches for Naturally Occurring Tumor-Reactive T Cells in Tumor. *Clin. Cancer Res.* **2014**, *20*, 44–55. [CrossRef]
18. Donia, M.; Andersen, R.; Kjeldsen, J.W.; Fagone, P.; Munir, S.; Nicoletti, F.; Andersen, M.H.; thor Straten, P.; Svane, I.M. Aberrant Expression of MHC Class II in Melanoma Attracts Inflammatory Tumor-Specific CD4 + T- Cells, Which Dampen CD8 + T-cell Antitumor Reactivity. *Cancer Res.* **2015**, *75*, 3747–3759. [CrossRef]
19. Westergaard, M.C.W.; Andersen, R.; Chong, C.; Kjeldsen, J.W.; Pedersen, M.; Friese, C.; Hasselager, T.; Lajer, H.; Coukos, G.; Bassani-Sternberg, M.; et al. Tumour-reactive T cell subsets in the microenvironment of ovarian cancer. *Br. J. Cancer* **2019**, *120*, 424–434. [CrossRef]
20. Andersen, R.; Westergaard, M.C.W.; Kjeldsen, J.W.; Muller, A.; Pedersen, N.W.; Hadrup, S.R.; Met, O.; Seliger, B.; Kromann-Andersen, B.; Hasselager, T.; et al. T-cell responses in the microenvironment of primary renal cell carcinoma-implications for adoptive cell therapy. *Cancer Immunol. Res.* **2018**, *6*. [CrossRef]
21. Nielsen, M.; Krarup-Hansen, A.; Hovgaard, D.; Petersen, M.M.; Loya, A.C.; Westergaard, M.C.W.; Svane, I.M.; Junker, N. In vitro 4-1BB stimulation promotes expansion of CD8+ tumor-infiltrating lymphocytes from various sarcoma subtypes. *Cancer Immunol. Immunother.* **2020**. [CrossRef]
22. Andersen, R.; Borch, T.H.; Draghi, A.; Gokuldass, A.; Rana, A.H.M.; Pedersen, M.; Nielsen, M.; Kongsted, P.; Kjeldsen, J.W.; Westergaard, C.W.M.; et al. T cells isolated from patients with checkpoint inhibitor-resistant melanoma are functional and can mediate tumor regression. *Ann. Oncol.* **2018**, *29*, 1575–1581. [CrossRef] [PubMed]
23. Qi, Q.; Liu, Y.; Cheng, Y.; Glanville, J.; Zhang, D.; Lee, J.-Y.; Olshen, R.A.; Weyand, C.M.; Boyd, S.D.; Goronzy, J.J. Diversity and clonal selection in the human T-cell repertoire. *Proc. Natl. Acad. Sci. USA* **2014**, *111*, 13139–13144. [CrossRef] [PubMed]
24. Poschke, I.C.; Hassel, J.C.; Rodriguez Ehrenfried, A.; Lindner, K.A.M.; Heras-Murillo, I.; Appel, L.M.; Lehmann, J.; Lövgren, T.; Wickström, S.L.; Lauenstein, C.; et al. The outcome of ex vivo TIL expansion is highly influenced by spatial heterogeneity of the tumor T-cell repertoire and differences in intrinsic in vitro growth capacity between T-cell clones. *Clin. Cancer Res.* **2020**. [CrossRef] [PubMed]
25. Tirosh, I.; Izar, B.; Prakadan, S.M.; Ii, M.H.W.; Treacy, D.; Trombetta, J.J.; Rotem, A.; Rodman, C.; Lian, C.; Murphy, G.; et al. Dissecting the multicellular ecosystem of metastatic melanoma by single-cell RNA-seq. *Science* **2016**, *352*, 189–196. [CrossRef] [PubMed]
26. Jerby-Arnon, L.; Shah, P.; Cuoco, M.S.; Rodman, C.; Su, M.J.; Melms, J.C.; Leeson, R.; Kanodia, A.; Mei, S.; Lin, J.R.; et al. A Cancer Cell Program Promotes T Cell Exclusion and Resistance to Checkpoint Blockade. *Cell* **2018**, *175*, 984–997.e24. [CrossRef] [PubMed]
27. Sade-Feldman, M.; Yizhak, K.; Bjorgaard, S.L.; Ray, J.P.; de Boer, C.G.; Jenkins, R.W.; Lieb, D.J.; Chen, J.H.; Frederick, D.T.; Barzily-Rokni, M.; et al. Defining T Cell States Associated with Response to Checkpoint Immunotherapy in Melanoma. *Cell* **2018**, *175*, 998–1013.e20. [CrossRef]
28. Savas, P.; Virassamy, B.; Ye, C.; Salim, A.; Mintoff, C.P.; Caramia, F.; Salgado, R.; Byrne, D.J.; Teo, Z.L.; Dushyanthen, S.; et al. Single-cell profiling of breast cancer T cells reveals a tissue-resident memory subset associated with improved prognosis. *Nat. Med.* **2018**, *24*, 986–993. [CrossRef]
29. Guo, X.; Zhang, Y.; Zheng, L.; Zheng, C.; Song, J.; Zhang, Q.; Kang, B.; Liu, Z.; Jin, L.; Xing, R.; et al. Global characterization of T cells in non-small-cell lung cancer by single-cell sequencing. *Nat. Med.* **2018**, *24*, 978–985. [CrossRef]
30. Zheng, C.; Zheng, L.; Yoo, J.K.; Guo, H.; Zhang, Y.; Guo, X.; Kang, B.; Hu, R.; Huang, J.Y.; Zhang, Q.; et al. Landscape of Infiltrating T Cells in Liver Cancer Revealed by Single-Cell Sequencing. *Cell* **2017**, *169*, 1342–1356.e16. [CrossRef]
31. Young, M.D.; Mitchell, T.J.; Vieira Braga, F.A.; Tran, M.G.B.; Stewart, B.J.; Ferdinand, J.R.; Collord, G.; Botting, R.A.; Popescu, D.M.; Loudon, K.W.; et al. Single-cell transcriptomes from human kidneys reveal the cellular identity of renal tumors. *Science* **2018**, *361*, 594–599. [CrossRef]
32. Zhang, L.; Yu, X.; Zheng, L.; Zhang, Y.; Li, Y.; Fang, Q.; Gao, R.; Kang, B.; Zhang, Q.; Huang, J.Y.; et al. Lineage tracking reveals dynamic relationships of T cells in colorectal cancer. *Nature* **2018**, *1*. [CrossRef] [PubMed]
33. Osipov, A.; Lim, S.J.; Popovic, A.; Azad, N.S.; Laheru, D.A.; Zheng, L.; Jaffee, E.M.; Wang, H.; Yarchoan, M. Tumor Mutational Burden, Toxicity and Response of Immune Checkpoint Inhibitors (ICIs) Targeting PD(L)1, CTLA-4, and Combination: A Meta-Regression Analysis. *Clin. Cancer Res.* **2020**. [CrossRef] [PubMed]

34. Yarchoan, M.; Hopkins, A.; Jaffee, E.M. Tumor Mutational Burden and Response Rate to PD-1 Inhibition. *N. Engl. J. Med.* **2017**, *377*, 2500–2501. [CrossRef] [PubMed]
35. Alexandrov, L.B.; Nik-Zainal, S.; Wedge, D.C.; Aparicio, S.A.; Behjati, S.; Biankin, A.V.; Bignell, G.R.; Bolli, N.; Borg, A.; Børresen-Dale, A.-L.; et al. Signatures of mutational processes in human cancer. *Nature* **2013**, *500*, 415–421. [CrossRef]
36. Lee, J.S.; Ruppin, E. Multiomics Prediction of Response Rates to Therapies to Inhibit Programmed Cell Death 1 and Programmed Cell Death 1 Ligand 1. *JAMA Oncol.* **2019**, *20892*, 1–5. [CrossRef] [PubMed]
37. Andersen, R.; Donia, M.; Ellebæk, E.; Borch, T.H.; Kongsted, P.; Iversen, T.Z.; Hölmich, L.R.; Hendel, H.W.; Met, Ö.; Andersen, M.H.; et al. Long-lasting complete responses in patients with metastatic melanoma after adoptive cell therapy with tumor-infiltrating lymphocytes and an attenuated IL-2 regimen. *Clin. Cancer Res.* **2016**. [CrossRef] [PubMed]
38. Dudley, M.E.; Gross, C.A.; Somerville, R.P.T.; Hong, Y.; Schaub, N.P.; Rosati, S.F.; White, D.E.; Nathan, D.; Restifo, N.P.; Steinberg, S.M.; et al. Randomized Selection Design Trial Evaluating CD8+-Enriched Versus Unselected Tumor-Infiltrating Lymphocytes for Adoptive Cell Therapy for Patients With Melanoma. *J. Clin. Oncol.* **2013**, *31*, 2152. [CrossRef]
39. Rosenberg, S.A.; Yang, J.C.; Topalian, S.L.; Schwartzentruber, D.J.; Weber, J.S.; Parkinson, D.R.; Seipp, C.A.; Einhorn, J.H.; White, D.E. Treatment of 283 consecutive patients with metastatic melanoma or renal cell cancer using high-dose bolus interleukin 2. *JAMA* **1994**, *271*, 907–913. [CrossRef]
40. Ahmadzadeh, M.; Felipe-Silva, A.; Heemskerk, B.; Powell, D.J.; Wunderlich, J.R.; Merino, M.J.; Rosenberg, S.A. FOXP3 expression accurately defines the population of intratumoral regulatory T cells that selectively accumulate in metastatic melanoma lesions. *Blood* **2008**, *112*, 4953–4960. [CrossRef]
41. Ellebaek, E.; Iversen, T.Z.; Junker, N.; Donia, M.; Engell-Noerregaard, L.; Met, O.; Hölmich, L.R.; Andersen, R.S.; Hadrup, S.R.; Andersen, M.H.; et al. Adoptive cell therapy with autologous tumor infiltrating lymphocytes and low-dose Interleukin-2 in metastatic melanoma patients. *J. Transl. Med.* **2012**, *10*, 169. [CrossRef]
42. Pedersen, M.; Westergaard, M.C.W.; Milne, K.; Nielsen, M.; Borch, T.H.; Poulsen, L.G.; Hendel, H.W.; Kennedy, M.; Briggs, G.; Ledoux, S.; et al. Adoptive cell therapy with tumor-infiltrating lymphocytes in patients with metastatic ovarian cancer: A pilot study. *Oncoimmunology* **2018**, *7*, e1502905. [CrossRef] [PubMed]
43. Kverneland, A.H.; Pedersen, M.; Westergaard, M.C.W.; Nielsen, M.; Borch, T.H.; Olsen, L.R.; Aasbjerg, G.; Santegoets, S.J.; van der Burg, S.H.; Milne, K.; et al. Adoptive cell therapy in combination with checkpoint inhibitors in ovarian cancer. *Oncotarget* **2020**, *11*, 2092–2105. [CrossRef] [PubMed]
44. Fischer, A.H.; Zhao, C.; Li, Q.K.; Gustafson, K.S.; Eltoum, I.-E.E.; Tambouret, R.; Benstein, B.; Savaloja, L.C.; Kulesza, P. The cytologic criteria of malignancy. *J. Cell. Biochem.* **2010**, *110*, 795–811. [CrossRef] [PubMed]
45. Donia, M.; Hansen, M.; Sendrup, S.L.; Iversen, T.Z.; Ellebæk, E.; Andersen, M.H.; Straten, P.T.; Svane, I.M. Methods to improve adoptive T-cell therapy for melanoma: IFN-γ enhances anticancer responses of cell products for infusion. *J. Invest. Dermatol.* **2013**, *133*, 545–552. [CrossRef]
46. Donia, M.; Junker, N.; Ellebaek, E.; Andersen, M.H.; Straten, P.T.; Svane, I.M. Characterization and comparison of "Standard" and "Young" tumor infiltrating lymphocytes for adoptive cell therapy at a Danish Translational Research Institution. *Scand. J. Immunol.* **2011**, *75*, 157–167. [CrossRef]
47. Junker, N.; Andersen, M.H.; Wenandy, L.; Dombernowsky, S.L.; Kiss, K.; Sørensen, C.H.; Therkildsen, M.H.; Von Buchwald, C.; Andersen, E.; Straten, P.T.; et al. Bimodal ex vivo expansion of T cells from patients with head and neck squamous cell carcinoma: A prerequisite for adoptive cell transfer. *Cytotherapy* **2011**, *13*, 822–834. [CrossRef]
48. Draghi, A.; Gokuldass, A.; Chamberlain, C.A.; Radic, H.D.; Svane, I.M.; Donia, M. Combined Detection of CD137 and Type 1 Functions Improves Identification and Characterization of the Activated T Lymphocyte Repertoire—Annals of Oncology. Available online: https://www.annalsofoncology.org/article/S0923-7534(20)34468-9/abstract (accessed on 4 May 2020).
49. Nowak, A.; Lock, D.; Bacher, P.; Hohnstein, T.; Vogt, K.; Gottfreund, J.; Giehr, P.; Polansky, J.K.; Sawitzki, B.; Kaiser, A.; et al. CD137+CD154- expression as a regulatory T cell (Treg)-specific activation signature for identification and sorting of stable human tregs from in vitro expansion cultures. *Front. Immunol.* **2018**, *9*, 199. [CrossRef]

50. Betts, M.R.; Koup, R.A. Detection of T-cell degranulation: CD107a and b. *Methods Cell Biol.* **2004**, *2004*, 497–512. [CrossRef]
51. Lauss, M.; Donia, M.; Harbst, K.; Andersen, R.; Mitra, S.; Rosengren, F.; Salim, M.; Vallon-Christersson, J.; Törngren, T.; Kvist, A.; et al. Mutational and putative neoantigen load predict clinical benefit of adoptive T cell therapy in melanoma. *Nat. Commun.* **2017**, *8*. [CrossRef]
52. Tran, E.; Turcotte, S.; Gros, A.; Robbins, P.F.; Lu, Y.-C.; Dudley, M.E.; Wunderlich, J.R.; Somerville, R.P.; Hogan, K.; Hinrichs, C.S.; et al. Cancer immunotherapy based on mutation-specific CD4+ T cells in a patient with epithelial cancer. *Science* **2014**, *344*, 641–645. [CrossRef]
53. Tran, E.; Robbins, P.F.; Lu, Y.-C.; Prickett, T.D.; Gartner, J.J.; Jia, L.; Pasetto, A.; Zheng, Z.; Ray, S.; Groh, E.M.; et al. T-Cell Transfer Therapy Targeting Mutant KRAS in Cancer. *N. Engl. J. Med.* **2016**, *375*, 2255–2262. [CrossRef] [PubMed]
54. Zacharakis, N.; Chinnasamy, H.; Black, M.; Xu, H.; Lu, Y.-C.; Zheng, Z.; Pasetto, A.; Langhan, M.; Shelton, T.; Prickett, T.; et al. Immune recognition of somatic mutations leading to complete durable regression in metastatic breast cancer. *Nat. Med.* **2018**, *24*, 724–730. [CrossRef] [PubMed]
55. Stevanovic, S.; Draper, L.M.; Langhan, M.M.; Campbell, T.E.; Kwong, M.L.; Wunderlich, J.R.; Dudley, M.E.; Yang, J.C.; Sherry, R.M.; Kammula, U.S.; et al. Complete Regression of Metastatic Cervical Cancer After Treatment With Human Papillomavirus-Targeted Tumor-Infiltrating T Cells. *J. Clin. Oncol.* **2015**, *33*, 1543. [CrossRef] [PubMed]

Publisher's Note: MDPI stays neutral with regard to jurisdictional claims in published maps and institutional affiliations.

© 2020 by the authors. Licensee MDPI, Basel, Switzerland. This article is an open access article distributed under the terms and conditions of the Creative Commons Attribution (CC BY) license (http://creativecommons.org/licenses/by/4.0/).

Article

ADCC against MICA/B Is Mediated against Differentiated Oral and Pancreatic and Not Stem-Like/Poorly Differentiated Tumors by the NK Cells; Loss in Cancer Patients due to Down-Modulation of CD16 Receptor

Kawaljit Kaur [1,†], Tahmineh Safaie [1,†], Meng-Wei Ko [1], Yuhao Wang [1] and Anahid Jewett [1,2,*]

[1] Division of Oral Biology and Oral Medicine, School of Dentistry and Medicine, Los Angeles, CA 90095, USA; drkawalmann@g.ucla.edu (K.K.); tahmineh19521@g.ucla.edu (T.S.); mengwei@g.ucla.edu (M.-W.K.); yuhaowang@ucla.edu (Y.W.)
[2] The Jonsson Comprehensive Cancer Center, UCLA School of Dentistry and Medicine, Los Angeles, CA 90095, USA
* Correspondence: ajewett@ucla.edu; Tel.: +1-310-206-3970; Fax: +1-310-794-7109
† Equal contribution by these authors.

Citation: Kaur, K.; Safaie, T.; Ko, M.-W.; Wang, Y.; Jewett, A. ADCC against MICA/B Is Mediated against Differentiated Oral and Pancreatic and Not Stem-Like/Poorly Differentiated Tumors by the NK Cells; Loss in Cancer Patients due to Down-Modulation of CD16 Receptor. *Cancers* **2021**, *13*, 239. https://doi.org/10.3390/cancers13020239

Received: 10 December 2020
Accepted: 5 January 2021
Published: 11 January 2021

Publisher's Note: MDPI stays neutral with regard to jurisdictional claims in published maps and institutional affiliations.

Copyright: © 2021 by the authors. Licensee MDPI, Basel, Switzerland. This article is an open access article distributed under the terms and conditions of the Creative Commons Attribution (CC BY) license (https://creativecommons.org/licenses/by/4.0/).

Simple Summary: Natural Killer cells are known to eliminate tumors directly or via antibody dependent cellular cytotoxicity. The complete modes and mechanisms of such killings are yet to be delineated. It is also unclear at what stages of tumor differentiation NK cells are capable of mediating the two modes of tumor killing. In this report we provide evidence that NK cells mediate killing of both stem-like/poorly differentiated tumors and well-differentiated tumors via direct cytotoxicity and antibody dependent cellular cytotoxicity, respectively. By using antibodies to MICA/B, EGFR and PDL1 surface receptors expressed on well-differentiated but not on stem-like/poorly differentiated tumors we demonstrate significant NK cell mediated antibody dependent cellular cytotoxicity in the absence of direct killing. In addition, our results suggested the possibility of CD16 receptors mediating both direct cytotoxicity and antibody dependent cellular cytotoxicity, resulting in the competitive use of these receptors in either direct killing or antibody dependent cellular cytotoxicity.

Abstract: Tumor cells are known to upregulate major histocompatibility complex-class I chain related proteins A and B (MICA/B) expression under stress conditions or due to radiation exposure. However, it is not clear whether there are specific stages of cellular maturation in which these ligands are upregulated or whether the natural killer (NK) cells differentially target these tumors in direct cytotoxicity or antibody-dependent cell cytotoxicity (ADCC). We used freshly isolated primary and osteoclast (OCs)-expanded NK cells to determine the degree of direct cytotoxicity or of ADCC using anti-MICA/B monoclonal antibodies (mAbs) against oral stem-like/poorly-differentiated oral squamous cancer stem cells (OSCSCs) and Mia PaCa-2 (MP2) pancreatic tumors as well as their well-differentiated counterparts: namely, oral squamous carcinoma cells (OSCCs) and pancreatic PL12 tumors. By using phenotypic and functional analysis, we demonstrated that OSCSCs and MP2 tumors were primary targets of direct cytotoxicity by freshly isolated NK cells and not by ADCC mediated by anti-MICA/B mAbs, which was likely due to the lower surface expression of MICA/B. However, the inverse was seen when their MICA/B-expressing differentiated counterparts, OSCCs and PL12 tumors, were used in direct cytotoxicity and ADCC, in which there was lower direct cytotoxicity but higher ADCC mediated by the NK cells. Differentiation of the OSCSCs and MP2 tumors by NK cell-supernatants abolished the direct killing of these tumors by the NK cells while enhancing NK cell-mediated ADCC due to the increased expression of MICA/B on the surface of these tumors. We further report that both direct killing and ADCC against MICA/B expressing tumors were significantly diminished by cancer patients' NK cells. Surprisingly, OC-expanded NK cells, unlike primary interleukin-2 (IL-2) activated NK cells, were found to kill OSCCs and PL12 tumors, and under these conditions, we did not observe significant ADCC using anti-MICA/B mAbs, even though the tumors expressed a higher surface expression of MICA/B. In addition, differentiated tumor cells also expressed higher levels of surface epidermal growth factor receptor (EGFR) and programmed death-ligand 1(PDL1) and were more susceptible to NK cell-mediated ADCC in the presence of anti-EGFR and anti-PDL1 mAbs compared to their stem-like/poorly differentiated counterparts. Overall, these results suggested the possibility of CD16 receptors mediating both direct cytotoxicity and ADCC, resulting in the competitive use of these receptors in either direct killing or ADCC, depending on the differentiation status of tumor cells and the stage of maturation and activation of NK cells.

Keywords: NK cells; cancer stem cells (CSCs); antibody-dependent cellular cytotoxicity (ADCC); differentiation; humanized-BLT mice; cytotoxicity; IFN-γ; osteoclasts; MICA/B mAb

1. Introduction

Natural killer (NK) cells were first discovered as a functional cell type in 1970 and were named by Kiessling et al. in 1975 [1]. NK cells were so named for their effector functions, which include direct natural cytotoxicity, antibody-dependent cellular cytotoxicity (ADCC), as well as the secretion of inflammatory cytokines and chemokines which indirectly regulate the functions of other immune cells [2,3]. Conventional human NK cells are identified by the expression of CD16 and CD56 and by the lack of surface CD3 receptor expression [4]. NK cells mediate their functions through several important activating and inhibitory cell receptors such as CD16, NKG2D, natural cytotoxicity receptors (NCR), killer immunoglobulin-like receptors (KIR), and the NKG2 family of receptors, which form heterodimers with CD94 [5–7]. The balance between activating and inhibitory signals that NK cells receive through the surface receptors determines their functional fate [5]. As such, activated NK cells are able to recognize and lyse tumor cells expressing certain surface receptors without prior antigenic sensitization [8,9]. Many tumors, especially differentiated tumors, express major histocompatibility complex-class I (MHC-class I) chain related proteins A and B (MICA/B), which mark them for elimination by the NK cells [10–13]. However, tumor cells can successfully evade detection by NK cells by shedding MICA/B [11,14]. Moreover, differentiated tumors were also found to have higher expression of epidermal growth factor receptor (EGFR) [15,16] and programmed death-ligand 1(PDL1) [17,18].

Studies have shown that NK cell-mediated ADCC can be exploited as an important cancer treatment [19]. NK cell-mediated ADCC is triggered when FcγRIIIA (CD16) binds to the Fc region of antibodies bound to their cognate antigens expressed on target cells. This binding induces the directed exocytosis of granzyme- and perforin-containing granules that then lyse the target cells [20,21]. Thus, CD16 is a major FcγR on NK cells and is crucial for activating ADCC activity in NK cells [22–25]. NKG2D is an activating surface receptor of NK cells which in conjunction with CD16 influences NK cell function [26–28]. In addition, NKG2D was found to play a significant role in tumor rejection and tumor immunosurveillance through binding to MICA/B, which are among the ligands binding to NKG2D receptors [29–32]. However, tumor-associated NK cells are refractory to CD16 receptor stimulation, resulting in diminished ADCC against autologous tumor cells [33]. Moreover, ADCC was also found to be impaired in cancer patients [34–37].

We have previously demonstrated that NK cells secrete elevated levels of cytokines, particularly interferon gamma (IFN-γ) and tumor necrosis factor alpha (TNF-α), in the presence of decreased cytotoxicity when CD16 receptors are triggered on their surface. We termed this functional stage of NK cells as "split anergy". As indicated, NK cells become split-anergized upon CD16 receptor crosslinking or during interactions with cancer stem cells (CSCs) or undifferentiated cells [38,39]. Cytokines secreted by split-anergized NK cells play an important role in mediating tumor cell differentiation [38,40,41]. It was previously demonstrated that decreased NK cell counts, suppressed NK function, and the down-modulation of NK cell surface receptors in the peripheral blood and the tumor microenvironment were associated with poor prognoses in cancer patients [42–60]. Our previous work also illustrated that NK cells of cancer patients and of tumor-bearing humanized-bone marrow/liver/thymus (BLT) mice mediated less cytotoxicity against cancer cells and secreted lower levels of cytokines [61–63].

This study explored the different levels of NK cell-mediated IFN-γ secretion, direct cytotoxicity, ADCC, and the surface receptor expression on NK cells from healthy individuals and those of cancer patients. Next, we determined the surface expression of MICA/B on CSCs and their differentiated counterparts. Finally, we elucidated the differences between the NK cell-mediated ADCC between freshly isolated NK cells, osteoclast (OC)-expanded supercharged NK cells, and NK92 tumors transfected with CD16 receptors. NK cell-mediated ADCC against MICA/B bearing differentiated tumor cells in the presence of MICA/B antibody was compared to that mediated by the antibodies against EGFR and PDL1.

2. Results

2.1. Cancer Patients' NK Cells Exhibit Decreased Direct Killing and NK Cell-Mediated ADCC Compared to Healthy Individuals' NK Cells

We determined NK cell function in cancer patients using NK cell-mediated cytotoxicity and IFN-γ secretion. We found that cancer patients' NK cells mediated significantly lower levels of cytotoxicity (Figure 1A and Figure S1) and secreted lower amounts of IFN-γ when compared to healthy individuals' NK cells (Figure 1B and S2). Decreased IFN-γ secretion by patients' NK cells was also seen in the presence of CSCs (Figure S2). Then, we used interleukin-2 (IL-2)-treated NK cells from cancer patients and healthy individuals as effectors to target tumors in the absence and presence of anti-MICA/B monoclonal antibodies (mAbs). We observed NK cells mediated ADCC against differentiated tumors (Figure 1D,E,G,H) but not against stem-like tumors (Figure 1C,F). However, very little or no NK cell-mediated ADCC against both oral (Figure 1C–E) and pancreatic tumor cells (Figure 1F–H) was seen from cancer patients' NK cells. Next, we analyzed the expression of NK cell surface receptors isolated from healthy individuals and cancer patients. We detected a lower expression of Nkp44, CD94, NKG2D, and KIR2, and a higher expression of Nkp30, Nkp46, and KIR3 on CD16+ NK cells from cancer patients (Figure 1I–K and Figure S3). These data indicate that cancer patients' NK cells express lower levels of surface receptors important in ADCC and substantially decreased cytotoxic activity against tumors when compared to those from healthy individuals' NK cells.

2.2. Differentiated Tumor Cells Expressed Higher Levels of Surface MICA/B and Were More Susceptible to NK Cell-Mediated ADCC in the Presence of Anti-MICA/B mAb Compared to Their Stem-Like/Poorly Differentiated Counterparts

We have previously demonstrated that IFN-γ secreted by IL-2+anti-CD16 mAb-treated NK cells promotes tumor differentiation [18,64]. Therefore, we used the supernatants from IL-2+anti-CD16 mAb-treated NK cells to differentiate CSCs as described in the Materials and Methods section. We first investigated the surface expression of MICA/B on stem-like oral stem-like/poorly-differentiated oral squamous cancer stem cells (OSCSCs) and MP2 tumors, differentiated oral squamous carcinoma cells (OSCCs) and PL12 tumors, and NK cell-differentiated OSCSCs and MP2 tumors. OSCCs and the NK cell-differentiated OSCSCs expressed higher levels of MICA/B when compared to their stem-like counterparts OSCSCs (Figure 2A). Similarly, the differentiated pancreatic PL12 tumors and the NK cell-differentiated MP2 tumors expressed higher levels of MICA/B when compared to their stem-like counterparts MP2 tumors (Figure 2B).

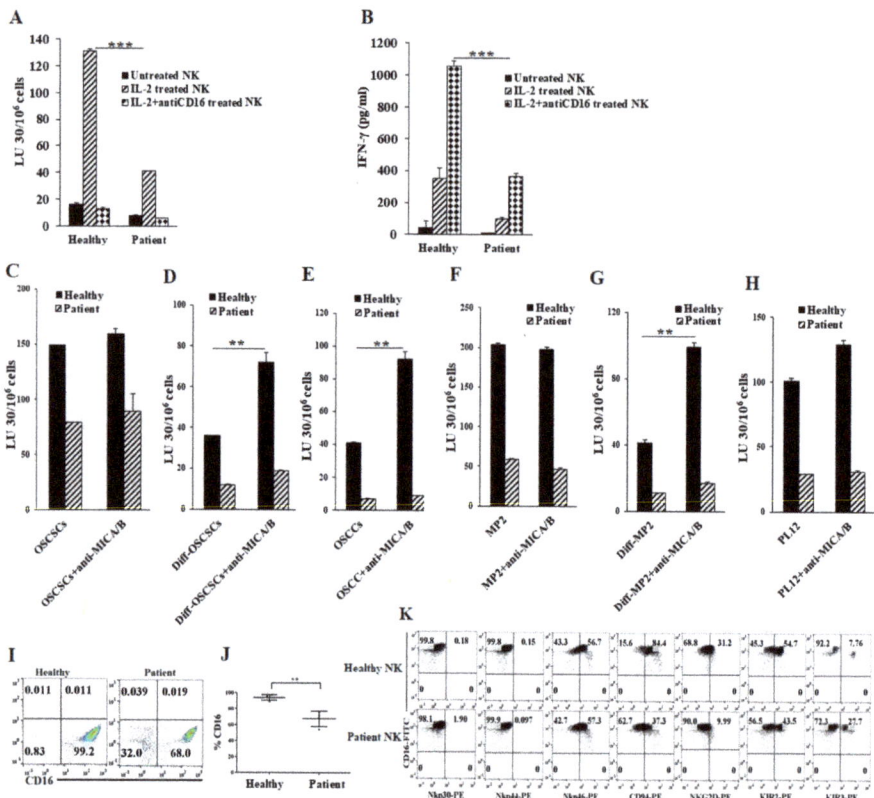

Figure 1. Cancer patients' natural killer (NK) cells exhibit lower functional activity and NK cell-mediated antibody-dependent cell cytotoxicity (ADCC) when compared to healthy individuals' NK cells. Purified NK cells (1 × 10^6 cells/mL) from healthy individuals and pancreatic cancer patients were left untreated, treated with interleukin-2 (IL-2) (1000 U/mL), or treated with a combination of IL-2 (1000 U/mL) and anti-CD16 mAb (3 µg/mL) for 18 h and were added to ^{51}Cr-labeled oral stem-like/poorly-differentiated oral squamous cancer stem cells (OSCSCs) at various effector-to-target ratios. NK cell-mediated cytotoxicity was measured using a standard 4-h ^{51}Cr release assay against OSCSCs. The lytic units (LU) 30/10^6 cells were determined using the inverse number of NK cells required to lyse 30% of target cells × 100 (**A**) *** (p value < 0.001). NK cells were isolated and prepared as described in Figure 1A for 18 h before the supernatants were harvested and the levels of IFN-γ secretion were determined using single ELISA (**B**) *** (p value < 0.001). One of 20 experiments is shown in Figure 1A,B. Purified NK cells (1 × 10^6 cells/mL) from healthy individuals and cancer patients were treated with IL-2 (1000 U/mL) for 18 h and were used as effectors in ^{51}Cr release assay. OSCSCs and Mia PaCa-2 (MP2) tumors were differentiated as described in the Materials and Methods. OSCSCs (**C**), NK cell-differentiated OSCSCs (**D**), oral squamous carcinoma cells (OSCCs) (**E**), MP2 (**F**), NK cell-differentiated MP2 (**G**), and PL12 cells (**H**) were labeled with ^{51}Cr for an hour, after which cells were washed to remove unbound ^{51}Cr. Then, ^{51}Cr-labeled tumor cells were left untreated or treated with anti-major histocompatibility complex-class I chain related proteins A and B (MICA/B) monoclonal antibodies (mAbs) (5 µg/mL) for 30 min. The unbounded antibodies were washed away, and the cytotoxicity against the tumor cells was determined using a standard 4-h ^{51}Cr release assay. LU 30/10^6 cells were determined as described in Materials and Methods (**C–H**) ** (p value 0.001–0.01). The surface expression levels of CD16 of freshly purified NK cells from healthy individuals and cancer patients were analyzed using flow cytometry. IgG2 isotype antibodies were used as controls (n = 4) (**I,J**) ** (p value 0.001–0.01). Freshly purified NK cells from healthy individuals and cancer patients were analyzed for the surface expression levels of CD16, Nkp30, Nkp44, Nkp46, CD94, NKG2D, KIR2, and KIR3 using flow cytometry. IgG2 isotype antibodies were used as controls (**K**). One of eight representative experiments is shown in Figure 1K.

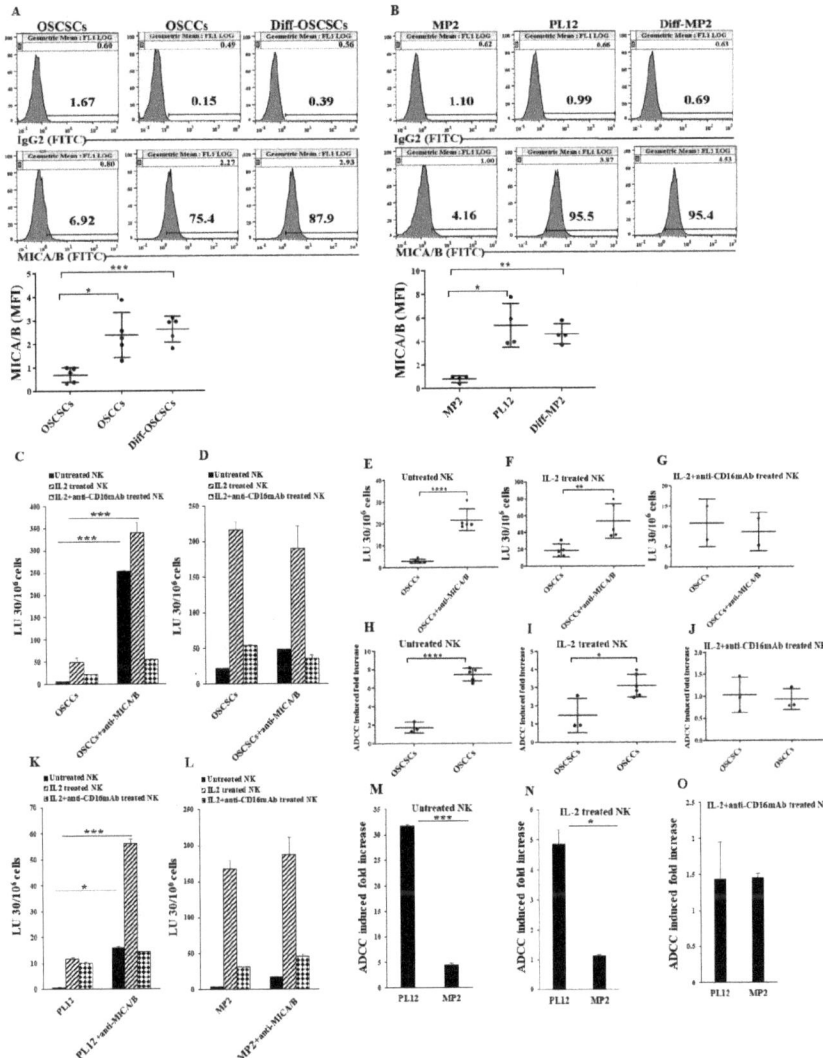

Figure 2. Differentiated tumors expressed higher surface levels of MICA/B and were more susceptible to NK cell-mediated cytotoxicity when compared to their stem-like counterparts in the presence of anti-MICA/B mAbs. OSCSCs were differentiated as described in the Materials and Methods. The surface expression levels of MICA/B on OSCSCs, OSCCs, and NK cell-differentiated OSCSCs were assessed using flow cytometric analysis. IgG2 isotype antibodies were used as controls ($n = 5$) (**A**). MP2 were differentiated as described in the Materials and Methods. The surface expression levels of MICA/B on MP2, PL12, and NK cell-differentiated MP2 were assessed using flow cytometric analysis. IgG2 isotype antibodies were used as controls ($n = 4$) (**B**). Freshly purified NK cells were left untreated, treated with IL-2 (1000 U/mL), or treated with a combination of IL-2 and anti-CD16 mAb (3 µg/mL) for 18 h and were used as effectors against OSCCs (**C**) and OSCSCs (**D**) to measure NK cell-mediated ADCC as described in Materials and Methods ($n = 6$) (**C,D**). The NK cell-mediated ADCC was measured using untreated ($n = 5$) (**E**), IL-2 treated ($n = 5$) (**F**), and IL-2 + anti-CD16 mAb treated ($n = 2$) (**G**) NK cells as effectors against target OSCCs (**E–G**). Fold increase in ADCC against OSCCs and OSCSCs mediated by untreated (**H**) ($n = 5$), IL-2 treated (**I**) ($n = 5$), and IL-2+anit-CD16 mAbs (**J**) ($n = 3$) were calculated. NK cells were prepared as described in Figure 2C and were used as effectors to measure NK cell-mediated ADCC against PL12 (**K**) and MP2 tumors (**L**). Fold increases in NK cell-mediated ADCC against MP2 and PL12 tumors by untreated (**M**), IL-2 treated (**N**), and IL-2+anti-CD16 mAb treated (**O**) NK cells were calculated. **** (p value < 0.0001), *** (p value < 0.001), ** (p value 0.001–0.01), * (p value 0.01–0.05)

Our previous studies also demonstrated that CSCs/poorly differentiated tumors are excellent targets of NK cell-mediated cytotoxicity, whereas their differentiated counterparts are significantly more resistant [64–66]. Here, we evaluated NK cell-mediated cytotoxicity against CSCs/poorly differentiated and well-differentiated tumor cells treated with monoclonal antibodies specific for MICA/B. We found that susceptibility to NK cell-mediated cytotoxicity increased significantly against anti-MICA/B mAb-treated OSCCs (Figure 2C,E,F), while anti-MICA/B mAb-treated OSCSCs remained relatively unchanged when compared to the killing of target cells in the absence of MICA/B antibodies (Figure 2D). Significantly higher levels of fold increase in NK cell-mediated ADCC were seen against OSCCs using MICA/B antibodies as compared to those in the absence of antibody, whereas slight increases in ADCC could be seen against OSCSCs when untreated or IL-2 treated NK cells were used as effectors (Figure 2H,I). The addition of antibody to CD16 receptor abolished the increase in ADCC against OSCCs (Figure 2G,J). Consistent with our findings in oral tumor cells, we observed higher levels of NK cell-mediated ADCC against anti-MICA/B mAb-treated pancreatic PL12 cells when compared to MP2 tumors (Figure 2K–O).

2.3. Differentiated Tumor Cells Treated with Anti-MICA/B mAb Triggered Increased IFN-γ Secretion by NK Cells

We have previously demonstrated that NK cells secrete higher levels of IFN-γ when co-cultured with CSCs/poorly differentiated tumor cells than with the differentiated tumors [64]. To assess the effect of anti-MICA/B mAbs-induced NK cell activation, we treated OSCSCs and OSCCs with or without anti-MICA/B mAbs and co-cultured them with NK cells. As expected, we did not detect any changes in IFN-γ secretion in cultures containing untreated NK cells (Figure 3A). When NK cells were treated with IL-2, we observed an increase in IFN-γ secretion in response to anti-MICA/B mAb-treated OSCCs as compared to untreated OSCCs (Figure 3B,D). There was no or a slight increase in IFN-γ secretion when IL-2-treated NK cells were co-cultured with anti-MICA/B mAb-treated OSCSCs as compared to untreated OSCSCs (Figure 3B,D). Consistent with our previous findings, the levels of IFN-γ produced by IL-2+anti-CD16 mAb treated NK cells were plateaued (Figure 3C). Therefore, we were not able to detect a noticeable difference in the concentrations of IFN-γ when untreated or anti-MICA/B mAb-treated OSCCs or OSCSCs were co-cultured with IL-2+anti-CD16 mAb-treated NK cells (Figure 3C).

2.4. Higher Levels of NK Cell-Mediated ADCC Were Seen in Freshly Isolated Primary NK Cells When Compared to OC-Expanded NK Cells

We have previously shown that osteoclast (OC)-expanded NK cells exhibit a greater potential to directly kill tumor cells while also expressing lower levels of CD16 receptors when compared to freshly isolated primary NK cells [61]. Thus, we tested surface expressions and NK cell-mediated ADCC from both freshly isolated primary NK cells and those expanded by the OCs. Primary NK cells expressed higher levels of CD16 but low no or low levels of NKG2D on their surface compared to OC-expanded NK cells (Figure S4A,B). High levels of NK cell-mediated ADCC were seen in IL-2-treated primary NK cells, but very little or no NK cell-mediated ADCC was seen in OC-expanded NK cells (Figure 4A,B and Figure S5). Next, we determined the effects of CD16 cross-linking on primary and OC-expanded NK cells. As expected, in primary NK cells, NK cell-mediated cytotoxicity decreased when they were treated with IL-2+anti-CD16 mAb as compared to IL-2-activated NK cells (Figure 4C and Figure S5A). However, we observed no significant differences between the levels of NK cell-mediated cytotoxicity from IL-2+anti-CD16 mAb-treated and IL-2-activated OC-expanded NK cells (Figure 4C and Figure S5B). Similarly, we observed higher levels of NK cell-mediated ADCC by primary NK cells in comparison to OC-expanded NK cells when PL12 tumors were used as targets (Figure S5C).

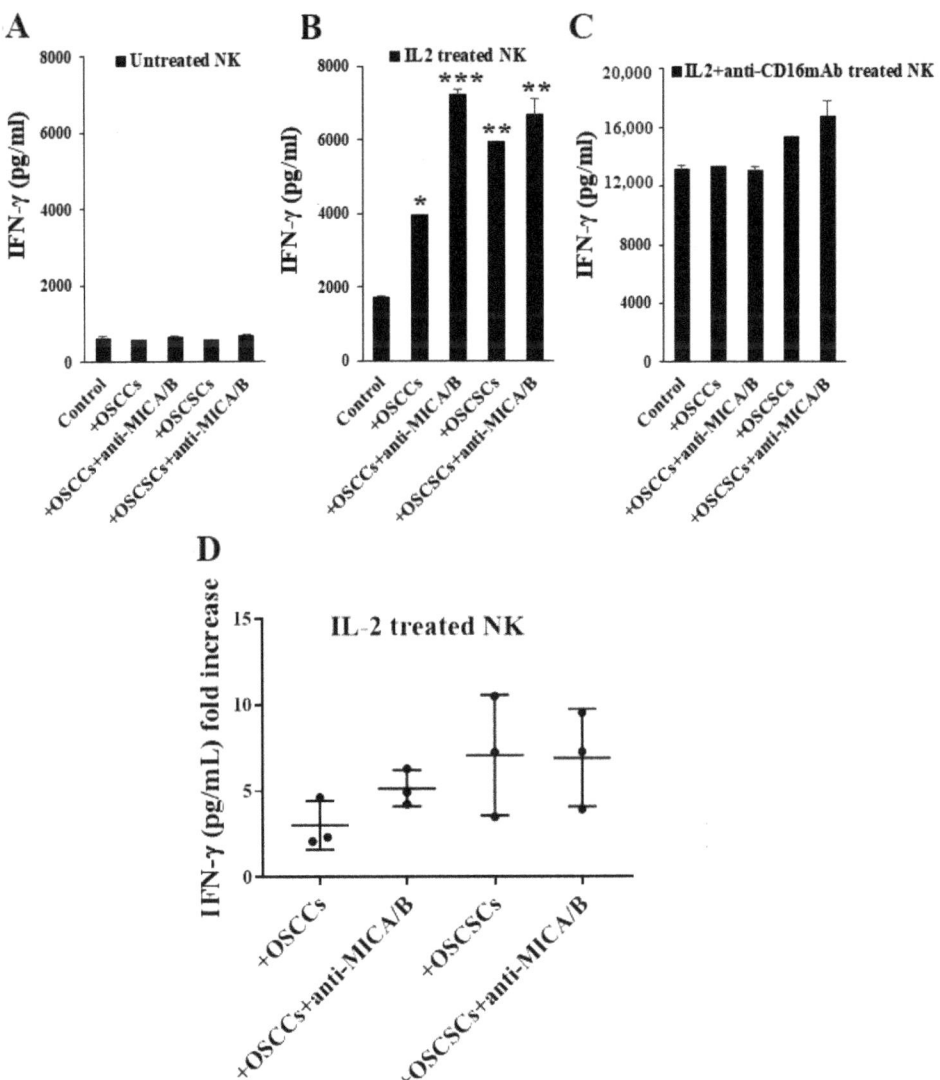

Figure 3. Differentiated tumor cells treated with anti-MICA/B mAb increased IFN-γ secretion of IL-2 activated NK cells. Freshly purified NK cells from healthy individuals were left untreated (**A**), treated with IL-2 (1000 U/mL) (**B**), or treated with a combination of IL-2 and anti-CD16 mAb (3 µg/mL) (**C**) for 18 h. OSCCs and OSCSCs were treated with anti-MICA/B mAb (5 µg/mL) for 18 h, excess unbounded antibodies were removed, and the tumor cells were co-cultured with NK cells. The supernatants were harvested from the co-cultures after 24 h, and the concentrations of IFN-γ were determined using single ELISA ($n = 2$) (**A–C**). The ratios of IFN-γ secretion of IL-2 activated NK cells induced by untreated or anti-MICA/B mAb treated oral tumor cells (OSCCs and OSCSCs) were determined as fold increase ($n = 3$) (**D**). *** (p value < 0.001), ** (p value 0.001–0.01), * (p value 0.01–0.05).

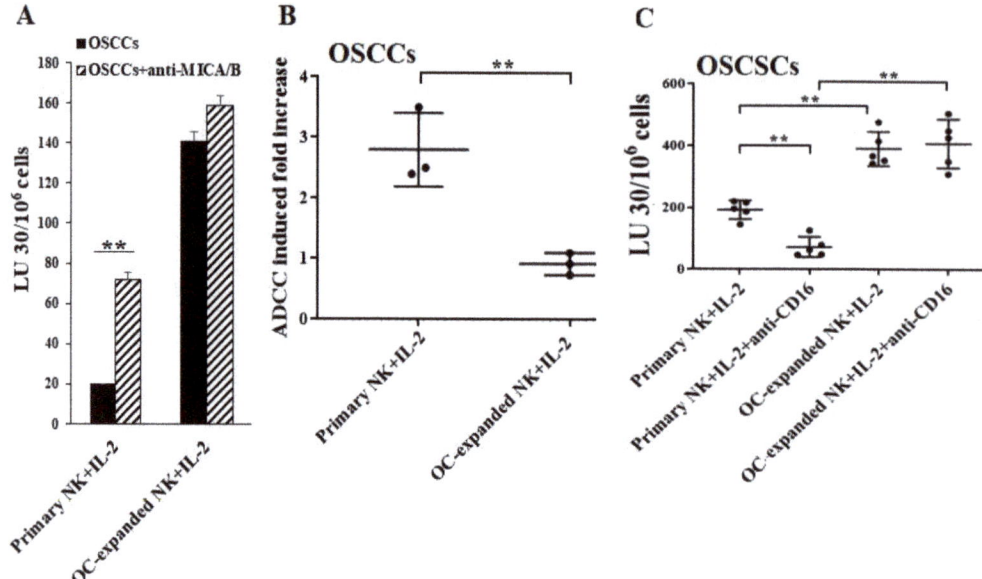

Figure 4. Higher levels of NK cell-mediated ADCC were seen in freshly isolated primary NK cells vs. in osteoclast (OC)-expanded NK cells. OCs were generated and OC-expanded NK cells were prepared as described in the Materials and Methods. Freshly purified primary and OC-expanded NK cells were both treated with IL-2 (1000 U/mL) for 18 h and were used as effector cells to measure NK cell-mediated ADCC against OSCCs as described in Materials and Methods (**A**). Fold increases in ADCC were calculated (**B**). Freshly purified primary and OC-expanded NK cells were treated with IL-2 (1000 U/mL) or a combination of IL-2 (1000 U/mL) and anti-CD16 mAbs (3 μg/mL) for 18 h before being used as effector cells to measure NK cell-mediated ADCC against OSCSCs as described in the Materials and Methods ($n = 5$) (**C**). ** (p value 0.001–0.01).

2.5. Higher Levels of NK Cell-Mediated Direct Cytotoxicity and ADCC by Freshly Isolated NK Cells When Compared to Either Parental NK92 Tumors and Those Expressing CD16 Receptor

Next, we explored the CD16 surface expression and the functional differences between primary human peripheral blood-derived NK cells, parental NK92, and the CD16-expressing NK92 tumors (NK92-176V). Primary NK cells exhibited a high surface expression of CD16 receptor as compared to NK92 or NK92-176V (Figure S4C). We have also found lower levels of NK cell-mediated cytotoxicity and ADCC mediated by the NK92 parental and NK92-176V tumors than by primary NK cells when tested against untreated or anti-MICA/B mAb-treated PL12 tumors (Figure 5A). The levels of NK92 and NK92-176V-mediated direct cytotoxicity were lower than those of primary NK cells against MP2 tumors (Figure 5B). NK92-176V cells mediated higher levels of cytotoxicity against both PL12 and MP2 cells when compared to NK92 cells, although the effects of anti-MICA/B mAbs were not pronounced in either group (Figure 5).

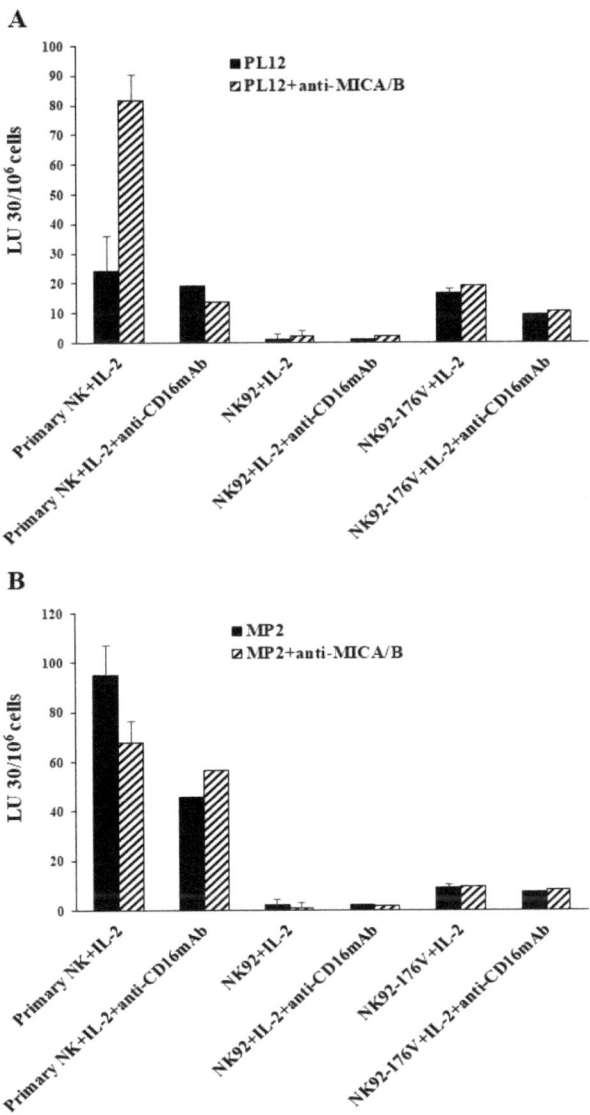

Figure 5. Higher levels of NK cell-mediated ADCC were seen in freshly isolated primary NK cells vs. in NK92 and NK92-176V. Freshly purified primary NK cells, NK92, and NK92-176V cells were either treated with IL-2 (1000 U/mL) or a combination of IL-2 and anti-CD16 mAbs (3 µg/mL) for 18 h and were used as effector cells to measure NK cell-mediated ADCC against PL12 (**A**) and MP2 tumors (**B**), as described in Materials and Methods.

2.6. Differentiated Tumor Cells Expressed Higher Levels of Surface EGFR and PDL-1 and Were More Susceptible to NK Cell-Mediated ADCC in the Presence of Anti-EGFR and Anti-PDL1 mAbs Compared to Their Stem-Like/Poorly Differentiated Counterparts

In addition to increase in MICA/B ligands on OSCCs, PL12, and NK-differentiated OSCSCs, we also observed higher levels of epidermal growth factor receptor (EGFR) (Figure 6A) and PDL1 receptors (Figure S6B,D) on these tumors when compared to OSC-SCs and MP2 tumors. Indeed, unlike OSCSCs, signaling through EGFR on OSCCs was able to increase the expression of phospho-STAT3 (signal transducer and activator of transcription 3) (Figure S6A). IL-2-treated NK cells mediated significant ADCC against OSCCs in the presence of anti-EGFR mAbs, whereas these antibodies inhibited IL-2-treated NK cell-mediated ADCC against OSCSCs tumors (Figure 6B). Similarly, NK cells mediated significant ADCC against OSCCs, NK-differentiated OSCSCs, and PL12 in the presence of anti-PDL1 mAbs, whereas these antibodies inhibited NK cell-mediated ADCC against OSCSCs and MP2 tumors (Figure S6C,E). Taken together, the data indicated that well-differentiated tumor cells expressed MICA/B, EGFR, and PDL1 and that the treatment of these tumors with their respective antibodies mediated ADCC by the NK cells.

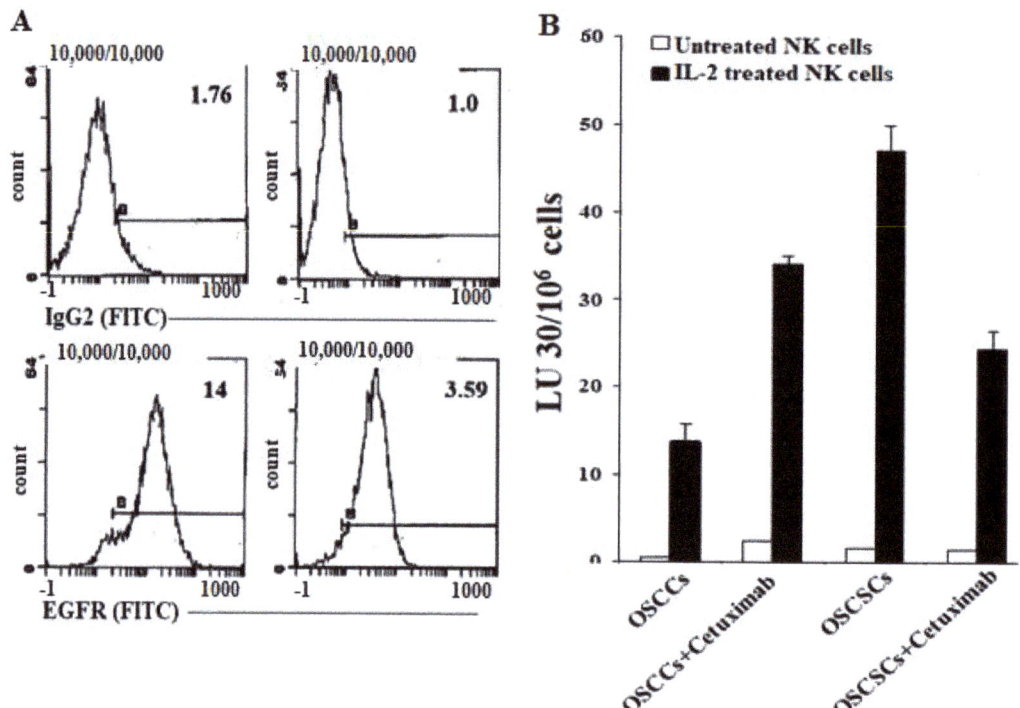

Figure 6. Higher levels of epidermal growth factor receptor (EGFR) surface expressions and NK cell-mediated ADCC in OSCCs in comparison to OSCSCs. The levels of EGFR expression were determined on OSCCs and OSCSCs using surface staining with Cetuximab (**A**). Numbers in the histograms represent mean fluorescence intensity (MFI). Purified NK cells (1×10^6 cells/mL) were left untreated or treated with IL-2 (1000 U/mL) for 18 h and used against untreated and Cetuximab-treated OSCCs and OSCSCs in a 4-h ^{51}Cr release assay. The lytic units (LU) $30/10^6$ cells were determined using the inverse number of NK cells required to lyse 30% of target cells × 100 (**B**).

3. Discussion

NK cells mediate their cytotoxic function against tumors through direct cytotoxicity and ADCC. A great number of antibodies made against distinct receptors on a variety of tumor cells are currently in clinical use. These antibodies not only inhibit various tumor cell functions by targeting specific receptors, they also guide NK cells to the targeted tumor cells to affect ADCC. Unfortunately, cancer patients who have dysfunctional NK cells also lack ADCC, as shown in this study, and the defect is in great part due to the decrease in CD16 expression in cancer patients. In a number of previous studies, it was shown that Adam 17 is an important enzyme that regulates the shedding of the CD16 receptor, and it can inhibit shedding, thereby increasing NK cell function [67,68]. Whether Adam 17 or any other effective enzyme could increase or restore NK cell-mediated ADCC in cancer patients awaits future investigations.

We have identified and characterized several oral and pancreatic tumor cell lines in different stages of differentiation [18,63,69,70]. Our previous studies have established OSCSCs as oral and MP2 as pancreatic stem-like/poorly differentiated tumors and OSCCs and PL12 tumors as well-differentiated oral and pancreatic tumors, respectively using CD44, CD54, MHC-class I, and PD-L1 surface antigens [62,70]. In this study, we established that the levels of MICA/B expressions were higher on the surfaces of well-differentiated tumor cells than on stem-like/poorly differentiated tumor cells. In accordance, untreated and IL-2-treated primary NK cells mediated significantly higher ADCC against anti-MICA/B mAb-treated OSCCs and PL12 tumors, and the addition of anti-CD16 mAbs to IL-2-activated NK cells abolished the increase in anti-MICA/B mAb-mediated ADCC (Figure 2C,E–G,K). On the other hand, no significant NK cell-mediated ADCC could be seen against OSCSCs or MP2 tumors that express no or lower levels of MICA/B, albeit NK cells mediated significant direct cytotoxicity against these tumors (Figure 2D,L). When OSCSCs or MP2 tumors were differentiated by NK supernatants and the levels of direct cytotoxicity and ADCC were measured, a significant increase in ADCC in the presence of negligible direct cytotoxicity (Figure 1D,G) was observed, which correlated with the increase in the surface expressions of MICA/B on the NK cell-differentiated OSCSCs and MP2 tumors. No increase in the levels of NK cell-mediated ADCC could be observed using patients' NK cells, indicating a severe inhibition of ADCC. This is surprising, since even though there is a substantial decrease in CD16 expression on the surface of patients' NK cells, there still exists a portion of the NK cells with decent levels of CD16 receptors. Whether there is also a functional deficiency of CD16 receptors in regard to ADCC in addition to decreased expression of this receptor on patients' NK cells will require further studies.

The fold increase in ADCC was higher by untreated NK cells than those treated with IL-2. However, IFN-γ secretion was only induced in IL-2-treated NK cells during ADCC and not by the untreated NK cells, even though they mediated higher levels of ADCC. These results indicated the differential regulation of ADCC and IFN-γ secretion. In addition, untreated NK cells mediated no or slight direct killing in the majority of tumors tested. When NK cells were activated with IL-2, the levels of direct cytotoxicity increased, but the fold increase in ADCC was lower against OSCCs and PL12 tumors than those displayed by untreated NK cells (Figure 2I,N). These results indicated that once IL-2 triggers direct cytotoxicity, the levels of ADCC decreases, potentially indicating competition for CD16 receptors by the tumor cell ligands for direct killing as well as ADCC-mediated killing.

Unlike IL-2-activated primary NK cells, OC-expanded NK cells were found to mediate direct cytotoxicity against differentiated OSCCs and PL12 tumors (Figure 4A and Figure S5B,C). When these tumors were used as targets of OC-expanded NK cells, no significant increase in ADCC could be observed; however, there were significant levels of direct cytotoxicity. Although there was a down-modulation of CD16 receptor on OC-expanded NK cells, the remaining amounts of CD16 were presumably sufficient to mediate ADCC. This observation further reinforces the possibility that putative tumor ligands or Fc regions of antibodies are likely engaged in a competitive manner in CD16 receptor binding, thereby decreasing the levels of ADCC while increasing cytotoxicity or vice versa. Indeed, OC-expanded NK cells have increased levels of NKG2D, which could directly bind to MICA/B and mediate cytotoxicity [61].

Primary NK cells exhibited a greater potential for ADCC than did the CD16-expressing NK92 cells, although CD16-expressing NK92 cells could also mediate ADCC to a much lower extent against MICA/B ligands. These differences can be due to the density of CD16 expression on primary NK cells as well as the superb ability of these cells to mediate direct cytotoxicity as well as ADCC. Indeed, NK92 cells do not mediate significant direct killing against OSCSCs and MP2 tumors ([71], and Figure 5B).

Similar to MICA/B, the levels of EGFR and PDL1 are increased on NK-differentiated OSCSCs and well-differentiated OSCCs and PL-12 tumors but not on OSCSCs or MP2

CSCs/poorly differentiated tumors (Figure 6A and Figure S6A,B,D). Accordingly, the levels of NK cell-mediated ADCC using these antibodies were increased in NK-differentiated OSCSCs and in well differentiated OSCCs and PL-12 tumors. In contrast, there was a slight change or a decrease in direct cytotoxicity when antibodies were used in the cultures of NK cells with OSCSC and MP2 oral and pancreatic tumors. These results are in agreement with the findings obtained using anti-MICA/B mAbs. It is possible that a lack of expression of many key receptors on OSCSCs and MP2 tumors, as seen in this study, is a potential underlying mechanism for their aggressiveness. In addition, the lack of such receptor expression is likely to shield these tumors from receiving signals, which could potentially control their growth and expansion.

Overall, there is a possibility that CD16 receptors can be used in both direct cytotoxicity and in ADCC, resulting in competition for the use of the receptors depending on the differentiation status of the tumor cells and the stage of maturation of the NK cells. In fact, ligands other than the Fc portion of antibodies have previously been identified that can bind to the CD16 receptor [72–74]. Delineation of the ligands used for binding to CD16 and induction of direct cytotoxicity versus those used for ADCC should provide the basis for novel treatment strategies. Indeed, the addition of anti-CD16 receptor antibody also inhibits direct cytotoxicity as well as inhibition of ADCC.

4. Materials and Methods

4.1. Cell Lines, Reagents, and Antibodies

Oral squamous carcinoma cells (OSCCs) and oral squamous carcinoma stem cells (OSCSCs) were isolated from patients with tongue tumors at UCLA [64,75]. OSCCs, OSCSCs, and K562 tumors were cultured in RPMI 1640 (Life Technologies, Los Angeles, CA, USA) supplemented with 10% fetal bovine serum (FBS) (Gemini Bio-Product, West Sacramento, CA, USA). Mia PaCa-2 (MP2) and PL12 human pancreatic cell lines were provided by Drs. Guido Eibl and Nicholas Cacalano (UCLA David Geffen School of Medicine, Los Angeles, CA, USA). MP2 and PL12 cells were cultured in dulbecco's modified eagle medium (DMEM) supplemented with 10% FBS and 2% penicillin–streptomycin (Gemini Bio-Products, West Sacramento, CA, USA). RPMI 1640 supplemented with 10% FBS was used to culture human NK cells. Alpha-minimum essential medium (α-MEM) (Life Technologies, Los Angeles, CA, USA) supplemented with 10% FBS was used for osteoclasts (OCs) cultures. Macrophage colony-stimulating factor (M-CSF), anti-CD16 mAb, and flow cytometric antibodies were purchased from Biolegend (San Diego, CA, USA). Receptor activator of nuclear factor kappa-B ligand (RANKL) was purchased from PeproTech (Cranbury, NJ, USA), and recombinant human IL-2 was obtained from NIH-BRB. Anti-MICA/B mAbs used for ADCC were a generous gift from Dr. Jennifer Wu (Feinberg School of Medicine, Northwestern University, Evanston, IL, USA). NK92 was obtained from ATCC (Baltimore, MD, USA) and maintained in Alpha-MEM medium without ribonucleosides and deoxyribonucleosides supplemented with 2 mM L-glutamine, 1.5 g/L sodium bicarbonate, 0.2 mM inositol, 0.1 mM 2-mercaptoethanol, 0.02 mM folic acid, 100–200 U/mL rh-IL-2, 10% horse serum, and 10% FBS. NK92-176V (CD16^{high} transfected NK92) was a generous gift from Dr. Kerry Campbell (FOX Chase Cancer Center). AJ2 is a combination of eight different strains of Gram-positive probiotic bacteria (*Streptococcus thermophiles, Bifidobacterium longum, Bifidobacterium breve, Bifidobacterium infantis, Lactobacillus acidophilus, Lactobacillus plantarum, Lactobacillus casei*, and *Lactobacillus bulgaricus*) selected for its superior ability to optimally induce the secretion of both pro-inflammatory and anti-inflammatory cytokines from NK cells [18]. RPMI 1640 supplemented with 10% FBS was used to re-suspend AJ2. Human ELISA kits for IFN-γ were purchased from Biolegend (San Diego, CA, USA). Phosphate-buffered saline (PBS) and bovine serum albumin (BSA) were purchased from Life Technologies (Los Angeles, CA, USA).

4.2. Purification of Human NK Cells and Monocytes

Written informed consents approved by the UCLA Institutional Review Board (IRB) were obtained from healthy donors and cancer patients, and all procedures were approved by the UCLA-IRB. Peripheral blood mononuclear cells (PBMCs) were isolated from peripheral blood as previously described [76]. Briefly, PBMCs were obtained after Ficoll-hypaque centrifugation and were used to isolate NK cells and monocytes using the EasySep® Human NK cell and EasySep® Human Monocytes enrichment kits, respectively, purchased from Stem Cell Technologies (Vancouver, BC, Canada). Isolated NK cells and monocytes were stained with anti-CD16 and anti-CD14 antibodies, respectively, to measure the cell purity using flow cytometric analysis.

4.3. NK Cells Induced Differentiation of OSCSCs and MP2 Tumors

Human NK cells were purified from healthy individuals' PBMCs as described above. NK cells were treated with a combination of IL-2 (1000 U/mL) and anti-CD16mAbs (3 µg/mL) for 18 h, after which the supernatant was harvested and the levels of IFN-γ were assessed using single ELISA and later used in differentiation experiments. The differentiation of OSCSCs and MP2 cells was conducted with an average total of 2000 to 3500 pg and 5000 to 7000 pg of IFN-γ from IFN-γ containing supernatants, respectively, over a 7-day period, as previously described [64]. Initially, 1×10^6 tumor cells were cultured and treated with NK supernatant for differentiation as described, after which tumor cells were rinsed with 1× PBS, detached, and used for experiments.

4.4. Generation of Human OCs

To generate OCs, monocytes were cultured in alpha-MEM media supplemented with M-CSF (25 ng/mL) for 21 days and RANKL (25 ng/mL) from day 6 to 21 days. The media were replenished every three days.

4.5. Sonication of Probiotic Bacteria (AJ2)

AJ2 bacteria were weighed and re-suspended in RPMI 1640 medium containing 10% FBS at a concentration of 10 mg/mL. The bacteria were thoroughly vortexed and sonicated on ice for 15 s at six to eight amplitudes. Then, sonicated samples were incubated for 30 s on ice, and the cycle was repeated for five rounds. After every five rounds of sonication, the samples were examined under the microscope until at least 80% of bacterial walls were lysed. It was determined that approximately 20 rounds of sonication/incubation on ice were necessary to achieve complete sonication. Finally, the sonicated AJ2 (sAJ2) were aliquoted and stored at −80 °C until use.

4.6. Expansion of NK Cells

Purified human NK cells were activated with rh-IL-2 (1000 U/mL) and anti-CD16 mAbs (3 µg/mL) for 18–20 h before they were co-cultured with osteoclasts (OCs) and sAJ2 (OCs:NK:sAJ2; 1:2:4) in RPMI 1640 medium containing 10% FBS. The media were refreshed every three days with RPMI complete medium containing rh-IL-2 (1500 U/mL).

4.7. Enzyme-Linked Immunosorbent Assays (ELISAs)

Single ELISAs were performed as previously described [76]. To analyze and obtain the cytokine and chemokine concentration, a standard curve was generated by either two- or three-fold dilutions of recombinant cytokines provided by the manufacturer.

4.8. ^{51}Cr release Cytotoxicity Assay

The ^{51}Cr release cytotoxicity assay was performed as previously described [77]. Briefly, different numbers of effector cells were incubated with ^{51}Cr–labeled target cells. After a 4-h incubation period, the supernatants were harvested from each sample, and the released radioactivity was counted using a gamma counter. The percentage specific cytotoxicity was calculated as follows:

$$\%\text{cytotoxicity} = \frac{\text{Experimental cpm} - \text{spontaneous cpm}}{\text{Total cpm} - \text{spontaneous cpm}} \quad (1)$$

where LU $30/10^6$ is calculated by using the inverse of the number of effector cells needed to lyse 30% of tumor target cells × 100.

4.9. Antibody-Dependent Cell-Mediated Cytotoxicity (ADCC) Measurements

Tumor cells (target cells) were ^{51}Cr-labeled and were incubated for an hour, after which unbound ^{51}Cr was washed. Then, cells (1×10^6 cells/mL) were treated with anti-MICA/B mAbs (5 µg/mL) or anti-EGFR mAbs (10 µg/mL) or anti-PDL1 mAbs (20 µg/mL) for 30 min and washed with medium to remove excess unbound antibodies. Then, antibody-treated ^{51}Cr-labeled cells were cultured with effector cells at various effector to target ratios, and the cytotoxicity against tumor cells was assessed using the ^{51}Cr release cytotoxicity assay as described above.

4.10. Surface Staining Assay

For surface staining, the cells were washed twice using 1%BSA/PBS. Predetermined optimal concentrations of specific human monoclonal antibodies were added to 1×10^4 cells in 50 µL of 1%BSA/PBS and were incubated on ice for 30 min. Thereafter, cells were washed in 1%BSA/PBS and brought to 500 µL with 1%BSA/PBS. Flow cytometric analysis was performed using the Beckman Coulter Epics XL cytometer (Brea, CA, USA), and the results were analyzed in the FlowJo vX software (Ashland, OR, USA).

4.11. Western Blot

OSCCs and OSCSCs tumor cells were lysed in a lysis buffer containing 50 mM Tris-HCL (pH 7.4), 150 mM NaCl, 1% Nonidet P-40 (v/v), 1 mM sodium orthovanadate, 0.5 mM ethylenediaminetetraactetic acid (EDTA), 10 mM NaF, 2 mM phenylmethylsulfonyl fluoride (PMSF), 10 µg/mL leupeptin, and 2 U/mL aprotinin for 15 min on ice. Then, the samples were sonicated for 3 s. The tumor cell lysates were centrifuged at 14,000 rpm for 10 min, and the supernatants were removed and the levels of protein were quantified by the Bradford method. The cell lysates were denatured by boiling in 5× SDS sample buffer. Equal amounts of cell lysates were loaded onto 10% SDS-PAGE and transferred onto Immobilon-P membranes (Millipore, Billerica, MA, USA). The membranes were blocked with 5% non-fat milk in PBS plus 0.1% Tween-20 for 1 h. Primary antibodies at the predetermined dilutions were added for 1 h at room temperature. Then, membranes were incubated with 1:1000 dilution of horseradish peroxidase-conjugated secondary antibody. Blots were developed by enhanced chemiluminescence (ECL purchased from Pierce Biotechnology, Rockford, IL, USA).

4.12. Statistical Analyses

All statistical analyses were performed using the GraphPad Prism-8 software. An unpaired or paired, two-tailed Student's t-test was performed for experiments with two groups. One-way ANOVA with a Bonferroni post-test was used to compare different groups for experiments with more than two groups. (n) denotes the number of human donors or the number of samples for each experimental condition. Duplicate or triplicate samples were used in the in vitro studies for assessment. The following symbols represent the levels of statistical significance within each analysis: **** (p value < 0.0001), *** (p value < 0.001), ** (p value 0.001–0.01), * (p value 0.01–0.05)

5. Conclusions

In this study we provided evidence that NK cells mediate lysis of both stem-like/poorly differentiated tumors and well-differentiated tumors via direct cytotoxicity and ADCC, respectively. By using antibodies to MICA/B, EGFR and PDL1 surface receptors expressed on well-differentiated but not on stem-like/poorly differentiated tumors we demonstrated significant NK cell mediated ADCC in the absence of direct killing. In addition, our results suggested the possibility of CD16 receptors mediating both direct cytotoxicity and ADCC, resulting in the competitive use of these receptors in either direct killing or antibody dependent cellular cytotoxicity, depending on the differentiation status of tumor cells and the stage of maturation and activation of NK cells. Thus, these two modes of NK cell mediated killing ensures that both CSC/poorly differentiated and well-differentiated tumors are eliminated.

Supplementary Materials: The following are available online at https://www.mdpi.com/2072-6694/13/2/239/s1, This study was conducted according to UCLA Institutional Review Board (IRB). Figure S1: Cancer patients' NK cells exhibit lower cytotoxicity against K562 tumors, Figure S2: Cancer patients' NK cells secreted low IFN-γ secretions when compared to healthy individuals' NK cells, Figure S3: NK cells' surface expression on healthy individual and cancer patients, Figure S4: Surface expression levels of CD16 on primary NK cells, OC-expanded NK cells and NK92 cell lines, and NKG2D surface expression level on primary NK and OC-expanded NK cells, Figure S5: Higher levels of NK cell-mediated ADCC was seen in freshly isolated primary NK cells vs. OC-expanded NK cells, Figure S6. Higher surface expression of EGF and B7H1, and NK cell-mediated ADCC in differentiated tumors in comparison to their stem-like counterparts.

Author Contributions: K.K. prepared study design, performed experiments, data analysis, figures preparation, manuscript writing/editing. T.S. performed experiments and edited manuscript. M.-W.K. performed supporting experiments and edited manuscript. Y.W. edited manuscript. A.J. arranged funding, and worked with K.K. for study design, data analysis, figures preparation, manuscript writing/editing. All authors have read and agreed to the published version of the manuscript.

Funding: This research was funded by NIH-NIDCR RO1-DE022552; RO1 DE12880; UCLA Academic senate grant and School of Dentistry Seed grant

Institutional Review Board Statement: This study was conducted according to UCLA Institutional Review Board (IRB). Protocol ID: IRB#11-000781. Approval Date: 2 December 2020.

Informed Consent Statement: Written informed consents approved by the UCLA Institutional Review Board (IRB) were obtained from healthy donors and cancer patients, and all procedures were approved by the UCLA-IRB.

Data Availability Statement: The data presented in this study are available in the article or supplementary materials of "ADCC against MICA/B is Mediated against Differentiated Oral and Pancreatic and Not Stem-Like/Poorly Differentiated Tumors by the NK Cells; Loss in Cancer Patients due to Down-Modulation of CD16 Receptor".

Conflicts of Interest: The authors declare that the research was conducted in the absence of any commercial or financial relationships that could be construed as a potential conflict of interest.

Abbreviations

NK cells	Natural killer cells
MICA/B	Major histocompatibility complex-class I chain related proteins A and B
ADCC	Antibody-dependent cellular cytotoxicity (ADCC)
Hu-BLT	Humanized-bone marrow/liver/thymus
MP2	Mia PaCa-2 pancreatic cancer stem cells
OSCSCs	Oral squamous cancer stem cells
OSCCs	Oral squamous carcinoma cells
MHC-Class I	Major histocompatibility complex molecule class I
IFN-γ	Interferon-gamma
TNF-α	Tumor necrosis factor-α
CSCs	Cancer stem cells
rhIL-2	Recombinant human IL-2

References

1. Kiessling, R.; Klein, E.; Pross, H.; Wigzell, H. "Natural" killer cells in the mouse. II. Cytotoxic cells with specificity for mouse Moloney leukemia cells. Characteristics of the killer cell. *Eur. J. Immunol.* **1975**, *5*, 117–121. [CrossRef] [PubMed]
2. Fildes, J.E.; Yonan, N.; Leonard, C.T. Natural killer cells and lung transplantation, roles in rejection, infection, and tolerance. *Transpl. Immunol.* **2008**, *19*, 1–11. [CrossRef] [PubMed]
3. Farag, S.S.; Caligiuri, M.A. Human natural killer cell development and biology. *Blood Rev.* **2006**, *20*, 123–137. [CrossRef] [PubMed]
4. Cooper, M.A.; Fehniger, T.A.; Caligiuri, M.A. The biology of human natural killer-cell subsets. *Trends Immunol.* **2001**, *22*, 633–640. [CrossRef]
5. Pegram, H.J.; Andrews, D.M.; Smyth, M.J.; Darcy, P.K.; Kershaw, M.H. Activating and inhibitory receptors of natural killer cells. *Immunol. Cell Biol.* **2011**, *89*, 216–224. [CrossRef]
6. Gogali, F.; Paterakis, G.; Rassidakis, G.Z.; Liakou, C.I.; Liapi, C. CD3(-)CD16(-)CD56(bright) immunoregulatory NK cells are increased in the tumor microenvironment and inversely correlate with advanced stages in patients with papillary thyroid cancer. *Thyroid Off. J. Am. Thyroid Assoc.* **2013**, *23*, 1561–1568. [CrossRef] [PubMed]
7. Lopez-Cobo, S.; Pieper, N.; Campos-Silva, C.; Garcia-Cuesta, E.M.; Reyburn, H.T.; Paschen, A.; Vales-Gomez, M. Impaired NK cell recognition of vemurafenib-treated melanoma cells is overcome by simultaneous application of histone deacetylase inhibitors. *Oncoimmunology* **2018**, *7*, e1392426. [CrossRef]
8. Morel, P.A.; Ernst, L.K.; Metes, D. Functional CD32 molecules on human NK cells. *Leuk. Lymphoma* **1999**, *35*, 47–56. [CrossRef]
9. Lanier, L.L.; Ruitenberg, J.J.; Phillips, J.H. Functional and biochemical analysis of CD16 antigen on natural killer cells and granulocytes. *J. Immunol.* **1988**, *141*, 3478–3485.
10. Jinushi, M.; Takehara, T.; Tatsumi, T.; Kanto, T.; Groh, V.; Spies, T.; Kimura, R.; Miyagi, T.; Mochizuki, K.; Sasaki, Y.; et al. Expression and role of MICA and MICB in human hepatocellular carcinomas and their regulation by retinoic acid. *Int. J. Cancer* **2003**, *104*, 354–361. [CrossRef]
11. Salih, H.R.; Rammensee, H.G.; Steinle, A. Cutting edge: Down-regulation of MICA on human tumors by proteolytic shedding. *J. Immunol.* **2002**, *169*, 4098–4102. [CrossRef] [PubMed]
12. Moncayo, G.; Lin, D.; McCarthy, M.T.; Watson, A.A.; O'Callaghan, C.A. MICA Expression Is Regulated by Cell Adhesion and Contact in a FAK/Src-Dependent Manner. *Front. Immunol.* **2016**, *7*, 687. [CrossRef] [PubMed]
13. Thompson, T.W.; Kim, A.B.; Li, P.J.; Wang, J.; Jackson, B.T.; Huang, K.T.H.; Zhang, L.; Raulet, D.H. Endothelial cells express NKG2D ligands and desensitize antitumor NK responses. *Elife* **2017**, *6*. [CrossRef] [PubMed]
14. Zhao, Y.; Chen, N.; Yu, Y.; Zhou, L.; Niu, C.; Liu, Y.; Tian, H.; Lv, Z.; Han, F.; Cui, J. Prognostic value of MICA/B in cancers: A systematic review and meta-analysis. *Oncotarget* **2017**, *8*, 96384–96395. [CrossRef]
15. Lin, S.R.; Wen, Y.C.; Yeh, H.L.; Jiang, K.C.; Chen, W.H.; Mokgautsi, N.; Huang, J.; Chen, W.Y.; Liu, Y.N. EGFR-upregulated LIFR promotes SUCLG2-dependent castration resistance and neuroendocrine differentiation of prostate cancer. *Oncogene* **2020**, *39*, 6757–6775. [CrossRef]
16. Eriksen, J.G.; Steiniche, T.; Askaa, J.; Alsner, J.; Overgaard, J. The prognostic value of epidermal growth factor receptor is related to tumor differentiation and the overall treatment time of radiotherapy in squamous cell carcinomas of the head and neck. *Int. J. Radiat. Oncol. Biol. Phys.* **2004**, *58*, 561–566. [CrossRef]
17. Kozlowska, A.K.; Tseng, H.C.; Kaur, K.; Topchyan, P.; Inagaki, A.; Bui, V.T.; Kasahara, N.; Cacalano, N.; Jewett, A. Resistance to cytotoxicity and sustained release of interleukin-6 and interleukin-8 in the presence of decreased interferon-γ after differentiation of glioblastoma by human natural killer cells. *Cancer Immunol. Immunother.* **2016**, *65*, 1085–1097. [CrossRef]
18. Bui, V.T.; Tseng, H.C.; Kozlowska, A.; Maung, P.O.; Kaur, K.; Topchyan, P.; Jewett, A. Augmented IFN-γ and TNF-α Induced by Probiotic Bacteria in NK Cells Mediate Differentiation of Stem-Like Tumors Leading to Inhibition of Tumor Growth and Reduction in Inflammatory Cytokine Release; Regulation by IL-10. *Front. Immunol.* **2015**, *6*, 576. [CrossRef]
19. Alderson, K.L.; Sondel, P.M. Clinical Cancer Therapy by NK Cells via Antibody-Dependent Cell-Mediated Cytotoxicity. *J. Biomed. Biotechnol.* **2011**, *2011*, 379123. [CrossRef]
20. Wang, W.; Erbe, A.K.; Hank, J.A.; Morris, Z.S.; Sondel, P.M. NK Cell-Mediated Antibody-Dependent Cellular Cytotoxicity in Cancer Immunotherapy. *Front. Immunol.* **2015**, *6*, 368. [CrossRef]
21. Krzewski, K.; Coligan, J.E. Human NK cell lytic granules and regulation of their exocytosis. *Front. Immunol.* **2012**, *3*, 335. [CrossRef] [PubMed]
22. Yeap, W.H.; Wong, K.L.; Shimasaki, N.; Teo, E.C.Y.; Quek, J.K.S.; Yong, H.X.; Diong, C.P.; Bertoletti, A.; Linn, Y.C.; Wong, S.C. CD16 is indispensable for antibody-dependent cellular cytotoxicity by human monocytes. *Sci. Rep.* **2016**, *6*, 34310. [CrossRef] [PubMed]
23. Bhatnagar, N.; Ahmad, F.; Hong, H.S.; Eberhard, J.; Lu, I.N.; Ballmaier, M.; Schmidt, R.E.; Jacobs, R.; Meyer-Olson, D. FcgammaRIII (CD16)-mediated ADCC by NK cells is regulated by monocytes and FcgammaRII (CD32). *Eur. J. Immunol.* **2014**, *44*, 3368–3379. [CrossRef] [PubMed]
24. Oboshi, W.; Watanabe, T.; Matsuyama, Y.; Kobara, A.; Yukimasa, N.; Ueno, I.; Aki, K.; Tada, T.; Hosoi, E. The influence of NK cell-mediated ADCC: Structure and expression of the CD16 molecule differ among FcgammaRIIIa-V158F genotypes in healthy Japanese subjects. *Hum. Immunol.* **2016**, *77*, 165–171. [CrossRef]
25. Nimmerjahn, F.; Ravetch, J.V. Fcgamma receptors as regulators of immune responses. *Nat. Rev. Immunol.* **2008**, *8*, 34–47. [CrossRef] [PubMed]

26. Parsons, M.S.; Richard, J.; Lee, W.S.; Vanderven, H.; Grant, M.D.; Finzi, A.; Kent, S.J. NKG2D Acts as a Co-Receptor for Natural Killer Cell-Mediated Anti-HIV-1 Antibody-Dependent Cellular Cytotoxicity. *Aids Res. Hum. Retrovir.* **2016**, *32*, 1089–1096. [CrossRef]
27. Di Modica, M.; Sfondrini, L.; Regondi, V.; Varchetta, S.; Oliviero, B.; Mariani, G.; Bianchi, G.V.; Generali, D.; Balsari, A.; Triulzi, T.; et al. Taxanes enhance trastuzumab-mediated ADCC on tumor cells through NKG2D-mediated NK cell recognition. *Oncotarget* **2016**, *7*, 255–265. [CrossRef]
28. Märklin, M.; Hagelstein, I.; Koerner, S.P.; Rothfelder, K.; Pfluegler, M.S.; Schumacher, A.; Grosse-Hovest, L.; Jung, G.; Salih, H.R. Bispecific NKG2D-CD3 and NKG2D-CD16 fusion proteins for induction of NK and T cell reactivity against acute myeloid leukemia. *J. Immunother. Cancer* **2019**, *7*, 143. [CrossRef]
29. Raulet, D.H.; Gasser, S.; Gowen, B.G.; Deng, W.; Jung, H. Regulation of Ligands for the NKG2D Activating Receptor. *Annu. Rev. Immunol.* **2013**, *31*, 413–441. [CrossRef]
30. Li, P.; Morris, D.L.; Willcox, B.E.; Steinle, A.; Spies, T.; Strong, R.K. Complex structure of the activating immunoreceptor NKG2D and its MHC class I-like ligand MICA. *Nat. Immunol.* **2001**, *2*, 443–451. [CrossRef]
31. Cerwenka, A.; Baron, J.L.; Lanier, L.L. Ectopic expression of retinoic acid early inducible-1 gene (RAE-1) permits natural killer cell-mediated rejection of a MHC class I-bearing tumor in vivo. *Proc. Natl. Acad. Sci. USA* **2001**, *98*, 11521–11526. [CrossRef] [PubMed]
32. Groh, V.; Wu, J.; Yee, C.; Spies, T. Tumour-derived soluble MIC ligands impair expression of NKG2D and T-cell activation. *Nature* **2002**, *419*, 734–738. [CrossRef] [PubMed]
33. Carlsten, M.; Norell, H.; Bryceson, Y.T.; Poschke, I.; Schedvins, K.; Ljunggren, H.G.; Kiessling, R.; Malmberg, K.J. Primary human tumor cells expressing CD155 impair tumor targeting by down-regulating DNAM-1 on NK cells. *J. Immunol.* **2009**, *183*, 4921–4930. [CrossRef] [PubMed]
34. Kono, K.; Takahashi, A.; Ichihara, F.; Sugai, H.; Fujii, H.; Matsumoto, Y. Impaired antibody-dependent cellular cytotoxicity mediated by herceptin in patients with gastric cancer. *Cancer Res.* **2002**, *62*, 5813–5817. [PubMed]
35. Kawaguchi, Y.; Kono, K.; Mimura, K.; Sugai, H.; Akaike, H.; Fujii, H. Cetuximab induce antibody-dependent cellular cytotoxicity against EGFR-expressing esophageal squamous cell carcinoma. *Int. J. Cancer* **2007**, *120*, 781–787. [CrossRef] [PubMed]
36. Watanabe, M.; Kono, K.; Kawaguchi, Y.; Mizukami, Y.; Mimura, K.; Maruyama, T.; Izawa, S.; Fujii, H. NK cell dysfunction with down-regulated CD16 and up-regulated CD56 molecules in patients with esophageal squamous cell carcinoma. *Dis. Esophagus* **2010**, *23*, 675–681. [CrossRef]
37. Nakajima, T.; Okayama, H.; Ashizawa, M.; Noda, M.; Aoto, K.; Saito, M.; Monma, T.; Ohki, S.; Shibata, M.; Takenoshita, S.; et al. Augmentation of antibody-dependent cellular cytotoxicity with defucosylated monoclonal antibodies in patients with GI-tract cancer. *Oncol. Lett.* **2018**, *15*, 2604–2610. [CrossRef]
38. Tseng, H.C.; Cacalano, N.; Jewett, A. Split anergized Natural Killer cells halt inflammation by inducing stem cell differentiation, resistance to NK cell cytotoxicity and prevention of cytokine and chemokine secretion. *Oncotarget* **2015**, *6*, 8947–8959. [CrossRef]
39. Jewett, A.; Teruel, A.; Romero, M.; Head, C.; Cacalano, N. Rapid and potent induction of cell death and loss of NK cell cytotoxicity against oral tumors by F(ab')2 fragment of anti-CD16 antibody. *Cancer Immunol. Immunother.* **2008**, *57*, 1053–1066. [CrossRef]
40. Bonavida, B.; Lebow, L.T.; Jewett, A. Natural killer cell subsets: Maturation, differentiation and regulation. *Nat. Immun.* **1993**, *12*, 194–208.
41. Kaur, R.; Nanut, M.P.; Ko, M.W.; Safaie, T.; Kos, J.; Jewett, A. Natural killer cells target and differentiate cancer stem-like cells/undifferentiated tumors: Strategies to optimize their growth and expansion for effective cancer immunotherapy. *Curr. Opin. Immunol.* **2018**, *51*, 170–180. [CrossRef] [PubMed]
42. Vela-Ojeda, J.; Esparza, M.A.G.; Majluf-Cruz, A.; Garcia-Chavez, J.; Montiel-Cervantes, L.A.; Reyes-Maldonado, E.; Hernandez-Caballero, A.; Rodriguez-Gonzalez, M.G. Post-treatment improvement of NK cell numbers predicts better survival in myeloma patients treated with thalidomide-based regimens. *Int. J. Hematol.* **2019**, *110*, 306–312. [CrossRef] [PubMed]
43. Tang, Y.-P.; Xie, M.-Z.; Li, K.-Z.; Li, J.-L.; Cai, Z.-M.; Hu, B.-L. Prognostic value of peripheral blood natural killer cells in colorectal cancer. *BMC Gastroenterol.* **2020**, *20*, 31. [CrossRef] [PubMed]
44. Levy, E.M.; Roberti, M.P.; Mordoh, J. Natural killer cells in human cancer: From biological functions to clinical applications. *J. Biomed. Biotechnol.* **2011**, *2011*, 676198. [CrossRef]
45. Kuss, I.; Hathaway, B.; Ferris, R.L.; Gooding, W.; Whiteside, T.L. Decreased absolute counts of T lymphocyte subsets and their relation to disease in squamous cell carcinoma of the head and neck. *Clin. Cancer Res.* **2004**, *10*, 3755–3762. [CrossRef]
46. Kim, J.W.; Tsukishiro, T.; Johnson, J.T.; Whiteside, T.L. Expression of pro- and antiapoptotic proteins in circulating CD8+ T cells of patients with squamous cell carcinoma of the head and neck. *Clin. Cancer Res.* **2004**, *10*, 5101–5110. [CrossRef]
47. Burke, S.; Lakshmikanth, T.; Colucci, F.; Carbone, E. New views on natural killer cell-based immunotherapy for melanoma treatment. *Trends Immunol.* **2010**, *31*, 339–345. [CrossRef]
48. Larsen, S.K.; Gao, Y.; Basse, P.H. NK cells in the tumor microenvironment. *Crit. Rev. Oncog.* **2014**, *19*, 91–105. [CrossRef]
49. Nolibe, D.; Poupon, M.F. Enhancement of pulmonary metastases induced by decreased lung natural killer cell activity. *J. Natl. Cancer Inst.* **1986**, *77*, 99–103.
50. Imai, K.; Matsuyama, S.; Miyake, S.; Suga, K.; Nakachi, K. Natural cytotoxic activity of peripheral-blood lymphocytes and cancer incidence: An 11-year follow-up study of a general population. *Lancet* **2000**, *356*, 1795–1799. [CrossRef]
51. Harning, R.; Koo, G.C.; Szalay, J. Regulation of the metastasis of murine ocular melanoma by natural killer cells. *Investig. Ophthalmol. Vis. Sci.* **1989**, *30*, 1909–1915.

52. Coca, S.; Perez-Piqueras, J.; Martinez, D.; Colmenarejo, A.; Saez, M.A.; Vallejo, C.; Martos, J.A.; Moreno, M. The prognostic significance of intratumoral natural killer cells in patients with colorectal carcinoma. *Cancer* 1997, 79, 2320–2328. [CrossRef]
53. Bruno, A.; Ferlazzo, G.; Albini, A.; Noonan, D.M. A think tank of TINK/TANKs: Tumor-infiltrating/tumor-associated natural killer cells in tumor progression and angiogenesis. *J. Natl. Cancer Inst.* 2014, 106, dju200. [CrossRef] [PubMed]
54. Gross, E.; Sunwoo, J.B.; Bui, J.D. Cancer immunosurveillance and immunoediting by natural killer cells. *Cancer J.* 2013, 19, 483–489. [CrossRef] [PubMed]
55. Mirjacic Martinovic, K.M.; Babovic, N.; Dzodic, R.R.; Jurisic, V.B.; Tanic, N.T.; Konjevic, G.M. Decreased expression of NKG2D, NKp46, DNAM-1 receptors, and intracellular perforin and STAT-1 effector molecules in NK cells and their dim and bright subsets in metastatic melanoma patients. *Melanoma Res.* 2014, 24, 295–304. [CrossRef] [PubMed]
56. Gubbels, J.A.; Felder, M.; Horibata, S.; Belisle, J.A.; Kapur, A.; Holden, H.; Petrie, S.; Migneault, M.; Rancourt, C.; Connor, J.P.; et al. MUC16 provides immune protection by inhibiting synapse formation between NK and ovarian tumor cells. *Mol. Cancer* 2010, 9, 11. [CrossRef] [PubMed]
57. Balsamo, M.; Scordamaglia, F.; Pietra, G.; Manzini, C.; Cantoni, C.; Boitano, M.; Queirolo, P.; Vermi, W.; Facchetti, F.; Moretta, A.; et al. Melanoma-associated fibroblasts modulate NK cell phenotype and antitumor cytotoxicity. *Proc. Natl. Acad. Sci. USA* 2009, 106, 20847–20852. [CrossRef]
58. Castriconi, R.; Cantoni, C.; Della Chiesa, M.; Vitale, M.; Marcenaro, E.; Conte, R.; Biassoni, R.; Bottino, C.; Moretta, L.; Moretta, A. Transforming growth factor beta 1 inhibits expression of NKp30 and NKG2D receptors: Consequences for the NK-mediated killing of dendritic cells. *Proc. Natl. Acad. Sci. USA* 2003, 100, 4120–4125. [CrossRef]
59. Pietra, G.; Manzini, C.; Rivara, S.; Vitale, M.; Cantoni, C.; Petretto, A.; Balsamo, M.; Conte, R.; Benelli, R.; Minghelli, S.; et al. Melanoma cells inhibit natural killer cell function by modulating the expression of activating receptors and cytolytic activity. *Cancer Res.* 2012, 72, 1407–1415. [CrossRef]
60. Krockenberger, M.; Dombrowski, Y.; Weidler, C.; Ossadnik, M.; Honig, A.; Hausler, S.; Voigt, H.; Becker, J.C.; Leng, L.; Steinle, A.; et al. Macrophage migration inhibitory factor contributes to the immune escape of ovarian cancer by down-regulating NKG2D. *J. Immunol.* 2008, 180, 7338–7348. [CrossRef]
61. Kaur, K.; Cook, J.; Park, S.H.; Topchyan, P.; Kozlowska, A.; Ohanian, N.; Fang, C.; Nishimura, I.; Jewett, A. Novel Strategy to Expand Super-Charged NK Cells with Significant Potential to Lyse and Differentiate Cancer Stem Cells: Differences in NK Expansion and Function between Healthy and Cancer Patients. *Front. Immunol.* 2017, 8, 297. [CrossRef] [PubMed]
62. Kaur, K.; Kozlowska, A.K.; Topchyan, P.; Ko, M.W.; Ohanian, N.; Chiang, J.; Cook, J.; Maung, P.O.; Park, S.H.; Cacalano, N.; et al. Probiotic-Treated Super-Charged NK Cells Efficiently Clear Poorly Differentiated Pancreatic Tumors in Hu-BLT Mice. *Cancers* 2019, 12, 63. [CrossRef] [PubMed]
63. Kaur, K.; Topchyan, P.; Kozlowska, A.K.; Ohanian, N.; Chiang, J.; Maung, P.O.; Park, S.H.; Ko, M.W.; Fang, C.; Nishimura, I.; et al. Super-charged NK cells inhibit growth and progression of stem-like/poorly differentiated oral tumors in vivo in humanized BLT mice; effect on tumor differentiation and response to chemotherapeutic drugs. *Oncoimmunology* 2018, 7, e1426518. [CrossRef] [PubMed]
64. Tseng, H.C.; Bui, V.; Man, Y.G.; Cacalano, N.; Jewett, A. Induction of Split Anergy Conditions Natural Killer Cells to Promote Differentiation of Stem Cells through Cell-Cell Contact and Secreted Factors. *Front. Immunol.* 2014, 5, 269. [CrossRef] [PubMed]
65. Jewett, A.; Man, Y.-G.; Tseng, H.C. Dual Functions of Natural Killer Cells in Selection and Differentiation of Stem Cells; Role in Regulation of Inflammation and Regeneration of Tissues. *J. Cancer* 2013, 4, 12–24. [CrossRef] [PubMed]
66. Tseng, H.-C.; Arasteh, A.; Paranjpe, A.; Teruel, A.; Yang, W.; Behel, A.; Alva, J.A.; Walter, G.; Head, C.; Ishikawa, T.-O.; et al. Increased Lysis of Stem Cells but Not Their Differentiated Cells by Natural Killer Cells; De-Differentiation or Reprogramming Activates NK Cells. *PLoS ONE* 2010, 5. [CrossRef]
67. Wiernik, A.; Foley, B.; Zhang, B.; Verneris, M.R.; Warlick, E.; Gleason, M.K.; Ross, J.A.; Luo, X.; Weisdorf, D.J.; Walcheck, B.; et al. Targeting natural killer cells to acute myeloid leukemia in vitro with a CD16 x 33 bispecific killer cell engager and ADAM17 inhibition. *Clin. Cancer Res.* 2013, 19, 3844–3855. [CrossRef]
68. Jing, Y.; Ni, Z.; Wu, J.; Higgins, L.; Markowski, T.W.; Kaufman, D.S.; Walcheck, B. Identification of an ADAM17 cleavage region in human CD16 (FcγRIII) and the engineering of a non-cleavable version of the receptor in NK cells. *PLoS ONE* 2015, 10, e0121788. [CrossRef]
69. An, B.C.; Hong, S.; Park, H.J.; Kim, B.K.; Ahn, J.Y.; Ryu, Y.; An, J.H.; Chung, M.J. Anti-Colorectal Cancer Effects of Probiotic-Derived p8 Protein. *Genes* 2019, 10, 624. [CrossRef]
70. Kozlowska, A.K.; Topchyan, P.; Kaur, K.; Tseng, H.C.; Teruel, A.; Hiraga, T.; Jewett, A. Differentiation by NK cells is a prerequisite for effective targeting of cancer stem cells/poorly differentiated tumors by chemopreventive and chemotherapeutic drugs. *J. Cancer* 2017, 8, 537–554. [CrossRef]
71. Magister, Š.; Tseng, H.C.; Bui, V.T.; Kos, J.; Jewett, A. Regulation of split anergy in natural killer cells by inhibition of cathepsins C and H and cystatin F. *Oncotarget* 2015, 6, 22310–22327. [CrossRef] [PubMed]
72. Pazina, T.; James, A.M.; MacFarlane, A.W.t.; Bezman, N.A.; Henning, K.A.; Bee, C.; Graziano, R.F.; Robbins, M.D.; Cohen, A.D.; Campbell, K.S. The anti-SLAMF7 antibody elotuzumab mediates NK cell activation through both CD16-dependent and -independent mechanisms. *Oncoimmunology* 2017, 6, e1339853. [CrossRef] [PubMed]
73. Aramburu, J.; Azzoni, L.; Rao, A.; Perussia, B. Activation and expression of the nuclear factors of activated T cells, NFATp and NFATc, in human natural killer cells: Regulation upon CD16 ligand binding. *J. Exp. Med.* 1995, 182, 801–810. [CrossRef] [PubMed]
74. Tamm, A.; Schmidt, R.E. The binding epitopes of human CD16 (Fc gamma RIII) monoclonal antibodies. Implications for ligand binding. *J. Immunol.* 1996, 157, 1576–1581. [PubMed]

75. Tseng, H.C.; Inagaki, A.; Bui, V.T.; Cacalano, N.; Kasahara, N.; Man, Y.G.; Jewett, A. Differential Targeting of Stem Cells and Differentiated Glioblastomas by NK Cells. *J. Cancer* **2015**, *6*, 866–876. [CrossRef]
76. Jewett, A.; Bonavida, B. Target-induced inactivation and cell death by apoptosis in a subset of human NK cells. *J. Immunol.* **1996**, *156*, 907–915. [PubMed]
77. Jewett, A.; Wang, M.Y.; Teruel, A.; Poupak, Z.; Bostanian, Z.; Park, N.H. Cytokine dependent inverse regulation of CD54 (ICAM1) and major histocompatibility complex class I antigens by nuclear factor kappaB in HEp2 tumor cell line: Effect on the function of natural killer cells. *Hum. Immunol.* **2003**, *64*, 505–520. [CrossRef]

Article

Co-Expression of IL-7 Improves NKG2D-Based CAR T Cell Therapy on Prostate Cancer by Enhancing the Expansion and Inhibiting the Apoptosis and Exhaustion

Cong He [†], Ying Zhou [†], Zhenlong Li, Muhammad Asad Farooq, Iqra Ajmal, Hongmei Zhang, Li Zhang, Lei Tao, Jie Yao, Bing Du, Mingyao Liu * and Wenzheng Jiang *

Shanghai Key Laboratory of Regulatory Biology, School of Life Sciences, East China Normal University, Shanghai 200241, China; hecong0126@126.com (C.H.); zoeyzhouying@126.com (Y.Z.); lzlecnu@163.com (Z.L.); asadfarooq601@yahoo.com (M.A.F.); iqraasad1263@yahoo.com (I.A.); zhanghongmei0501@126.com (H.Z.); 15221152917@163.com (L.Z.); taol1988@126.com (L.T.); ecnuyaojie@163.com (J.Y.); bdu@bio.ecnu.edu.cn (B.D.)
* Correspondence: myliu@bio.ecnu.edu.cn (M.L.); wzjiang@bio.ecnu.edu.cn (W.J.); Tel.: +86-21-54341035 (W.J.); Fax: +86-21-54341006 (W.J.)
† These authors contributed equally to this work.

Received: 14 June 2020; Accepted: 18 July 2020; Published: 20 July 2020

Abstract: Chimeric antigen receptor (CAR) T-cell therapy is a promising approach in treating solid tumors but the therapeutic effect is limited. Prostate cancer is a typical solid malignancy with invasive property and a highly immunosuppressive microenvironment. Ligands for the NKG2D receptor are primarily expressed on many cancer cells, including prostate cancer. In this study, we utilized NKG2D-based CAR to treat prostate cancer, and improved the therapeutic effect by co-expression of IL-7. The results showed that NKG2D-CAR T cells performed significantly increased cytotoxicity against prostate cancer compared to non-transduced T cells in vitro and in vivo. Moreover, the introduction of the *IL-7* gene into the NKG2D-CAR backbone enhanced the production of IL-7 in an antigen-dependent manner. NKG2DIL7-CAR T cells exhibited better antitumor efficacy at 16 h and 72 h in vitro, and inhibited tumor growth in xenograft models more effectively. In mechanism, enhanced proliferation and Bcl-2 expression in CD8[+] T cells, decreased apoptosis and exhaustion, and increased less-differentiated cell phenotype may be the reasons for the improved persistence and survival of NKG2DIL7-CAR T cells. In conclusion, these findings demonstrated that NKG2D is a promising option for CAR T-cell therapy on prostate cancer, and IL-7 has enhanced effect on NKG2D-based CAR T-cell immunotherapy, providing a novel adoptive cell therapy for prostate cancer either alone or in combination with IL-7.

Keywords: NKG2D; CAR T; IL-7; prostate cancer; cell therapy

1. Introduction

The application of genetic redirection of T lymphocytes with chimeric antigen receptors (CARs) in cancers has re-energized the field of cancer immunotherapy. The tremendous success of CAR T cells in eradicating CD19-expressing acute and chronic B cell leukemias has attracted more attention in applying CARs to solid tumors [1,2]. However, several limitations need to be resolved to extend the success to solid tumors [3]. The key limitations are the accumulation and survival of transferred CAR T cells in the immunosuppressive tumor microenvironment, which can impair the proliferative ability and in vivo persistence of the infused T cells [4].

Prostate cancer is the second most frequent cancer among males, leading to a huge burden of incidence and mortality in the world [5]. Once symptoms appear, it is mostly diagnosed as a progressive prostate cancer, and approximately one in five patients of metastatic castration-resistant prostate cancer (mCRPC) dies annually [6]. Given that metastatic prostate cancer is associated with an unfavorable prognosis and poses enormous therapeutic challenges, any novel strategy of effective treatment developed for this advanced disease is a top priority for the scientists nowadays [7].

Ligands for the natural killer group 2 member D (NKG2D), an NK-cell activating receptor [8], are primarily expressed on most types of tumor cells, including hematological and solid tumors, but are normally absent or expressed in low levels on healthy tissues [9,10]. Several kinds of NKG2D-based CARs have been developed and their extensive therapeutic effects on various tumors have been studied [11]. Here, we identified NKG2DLs-expressing prostate cancer as being susceptible to NKG2D-CAR T cell-mediated attack, providing a new strategy for effective treatment.

NKG2D-CAR T cells have shown safety in clinical trials for treating patients with multiple hematological and solid tumors, and although complete alleviation has been achieved in selected patients, higher efficacy still remains desirable. The limited expansion of CAR T cells in vivo is one of the obstacles that need to be overcome to boost clinical efficacy [12]. Transgenic expression of growth-promoting cytokines (e.g., IL-15, IL-12) in CAR T cells represents a strategy to support the long-term expansion and persistence [13,14]. IL-7 has been applied to augment the T-cell antitumor immune response as a T-cell growth factor [15,16]. Therefore, we further modified NKG2D-CAR T cells to secrete IL-7, a stimulatory cytokine known to improve the proliferation and survival of T cells [17,18]. Our study demonstrated that NKG2D-CAR T-cell treatment effectively inhibited the growth of prostate cancer. Furthermore, transgenic expression of IL-7 enhanced the proliferative, persistence and anti-prostate cancer activity of NKG2D-CAR T cells in vitro and in vivo.

2. Results

2.1. NKG2D-CAR T cells Effectively Recognize and Lyse NKG2DLs$^+$ Prostate Cancer Cell Lines In Vitro and Co-Expression of IL-7 Enhances Its Activation and Function

We synthesized NKG2D-CAR construct consisting of the extracellular part of NKG2D, linked to the intracellular signaling domains of 4-1BB and CD3ζ molecule via a CD8α hinge-transmembrane domain (Upper panel of Figure 1a). CD3/CD28-activated T cells separated from healthy donors were transduced with lentivirus expressing NKG2D-CAR, and the transduction efficiency was identified by staining of the anti-human NKG2D antibody. To determine whether NKG2D-CAR T cells can recognize and lyse prostate cancer cells such as PC-3, DU 145 and C4-2, which expresses a high level of NKG2DLs (Additional file 1: Figure S1), the cytotoxicity of NKG2D-CAR T cells against the NKG2DLs$^+$ prostate cancer cells was determined by detection of the apoptosis of the targeT cells. The results showed that NKG2D-CAR T cells exhibited significant cytolytic activity against several prostate cancer cell lines in an E:T ratio-dependent manner, but had no killing effect on the NKG2DLs$^-$ cell line B16-F10 (Figure 1b). Therefore, NKG2D-based CAR T cells could specifically and efficiently kill NKG2DLs$^+$ prostate cancer cells in vitro.

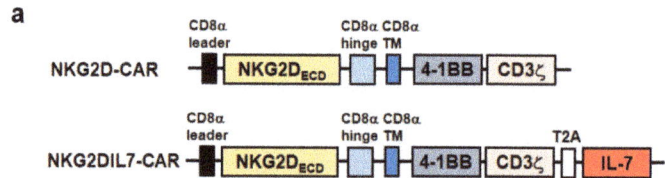

Figure 1. *Cont.*

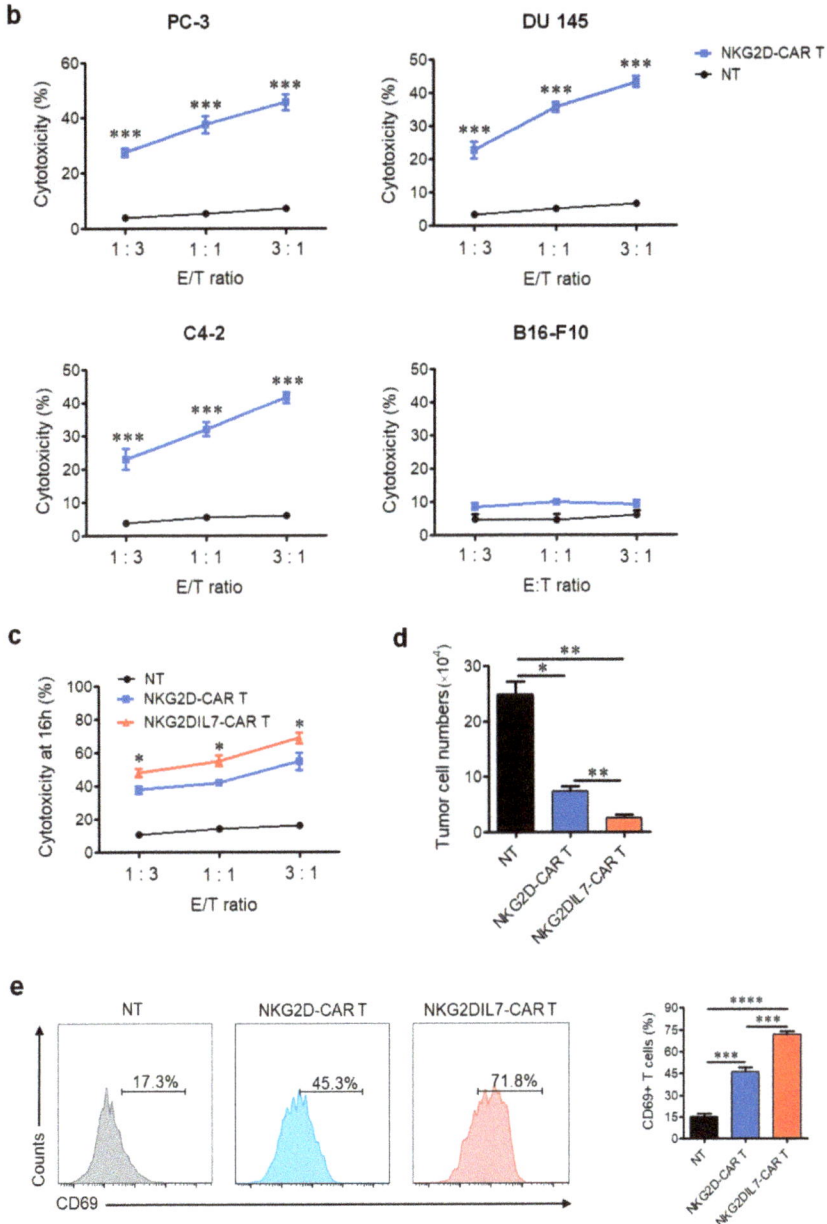

Figure 1. *Cont.*

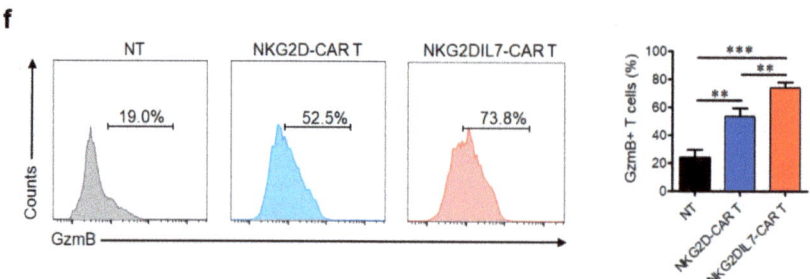

Figure 1. NKG2DIL7-chimeric antigen receptor (CAR T cells display enhanced antitumor activity in vitro. (**a**) Domain architecture of engineered NKG2D-CAR and NKG2DIL7-CAR constructs. (**b**) Cytotoxic activity of NKG2D-CAR or non-transduced T cells against prostate cancer cell lines was determined by Annexin-V staining. B16-F10 melanoma cells served as negative target cell control. The effector cells were co-cultured for 4 h with targeT cells at E:T ratio of 1:3, 1:1 and 3:1, respectively. (**c,d**) Cytotoxic assays were determined by Annexin-V staining at 16 h of co-culture of NKG2D-CAR or NKG2DIL7-CAR T with PC-3 at E:T ratios of 3:1,1:1 and 1:3. E (**c**), PC-3 tumor cell viability assay was performed after 72 h of co-culture with non-transduced T (NT), NKG2D-CAR or NKG2DIL7-CAR T cells at E:T ratio of 3:1 (**d**). (**e,f**) Flow cytometric analysis of CD69 and granzyme B in T cells after the stimulation of tumor cells. * $p < 0.05$, ** $p < 0.01$, *** $p < 0.001$, **** $p < 0.0001$. Data are representative of greater than three independent experiments.

To generate NKG2D-CAR T cells expressing IL-7 (NKG2DIL7-CAR T cells), we modified T cells with a lentivirus vector encoding *IL-7* on NKG2D-CAR backbone as shown in the schematic diagram (Lower panel of Figure 1a). IL-7 linked to NKG2D-CAR with 2A peptide could be secreted outside the cells. T cells separated from PBMCs were transduced with lentivirus NKG2DIL7-CAR. Transduction efficiency was determined by FACS analysis (Additional file 2: Figure S2). To determine the killing ability of NKG2DIL7-CAR T cells against prostate cancer cells, prostate cancer cell line PC-3 was used as targeT cells and the cytotoxicity assay was performed at different E:T. The results showed that both of two CAR T cells had a significant cytotoxic effect on PC-3 cells and NKG2DIL7-CAR T cells exhibited better antitumor efficacy than conventional NKG2D-CAR T cells at 16 h (Figure 1c) and 72 h (Figure 1d), demonstrating that the killing capacity of NKG2D-based CAR T cells could be enhanced by co-expressing of IL-7.

We next explored the expression of CD69, a sensitive activation marker for T-cell function [19,20]. A higher level of CD69-positive cells was observed in both types of CAR T cells compared to non-transduced T cells in response to PC-3 tumor cells. However, a higher level of CD69 expression was detected in NKG2DIL7-CAR T cells (Figure 1e).

Furthermore, granzyme B is also pivotal for cytolytic function of CAR T cells [21,22]. The results demonstrated that NKG2D-CAR T cells produced more granzyme B than non-transduced T cells when co-cultured with targeT cells and transgenic expression of IL-7 into conventional NKG2D-CAR T cells could significantly enhance the expression of granzyme B (Figure 1f).

2.2. Co-Expression of IL-7 Enhances the Proliferation of NKG2D-CAR T cells

To validate the expression of IL-7, NKG2DIL7-CAR, NKG2D-CAR and non-transduced T cells were cultured in media with or without tumor cells for 24 h. The supernatants were collected to determine the secretion of IL-7. We found that NKG2DIL7-CAR T cells produced a relatively greater amount of IL-7 compared with conventional CAR T cells in the absence of a tumor (Figure 2a). Surprisingly, a robust increase of IL-7 expression was observed in NKG2DIL7-CAR T cells when co-cultured with PC-3 cells. These results indicated that the production of IL-7 was dependent on the presence of target cells.

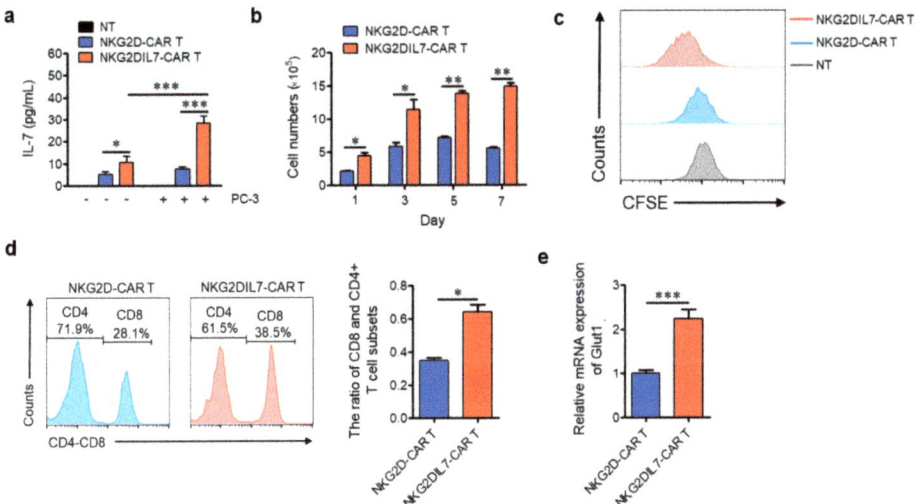

Figure 2. Co-expression of IL-7 enhances the proliferation of NKG2D-CAR T cells. (**a**) NKG2D-CAR or NKG2DIL7-CAR T cells were cultured in the absence or presence of PC-3 tumor cells at E:T ratio of 3:1 for 24 h without any exogenous cytokines, and the co-culture supernatants were detected for concentrations of IL-7 by ELISA. (**b**) Expansion of NKG2D-CAR and NKG2DIL7-CAR T cells after stimulation with tumor cells. The number of initial CAR T cells was 2.5×10^5, and cell numbers were measured by Vi-CELL every other day. (**c**) Non-transduced, NKG2D-CAR and NKG2DIL7-CAR T cells were labeled with 5(6)-carboxyfluorescein diacetate succinimidyl ester (CFSE) before being stimulated by PC-3 tumor cells, the dilution of CFSE was determined by flow cytometry after 7 days of co-culture. (**d**) The flow cytometric analysis of the percentage and ratio of $CD8^+$ and $CD4^+$ T cells in vitro on 7th day after stimulation, the initial CD4 and CD8 percentages were the same. (**e**) Effector and targeT cells were co-cultured at E:T ratio of 2:1 for 24 h, and the expression of Glut1 was measured by quantitative real-time PCR. * $p < 0.05$, ** $p < 0.01$, *** $p < 0.001$. Data are representative of 4 donors and have been gated on $NKG2D^+$ T cells.

To define the effect of IL-7 on the proliferation of NKG2D-CAR T cells, non-transduced T cells and two types of transduced CAR T cells were stimulated with targeT cells (PC-3). Cell numbers were measured by Vi-CELL every other day and CFSE-based assay was performed to assess the proliferation ability. The results showed that co-expression of IL-7 induced significantly greater expansion of the numbers of CAR T cells when compared with its counterpart on day 1, day 3 until day 7 (Figure 2b). Moreover, NKG2DIL7-CAR T cells displayed greater proliferation potential after 7 days of culture compared with NKG2D-CAR and non-transduced T cells (Figure 2c). The data of T cell subsets of $CD4^+$ and $CD8^+$ T cells revealed a pronounced increase in $CD8^+$ T cell numbers, resulting in an obvious increase in CD8/CD4 ratio in NKG2DIL7-CAR transduced T cells compared with NKG2D-CAR T cells (Figure 2d).

IL-7 can promote glucose transporter (Glut1) trafficking and glucose uptake to support cell survival [23]. To determine whether overexpression of IL-7 likewise affected Glut1 expression in NKG2D-CAR T cells, CAR T cells were stimulated with antigen cells for 24 h and mRNA level of Glut1 was detected (Figure 2e). Compared with NKG2D-CAR T cells, a higher mRNA level of Glut1 was detected in NKG2DIL7-CAR T cells, demonstrating that higher Glut1 transport may be an important factor for IL-7 to improve the survival of NKG2D-CAR T cells.

2.3. Transgenic Expression of IL-7 Reduces the Apoptosis of NKG2D-CAR T cells

The role of IL-7 in survival and anti-apoptosis was verified in several studies previously [17,24]. To figure out whether co-expression of IL-7 enhanced the anti-apoptosis ability of CAR T cells, NKG2D-CAR and NKG2DIL7-CAR T cells were cocultured with PC-3 tumor cells without IL-2 for 7 days and the apoptosis cells was detected by Annexin-V/7AAD staining. The results showed that the cells apoptotic rate in NKG2DIL7-CAR T cells (36 ± 8%) was significantly lower than NKG2D-CAR T cells (81.1 ± 15%) (Figure 3a).

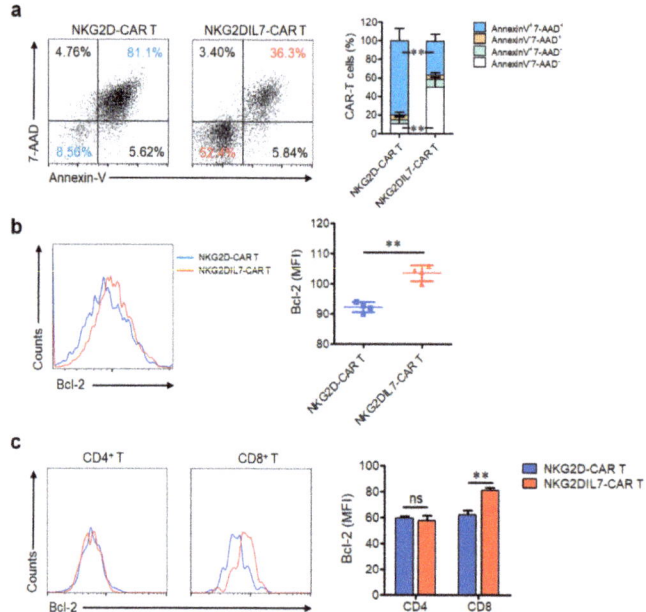

Figure 3. Co-expression of IL-7 reduces the apoptosis of NKG2D-CAR T cells. (**a**) After NKG2D-CAR and NKG2DIL7-CAR T cells were cocultured with PC-3 tumor cells for 7 days, the cells were stained with Annexin-V/7-AAD and the apoptosis was detected by FACS. The statistical analysis was shown in the right panel. (**b**) Flow cytometric analysis of Bcl-2 protein expression in CAR T cells 7 days after tumor cell stimulation. (**c**) The Bcl-2 expression in $CD4^+$ and $CD8^+$ CAR T cells was detected by FACS 7 days after tumor cell stimulation. ** $p < 0.01$. Data are representative of 4 donors and have been gated on $NKG2D^+$ T cells.

Bcl-2 is a downstream protein of IL-7-mediated signal pathways [24,25]. The up-regulation of Bcl-2 is related to improved anti-apoptosis and survival ability [26]. FACS analysis of Bcl-2 expression illustrated that NKG2DIL7-CAR T cells expressed significantly higher amounts of anti-apoptotic protein Bcl-2 compared with NKG2D-CAR T cells (Figure 3b). In addition, further analysis of Bcl-2 expression in T cell subsets showed that there was no difference in $CD4^+$ T cells, but a significantly higher expression was observed in $CD8^+$ NKG2DIL7-CAR T cells (Figure 3c).

2.4. IL-7 Preserves Less Differentiated Cell Phenotype and Inhibits the Exhaustion of CAR T cells

T cells usually exist several phenotypes, which have distinct proliferation, survival, and effector capabilities [27–29]. To determine whether overexpression of IL-7 influenced the differentiation of CAR T cells, NKG2D-CAR and NKG2DIL7-CAR T cells were cultured for 14 days, and were stained with CD45RA and CCR7 to analyze T cell subsets. A comparable proportion of $CCR7^+CD45RA^+$ subset in $CD4^+$ T cells, but a remarkable increased CCR7 and CD45RA double-positive cells in $CD8^+$ T cells were observed after co-expression of IL-7 (Figure 4a,b). T cells that express CD45RA and CCR7 are a group of less differentiated cells, which correlate with CAR-T cell expansion, survival and long-term persistence [30]. Our results suggested that IL-7 could preserve a less differentiated phenotype of CD8+ T cells, which might be beneficial for the future clinical application of CAR T cell therapy.

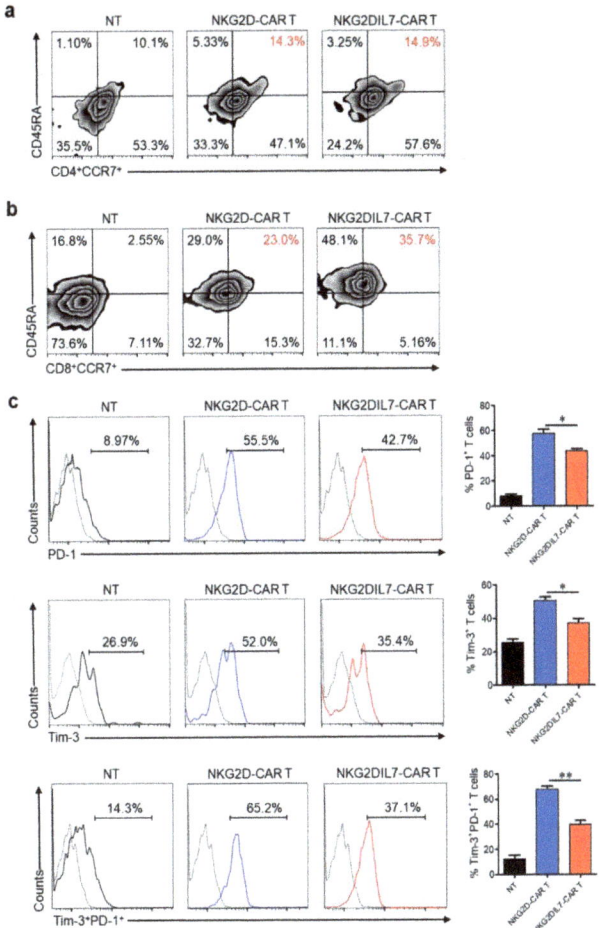

Figure 4. IL-7 increases CAR T cells with the less differentiated phenotype and inhibits CAR T-cell exhaustion. (**a,b**) T-cell phenotype based on CD45RA and CCR7 expression in CD4+ (**a**) and CD8+ (**b**) CAR T cells was analyzed by FACS after 14 days of culture. (**c**), Surface expression of PD-1 and Tim-3 on CAR T cells was analyzed by FACS after 7 days of co-culture with tumor cells. * $p < 0.05$, ** $p < 0.01$. Data are representative of four T-cell lines.

Cancer cells can inhibit the functions of T cells by expressing the ligands of inhibitory receptors such as PD-1 and Tim-3 [31,32], leading to T cell exhaustion. After 7 days of co-cultured with tumor cells, NKG2DIL7-CAR T cells exhibited lower expression of PD-1 and Tim-3, especially in Tim3+ PD1+ expression (Figure 4c), suggesting that IL-7 could prevent CAR-T cell exhaustion and protect T cells from deleterious immunosuppressive actions of tumor cells.

2.5. NKG2D-CAR T cells Expressing IL-7 Have Improved Antitumor Activity against Xenograft Prostate Tumor Model

Finally, we evaluated the ability of two types of CAR T cells against PC-3 prostate tumor cells in a xenograft mouse model. NSG mice were engrafted s.c. with 2×10^6 tumor cells. Once the tumor volumes had reached approximately 150–200 mm^3 size, non-transduced T cells, NKG2D-CAR and NKG2DIL7-CAR T cells (1×10^7 cells/mouse) were injected i.v. into the tumor-bearing mice (Figure 5a). All the mice of NT group had to be euthanized due to large tumor volume on day 18, NKG2D-CAR T cells produced remarkable antitumor ability in vivo compared with NT group, and the survival rate of tumor-bearing mice was more than 80% (Figure 5b). Furthermore, the tumor volume and weight of the group treated with NKG2D-CAR T cells reduced significantly compared to the control group. Interestingly, the tumor volume and weight were lower in the group treated with NKG2DIL7-CAR T cells than that with NKG2D-CAR T cells (Figure 5c,d), demonstrating that NKG2D-CAR T-cell therapy on prostate cancer could be an efficient method and introduction of IL-7 into NKG2D-CAR T cells can enhance antitumor ability in vivo.

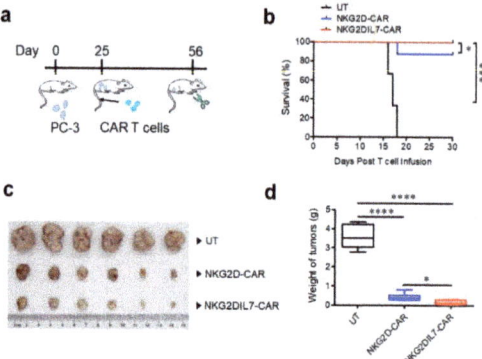

Figure 5. Engineering expression of IL-7 enhances the antitumor activity of NKG2D-CAR T cells in the xenograft prostate tumor model. (**a**) A schematic of in vivo experiment. (**b**) Kaplan-Meier survival analysis of PC-3 challenged mice after treatment with CAR-T cells. (**c,d**) The photograph of residual tumors (**c**) and the tumor weight (**d**) of the mice treated with NKG2D-CAR and NKG2DIL7-CAR T cells at the endpoint of the experiments. * $p < 0.05$, *** $p < 0.001$, **** $p < 0.0001$, $n = 6$.

The sera of mice were collected to determine the IL-7 cytokine level, the results showed that a high level of IL-7 was detected in the NKG2DIL7-CAR treatment group but not in the other two groups (Figure 6a). T cells in the blood of mice were analyzed by flow cytometry, the data indicated that the proportion of T cells in the group treated with NKG2DIL7-CAR T cells was significantly higher than that with NKG2D-CAR T cells (Figure 6b). The proportion of CD4$^+$ and CD8$^+$ T cells in blood was also analyzed by FACS and more CD8$^+$ T cells have been detected in the mice treated with NKG2DIL7-CAR T cells (Figure 6c).

To analyze the persistence and accumulation of CAR T cells at the tumor site, the tumors were excised after the treatment and the histopathological analysis was performed. H&E staining results showed that there were more T cells in the tumor sections of the mice treated with NKG2DIL7-CAR T cells compared with the conventional CAR-T group (Figure 6d). Furthermore, immunohistochemistry staining analysis indicated that there were more CD8$^+$ T cells in the tumor sections of the NKG2DIL7-CAR treatment group (Figure 6e), all of which were consistent with the results in vitro.

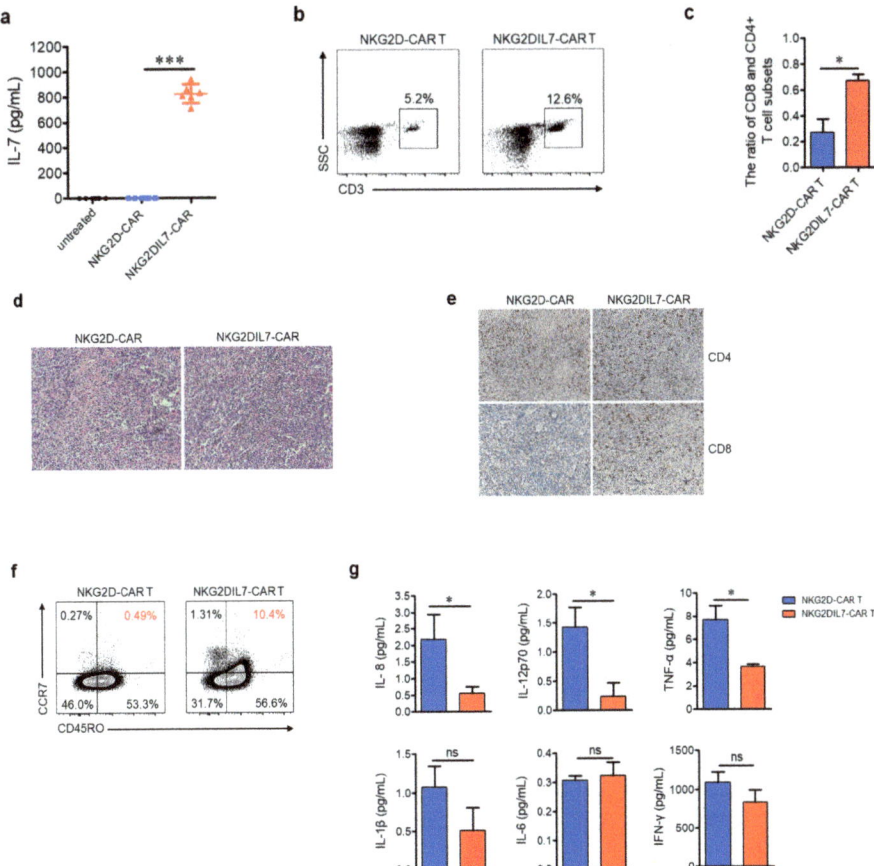

Figure 6. Engineering expression of IL-7 increases T cell accumulation and central memory-T cell subsets in the tumor site. (**a**) The level of IL-7 in sera of mice was detected by ELISA. (**b,c**) Detection of human T cells and the subsets of CD4$^+$ and CD8$^+$ T cells in the blood 25 days after CAR T cells adoptive transfer. (**d,e**) Tumor tissues were removed from the mice 25 days after treatment with CAR-T cells, and each tissue was divided into two parts. One part was stained with H&E (**d**), the other part was used for immunohistochemistry (IHC) (**e**). For IHC, combinations of anti-CD4 antibody and anti-CD8 antibody. The images were obtained under ×100 magnification. (**f**) Detection of Tcm phenotype of CAR T cells in the tumor site at the endpoint of the experiments. (**g**) The levels of IL-1β, IL-6, IL-8, IL12p70, IFN-γ and TNF-α in mouse serum were evaluated by a CBA kit. ns, not significant, * $p < 0.05$, *** $p < 0.001$. Data shown are representative of 6 mice per group from 2 independent experiments.

To address whether IL-7 affected the T cell phenotype in vivo, CD45RO and CCR7 on T cells, the markers of central memory T cells (Tcm), were examined by flow cytometry. The results revealed a higher proportion of Tcm-phenotypic cells in the tumor tissue from the mice treated with NKG2DIL7-CAR T cells (Figure 6f), suggesting that IL-7 might be beneficial to the formation of central memory T cells. Moreover, there were fewer cytokines such as IL-8, IL-12, TNF-α in the blood from the mice treated with NKG2DIL7-CAR T cells (Figure 6g). Collectively, our in vivo results elucidated that NKG2D-based CAR T cells could effectively kill prostate cancer cells, and co-expression of IL-7 could enhance the antitumor function of NKG2D-CAR T cells against prostate cancer.

3. Discussion

For treating B-cell malignancies, the second and third-generation CAR T cells have gained much success [14,33], but its efficacy to treat solid tumors still remains insufficient [3,4,34], particularly due to poor proliferation and survival of CAR T cells in vivo. NKG2D-based CAR T cells for immunotherapies have been reported to be promising for targeting NKG2D ligand-positive cancers [11,35]. In addition, NKG2D ligands are also expressed in tumor blood vessels, myeloid cells, immunosuppressive cells (such as Tregs and MDSCs) and endothelial cells in the tumor microenvironment [36], indicating that NKG2D-CAR could not only target tumors but also other cells in the microenvironment that promote tumor progression. To the best of our knowledge, there were few reports of targeting prostate cancer with NKG2D-CAR T cells. Here, we found that NKG2D ligands were highly expressed on human prostate cancer cell lines, and our designed NKG2D-based CAR T cells could effectively recognize and lyse NKG2D ligand-positive prostate cancer cells.

CAR T-cell immunotherapy for prostate cancer is extremely promising [37,38], but a major challenge that needs to be addressed is enhancing the proliferation and survival of CAR T cells in the highly immunosuppressive microenvironment. Previous studies have characterized that incorporation of CD137 (4-1BB) signaling domain into CARs rather than CD28 domain can improve the persistence and antitumor ability in vivo [39–41]. Nevertheless, in the highly immunosuppressive microenvironment of metastatic prostate cancer, the second generation of CAR with a 4-1BB signaling domain may not provide sufficient co-stimulation for T cells. Therefore, co-expression of a cytokine such as IL-7 in CAR T cells may be a valid way to improve the persistence and antitumor activity of CAR T cells in vitro and in vivo.

In the present study, there was no major difference in IL-7 production between NKG2D-CAR and NKG2DIL7-CAR T cells in the absence of tumor, but when co-cultured with PC-3 cells, IL-7 expression in NKG2DIL7-CAR-transduced T cells increased vigorously. Thus, our current approach provided cytokine IL-7 directly to the tumor site, which could possibly avoid toxicities reported with high dose systemic cytokine administration [42,43]. Previously, it has been reported that the constitutive expression of IL-7 has led to increased T-cell accumulation in vitro by enhancing T cell proliferation and survival. The author demonstrated that overall enhanced tumor rejection was due to improved T cell expansion rather than upregulation of effector function [44]. In another study, it has been reported that transgenic expression of IL-7 could also enhance the effector function of CAR T cells both in vitro and in vivo [45]. Here we validated that transgenic expression of IL-7 could increase the effector function as well as promote the expansion at the tumor site. The expression of T cell exhaustion markers such as PD-1 and TIM3 were also downregulated by IL-7 transgenic expression, which also pointed towards enhanced T cell effector function.

NKG2DIL7-CAR T cells exhibited greater cell numbers and cell viability than conventional NKG2D-CAR T cells on day 7. Our analysis indicated that co-expression of IL-7 could enhance both proliferation and survival of NKG2D-CAR T cells. Elevated expression of Bcl-2, an anti-apoptotic protein [46], and Glut1 in NKG2DIL7-CAR T cells further supported our claim. In vitro expression analysis of Bcl-2 in CD4 and CD8 populations further revealed that expansion in T cell numbers was mainly due to a rise in $CD8^+$ T cells, but not $CD4^+$ T cells. Less differentiated $CD8^+$ T cell subsets i.e., naïve and T memory stem cells have been recognized recently to be critical for expansion, survival and long-term persistence in vivo [30,47]. Interestingly, in the present study $CD8^+$ T cells displayed $CD45RA^+$ $CCR7^+$ phenotype in vitro, which was also beneficial for in vivo antitumor efficacy.

Consistent with in vitro results, there were more CAR T cells, especially $CD8^+$ T cells, in peripheral blood from the mice treated with NKG2DIL7-CAR T cells, which might be due to a more plentiful $CD8^+$ population after in vitro culture. Moreover, the histological analysis indicated that CAR T cells could infiltrate into the tumor tissues and there was more T cell infiltration, especially $CD8^+$ T cells, in tumor tissues treated with NKG2DIL7-CAR T cells. Increased $CD8^+$ T cells will be more effective for adoptive T cell immunotherapy in prostate cancer patients [48].

We found increased expression of CD45RO and CCR7 with transgenic expression of IL-7 in vivo, representing central memory T cell (Tcm) phenotype. As we know, Tcm is beneficial for adoptive T cell transfer because it provides instant antitumor immunity to patients and endows them with immune memory to prevent cancer recurrence [28]. Interestingly, there were fewer cytokines such as IL-8, IL-12, TNF-α in the blood of the mice treated with NKG2DIL7-CAR T cells, indicating that NKG2DIL7-CAR T therapy was relatively safe.

The limitation in our study is that human NKG2D-based CAR we used does not recognize murine NKG2D ligands, so the associated potential toxicity in the current mouse model could not be assessed. Besides, the effects of IL-7 for NKG2D CAR T cell functions were evaluated in the absence of Tregs, MDSCs, M2-macrophages and immune checkpoint molecules which created an immune-suppressive environment in cancer tissues. In our study, there is no evidence that IL-7 plays a role in immuno-suppressive microenvironment in prostate cancer tissues of the patients. However, several reports have shown that IL-7 is able to antagonize the immunosuppressive network to improve immune function on cancer cells [49]. For example, Treg cells, which have low expression of CD127 and high level of CD132 on their surface, accumulate in the tumor microenvironment and inhibit immune responses. IL-7 can directly abrogate the Treg-mediated suppression of effector T cell proliferation [50] and decrease the population of Tregs in the lung cancer model [51]. Therefore, we suppose that NKG2D-CAR T cells expressing IL-7 would have the capacity to persist in the immuno-suppressive microenvironment in prostate cancer tissues and induce potent antitumor immunity in patients, but more evidence is needed.

4. Materials and Methods

4.1. Cell Lines and Culture

The immortalized human embryonic kidney (HEK) -293T cell line was purchased from the American Type Culture Collection (ATCC, Manassas, VA, USA) and used for lentivirus packaging. Prostate cancer cell lines (PC-3, DU 145 and C4-2) were kindly provided by Dr. Zhengfang Yi (East China Normal University, Shanghai, China). The mouse melanoma cell line, B16-F10, was used as an antigen-negative control. The cell lines were cultured and maintained in DMEM (GIBCO, Waltham, MA, USA) supplemented with 10% heat-inactivated FBS, 100 IU/mL penicillin, and 100 mg/mL streptomycin at 37 °C in a humidified atmosphere containing 5% CO_2.

4.2. CAR Construction and Lentivirus Production

We constructed two different human codon-optimized second-generation CARs, which have a specificity against NKG2D ligands on tumors. NKG2D-CAR lentiviral vector consisted of the extracellular portion of human NKG2D (aa 82-216), linked to a CD8α hinge-transmembrane domain along with CD3ζ and 4-1BB signaling domains as described previously [52,53]. Human IL-7 gene (GenBank NM_000880.4) was fused with the NKG2D-CAR by foot-and-mouth disease virus 2A ribosomal skipping sequence to generate NKG2DIL7-CAR.

High titer replication-defective lentivirus particles were generated by using HEK-293T lentivirus packaging cell line along with packaging plasmid vectors (psPAX2, pMD2.G). On the day before transfection, HEK-293T cells were seeded in a 10 cm culture dish and when the cell fusion rate reached to 80–90%, CAR-encoding vector, psPAX2 and pMD2.G were transfected into HEK-293T cells at a ratio of 5:5:3 with the help of polyethyleneimine (MW 25000) (Polysciences, Warrington, PA, USA) transfection system [54]. Supernatant harvested at 48 h and 72 h post-transfection were concentrated by ultracentrifugation (Beckman Coulter, Brea, CA, USA) for 2.5 h at 25,000 rpm at 4 °C. Viruses were aliquoted and stored at −80 °C until used for experiments. All the experiments performed in this study were from the concentrated virus stock.

4.3. T-Cell Isolation, Modification and Culture

Peripheral blood mononuclear cells (PBMCs) were isolated from healthy volunteer donor cord blood after informed consent under the protocol approved by East China Normal University Internal Review Board. Primary human $CD4^+$ and $CD8^+$ T cells were positively selected from PBMCs with the CD4 and CD8 MicroBeads (Miltenyi, Bergisch Gladbach, Germany) following the manufacturer's instructions. Then, $CD4^+$ and $CD8^+$ T cells were mixed and activated for 48 h using the T Cell TransAct™ (Miltenyi) with the recommended titer of 1:100. Activated T cells were transduced with lentiviruses expressing NKG2D-CAR or NKG2DIL7-CAR supplemented with polybrene (10 µg/mL) and centrifuged for 1 h at 1800 rpm, 28 °C, and then incubated overnight. After 12 h of post-transduction, T cells were washed and cultured in X-VIVO™ 15 medium (Lonza, Switzerland) in the presence of human recombinant IL-2 (rhIL-2, 50 U/mL) and incubated for 2 days. Transduction efficiency was monitored by flow cytometry with APC anti-human NKG2D antibody staining. No exogenous cytokines were added in subsequent experiments.

4.4. Flow Cytometry and Antibodies

The following cell surface fluorochrome-conjugated monoclonal antibodies were used to detect T cells phenotype: FITC anti-human CD4, PE/Cy7 anti-human CD8, PE anti-human CD3 (Biolegend, San Diego, CA, USA) and APC anti-human NKG2D (eBioscience). T cell activity was determined by staining the surface APC anti-human CD69, intracellular PE anti-human IFN-γ (BD Biosciences, San Diego, CA, USA), PE anti-human GzmB (eBioscience, San Diego, CA, USA) and PE anti-human Bcl-2 (Biolegend) antibodies. PE anti-human PD-1 and APC anti-human Tim-3 were purchased from Biolegend. FITC anti-human CD45RA, Percp-cy5.5 anti-human CCR7 and PE anti-human CD45RO were purchased from Biolegend to determine T cell subsets. APC Annexin-V and 7-Aminoactinomycin D (7-ADD) from BD Biosciences were used for apoptosis staining. APC mouse IgG1, κ isotype, PE mouse IgG1 isotype, PE/Cy7 mouse IgG2a, κ isotype, FITC mouse IgG1 isotype, PE/Cy7 mouse IgG1, κ isotype (Biolegend) were used as controls. Flow cytometry was performed on the BD LSRFortessa flow cytometer, and data were analyzed using FlowJo Version 10 software.

4.5. Cytotoxicity Assays

To evaluate the cytotoxicity of CAR T cells, prostate cancer cell lines were used as targeT cells when the viability was >95% on the Vi-CELL counting machine (Beckman Coulter). The targeT cells were labeled with CFSE (eBioscience), and co-cultured with NKG2D-CAR T cells or NKG2DIL7-CAR T cells at an effector: target ratio of 3:1, 1:1 and 1:3. No exogenous cytokines were added. Cytotoxicity was measured as the percentage of apoptotic target cells.

Additional cytotoxicity of NKG2D-CAR and NKG2DIL7-CAR T cells was measured by the Vi-CELL counting machine (Beckman Coulter). After the targeT cells were co-cultured with effector cells at an effector: target ratio of 3:1 for 72 h, the numbers of alive cells were counted.

4.6. Cytokine Assay

To measure the production level of IL-7, non-transduced T cells, NKG2D-CAR or NKG2DIL7-CAR T cells (3×10^5 cells per well) were cultured in 24-well plates with and without $NKG2DLs^+$ PC-3 cells in an effector to target ratio of 2:1 without the addition of exogenous cytokines. After 24 h of co-culture, supernatants were harvested to measure IL-7 with enzyme-linked immunosorbent assay kit (Invitrogen, Carlsbad, CA, USA).

To detect the in vivo level of cytokine production, the mouse blood was collected and clotted at 4 °C, the sera were used to determine INF-γ, TNF-α, IL-6, IL-8, IL-1β and IL-12p70 with Human Inflammatory Cytokine Kit (BD Biosciences) according to the manufacturer's instruction.

4.7. T-Cell Proliferation, Survival and Apoptosis Assay

To assess T-cell growth, survival and apoptosis upon antigen exposure, non-transduced T cells, NKG2D-CAR or NKG2DIL7-CAR T cells were labeled with CFSE (2 µM) and co-cultured with the targeT cells PC-3 in a ratio E:T (3:1) without the addition of exogenous cytokines. After 7 days of co-culture, CFSE dilution was measured for T cell division. Annexin-V/7-amino-actinomycin (7-ADD) staining was used to determine the apoptotic rate and the expression of Glut-1, Bcl-2 and exhaustion markers such as PD-1 and Tim-3 was detected by FACS analysis.

4.8. Quantitative Real-Time PCR

Effector and targeT cells were co-cultured in E:T (2:1) for 24 h and the total RNAs were extracted from cultured cells using Trizol reagent (Takara) and reverse-transcribed using Reverse Transcription Kit (Prime Script First Strand cDNA Synthesis kit R047A, Takara). Reverse-transcribed single-stranded DNA was subjected to quantitative real-time PCR (q-PCR) using SYBR green master mix and amplified on Light Cycle (Agilent). Each experiment was performed in a duplicate manner and relative expression was calculated using the $2^{-\Delta\Delta Ct}$ change-in-cycling-threshold method with GAPDH as a reference. Primers were designed using primer 5 as follows: Glut1, sense, 5′-ATTGGCTCCGGTATCGTCAAC-3′, antisense, 5′-GCTCAGATAGGACATCCAGGGTA-3′; GAPDH, sense, 5′-AGGTCGGTGTGAACGGATTTG-3′, antisense, 5′-TGTAGACCATGTAGTTGAGGTCA-3′.

4.9. Xenograft Model

Female 6- to 8-week-old NOD/SCID/γ-chain$^{-/-}$ (NSG) mice, purchased from Beijing Biocytogen Co., Ltd., were raised, treated and maintained in a non-specific pathogenic environment under the approval of the Animal ethics Committee of East China Normal University. To establish a prostate cancer model, mice were inoculated subcutaneously with 2×10^6 PC-3 cells on its right flank and defined it as day 0. Mice were observed regularly, and their tumor dimensions were measured with calipers. When the tumor burden reached approximately 150–200 mm^3, animals were randomly divided into three study groups ($n = 6$) and injected intravenously (i.v.) with 200 µL of T cells, NKG2D-CAR and NKG2DIL7-CAR T cells (1×10^7 cells/mouse). No exogenous cytokines were injected in the mice. The magnitudes of tumors were measured by caliper and the volume of tumors was calculated using formula V = π/6 × (length × width^2), where length is the largest longitudinal diameter and width is the largest transverse diameter. Mouse survival was observed and mice were sacrificed when the tumor burden reached a size of 1500–2000 mm^3.

4.10. Histopathological Analysis

To assess histopathological changes, tumor tissues were fixed with 4% paraformaldehyde and embedded in paraffin wax. Tissues were sliced into 4 µm-thick tumor sections and then stained with hematoxylin/eosin (H&E) for visualization of the tissue structure. For the IHC staining assay, the tissue sections were deparaffinated and incubated with 5% bovine serum albumin for 60 min. Then the tissue sections were incubated with anti-CD3 antibody (1:100, Servicebio, GB130144-M), anti-CD4 antibody (1:100, Servicebio, GB13064) or anti-CD8 antibody (1:100, Servicebio, GB13068) for 2 h, respectively. After incubation, 3% H_2O_2 was used to eliminate the activity of endogenous peroxidase and Goat anti-rabbit lgG was used as a secondary antibody. 3,3′-diaminobenzidinetetrahydrochloride (DAB-4HCl) was used to visualize the CD3, CD4 and CD8 expression. Images were acquired using a Nikon Eclipse 80i microscope (Nikon, Badhoevedorp, The Netherlands). Tissue evaluation was performed by two independent examiners and semi quantitated by image J Software.

4.11. Statistical Analysis

The data were reported as mean ± SD. Statistical analysis was performed using unpaired Student's t-test or ANOVA. Survival was plotted using a Kaplan–Meier survival curve and statistical significance was determined by the Log-rank (Mantel–Cox) test. Prism software version 6.0 (GraphPad) was used for statistical calculation and *p*-value < 0.05 was accepted as indicating a significant difference.

4.12. Ethics Approval and Consent to Participate

All fresh blood was collected under a protocol approved by the Ethics Committee of East China Normal University, following written informed consent. All animal studies were approved by the Institutional Animal Care and Use Committee of East China Normal University (Approval No. m20170224).

5. Conclusions

In conclusion, our study emphasized the potential of NKG2D-based CAR T cells as a promising therapeutic option for prostate cancer. The incorporation of stimulation cytokine IL-7 could further improve the expansion and antitumor effects of NKG2D-CAR T cells and increase its potential clinical applicability. In addition, the strategy of co-expression of IL-7 can also be used to modify other targeted CAR T cells and to the treatment of other solid tumors.

Supplementary Materials: The following are available online at http://www.mdpi.com/2072-6694/12/7/1969/s1, Figure S1: Expression level of NKG2D ligands. Q-PCR (a) and flow cytometry analysis (b) of the expression of NKG2D ligands on prostate cancer cell lines of PC-3, DU 145 and C4-2. The same cells stained with isotype antibody were used for gating (black). Data shown are representatives of experiments with similar results, Figure S2: Expression level of NKG2D CAR. Human T cells were transduced with lentiviruses expressing NKG2D-CAR or NKG2DIL7-CAR and the representative flow cytometry plot was indicated. Gating was based on the same cells stained with isotype-matched antibody. Data shown are representatives of experiments with similar results.

Author Contributions: C.H. and W.J. conceived and designed the study. C.H., Y.Z., Z.L., H.Z., L.Z., L.T., J.Y., B.D. and W.J. analyzed and interpreted the data. C.H., Y.Z., H.Z., L.Z., L.T. and J.Y. performed the experiments. C.H., M.A.F., I.A., M.L. and W.J. wrote and revised the manuscript. All authors read and approved the final manuscript.

Funding: This work was supported by National Natural Science Foundation of China (81771306, 81072459, 81830083), National Key Research and Development Program of China (2016YFC1200400), Program for New Century Excellent Talents in University (NCET-12-0179), Science and Technology Commission of Shanghai Municipality (201409002900,14140904200). ECNU Public Platform for innovation (011).

Acknowledgments: We would like to thank Ying Zhang for her technological assistance. We also acknowledge Shanghai Bioray Laboratories Inc. and the Flow Cytometry Core Facility.

Conflicts of Interest: The authors declare no potential conflict of interest.

Abbreviations

7-AAD	7-Aminoactinomycin D;
CAR	chimeric antigen receptor;
E:T ratio	effector-to-target ratio;
Glut1	glucose transporter 1;
GzmB	granzyme B;
IFN-γ	interferon gamma;
mCRPC	metastatic castration-resistant prostate cancer;
NKG2D	natural killer group 2 member D;
NKG2DLs	NKG2D ligands;
NT	non-transduced T cells;
PBMC	peripheral blood mononuclear cells;
q-PCR	quantitative real-time PCR;
TNF-α	tumor necrosis factor alpha.

References

1. Ahmed, N.; Brawley, V.S.; Hegde, M.; Robertson, C.; Ghazi, A.; Gerken, C.; Liu, E.; Dakhova, O.; Ashoori, A.; Corder, A.; et al. Human Epidermal Growth Factor Receptor 2 (HER2) -Specific Chimeric Antigen Receptor-Modified T cells for the Immunotherapy of HER2-Positive Sarcoma. *J. Clin. Oncol.* **2015**, *33*, 1688–1696. [CrossRef] [PubMed]
2. Feng, K.; Guo, Y.; Dai, H.; Wang, Y.; Li, X.; Jia, H.; Han, W. Chimeric antigen receptor-modified T cells for the immunotherapy of patients with EGFR-expressing advanced relapsed/refractory non-small cell lung cancer. *Sci. China Life Sci.* **2016**, *59*, 468–479. [CrossRef] [PubMed]
3. Kakarla, S.; Gottschalk, S. CAR T cells for solid tumors: Armed and ready to go? *Cancer J.* **2014**, *20*, 151–155. [CrossRef] [PubMed]
4. Rabinovich, G.A.; Gabrilovich, D.; Sotomayor, E.M. Immunosuppressive strategies that are mediated by tumor cells. *Annu. Rev. Immunol.* **2007**, *25*, 267–296. [CrossRef] [PubMed]
5. Siegel, R.L.; Miller, K.D.; Jemal, A. Cancer statistics, 2020. *CA Cancer J. Clin.* **2020**, *70*, 7–30. [CrossRef]
6. Dong, L.; Zieren, R.C.; Xue, W.; de Reijke, T.M.; Pienta, K.J. Metastatic prostate cancer remains incurable, why? *Asian J. Urol.* **2019**, *6*, 26–41. [CrossRef]
7. Comiskey, M.C.; Dallos, M.C.; Drake, C.G. Immunotherapy in Prostate Cancer: Teaching an Old Dog New Tricks. *Curr. Oncol. Rep.* **2018**, *20*, 75. [CrossRef]
8. Raulet, D.H.; Gasser, S.; Gowen, B.G.; Deng, W.; Jung, H. Regulation of ligands for the NKG2D activating receptor. *Annu. Rev. Immunol.* **2013**, *31*, 413–441. [CrossRef]
9. Sentman, C.L.; Meehan, K.R. NKG2D CARs as cell therapy for cancer. *Cancer J.* **2014**, *20*, 156–159. [CrossRef] [PubMed]
10. Salih, H.R.; Antropius, H.; Gieseke, F.; Lutz, S.Z.; Kanz, L.; Rammensee, H.; Steinle, A. Functional expression and release of ligands for the activating immunoreceptor NKG2D in leukemia. *Blood* **2003**, *102*, 1389–1396. [CrossRef] [PubMed]
11. Demoulin, B.; Cook, W.J.; Murad, J.; Graber, D.J.; Sentman, M.; Lonez, C.; Gilham, D.E.; Sentman, C.L.; Agaugue, S. Exploiting natural killer group 2D receptors for CAR T-cell therapy. *Future Oncol.* **2017**, *13*, 1593–1605. [CrossRef] [PubMed]
12. Baumeister, S.H.; Murad, J.; Werner, L.; Daley, H.; Negre, H.T.; Gicobi, J.K.; Schmucker, A.; Reder, J.; Sentman, C.L.; Gilham, D.E.; et al. Phase I Trial of Autologous CAR T cells Targeting NKG2D Ligands in Patients with AML/MDS and Multiple Myeloma. *Cancer Immunol. Res.* **2019**, *7*, 100–112. [CrossRef]
13. Koneru, M.; Purdon, T.J.; Spriggs, D.; Koneru, S.; Brentjens, R.J. IL-12 secreting tumor-targeted chimeric antigen receptor T cells eradicate ovarian tumors in vivo. *OncoImmunology* **2015**, *4*, e994446. [CrossRef] [PubMed]
14. Hoyos, V.; Savoldo, B.; Quintarelli, C.; Mahendravada, A.; Zhang, M.; Vera, J.; Heslop, H.E.; Rooney, C.M.; Brenner, M.K.; Dotti, G. Engineering CD19-specific T lymphocytes with interleukin-15 and a suicide gene to enhance their anti-lymphoma/leukemia effects and safety. *Leukemia* **2010**, *24*, 1160–1170. [CrossRef] [PubMed]
15. Jicha, D.L.; Mule, J.J.; Rosenberg, S.A. Interleukin 7 generates antitumor cytotoxic T lymphocytes against murine sarcomas with efficacy in cellular adoptive immunotherapy. *J. Exp. Med.* **1991**, *174*, 1511–1515. [CrossRef] [PubMed]
16. Rochman, Y.; Spolski, R.; Leonard, W.J. New insights into the regulation of T cells by gamma(c) family cytokines. *Nat. Rev. Immunol.* **2009**, *9*, 480–490. [CrossRef] [PubMed]
17. Jiang, Q.; Li, W.Q.; Aiello, F.B.; Mazzucchelli, R.; Asefa, B.; Khaled, A.R.; Durum, S.k. Cell biology of IL-7, a key lymphotrophin. *Cytokine Growth FR* **2005**, *16*, 513–533. [CrossRef]
18. Mackall, C.L.; Fry, T.J.; Gress, R.E. Harnessing the biology of IL-7 for therapeutic application. *Nat. Rev. Immunol.* **2011**, *11*, 330–342. [CrossRef]
19. Cibrian, D.; Sanchez-Madrid, F. CD69: From activation marker to metabolic gatekeeper. *Eur. J. Immunol.* **2017**, *47*, 946–953. [CrossRef]
20. Gonzalez-Amaro, R.; Cortes, J.R.; Sanchez-Madrid, F.; Martin, P. Is CD69 an effective brake to control inflammatory diseases? *Trends Mol. Med.* **2013**, *19*, 625–632. [CrossRef]

21. Devadas, S.; Das, J.; Liu, C.; Zhang, L.; Roberts, A.I.; Pan, Z.; Moore, P.A.; Das, G.; Shi, Y. Granzyme B is critical for T cell receptor-induced cell death of type 2 helper T cells. *Immunity* **2006**, *25*, 237–247. [CrossRef] [PubMed]
22. Voskoboinik, I.; Whisstock, J.C.; Trapani, J.A. Perforin and granzymes: Function, dysfunction and human pathology. *Nat. Rev. Immunol.* **2015**, *15*, 388–400. [CrossRef] [PubMed]
23. Wofford, J.A.; Wieman, H.L.; Jacobs, S.R.; Zhao, Y.; Rathmell, J.C. IL-7 promotes Glut1 trafficking and glucose uptake via STAT5-mediated activation of Akt to support T-cell survival. *Blood* **2008**, *111*, 2101–2111. [CrossRef] [PubMed]
24. Lin, J.; Zhu, Z.; Xiao, H.; Xiao, H.; Wakefield, M.R.; Ding, V.A.; Bai, Q.; Fang, Y. The role of IL-7 in Immunity and Cancer. *Anticancer Res.* **2017**, *37*, 963–967. [PubMed]
25. Soares, M.V.; Borthwick, N.J.; Maini, M.K.; Janossy, G.; Salmon, M.; Akbar, A.N. IL-7-dependent extrathymic expansion of CD45RA+ T cells enables preservation of a naive repertoire. *J. Immunol.* **1998**, *161*, 5909–5917.
26. Vogler, M.; Walter, H.S.; Dyer, M.J.S. Targeting anti-apoptotic BCL2 family proteins in haematological malignancies—From pathogenesis to treatment. *Br. J. Haematol.* **2017**, *178*, 364–379. [CrossRef]
27. Golubovskaya, V.; Wu, L. Different Subsets of T cells, Memory, Effector Functions, and CAR-T Immunotherapy. *Cancers* **2016**, *8*, 36. [CrossRef]
28. Yang, S.; Gattinoni, L.; Liu, F.; Ji, Y.; Yu, Z.; Restifo, N.P.; Rosenberg, S.A.; Morgan, R.A. In vitro generated anti-tumor T lymphocytes exhibit distinct subsets mimicking in vivo antigen-experienced cells. *Cancer Immunol. Immun.* **2011**, *60*, 739–749. [CrossRef]
29. Neeson, P.; Shin, A.; Tainton, K.M.; Guru, P.; Prince, H.M.; Harrison, S.J.; Peinert, S.; Smyth, M.J.; Trapani, J.A.; Kershaw, M.H.; et al. Ex vivo culture of chimeric antigen receptor T cells generates functional CD8+ T cells with effector and central memory-like phenotype. *Gene Ther.* **2010**, *17*, 1105–1116. [CrossRef]
30. Xu, Y.; Zhang, M.; Ramos, C.A.; Durett, A.; Liu, E.; Dakhova, O.; Liu, H.; Creighton, C.J.; Gee, A.P.; Heslop, H.E.; et al. Closely related T-memory stem cells correlate with in vivo expansion of CAR.CD19-T cells and are preserved by IL-7 and IL-15. *Blood* **2014**, *123*, 3750–3759. [CrossRef]
31. Hou, L.; Jie, Z.; Liang, Y.; Desai, M.; Soong, L.; Sun, J. Type 1 interferon-induced IL-7 maintains CD8+ T-cell responses and homeostasis by suppressing PD-1 expression in viral hepatitis. *Cell Mol. Immunol.* **2015**, *12*, 213–221. [CrossRef]
32. Pellegrini, M.; Calzascia, T.; Toe, J.G.; Preston, S.P.; Lin, A.E.; Elford, A.R.; Shahinian, A.; Lang, P.A.; Lang, K.S.; Morre, M.; et al. IL-7 engages multiple mechanisms to overcome chronic viral infection and limit organ pathology. *Cell* **2011**, *144*, 601–613. [CrossRef]
33. Till, B.G.; Jensen, M.C.; Wang, J.; Qian, X.; Gopal, A.K.; Maloney, D.G.; Lindgren, C.G.; Lin, Y.; Pagel, J.M.; Budde, L.E.; et al. CD20-specific adoptive immunotherapy for lymphoma using a chimeric antigen receptor with both CD28 and 4-1BB domains: Pilot clinical trial results. *Blood* **2012**, *119*, 3940–3950. [CrossRef]
34. Li, W.; Song, X.; Jin, Y.; Li, F.; Yu, H.; Cao, C.; Jiang, Q. CARTs for Solid Tumors: Feasible or Infeasible? *Oncol. Res. Treat.* **2017**, *40*, 540–546. [CrossRef] [PubMed]
35. Han, Y.; Xie, W.; Song, D.G.; Powell, D.J., Jr. Control of triple-negative breast cancer using ex vivo self-enriched, costimulated NKG2D CAR T cells. *J. Hematol. Oncol.* **2018**, *11*, 92. [CrossRef] [PubMed]
36. Frazao, A.; Rethacker, L.; Messaoudene, M.; Avril, M.; Toubert, A.; Dulphy, N.; Caignard, A. NKG2D/NKG2-ligand pathway offers new opportunities in cancer treatment. *Front. Immunol.* **2019**, *10*, 661. [CrossRef]
37. Hillerdal, V.; Essand, M. Chimeric antigen receptor-engineered T cells for the treatment of metastatic prostate cancer. *BioDrugs* **2015**, *29*, 75–89. [CrossRef] [PubMed]
38. Zuccolotto, G.; Fracasso, G.; Merlo, A.; Montagner, I.M.; Rondina, M.; Bobisse, S.; Figini, M.; Cingarlini, S.; Colombatti, M.; Zanovello, P.; et al. PSMA-specific CAR-engineered T cells eradicate disseminated prostate cancer in preclinical models. *PLoS ONE* **2014**, *9*, e109427. [CrossRef] [PubMed]
39. Imai, C.; Mihara, K.; Andreansky, M.; Nicholson, I.C.; Pui, C.; Geiger, T.L.; Campana, D. Chimeric receptors with 4-1BB signaling capacity provoke potent cytotoxicity against acute lymphoblastic leukemia. *Leukemia* **2004**, *18*, 676–684. [CrossRef]
40. Milone, M.C.; Fish, J.D.; Carpenito, C.; Carroll, R.G.; Binder, G.K.; Teachey, D.; Samanta, M.; Lakhal, M.; Gloss, B.; Danet-Desnoyers, G.; et al. Chimeric receptors containing CD137 signal transduction domains mediate enhanced survival of T cells and increased antileukemic efficacy in vivo. *Mol. Ther.* **2009**, *17*, 1453–1464. [CrossRef]

41. Lee, H.W.; Nam, K.O.; Park, S.J.; Kwon, B.S. 4-1BB enhances CD8+ T cell expansion by regulating cell cycle progression through changes in expression of cyclins D and E and cyclin-dependent kinase inhibitor p27kip1. *Eur. J. Immunol.* **2003**, *33*, 2133–2141. [CrossRef]
42. Rosenberg, S.A.; Restifo, N.P.; Yang, J.C.; Morgan, R.A.; Dudley, M.E. Adoptive cell transfer: A clinical path to effective cancer immunotherapy. *Nat. Rev. Cancer* **2008**, *8*, 299–308. [CrossRef] [PubMed]
43. Dudley, M.E.; Rosenberg, S.A. Adoptive-cell-transfer therapy for the treatment of patients with cancer. *Nat. Rev. Cancer* **2003**, *3*, 666–675. [CrossRef] [PubMed]
44. Shum, T.; Omer, B.; Tashiro, H.; Kruse, R.L.; Wagner, D.l.; Parikh, K.; Yi, Z.; Sauer, T.; Liu, D.; Parihar, R.; et al. Constitutive Signaling from an Engineered IL7 Receptor Promotes Durable Tumor Elimination by Tumor-Redirected T cells. *Cancer Discov.* **2017**, *7*, 1238–1247. [CrossRef]
45. Markley, J.C.; Sadelain, M. IL-7 and IL-21 are superior to IL-2 and IL-15 in promoting human T cell-mediated rejection of systemic lymphoma in immunodeficient mice. *Blood* **2010**, *115*, 3508–3519. [CrossRef] [PubMed]
46. Vella, A.T.; Dow, S.; Potter, T.A.; Kappler, J.; Marrack, P. Cytokine-induced survival of activated T cells in vitro and in vivo. *Proc. Natl. Acad. Sci. USA* **1998**, *95*, 3810–3815. [CrossRef] [PubMed]
47. Gattinoni, L.; Lugli, E.; Ji, Y.; Pos, Z.; Paulos, C.M.; Quigley, M.F.; Almeida, J.R.; Gostick, E.; Yu, Z.; Carpenito, C.; et al. A human memory T cell subset with stem cell-like properties. *Nat. Med.* **2011**, *17*, 1290–1297. [CrossRef] [PubMed]
48. Fluxa, P.; Rojas-Sepulveda, D.; Gleisner, M.A.; Tittarelli, A.; Villegas, P.; Tapia, L.; Rivera, M.T.; Lopez, M.N.; Catan, F.; Uribe, M.; et al. High CD8(+) and absence of Foxp3(+) T lymphocytes infiltration in gallbladder tumors correlate with prolonged patients survival. *BMC Cancer* **2018**, *18*, 243. [CrossRef]
49. Pellegrini, M.; Calzascia, T.; Elford, A.R.; Shahinian, A.; Lin, A.E.; Dissanayake, D.; Dhanji, S.; Nguyen, L.T.; Gronski, M.A.; Morre, M.; et al. Adjuvant IL-7 antagonizes multiple cellular and molecular inhibitory networks to enhance immunotherapies. *Nat. Med.* **2009**, *15*, 528–536. [CrossRef]
50. Heninger, A.K.; Theil, A.; Wilhelm, C.; Petzold, C.; Huebel, N.; Kretschmer, K.; Bionifacio, E.; Monti, P. IL-7 abrogates suppressive activity of human CD4+CD25+FOXP3+ regulatory T cells and allows expansion of alloreactive and autoreactive T cells. *J. Immunol.* **2012**, *189*, 5649–5658. [CrossRef]
51. Andersson, A.; Yang, S.C.; Huang, M.; Zhu, L.; Kar, U.K.; Batra, R.K.; Elashoff, D.; Strieter, R.M.; Dubinett, S.M.; Sharma, S. IL-7 promotes CXCR3 ligand-dependent T cell antitumor reactivity in lung cancer. *J. Immunol.* **2009**, *182*, 6951–6958. [CrossRef] [PubMed]
52. Tao, K.; He, M.; Tao, F.; Xu, G.; Ye, M.; Zheng, Y.; Li, Y. Development of NKG2D-based chimeric antigen receptor-T cells for gastric cancer treatment. *Cancer Chemoth. Pharm.* **2018**, *82*, 815–827. [CrossRef] [PubMed]
53. Song, D.G.; Ye, Q.; Santoro, S.; Fang, C.; Best, A.; Powell, D.J., Jr. Chimeric NKG2D CAR-expressing T cell-mediated attack of human ovarian cancer is enhanced by histone deacetylase inhibition. *Hum. Gene Ther.* **2013**, *24*, 295–305. [CrossRef] [PubMed]
54. Kuroda, H.; Kutner, R.H.; Bazan, N.G.; Reiser, J. Simplified lentivirus vector production in protein-free media using polyethylenimine-mediated transfection. *J. Virol. Methods.* **2009**, *157*, 113–121. [CrossRef] [PubMed]

© 2020 by the authors. Licensee MDPI, Basel, Switzerland. This article is an open access article distributed under the terms and conditions of the Creative Commons Attribution (CC BY) license (http://creativecommons.org/licenses/by/4.0/).

Article

CD19-CAR-T Cells Bearing a KIR/PD-1-Based Inhibitory CAR Eradicate CD19⁺HLA-C1⁻ Malignant B Cells While Sparing CD19⁺HLA-C1⁺ Healthy B Cells

Lei Tao, Muhammad Asad Farooq, Yaoxin Gao, Li Zhang, Congyi Niu, Iqra Ajmal, Ying Zhou, Cong He, Guixia Zhao, Jie Yao, Mingyao Liu * and Wenzheng Jiang *

Shanghai Key Laboratory of Regulatory Biology, School of Life Sciences, East China Normal University, Shanghai 200241, China; 52161300026@stu.ecnu.edu.cn (L.T.); 52181300056@stu.ecnu.edu.cn (M.A.F.); 52181300027@stu.ecnu.edu.cn (Y.G.); 52171300027@stu.ecnu.edu.cn (L.Z.); 51171300098@stu.ecnu.edu.cn (C.N.); 52171300059@stu.ecnu.edu.cn (I.A.); 51181300130@stu.ecnu.edu.cn (Y.Z.); 52161300027@stu.ecnu.edu.cn (C.H.); 52191300040@stu.ecnu.edu.cn (G.Z.); 52201300031@stu.ecnu.edu.cn (J.Y.)

* Correspondence: myliu@bio.ecnu.edu.cn (M.L.); wzjiang@bio.ecnu.edu.cn (W.J.); Tel.: +86-21-54341035 (W.J.); Fax: +86-21-54341006 (W.J.)

Received: 3 August 2020; Accepted: 10 September 2020; Published: 13 September 2020

Simple Summary: CD19-targeted chimeric antigen receptor (CAR) T (CD19-CAR-T) cell therapy usually causes B cell aplasia because of "on-target off-tumor" toxicity. The aim of the study was to assess the concept that the introduction of an inhibitory CAR (iCAR) into CAR-T cells could alleviate the side effect of CD19-CAR-T cell therapy. The results showed that CD19-CAR-T cells with a novel KIR (killer inhibitory receptor) /PD-1 (programmed death receptor-1)-based inhibitory CAR (iKP-19-CAR-T) exhibited more naïve, less exhausted phenotypes and preserved a higher proportion of central memory T cells (T_{CM}). Furthermore, iKP-19-CAR-T cells exerted the similar level of cytotoxicity on CD19⁺HLA-C1⁻ Burkitt's lymphoma cells compared to CD19-CAR-T cells while sparing CD19⁺HLA-C1⁺ healthy human B cells both in vitro and in the xenograft model. Our data demonstrates that the KIR/PD-1-based inhibitory CAR can be a promising strategy to avoid B cell aplasia caused by CD19-CAR-T cell therapy.

Abstract: B cell aplasia caused by "on-target off-tumor" toxicity is one of the clinical side effects during CD19-targeted chimeric antigen receptor (CAR) T (CD19-CAR-T) cells treatment for B cell malignancies. Persistent B cell aplasia was observed in all patients with sustained remission, which increased the patients' risk of infection. Some patients even died due to infection. To overcome this challenge, the concept of incorporating an inhibitory CAR (iCAR) into CAR-T cells was introduced to constrain the T cells response once an "on-target off-tumor" event occurred. In this study, we engineered a novel KIR/PD-1-based inhibitory CAR (iKP CAR) by fusing the extracellular domain of killer cell immunoglobulin-like receptors (KIR) 2DL2 (KIR2DL2) and the intracellular domain of PD-1. We also confirmed that iKP CAR could inhibit the CD19 CAR activation signal via the PD-1 domain and CD19-CAR-T cells bearing an iKP CAR (iKP-19-CAR-T) exerted robust cytotoxicity in vitro and antitumor activity in the xenograft model of CD19⁺HLA-C1⁻ Burkitt's lymphoma parallel to CD19-CAR-T cells, whilst sparing CD19⁺HLA-C1⁺ healthy human B cells both in vitro and in the xenograft model. Meanwhile, iKP-19-CAR-T cells exhibited more naïve, less exhausted phenotypes and preserved a higher proportion of central memory T cells (T_{CM}). Our data demonstrates that the KIR/PD-1-based inhibitory CAR can be a promising strategy for preventing B cell aplasia induced by CD19-CAR-T cell therapy.

Keywords: CD19-CAR-T; B cell aplasia; KIR; PD-1; inhibitory CAR

1. Introduction

CD19-CAR-T (CD19-targeted chimeric antigen receptor T-) cells are the first cell therapy products for the treatment of relapsed or refractory B cell acute lymphoblastic leukemia (B-ALL) that were approved by US Food and Drug Administration (FDA) in 2017 [1,2]. Since then, CD19-CAR-T has brought a gigantic revolution in the field of immunotherapy because of the high percentage rate of complete remission (CR) in several blood-related malignancies [3–6]. While some challenges increase the risk of treatment failures, such as an "on-target off-tumor" adverse event instigating B cell aplasia, i.e., CD19-CAR-T cells kill all healthy B cells because of CD19 expressed in all B cells [7–9]. B cell aplasia contributes to hypogammaglobulinemia, which is one of the main factors leading to infection in patients [10]. Although these patients were administrated an intravenous immunoglobulin to maintain IgG levels, concomitant bacterial, viral, and fungal infections were still observed [11,12].

At present, some strategies have been developed to overcome the "on-target off-tumor" effect. Since the ideal specific target is almost non-existent in reality, it is a good idea to target structurally differentiated proteins. For example, the engineered CAR-T cells targeting the integrin β7 activated conformation was specifically effective against multiple myeloma (MM) without damaging normal hematopoietic cells [13]. Similarly, recognition of the Tn-glyco form of MUC1 by engineering CAR-T cells exhibited target-specific cytotoxicity to cellular adenocarcinoma [14]. According to another strategy, splitting 4-1BB domain and CD3ζ domain and fusing them together with two different single chain fragment variable regions (scFv) respectively, T cells would be entirely active if only two antigens were recognized at the same time [15]. Likewise, the same result was found using the "And gate" approach, in which logical control of CAR-T cells responses needed two different antigen engagements [16–18]. Although these strategies can reduce the incidence of the "on-target off-tumor" effect, they are not universal and not easy to implement.

In contrast to the strategies outlined earlier, inhibitory CAR (iCAR) is a versatile and implementable solution to subdue the "on-target off-tumor" effect by providing a negative signal to regulate T cell activation. In 2013, Fedorov et al. developed a PD-1-based iCAR strategy and they provided a proof of concept that CAR-T cells expressing an iCAR could discriminate between off target cells and target cells and functioned in a temporary and reversible manner [19].

KIRs are the most important inhibitory receptors expressed predominantly in NK cells and a small subset of T cells [20]. They can dampen the activation of NK cells after interacting with the human leukocyte antigen (HLA) ligands expressed on the surface of normal cells [21,22]. Tumor cells downregulate HLA to escape from the T cells immune surveillance [23–25]. Data from the Human Protein Atlas Database demonstrate that HLA-C is low or non-expressed on most tumor cell lines, but highly or moderately expressed in normal tissues. PD-1 is an inhibitory protein expressed in activated T cells to limit the excessive activation of T cells [26–28]. In this study, we engineered a novel iCAR consisting of the extracellular domain of KIR2DL2, CD8a hinge and transmembrane, and the intracellular domain of PD-1. This KIR/PD-1-based iCAR was termed as iKP CAR. We speculated that when KIR2DL2 recognized HLA-C1 on normal cells, iKP CAR would deliver a negative signal to inhibit T cells response via the PD-1 domain, meanwhile, iKP CAR did not work in the absence of HLA-C1 on tumor cells, so that iKP CAR could discriminate between normal cells (HLA-C1$^+$) and tumor cells (HLA-C1$^-$). Simultaneously, we hoped that CD19-CAR-T cells with an iKP CAR could eliminate CD19$^+$HLA-C1$^-$ malignant B cells, while reducing the damage to CD19$^+$HLA-C1$^+$ healthy B cells.

2. Results

2.1. iKP CAR Doesn't Affect CAR Expression, Viability, Proliferation and Subsets of CD19-CAR-T Cells

We designed iKP CAR by fusion of the extracellular domain of KIR2DL2 and the intracellular domain of PD-1 with CD8a hinge and transmembrane. Whereas, an iKP CAR with truncated PD-1 domain (named as iKPt CAR) was used as a negative control (upper panel of Figure 1A). Next, the commercially synthesized iKP/iKPt CAR was cloned into a vector expressing a CD19 CAR in which both CARs were separated by a T2A sequence (lower panel of Figure 1A). The bicistronic vector expressing CD19 CAR and iKP/iKPt CAR was used to package a lentivirus and transduced in Human Primary T cells from healthy donors to produce iKP-19-CAR-T/iKPt-19-CAR-T cells. Flow cytometry analysis showed that iKP-19-CAR-T/iKPt-19-CAR-T cells expressed analogous amounts of both CD19 CAR and iKP/iKPt CAR (Figure 1B), which ensured that iKP-19-CAR-T/iKPt-19-CAR-T cells would receive activation and suppression signals evenly. To study whether iKP/iKPt CAR affected characteristics of CD19-CAR-T cells, CD19-CAR-T and iKP-19-CAR-T/iKPt-19-CAR-T cells were cultured in X-VIVO media supplemented with 100 U/mL IL-2 for 14 days and the cells were analyzed by flow cytometry at different time points. We found that both CD19-CAR-T and iKP-19-CAR-T/iKPt-19-CAR-T cells displayed a similar expression level of CD19 CAR (Figure 1C), cell viability (Figure 1D), cell proliferation (Figure 1E) and proportion of CD8+ and CD4+ cells (Figure 1F). These results indicated that iKP CAR had no impact on CAR expression, viability, proliferation and subsets of CD19-CAR-T cells.

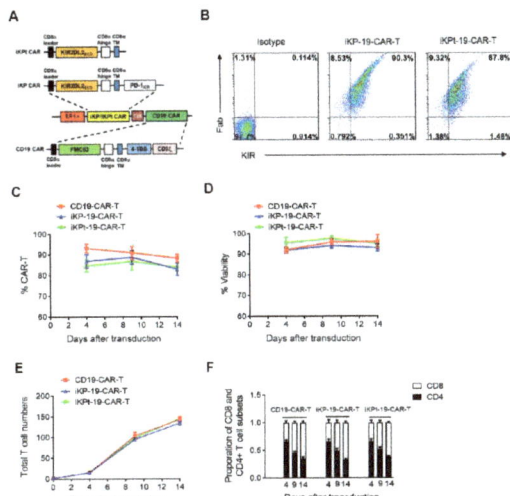

Figure 1. iKP CAR construction and expression. (**A**) Schematic diagram of the bicistronic vector expressing iKP/iKPt CAR and CD19 CAR. (**B–F**) iKP-19-CAR-T/iKPt-19-CAR-T cells or CD19-CAR-T cells were cultured for 14 days in X-VIVO media supplemented with 100U/mL IL-2. Representative iKP/iKPt CAR and CD19 CAR expression in iKP-19-CAR-T/iKPt-19-CAR-T cells were detected on day 4 by flow cytometry using PE-anti-human KIR antibody and Alexa Flour 647-anti-mouse F(ab')2 antibody (scFv of CD19 CAR is from mouse) ($n = 4$ different donors) (**B**). Detection of CD19 CAR-positive rate in iKP-19-CAR-T/iKPt-19-CAR-T and CD19-CAR-T on day 4, day 9 and day 14 by flow cytometry ($n = 4$ different donors) (**C**). Viability (**D**) or total cell numbers (**E**) of iKP-19-CAR-T/iKPt-19-CAR-T cells and CD19-CAR-T cells were also measured on day 4, day 9 and day 14 using Beckman Coulter counter ($n = 4$ different donors). Proportion of CD8+ and CD4+ T cell subsets in iKP-19-CAR-T/iKPt-19-CAR-T cells or CD19-CAR-T cells on day 4, day 9 and day 14 was measured using APC-anti-human CD8 antibody and PerCP-anti-human CD4 antibody ($n = 4$ different donors) (**F**). Three experiments were performed using PBMCs from each donor. Error bars represent ± SD.

2.2. iKP CAR Functions via PD-1 Signaling Upon Interacting with HLA-C1

To investigate whether iKP CAR could regulate the CD19 CAR signal through the intracellular PD-1 domain once it interacted with HLA-C1, Daudi cells (CD19$^+$HLA-C1$^-$) and Raji cells (CD19$^+$HLA-C1$^+$) were used as target cells and the presence of CD19 and HLA-C1 was analyzed by flow cytometry (Figure 2A). Next, CD19-CAR-T cells, iKP-19-CAR-T cells and iKPt-19-CAR-T cells were exposed to Daudi cells or Raji cells in RMPI-1640 medium after the CAR positive rate was unified. It was reported that PD-1 recruited SHP2 to dephosphorylate P-Zap70 to inhibit T cell activation [29,30]. In current study, the phosphorylated Zap70 (P-Zap70) was determined by flow cytometry six hours later. The results showed that the expression level of P-Zap70 in CD19-CAR-T cells, iKP-19-CAR-T cells, or iKPt-19-CAR-T cells was similar (Figure 2B) when exposed to Daudi cells, while the expression level of P-Zap70 in iKP-19-CAR-T cells was remarkably decreased compared to CD19-CAR-T cells or iKPt-19-CAR-T cells (Figure 2B) when exposed to Raji cells. The data indicated that in the absence of HLA-C1 (Daudi cells), iKP CAR would not affect the activation signal of CD19 CAR, however in the presence of HLA-C1 (Raji cells), iKP CAR would dephosphorylate P-Zap70 via intracellular PD-1 domain. Regardless of the presence of HLA-C1, iKPt CAR had no effect on the CD19 CAR activation signal, therefore we only compared the functional differences between iKP-19-CAR-T cells and CD19-CAR-T cells in further experiments.

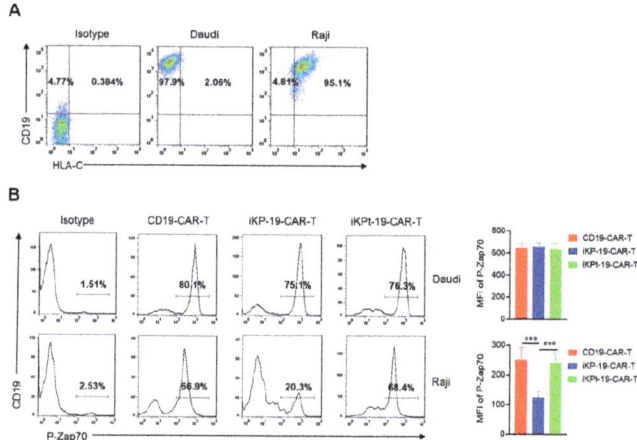

Figure 2. Dephosphorylating P-Zap70 by iKP CAR via intracellular PD-1 domain. (**A**) Flow cytometric analysis of CD19 and HLA-C1 expression in Daudi cells or Raji cells by using APC-anti-human CD19 and PE-anti-human HLA-C antibodies. (**B**) Expression analysis of P-Zap70 in different CAR-T cells by flow cytometry. iKP-19-CAR-T/iKPt-19-CAR-T cells and CD19-CAR-T cells were exposed to Daudi cells or Raji cells for 6 h at a 1:1 ratio in RPMI-1640 medium, stained with PE-anti-human P-Zap70 antibody and MFI of P-Zap70 was statistically analyzed ($n = 4$ different donors). All the experiments were conducted in triplicate manner using PBMCs from each donor. *** $p < 0.001$. Error bars represent ± SD. The CD19 CAR positive rate was unified using UT cells in all the co-culture experiments in this study.

2.3. iKP CAR Renders CD19-CAR-T Cells in Less Differentiated and Less Exhausted State Prior to Antigen Engagement

IL-2 activates T cells through PI3K-Akt-mTOR and MAPK signaling pathways [31,32], but high concentration of IL-2 in the media will cause excessive activation of T cells. PD-1 plays an opposite role to IL-2 also through PI3K-Akt-mTOR and MAPK signaling pathways to inhibit T cells activation and proliferation [33–35]. Since donor T cells expressed HLA-C1 as well (Supplementary Materials Figure S1), the interaction between iKP CAR and HLA-C1 on T cells could provide a negative PD-1 signal to suppress IL-2 induced T cells activation, which probably affected the properties of iKP-19-CAR-T cells different from CD19-CAR-T cells. Firstly, the differentiation status of CAR-T cells prior to antigen exposure was evaluated. Flow cytometry analysis for CCR7, CD45RO, and GzmB was performed in CD19-CAR-T cells and iKP-19-CAR-T cells. We found an increased CCR7 expression, decreased CD45RO and GzmB expression in iKP-19-CAR-T cells compared to CD19-CAR-T cells (Figure 3A). These results suggested that iKP-19-CAR-T cells were in a less differentiated state than CD19-CAR-T cells [35]. Furthermore, we observed that the percentage of T_{CM} (CD45RA$^-$CCR7$^+$) in iKP-19-CAR-T cells was higher than that in CD19-CAR-T cells (Figure 3B). Due to the high expression of Eomes and low expression of T-bet dedicated to T_{CM} [36,37], we analyzed these two transcription factors by flow cytometry and the results showed that iKP-19-CAR-T cells had a higher Eomes expression level and a lower T-bet expression level compared to CD19-CAR-T cells (Figure 3C,D). In addition, PD-1 expression in iKP-19-CAR-T cells was lower than that found in CD19-CAR-T cells (Figure 3E), while TIM-3 and LAG-3 expression showed no significant changes (data not shown). Therefore, our results demonstrated that the integration of iKP CAR into CD19-CAR-T cells leads to less differentiated and less exhausted T cell phenotypes.

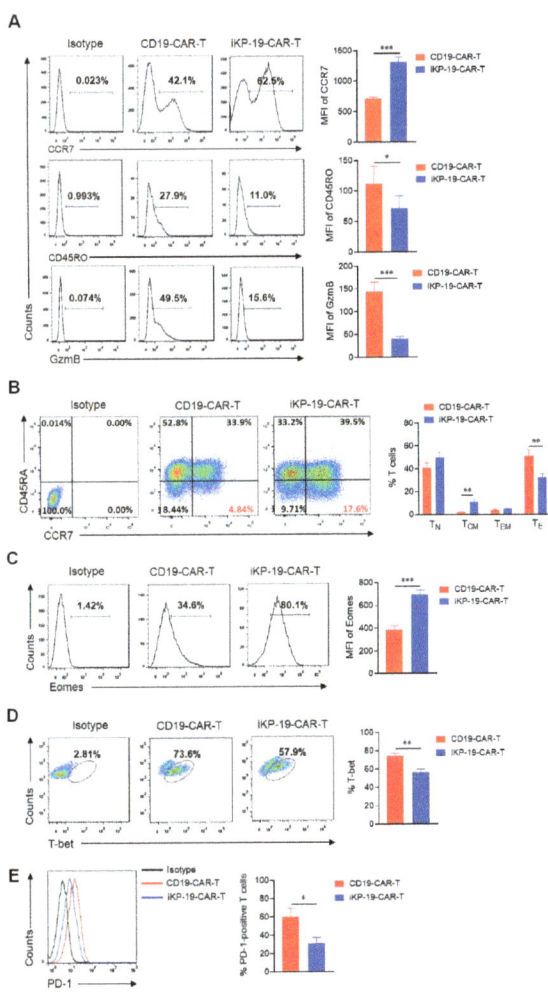

Figure 3. Characteristics of iKP-19-CAR-T cells and CD19-CAR-T cells. iKP-19-CAR-T cells or CD19-CAR-T cells were cultured for 10 days in X-VIVO media supplemented with 100 U/mL IL-2.

(**A**) The expression of T cell differentiation markers in CAR-T cells was analyzed by flow cytometry using PE/Cy7-anti-human CCR7 antibody, PE-anti-human-CD45RO antibody and PE-anti-human GzmB antibody ($n = 4$ different donors). (**B**) The frequency of naïve (T_N; CCR7$^+$CD45RA$^+$), T_{CM} (CCR7$^+$CD45RA$^-$), effector memory (T_{EM}; CCR7$^-$CD45RA$^-$) or effector (T_E; CCR7$^-$CD45RA$^+$) T cells were analyzed by flow cytometry using PE/Cy7-anti-human CCR7 antibody and FITC-anti-human CD45RA antibody ($n = 4$ different donors). (**C**) The expression of transcription factor Eomes in CAR-T cells was analyzed by flow cytometry using FITC-anti-human Eomes antibody ($n = 4$ different donors). (**D**) The expression of transcription factor T-bet in CAR-T cells was analyzed by flow cytometry using PE-anti-human T-bet antibody ($n = 4$ different donors). (**E**) The expression of T cell exhaustion marker PD-1 in CAR-T cells was analyzed by flow cytometry using FITC-anti-human PD-1 antibody ($n = 4$ different donors). All experiments were performed in triplicate manner using PBMCs from each donor and MFI or percentage was statistically analyzed. * $p < 0.05$, ** $p < 0.01$, *** $p < 0.001$. Error bars represent ± SD.

Although the KIR-PD-1 signal possibly attenuated the IL-2 signal, iKP CAR did not impair the cell proliferative capacity of CD19-CAR-T cells (Figure 1E), which meant that the T cells still had sufficient signal for proliferation. To investigate this, the expression level of P-Zap70 in iKP-19-CAR-T cells and CD19-CAR-T cells was analyzed by using flow cytometry. We found that there was no difference in the expression level of P-Zap70 between these two CAR-T cells (Supplementary Materials Figure S2).

2.4. CD19-CAR-T Cells Bearing an iKP CAR Eradicate CD19$^+$HLA-C1$^-$ Daudi Cells and Present Lower Cytotoxicity on CD19$^+$HLA-C1$^+$ Normal B Cells In Vitro

In order to study whether iKP-19-CAR-T cells could distinguish between malignant B cells and normal B cells in vitro, CD19$^+$HLA-C1$^+$ normal B cells from healthy donors were identified as one of the target cells (Figure 4A). Next iKP-19-CAR-T cells and CD19-CAR-T cells were co-cultured with Daudi cells or healthy B cells in RMPI-1640 medium for 6 h at a 1:1 ratio, respectively. T cell activation was assessed by using flow cytometry. The results revealed that the expression level of the activation marker CD69, the degranulation marker CD107a, and GzmB was similar in both iKP-19-CAR-T cells and CD19-CAR-T cells when co-culturing with Daudi cells. However, a significantly lower level of these molecules in iKP-19-CAR-T cells was observed when co-culturing with normal B cells (Figure 4B–D). Moreover, a lower level of P-Zap70 was observed in iKP-19-CAR-T cells compared to CD19-CAR-T cells when exposed to normal B cells (Figure 4E), but it had a similar expression level in both CAR-T cells when exposed to Daudi cells (Figure 2B). The data indicated that iKP CAR could constrain the activation of CD19-CAR-T cells effectively upon engagement of HLA-C1 on normal B cells.

To evaluate the cytotoxicity of iKP-19-CAR-T cells or CD19-CAR-T cells against different target cells, an LDH release assay was executed. At a different E:T ratio, iKP-19-CAR-T cells showed the same strong cytotoxicity on Daudi cells as CD19-CAR-T cells, a killing rate of almost 80% was observed at a 5:1 ratio (Figure 4F). However, the cytotoxicity of iKP-19-CAR-T cells on normal B cells was decreased dramatically compared to CD19-CAR-T cells, iKP CAR reduced the cytotoxicity by 51% at a 5:1 ratio, and the reduction effect was more pronounced at a lower E:T ratio (Figure 4F). The results suggested that the combination of iKP CAR and CD19 CAR could reduce the damage of CD19-CAR-T cells on CD19$^+$HLA-C1$^+$ normal B cells without decreasing the cytotoxicity on CD19$^+$HLA-C1$^-$ malignant B cells in vitro.

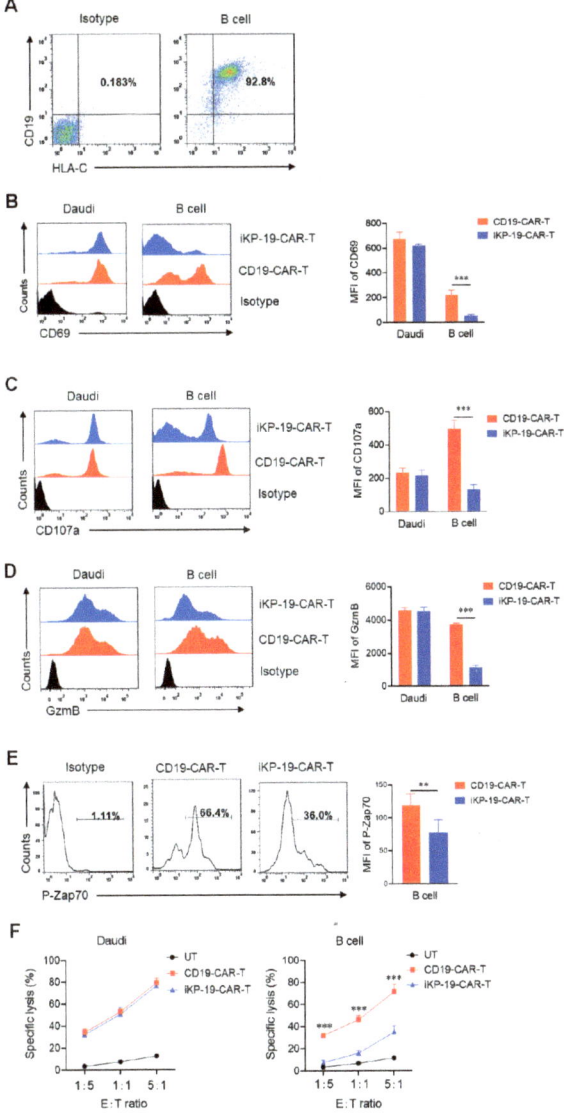

Figure 4. In vitro cytotoxicity of iKP-19-CAR-T cells against $CD19^+HLA$-$C1^-$ Daudi cells or $CD19^+HLA$-$C1^+$ normal B cells. (**A**) Representative CD19 and HLA-C1 expression in B cells. (**B**–**E**) After CD19 CAR rate was unified, iKP-19-CAR-T cells and CD19-CAR-T cells were co-cultured with Daudi cells or normal B cells at a 1:1 ratio for 6 h in RPMI-1640 medium. The expressions of activation marker CD69 (**B**), degranulation marker CD107a (**C**), GzmB (**D**), and signal molecule P-Zap70 (**E**) in CAR-T cells was detected by flow cytometry using PE-anti-human CD69 antibody, PE-anti-human CD107a antibody, PE-anti-human GzmB antibody and PE-anti-human P-Zap-70 antibody. MFI was statistically measured from three independent experiments (n = 4 different donors). (**F**) LDH assay was performed to test the cytotoxicity of iKP-19-CAR-T cells and CD19-CAR-T cells against Daudi cells or normal B cells after 6h co-culture at different E:T ratio (n = 4 different donors). All experiments were performed in triplicate manner using PBMCs of every donor. ** $p < 0.01$, *** $p < 0.001$. Error bars represent ± SD.

2.5. CD19-CAR-T Cells Bearing an iKP CAR Release Less Cytokines and Express Lower Exhaustion Markers during Lysing Malignant B Cells

Further, we measured the cytokines in media where CAR-T cells were cocultured with Daudi cells at a 1:1 ratio. The data showed that iKP-19-CAR-T cells released lower levels of cytokines including IL-6, IFN-γ and TNF-α compared to CD19-CAR-T cells (Figure 5A), which was beneficial to prevent cytokine release syndrome (CRS) [38,39]. Next, we tested the expression of surface markers on iKP-19-CAR-T cells or CD19-CAR-T cells and found that iKP-19-CAR-T cells expressed lower exhaustion markers of PD-1 and TIM-3 than CD19-CAR-T cells (Figure 5B). This data proved that CD19-CAR-T cells with an iKP CAR might have better properties than CD19-CAR-T cells.

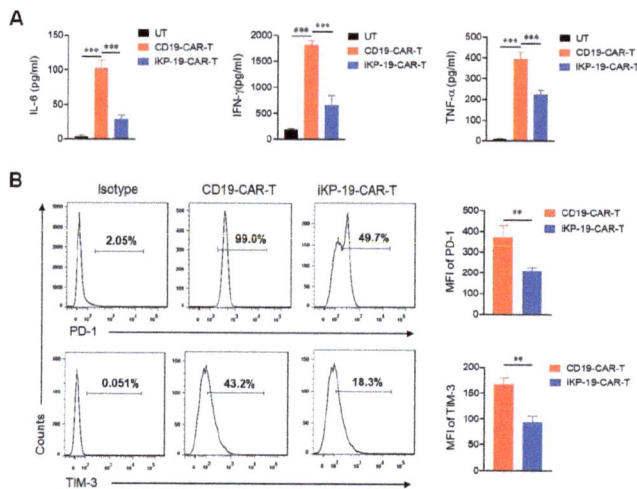

Figure 5. Cytokines release and exhaustion marker expression of iKP-19-CAR-T cells after coculture with Burkitt's lymphoma Daudi cells. iKP-19-CAR-T cells or CD19-CAR-T cells and Daudi cells were co-cultured in 96 well plate at a 1:1 ratio for 6 h in RPMI complete media. (**A**) Cell culture media were collected and IL-6, IFN-γ and TNF-α were determined by flow cytometry using CBA assay kit ($n = 4$ different donors). (**B**) CAR-T cells were collected and stained with FITC-anti-human PD-1 antibody and PE/Cy7-anti-human TIM-3 antibody to measure the expression of exhaustion markers PD-1 and TIM-3 ($n = 4$ different donors). MFI was statistically analyzed from three different experiments of each donor. ** $p < 0.01$, *** $p < 0.001$. Error bars represent ± SD.

2.6. CD19-CAR-T Cells Bearing an iKP CAR Discern CD19$^+$HLA-C1$^-$ Burkitt's Lymphoma Cell Line and CD19$^+$HLA-C1$^+$ Healthy B Cells In Vivo

To study the cytotoxicity of iKP-19-CAR-T cells on CD19$^+$HLA-C1$^-$ Burkitt's lymphoma cells or CD19$^+$HLA-C1$^+$ normal B cells in vivo, Daudi cells expressing luciferase were generated and were injected intravenously (i.v.) into six-week-old NOD-Prkdc^{em26Cd52}IL2rg^{em26Cd22}/Nju (NCG) mice via tail veins on day 0 and normal B cells were injected in the same way on day 6 (Figure 6A). The mice were divided into three separate groups ($n = 4$), and each group of mice received untransduced T cells (UT), CD19-CAR-T cells, or iKP-19-CAR-T cells intravenously. As shown in the IVIS imaging system, both iKP-19-CAR-T cells and CD19-CAR-T cells controlled B cell malignancy effectively as compared to UT cells (Figure 6B). When compared to the mice of the UT group, the total bioluminescence of tumors in mice of the iKP-19-CAR-T group or CD19-CAR-T group was decreased to a significantly lower level (Figure 6C). A 100% survival rate was recorded in those mice who received an iKP-19-CAR-T cell or CD19-CAR-T cell treatment on day 39, but all mice in the UT group died on day 30 (Figure 6D). On day 25, peripheral blood (PB) from mouse orbit was collected to analyze the persistence of normal B cells and cytokines release, respectively. Neither CD19$^+$HLA-C1$^-$ Daudi cells nor CD19$^+$HLA-C1$^+$ normal

B cells were detected in the CD19-CAR-T group (Figure 6E). However, certain quantities of normal B cells (1.54%) still existed but Daudi cells were not detected in the iKP-CD19-CAR-T group (Figure 6E). The results demonstrated that iKP-19-CAR-T cells could eliminate malignant B cells while still sparing normal B cells in vivo. More importantly, compared with CD19-CAR-T-treated mice, less cytokines such as IL-6, IFN-γband TNF-α in the sera from iKP-19-CAR-T-treated mice were detected (Figure 6F), which implied that iKP-19-CAR-T cells were safer than CD19-CAR-T cells. Furthermore, on day 32, we found that the mice treated with CD19-CAR-T cells had more T cell survival than the mice treated with iKP-19-CAR-T cells, but among the surviving T cells, the iKP-19-CAR-T-treated mice had a higher percentage of T_{CM} or less-differentiated cells (Figure 6G), which suggested that iKP-19-CAR-T cells would provide longer-term antitumor activities.

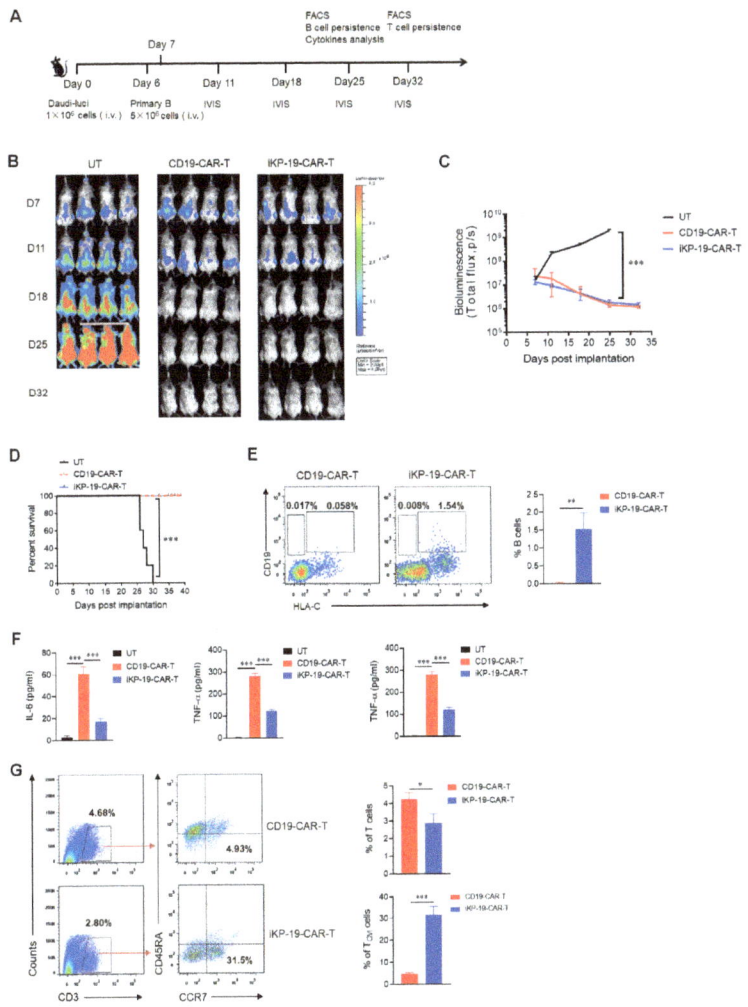

Figure 6. Controlling B cell malignancy effectively with sparing normal B cells in vivo of iKP-19-CAR-T cells. (**A**) Schematic representation of in-vivo experimental design. NCG mice (n = 4/group) were i.v. injected with 1×10^6 luciferase-expressed Daudi cells (Daudi-luc) on day 0. After 6 days, 5×10^6 normal B cells were administered. At day 7, 5×10^6 UT cells, CD19-CAR-T cells and iKP-19-CAR-T

cells were i.v. injected. IVIS imaging was performed to monitor tumor burden at day 11, 18, 25, 32. At day 25, normal B cells and cytokines in PB were analyzed. At day 32, T cells persistence was evaluated. (**B**) Representative bioluminescence images of Daudi-luc cells-derived tumor growth in the xenograft model. (**C**) Bioluminescence kinetics of Daudi-luc cells-derived tumor growth in the xenograft model. (**D**) Kaplan-Meier survival curve of mice. (**E**) Representative presence of Daui cells (CD19$^+$HLA-C1$^-$) or normal B cells (CD19$^+$HLA-C1$^+$) in mice at day 25 was determined by flow cytometry using APC-anti-human CD19 antibody and PE-anti-human HLA-C antibody. (**F**) Cytokine secretion of IL-6, IFN-γ and TNF-α in PB at day25 was measured by using CBA assay kit. (**G**) Flow cytometer analysis of total numbers of T cells and central memory T cells (T$_{CM}$) in different group of xenograft mice. T cells (CD3$^+$) or T$_{CM}$ (CCR7$^+$CD45RA$^-$) were detected from PB by using APC-anti CD3 antibody, FITC-anti-human CD45RA antibody and PE/Cy7-anti-human CCR7 antibody. All the experiments were performed with 4 mice per group. * $p < 0.05$, ** $p < 0.05$, *** $p < 0.001$. Error bars represent ± SD.

3. Discussion

CARs use scFv structure to recognize target antigens on cancer cells. The safety of CAR-T cells depends on the specificity of target antigens. However, most antigens are also expressed in normal cells, hence the "on-target off-tumor" effect is inevitable. This behavior of CAR-T cells causes severe side effects in the body systems expressing the target antigens [3,10,11,40]. Current clinical solutions are to use high-dose corticosteroid for treatment when "on-target off-tumor" events occur [41]. This immunosuppressive drug controls off-target toxicity at the cost of abolishing the T cells antitumor effect.

Fedorov et al. provided a model to elucidate the possibility to apply iCAR to regulate the function of CAR-T cells [19]. As we know, NK cells can discriminate between normal cells and abnormal cells that do not express adequate amounts of HLA such as cancer cells, virus-infected cells, etc [42]. Whether they are activated depends on the integrated signal of positive and negative signals. KIR is one of the important inhibitory receptors and exerts an inhibitory function to constrain NK cell response upon recognizing HLA on normal cells [43]. Based on the activation mechanism of NK cells, we fused the PD-1 intracellular signal domain and the extracellular recognition domain of KIR2DL2 to develop an iKP/PD-1-based iCAR (iKP CAR), and the data demonstrated that T cells co-expressing CD19 CAR and iKP CAR could discern between malignant B cells and normal B cells in vitro and in vivo.

Two factors are important for iCAR design. Firstly, the target of iCAR should be widely expressed on normal tissue cells, but rarely expressed on cancer cells. Obviously, HLA is an ideal target for iCAR. As HLA-C1 subtype has a high expression frequency in humans [44], we choose the extracellular domain of KIR2DL2 (whose ligand is HLA-C1) as the recognition domain of iCAR. Secondly, the intracellular signal domain should respond quickly and strongly in a transiently reversible manner once the extracellular recognition domain binds to the target. PD-1 is a powerful inhibitory molecule in T cells that dephosphorylates the TCR signal in a few hours after interacting with PD-L1 [45]. The dephosphorylation event is a dynamic and reversible process, which can ensure that T cell activity can restore during engagement of target cells. Therefore, we used the intracellular domain of PD-1 as the signaling domain of iCAR in the current study.

In our study, some characteristics of CD19-CAR-T cells were not affected by iKP CAR. Data showed that iKP-19-CAR-T cells had similar CAR transduction efficacy, cell viability, proliferation and CD8/CD4 ratio to CD19-CAR-T cells in X-VIVO media supplemented with IL-2. However, some characteristics were different. Prior to antigen engagement, iKP-19-CAR-T cells displayed lower differentiated (increased CCR7, decreased CD45RO and GzmB expression), less exhausted (lower PD-1 expression) phenotypes, and had an elevated proportion of T$_{CM}$ or less-differentiated cells. The reason might be that the negative signaling of iKP CAR suppressed the IL-2 signal and regulated the related gene expression (up-regulated Eomes expression and down-regulated T-bet expression), thereby inhibiting T cell differentiation and exhaustion simultaneously. A higher percentage of T$_{CM}$ or less differentiated cells in the peripheral blood from the mice treated with iKP-19-CAR-T cells had also been observed in vivo. Retrospective analysis from published CAR-T cell clinical studies had revealed that an elevated proportion of T_{CM} or less differentiated CAR-T cells provided superior antitumor efficacy [36]. Both in vitro and in vivo, we found that HLA-C1 on T cells did not reduce the toxicity of iKP-19-CAR-T cells on Daudi cells. Therefore, we speculated that iKP-19-CAR-T cells obtained an activation pattern similar to NK cells due to iCAR functioning in a temporary and reversible manner [19].

Subsequently, we demonstrated that the novel iKP CAR here had an ability to discern malignant B cells and normal B cells both in vitro and in vivo. Compared to CD19-CAR-T cells, iKP-19-CAR-T cells had an equivalent level of T cell activation, degranulation, and cytotoxic potential against Daudi cells, while all these were reduced significantly against normal B cells in vitro. Furthermore, we found that B cell malignancy could be controlled effectively by both CAR-T cells, but normal B cells were still detectable in the xenograft mice model treated with iKP-19-CAR-T cells while they could not be found in the mice treated with CD19-CAR-T cells. In contrast to CD19-CAR-T cells, the amount of CRS-associated cytokines IL-6, TNF-α and IFN-γ of iKP-19-CAR-T cells was decreased notably during lysing Daudi cells in vitro and this phenomenon was also observed in vivo, which indicated that iKP-19-CAR-T cells were safer than CD19-CAR-T cells. The result was similar to a CD19-CAR-T cells variant reported by Ying et al. [46]. This was possibly because iKP-19-CAR-T cells were not always activated, their activation was suppressed by the negative signaling provided by HLA-C1 expressed in normal B cells or T cells themselves. Therefore, the "missing self" activation mechanism like NK cells confers to iKP-19-CAR-T cells to control malignant B cells effectively and spare normal B cells.

"On-target off-tumor" toxicity has seriously limited the clinical application of CAR-T cells in the treatment of solid tumors. It has been reported that HER2-targeted CAR-T cell treatment for colon cancer caused a patient death because of CAR-T cells off target to pulmonary tissues [47]. In theory, iKP CAR can also reduce the "on-target off-tumor" toxicity of HER2-CAR-T cells to normal tissue cells. So far, our lab has detected the efficiency of iKP CAR in HER2-CAR-T cells therapy in vitro and in vivo, and similar results have been acquired (data not published).

In conclusion, we demonstrated a novel iKP CAR can recognize HLA-C1 and deliver an inhibitory signaling to T cells. T cells can be activated by CD19 CAR and kill malignant B cells because iKP CAR does not work (Figure 7A). Once it recognizes "self-HLA-C1" on normal B cells, iKP CAR will deliver an inhibitory signaling via the intracellular PD-1 domain to halt T cell activation mediated by CD19 CAR (Figure 7B). This "missing self" activation mechanism like NK cells confers to iKP-19-CAR-T to control malignant B cells effectively and spare normal B cells in vitro and in vivo. The effectiveness of iKP CAR in the human body needs to be verified in future clinical trials.

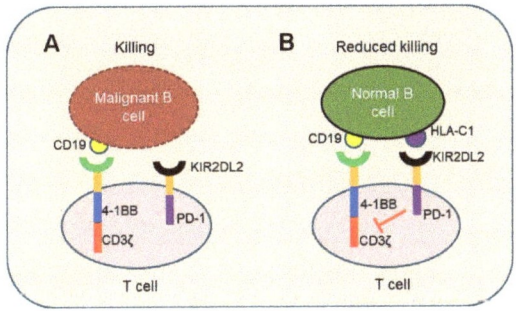

Figure 7. "Missing self" mechanism of iKP CAR. T cells co-expressing CD19 CAR and iKP CAR exploit "missing self" activation mechanism similar to NK cells. (**A**) T cells are activated by CD19 CAR and kill malignant B cells upon recognizing CD19 on malignant B cell tumors. (**B**) CD19 CAR activation signal is inhibited by PD-1 signal when iKP CAR is engaged to "self-HLA-C1" on normal B cells and iKP-19-CAR-T cells do not kill normal B cells.

4. Materials and Methods

4.1. Cell Lines

293T cell line was preserved in our lab and propagated in Dulbecco modified eagle medium (DMEM) (Invitrogen, Carlsbad, CA, USA) supplemented with 10% heat-inactivated fetal bovine serum (FBS) (Thermo Fisher Scientific, Waltham, MA, USA) and 1% penicillin/streptomycin (P/S) (Thermo Fisher Scientific, Waltham, MA, USA). The Burkitt's lymphoma Daudi cells and Raji cells were purchased from ATCC (Manassas, VA, USA) and maintained in RPIM-1640 medium (Invitrogen, Carlsbad, CA, USA) supplemented with 10% FBS and 1% P/S. Daudi cell line expressing Luciferase (Daudi-luc) was generated by stably transducing fire-fly luciferase in wild-type Daudi cells.

4.2. T-Cell and B-Cell Isolation

Human Peripheral blood was used to isolate peripheral blood mononuclear cells (PBMC) by density centrifugation method according to the manufacturer's guideline (Sigma, San Louis, MO, USA). Primary T cells were positively selected using a mixture of (1:1) anti-CD4 and anti-CD8 microbeads (Miltenyi, Koln, Germany) and normal B cells were negative selected by using B Cell Isolation Kit II (Miltenyi) from PBMCs according to the manufacturer's protocol. Both of them were cultured in X-VIVO^{TM} 15 serum free medium (LONZA, Basel, Switzerland) supplemented with 1% P/S, and stored in liquid nitrogen. All fresh blood was collected under a protocol approved by the Ethics Committee of East China Normal University (m20190315), following written informed consent.

4.3. iKP CAR/iKPt CAR Construction

iKP CAR was generated by linking the nucleotide sequence of extracellular domain of KIR2DL2 (AA22-245, Uniprot sequence ID P43627.) and intracellular domain of PD-1 (AA192-288, Uniprot sequence ID Q15116) with CD8a hinge and transmembrane nucleotide sequence. iKPt CAR was used as a negative control designed by truncating PD-1 domain of iKP CAR. Then, the commercially synthesized iKP CAR or iKPt CAR was cloned into a pCDH lentiviral vector expressing a CD19 CAR separated by a T2A sequence (GAGGGCAGAGGAAGTCTTCTAACATGCGGTGACGTGGAGGAGAATCCCGGCCCT).

4.4. Lentiviral Vector Production

Lentiviral supernatant was produced in the 293T packaging cell line according to the routine protocol [48]. In brief, 70% confluent 10 cm cell culture plates of 293T cells were co-transfected with 5 μg pCDH vector plasmid, 5 μg psPAX2 (Gag/pol/REV) and 3 μg pMD2.G (VSVG envelope) packaging plasmid using Lipofectamine 2000 transfecting reagent (Thermo Fisher Scientific, Waltham, MA, USA). Medium was replaced after 12-h transfection. The 48-h and 72-h viral supernatants were collected, combined and ultra-centrifuged at 25,000 rpm for 2 h to obtain concentrated lentivirus and stored at −80 °C for future use.

4.5. iKP-19-CAR-T/iKPt-19-CAR-T Cell Manufacture

After T cells were thawed and stimulated with CD3/CD28 Dynabeads (Miltenyi, Bergisch Gladbach, Germany) for 2 days, T cells were transduced with lentiviral vectors mentioned above at a MOI of 20, and maintained at 1×10^6 cells per ml in X-VIVO media with 100 U/mL of human IL-2 (Peprotech, Rocky Hill, NJ, USA) as described [44,49]. iKP-19-CAR-T/iKPt-19-CAR-T cells expansion was carried out for 14 days. Absolute cell counts and viability were obtained with a Coulter Counter (Beckman Coulter, Brea, CA, USA). The expression of CARs, T cell differentiation markers CCR7, CD45RO and GzmB, phenotype markers CCR7 and CD45RA, transcription factors Eomes and T-bet, exhaustion markers PD-1, TIM-3 and LAG-3, signal molecule P-Zap70 and $CD8^+/CD4^+$ T cells ratio were analyzed using flow cytometry.

4.6. Analysis of iKP CAR Function

Multiple sets of co-cultivation experiments were performed to study iKP CAR function. In every experiment, CD19 CAR-positive rate was unified using untransduced T cells (UT) and target cells were seeded at a density of 10^4 cells per well in 96 well plate in a triplicate manner. To verify whether iKP CAR performed PD-1 function after recognizing HLA-C1, CD19-CAR-T cells and iKP-19-CAR-T/iKPt-19-CAR-T cells were co-cultured with Daudi cells (CD19$^+$HLA-C1$^-$) or Raji cells (CD19$^+$HLA-C1$^+$) at 1:1 effector cells to target cells ratio (E:T) for 6 h, then the expression of P-Zap70 in different CAR-T cells was assayed by flow cytometry; In order to further investigate whether iKP CAR could suppress the activation of T cells, iKP-19-CAR-T cells and CD19-CAR-T cells were co-cultured with Daudi cells or normal B cells (CD19$^+$HLA-C1$^+$) at a 1:1 ratio for 6 h, then the expression level of CD69, CD107a, GzmB and P-Zap70 was determined by using flow cytometry.

4.7. Flow Cytometry

CD19-CAR-T cells or iKP-19-CAR-T cells were gated by CD19 CAR+, then the expression of the related molecules in CAR-T cells was analyzed. Especially, transcription factors T-bet and Eomes staining was performed using the FoxP3 TF Staining Buffer Set (eBioscience, San Diego, CA, USA) and intracellular proteins GzmB and P-Zap70 staining was performed using Fixation and Permeabilization Solution Kit (BD Biosciences, San Jose, CA, USA) according to the manufacturer's instruction respectively. In brief, 2×10^6 T cells or co-cultured cells were collected and washed twice with PBS, then cells were resuspended with 1 mL 1 × TF FIX/Perm Buffer per tube (or 250 µL BD 1 × Fixation/Permeabilization solution per tube) for 45 min at 4 °C to lyse nuclear membranes (or cell membranes). The supernatants were removed after centrifuging and cells were washed with 1 mL 1 × Perm/wash Buffer per tube (or 1 mL 1 × BD Perm/Wash Buffer per tube) twice. Next, cells were incubated with respective antibodies for 30 min at 4 °C and washed with PBS twice, protein expression levels were tested by flow cytometry. The following fluorescently-labeled monoclonal antibodies (mAbs) were used in this study: APC-anti-human CD19 (#555415, BD Biosciences, San Jose, CA, USA), PE-anti-human HLA-C (#566372, BD Biosciences, San Jose, CA, USA), PE-anti-human KIR (#556071, BD Biosciences, San Jose, CA, USA), PE-anti-human CD45RO (#561889, BD Biosciences, San Jose, CA, USA), PE-anti-human CD107a (#555801, BD Biosciences, San Jose, CA, USA), Alexa Flour 647-anti- mouse F(ab')2 antibody (#115-605-006, Jackson ImmunoResearch, West Grove, PA, USA), PerCP-anti-human CD4 (#317431, BioLegend, San Diego, CA, USA), APC-anti-human CD8 (#344722, BioLegend, San Diego, CA, USA), FITC-anti-human PD-1 (#621612, BioLegend, San Diego, CA, USA), PE/Cy7-anti-human TIM-3 (#345014, BioLegend, San Diego, CA, USA), PE-anti-human LAG-3 (#369306, BioLegend), PE/Cy7-anti-human CCR7 (#353226, BioLegend), FITC-anti-human CD45RA (#304106, BioLegend, San Diego, CA, USA), PE-anti-human CD69 (#310906, BioLegend, San Diego, CA, USA), APC-anti-human CD3 (#300312, BioLegend, San Diego, CA, USA), FITC-anti-human Eomes (#11-4877-41, eBioscience, San Diego, CA, USA), PE-anti-human P-Zap70 (#12-9006-4, eBioscience), PE-anti-human granzyme B (GzmB) (#MHGB04, Invitrogen) and PE-anti-human T-bet (#12-5825-82, eBioscience, San Diego, CA, USA). Isotype-matched, nonreactive fluorescently-labeled mAbs were always used as a fluorescence reference. LSRFortessa flow cytometer (BD Biosciences, San Jose, CA, USA) was used to acquire the cells and results were analyzed in FlowJo software (Tree Star Inc., San Carlos, CA, USA).

4.8. LDH Release Assay

The cytotoxicity of iKP-19 CAR-T cells or CD19-CAR-T cells against Daudi cells or normal B cells was evaluated by using a standard lactic dehydrogenase (LDH) release assay (Promega, Madison, WI, USA) as described earlier by Song et al. [50]. Briefly, target cells were seeded at a density of 10^4 cells per well in 96 well plate in a triplicate manner. The infection rate of the two CAR-T cells was adjusted to be similar, and an equal volume of effector cells and medium were added in order to make a different E:T ratio. After that 50 µL of each sample was transferred to plate in order to measure absorbance via plate reader (Thermo Fisher Scientific, Waltham, MA, USA). Results were calculated by using formulas provided by Promega.

4.9. Cytokine Assay

Cytokines such as IL-6, TNF-α and IFN-γ in supernatants collected from iKP-19-CAR-T cells and CD19-CAR-T cells against Daudi cells at a 1:1 ratio for 6 h or in sera from mice at day 25 were determined using a Cytometric Bead Array (CBA) assay kit (BD) according to the manufacturer's instruction.

4.10. In Vivo Daudi-Derived Xenograft Model

Six- to eight-week-old NOD-Prkdc^{em26Cd52}IL2rg^{em26Cd22}/Nju (NCG) mice (GemPharmatech, Nangjing, China) were bred and housed under pathogen-free conditions in the animal experiment facility of East China Normal University. Mice ($n = 4$/group) were I.V. injected with 1×10^6 Daudi-luc cells at day 0, followed by 5×10^6 normal B cells at day 6. At day 7, Mice were treated with 5×10^6 UT, CD19-CAR-T or iKP-19 CAR-T. Tumor burden was evaluated by bioluminescence (BLI) using Xenogen IVIS Imaging System with Living Image software (Xenogen Biosciences, Cranbury, NJ, USA). 150 mg/kg of D-luciferin (#115144-35-9, Merck, NJ, USA) was administered intraperitoneally to examine the tumor burden at specified time points. At day 25, Peripheral Blood (PB) of the mice was obtained from the eyelids. Daudi cell (CD19$^+$HLA-C1$^-$) and normal B cell (CD19$^+$HLA-C1$^+$) survival rate was analyzed by using flow cytometry, and cytokines release was determined by using CBA assay kits. The mice were euthanized at day 32, T cells (CD3$^+$) persistence, T$_{CM}$ (CCR7$^+$CD45RA$^-$) percentage, and exhaustion marker expression were analyzed by using flow cytometry. All procedures were performed in compliance with the institutional animal care committee of East China Normal University (m20190315).

4.11. Statistical Analysis

Statistical analysis was performed by using GraphPad prism software version 6 (La, Jolla, CA, USA). All of the in vitro experiments were performed in triplicate and the in vivo xenograft model contained 4 mice in each group. The data was analyzed by using a unpaired 2-tailed Student t test and the overall survival (OS) rate of the mice was determined by using a Mantle-Cox test. $p < 0.05$ was considered as statistically significant. The data is presented as mean ± SD.

5. Conclusions

The "on-target off-tumor" effect is a serious barrier to the clinical application of CAR-T cells. If CD19-CAR-T cells clear all healthy B cells, this will cause an infection in patients. We developed a KIR/PD-1-based inhibitory CAR (iKP CAR) and demonstrated that CD19-CAR-T cells bearing an iKP CAR could control B cell malignance effectively but spare healthy B cells both in vitro and in vivo. Furthermore, iKP-19-CAR-T cells exhibited a more naïve, less exhausted phenotypes and preserved a higher proportion of central memory T cells (T$_{CM}$). Our data support that iKP CAR can be developed into a clinically implementable and promising strategy to overcome "on-target off-tumor" toxicity.

Supplementary Materials: The following are available online at http://www.mdpi.com/2072-6694/12/9/2612/s1, Figure S1: HLA-C1 expression in primary T cells; Figure S2: The expression of P-Zap70 in CD19-CAR-T or iKP-19-CAR-T cells at day 10.

Author Contributions: L.T. and W.J. conceived and designed the study. L.T. and W.J. analyzed and interpreted the data. L.T., Y.G., C.N., L.Z., Y.Z., C.H., G.Z. and J.Y. performed the experiments. C.H., M.A.F., I.A., M.L. and W.J. wrote and revised the manuscript. All authors have read and agreed to the published version of the manuscript.

Funding: This work was supported by National Natural Science Foundation of China (81771306, 81072459, 81830083), Science and Technology Commission of Shanghai Municipality (201409002900, 14140904200), National Key Research and Development Program of China (2016YFC1200400).

Acknowledgments: We would like to thank Ying Zhang for her technological assistance. We also acknowledge Shanghai Bioray Laboratories Inc. and. ECNU Public Platform for innovation (011).

Conflicts of Interest: The authors declare no potential conflict of interest.

References

1. Thomas, X.; Paubelle, E. Tisagenlecleucel-T for the treatment of acute lymphocyticleukemia. *Expert. Opin. Biol. Ther.* **2018**, *18*, 1095–1106. [CrossRef]
2. Nair, R.; Neelapu, S.S. The promise of CAR T-cell therapy in aggressive B-cell lymphoma. *Best Pract. Res. Clin. Haematol.* **2018**, *31*, 293–298. [CrossRef] [PubMed]
3. Maude, S.L.; Frey, N.; Shaw, P.A.; Aplenc, R.; Barrett, D.M.; Bunin, N.J.; Chew, A.; Gonzalez, V.E.; Zheng, Z.; Lacey, S.F. Chimeric antigen receptor T cells for sustained remissions in leukemia. *N. Engl. J. Med.* **2014**, *371*, 1507–1517. [CrossRef] [PubMed]
4. Porter, D.L.; Hwang, W.-T.; Frey, N.V.; Lacey, S.F.; Shaw, P.A.; Loren, A.W.; Bagg, A.; Marcucci, K.T.; Shen, A.; Gonzalea, V. Chimeric antigen receptor T cells persist and induce sustained remissions in relapsed refractory chronic lymphocytic leukemia. *Sci. Transl. Med.* **2015**, *7*, 303ra139. [CrossRef]
5. Shah, N.N.; Fry, T.J. Mechanisms of resistance to CAR T cell therapy. *Nat. Rev. Clin. Oncol.* **2019**, *16*, 372–385. [CrossRef] [PubMed]
6. Garfall, A.L.; Maus, M.V.; Hwang, W.-T.; Lacey, S.F.; Mahnke, Y.D.; Melenhorst, J.J.; Zheng, Z.; Vogl, D.T.; Cohen, A.D.; Weiss, B.M. Chimeric antigen receptor T cells against CD19 for multiple myeloma. *N. Engl. J. Med.* **2015**, *373*, 1040–1047. [CrossRef] [PubMed]
7. Grupp, S.A.; Kalos, M.; Barrett, D.; Aplenc, R.; Porter, D.L.; Rheingold, S.R.; Teachey, D.T.; Chew, A.; Hauck, B.; Wright, J.F. Chimeric antigen receptor-modified T cells for acute lymphoid leukemia. *N. Engl. J. Med.* **2013**, *368*, 1509–1518. [CrossRef]
8. Kochenderfer, J.N.; Wilson, W.H.; Janik, J.E.; Dudley, M.E.; Stetler-Stevenson, M.; Feldman, S.A.; Maric, I.; Raffeld, M.; Nathan, D.-A.N.; Lanier, B.J. Eradication of B-lineage cells and regression of lymphoma in a patient treated with autologous T cells genetically engineered to recognize CD19. *Blood* **2010**, *116*, 4099–4102. [CrossRef]
9. Kochenderfer, J.N.; Dudley, M.E.; Feldman, S.A.; Wilson, W.H.; Spaner, D.E.; Maric, I.; Stevenson, M.S.; Phan, G.Q.; Hughes, M.S.; Sherry, R.M. B-cell depletion and remissions of malignancy along with cytokine-associated toxicity in clinical trial of anti-CD19 chimeric-antigen-receptor-transduced T cells. *Blood* **2012**, *119*, 2709–2720. [CrossRef]
10. Oluwole, O.O.; Davila, M.L. At the bedside: Clinical review of chimeric antigen receptor (CAR) T cell therapy for B cell malignancies. *J. Leukoc. Biol.* **2016**, *100*, 1265–1272. [CrossRef]
11. Pennell, C.A.; Barnum, J.L.; McDonald-Hyman, C.S.; Panoskaltsis-Mortari, A.; Riddle, M.J.; Xiong, Z.; Loschi, M.; Thangavelu, G.; Campbell, H.M.; Storlie, M.D. Human CD19-targeted mouse T cells induce B cell aplasia and toxicity in human CD19 transgenic mice. *Mol. Ther.* **2018**, *26*, 1423–1434. [CrossRef] [PubMed]
12. Hill, J.A.; Li, D.; Hay, K.A.; Green, M.L.; Cherian, S.; Chen, X.; Riddell, S.R.; Maloney, D.G.; Boeckh, M.; Turtle, C.J. Infectious complications of CD19-targeted chimeric antigen receptor-modified T-cell immunotherapy. *Blood* **2018**, *131*, 121–130. [CrossRef] [PubMed]
13. Hosen, N.; Matsunaga, Y.; Hasegawa, K.; Matsuno, H.; Nakamura, Y.; Makita, M.; Watanabe, K.; Yoshida, M.; Satoh, K.; Morimoto, S. The activated conformation of integrin β7 is a novel multiple myeloma-specific target for CAR T cell therapy. *Nat. Med.* **2017**, *23*, 1436–1443. [CrossRef] [PubMed]
14. Posey, A.D., Jr.; Schwab, R.D.; Boesteanu, A.C.; Steentoft, C.; Mandel, U.; Engels, B.; Stone, J.D.; Madsen, T.D.; Schreiber, K.; Haines, K.M. Engineered CAR T cells targeting the cancer-associated tn-glycoform of the membrane mucin muc1 control adenocarcinoma. *Immunity* **2016**, *44*, 1444–1454. [CrossRef] [PubMed]
15. Chen, C.; Li, K.; Jiang, H.; Song, F.; Gao, H.; Pan, X.; Shi, B.; Bi, Y.; Wang, H.; Wang, H. Development of T cells carrying two complementary chimeric antigen receptors against glypican-3 and asialoglycoprotein receptor 1 for the treatment of hepatocellular carcinoma. *Cancer Immunol. Immunother.* **2017**, *66*, 475–489. [CrossRef]
16. Roybal, K.T.; Rupp, L.J.; Morsut, L.; Walker, W.J.; Mcnally, K.A.; Park, J.S.; Lim, W.A. Precision tumor recognition by T cells with combinatorial antigen-sensing circuits. *Cell* **2016**, *164*, 770–779. [CrossRef]

17. Morsut, L.; Roybal, K.T.; Xiong, X.; Gordley, R.M.; Goyle, S.M.; Thomson, M.; Lim, W.A. Engineering customized cell sensing and response behaviors using synthetic notch receptors. *Cell* **2016**, *164*, 780–791. [CrossRef]
18. Roybal, K.T.; Williams, J.Z.; Morsut, L.; Rupp, L.J.; Kolinko, I.; Choe, J.H.; Walker, W.J.; McNally, K.A.; Lim, W.A. Engineering T cells with customized therapeutic response programs using synthetic notch receptors. *Cell* **2016**, *167*, 419–432. [CrossRef]
19. Fedorov, V.D.; Themeli, M.; Sadelain, M. PD-1- and CTLA-4-based inhibitory chimeric antigen receptors (iCARs) divert off-target immunotherapy responses. *Sci. Transl. Med.* **2013**, *5*, 215ra172. [CrossRef]
20. Martinet, L.; Smyth, M.J. Balancing natural killer cell activation through paired receptors. *Nat. Rev. Immunol.* **2015**, *15*, 243–254. [CrossRef]
21. Handgretinger, R.; Lang, P.; Andre, M.C. Exploitation of natural killer cells for the treatment of acute leukemia. *Blood* **2016**, *127*, 3341–3349. [CrossRef] [PubMed]
22. Bryceson, Y.T.; March, M.E.; Ljunggren, H.G.; Long, E.O. Activation, coactivation, and costimulation of resting human natural killer cells. *Immunol. Rev.* **2006**, *214*, 73–91. [CrossRef] [PubMed]
23. Khong, H.T.; Restifo, N.P. Natural selection of tumor variants in the generation of "tumor escape" phenotypes. *Nat. Immunol.* **2002**, *3*, 999–1005. [CrossRef]
24. Tsukahara, T.; Kawaguchi, S.; Torigoe, T.; Asanuma, H.; Nakazawa, E.; Shimozawa, K.; Nabeta, Y.; Kimura, S.; Kaya, M.; Nagoya, S. Prognostic significance of HLA class I expression in osteosarcoma defined by anti-pan HLA class I monoclonal antibody, EMR8-5. *Cancer Sci.* **2006**, *97*, 1374–1380. [CrossRef] [PubMed]
25. Mizukami, Y.; Kono, K.; Maruyama, T.; Watanabe, M.; Kawaguchi, Y.; Kamimura, K.; Fujii, H. Downregulation of HLA class I molecules in the tumor is associated with a poor prognosis in patients with oesophageal squamous cell carcinoma. *Br. J. Cancer* **2008**, *99*, 1462–1467. [CrossRef]
26. Pedoeem, A.; Azoulay-Alfaguter, I.; Strazza, M.; Silverman, G.J.; Mor, A. Programmed death-1 pathway in cancer and autoimmunity. *Clin. Immunol.* **2014**, *153*, 145–152. [CrossRef]
27. Riella, L.V.; Paterson, A.M.; Sharpe, A.H.; Chandraker, A. Role of the PD-1 pathway in the immune response. *Am. J. Transplant.* **2012**, *12*, 2575–2587. [CrossRef]
28. Chikuma, S. Basics of PD-1 in self-tolerance, infection, and cancer immunity. *Int. J. Clin. Oncol.* **2016**, *21*, 448–455. [CrossRef]
29. Gaud, G.; Lesourne, R.; Love, P.E. Regulatory mechanisms in T cell receptor signalling. *Nat. Rev. Immunol.* **2018**, *18*, 485–497. [CrossRef]
30. Wu, X.; Gu, Z.; Chen, Y.; Chen, B.; Chen, W.; Weng, L.; Liu, X. Application of PD-1 blockade in cancer immunotherapy. *Comput. Struct. Biotechnol. J.* **2019**, *17*, 661–674. [CrossRef]
31. Ellery, J.M.; Nicholls, P.J. Alternate signaling pathways from the interleukin-2 receptor. *Cytokine Growth Factor Rev.* **2002**, *13*, 27–40. [CrossRef]
32. Wang, L.H.; Kirken, R.A.; Erwin, R.A.; Yu, C.R.; Farrar, W.L. JAK3, STAT, and MAPK signaling pathways as novel molecular targets for the tyrphostin AG-490 regulation of IL-2-mediated T cell response. *J. Immunol.* **1999**, *162*, 3897–3904. [PubMed]
33. Sheppard, K.-A.; Fitz, L.J.; Lee, J.M.; Benander, C.; George, J.A.; Wooters, J.; Qiu, Y.; Jussif, J.M.; Carter, L.L.; Wood, C.R. PD-1 inhibits T-cell receptor induced phosphorylation of the ZAP70/CD3zeta signalosome and downstream signaling to PKCtheta. *FEBS Lett.* **2004**, *574*, 37–41. [CrossRef] [PubMed]
34. Patsoukis, N.; Brown, J.; Petkova, V.; Liu, F.; Li, L.; Boussiotis, V.A. Selective effects of PD-1 on Akt and Ras pathways regulate molecular compoents of the cell cycle and inhibit T cell proliferation. *Sci. Signal.* **2012**, *5*, ra46. [CrossRef]
35. Parry, R.V.; Chemnitz, J.M.; Frauwirth, K.A.; Lanfranco, A.R.; Braunstein, I.; Kobayashi, S.V.; Linsley, P.S.; Thompson, C.B.; Riley, J.L. CTLA-4 and PD-1 receptors inhibit T-cell activation by distinct mechanisms. *Mol. Cell. Biol.* **2005**, *25*, 9543–9553. [CrossRef]
36. Mahnke, Y.D.; Brodie, T.M.; Sallusto, F.; Roederer, M. The who's who of T-cell differentiation: Human memory T-cell subsets. *Eur. J. Immunol.* **2013**, *43*, 2797–2809. [CrossRef]
37. Joshi, N.S.; Cui, W.; Chandele, A.; Lee, H.K.; Urso, D.R.; Hagman, J.; Gapin, L.; Kaech, S.M. Inflammation directs memory precursor and short-lived effector CD8(+) T cells fates via the graded expression of T-bet transcription factor. *Immunity* **2007**, *27*, 281–295. [CrossRef]
38. Lee, D.W.; Levine, B.L.; Porter, D.L.; Louis, C.U.; Ahmed, N.; Jensen, M.; Grupp, S.A.; Mackall, C.L. Current concepts in the diagnosis and managment of cytokine release syndrome. *Blood* **2014**, *124*, 188–195. [CrossRef]

39. Kalos, M.; Levine, B.L.; Porter, D.L.; Katz, S.; Grupp, S.A.; Bagg, A.; June, C.H. T cells with chimeric antigen receptors have potent antitumor effects and can establish memory in patients with advanced leukemia. *Sci. Transl. Med.* **2011**, *3*, 95ra73. [CrossRef]
40. Lamers, C.H.; Sleijfer, S.; Vulto, A.G.; Kruit, W.H.; KLiffen, M.; Debets, R.; Gratama, J.W.; Stoter, G.; Oosterwijk, E. Treatment of metastatic renal cell carcinoma with autologous T-lymphocytes genetically retargeted against carbonic anhyfrase IX: First clinical experience. *J. Clin. Oncol.* **2006**, *24*, e20–e22. [CrossRef] [PubMed]
41. Akpek, G.; Lee, S.M.; Anders, V.; Vogelsang, G.B. A high-dose plus steroid regimen for controlling active chronic graft-versus-host disease. *Biol. Blood Marrow Transplant.* **2001**, *7*, 495–502. [CrossRef] [PubMed]
42. Karre, K.; Ljunggren, H.G.; Piontek, G.; Kiessling, R. Selective rejection of H-2-deficient lymphoma variants suggests alternative immune defence strategy. *Nature* **1986**, *319*, 675–678. [CrossRef] [PubMed]
43. Moretta, A.; Bottino, C.; Vitale, M.; Pende, D.; Biassoni, R.; Mingari, M.C.; Moretta, L. Receptors for HLA class I molecules in human natural killer cells. *Annu. Rev. Immunol.* **1996**, *14*, 619–648. [CrossRef] [PubMed]
44. Gomes-Silva, D.; Srinivasan, M.; Sharma, S.; Lee, C.M.; Wagner, D.L.; Davis, T.H.; Rouce, R.H.; Bao, G.; Brenner, M.K.; Mamonkin, M. CD7-edited T cells espressing a CD7-specific CAR for the therapy of T-cell malignancies. *Blood* **2017**, *130*, 285–296. [CrossRef]
45. Mary, E.K.; Manish, J.B.; Gordon, J.F.; Arlene, H.S. PD-1 and Its ligands in tolerance and immunity. *Annu. Rev. Immunol.* **2008**, *26*, 677–704.
46. Ying, Z.; Huang, X.F.; Xiang, X.; Liu, Y.; Kang, X.; Song, Y.; Guo, X.; Liu, H.; Ding, N.; Zhang, T. A safe and potent anti-CD19 CAR T cell therapy. *Nat. Med.* **2019**, *25*, 947–953. [CrossRef]
47. Morgan, R.A.; Yang, J.C.; Kitano, M.; Dudley, M.E.; Laurencot, C.M.; Rosenberg, S.A. Case report of a serious adverse event following the administration of T cells transduced with a chimeric antigen receptor recognizing ERBB2. *Mol. Ther.* **2010**, *18*, 843–851. [CrossRef]
48. Prieto, C.; Stam, R.W.; Agraz-Doblas, A.; Ballerini, P.; Camos, M.; Castano, J.; Marschalek, R.; Bursen, A.; Varela, I.; Bueno, C. Activated Kras cooperates with MLL-AF4 to promote extramedullary engraftment and migration of cord blood CD34+ HSPC but is insufficient to initiate leukemia. *Cancer Res.* **2016**, *76*, 2478–2489. [CrossRef]
49. Mamonkin, M.; Rouce, R.H.; Tashiro, H.; Brenner, M.K. A T-cell-directed chimeric antigen receptor for the selective treatment of T-cell malignancies. *Blood* **2015**, *126*, 983–992. [CrossRef] [PubMed]
50. Song, Y.; Tong, C.; Gao, Y.; Dai, H.; Guo, Y.; Zhao, X. Effective and persistent antitumor activity of HER2-directed CAR-T cells against gastric cancer cells in vitro and xenotransplanted tumors in vivo. *Protein Cell* **2018**, *9*, 867–878. [CrossRef]

© 2020 by the authors. Licensee MDPI, Basel, Switzerland. This article is an open access article distributed under the terms and conditions of the Creative Commons Attribution (CC BY) license (http://creativecommons.org/licenses/by/4.0/).

Article

Preclinical Evaluation of Recombinant Human IL15 Protein Fused with Albumin Binding Domain on Anti-PD-L1 Immunotherapy Efficiency and Anti-Tumor Immunity in Colon Cancer and Melanoma

Fei-Ting Hsu [1,†], Yu-Chang Liu [2,3,4,†], Chang-Liang Tsai [5,†], Po-Fu Yueh [1,6], Chih-Hsien Chang [5,7] and Keng-Li Lan [6,8,*]

1. Department of Biological Science and Technology, China Medical University, Taichung 406, Taiwan; sakiro920@mail.cmu.edu.tw (F.-T.H.); u409801001@ym.edu.tw (P.-F.Y.)
2. Department of Radiation Oncology, Chang Bing Show Chwan Memorial Hospital, Lukang, Changhua 505, Taiwan; kevinyc.liu@gmail.com
3. Department of Radiation Oncology, Show Chwan Memorial Hospital, Changhua 500, Taiwan
4. Department of Medical Imaging and Radiological Sciences, Central Taiwan University of Science and Technology, Taichung 406, Taiwan
5. Department of Biomedical Imaging and Radiological Sciences, National Yang Ming Chiao Tung University, Taipei 112, Taiwan; amos.tcl@ym.edu.tw (C.-L.T.); chchang@iner.gov.tw (C.-H.C.)
6. Institute of Traditional Medicine, School of Medicine, National Yang Ming Chiao Tung University, Taipei 112, Taiwan
7. Isotope Application Division, Institute of Nuclear Energy Research, Taoyuan 325, Taiwan
8. Department of Oncology, Taipei Veterans General Hospital, Taipei 112, Taiwan
* Correspondence: kengli@gmail.com or kllan@ym.edu.tw; Tel.: +886-2-2826-7000 (ext. 7121)
† Authors contribute equally.

Simple Summary: In this manuscript, we reported that a newly developed recombinant human IL15 fused with albumin binding domain (hIL15-ABD) showed superior biological half-life, pharmacokinetic and anti-tumor immunity than wild-type (WT) hIL15. Our hIL-15-ABD can effectively enhance anti-tumor efficacy of anti-PD-L1 on colon cancer and melanoma animal models. The anti-tumor potential of hIL-15-ABD was associated with tumor microenvironment (TME) regulation, including the activation of NK cells and CD8[+] T cells, the reduction of immunosuppressive cells (MDSCs and Tregs) and the suppression of immunosuppressive factors (IDO, FOXP3 and VEGF). In conclusion, our new hIL15-ABD combined with anti-PD-L1 antibody increased the activity of anti-tumor effector cells involved in both innate and adaptive immunities, decreased the TME's immunosuppressive cells, and showed greater anti-tumor effect than that of either monotherapy. We suggested hIL15-ABD as the potential complementary agent may effectively augment the therapeutic efficacy of anti-PD-L1 antibody in colon cancer and melanoma model.

Abstract: Anti-PD-L1 antibody monotherapy shows limited efficacy in a significant proportion of the patients. A common explanation for the inefficacy is a lack of anti-tumor effector cells in the tumor microenvironment (TME). Recombinant human interleukin-15 (hIL15), a potent immune stimulant, has been investigated in clinical trial with encouraging results. However, hIL15 is constrained by the short half-life of hIL15 and a relatively unfavorable pharmacokinetics profile. We developed a recombinant fusion IL15 protein composed of human IL15 (hIL15) and albumin binding domain (hIL15-ABD) and explored the therapeutic efficacy and immune regulation of hIL-15, hIL15-ABD and/or combination with anti-PD-L1 on CT26 murine colon cancer (CC) and B16-F10 murine melanoma models. We demonstrated that hIL15-ABD has significant inhibitory effect on the CT26 and B16-F10 tumor growths as compared to hIL-15. hIL-15-ABD not only showed superior half-life and pharmacokinetics data than hIL-15, but also enhance anti-tumor efficacy of antibody against PD-L1 via suppressive effect on accumulation of Tregs and MDSCs and activation of NK and CD8+T cells. Immune suppressive factors including VEGF and IDO were also decreased by combination treatment. hIL15-ABD combined with anti-PD-L1 antibody increased the activity of anti-tumor effector cells involved in both innate and adaptive immunities, decreased the TME's immunosuppressive cells, and showed greater anti-tumor effect than that of either monotherapy.

Keywords: PD-L1; IL15; colon cancer; melanoma; tumor microenvironment

1. Introduction

Active immune system possesses fighting ability against tumor development and progression. An example of this is cytotoxic T lymphocytes (CD8$^+$ T cells) and natural killer (NK) cells attacking tumor cells that can be elicited by antitumor immune signaling, resulting in tumor destruction [1,2]. However, tumors can escape immune surveillance through immunosuppressive tumor microenvironment (TME), restricting antitumor immunity. Immune checkpoints, which are composed of immunosuppressive molecule receptors or their ligands such as programed death receptor 1 (PD-1)/programed death ligand 1 (PD-L1) and cytotoxic T-lymphocyte-associated protein 4 (CTLA-4) modulate inactivation of CD8$^+$ T and nature killer (NK) cells, resulting in tumor immune evasion in TME. Evasion of immune surveillance is conductive to tumor survival and progression [3–5].

Immunotherapy, an innovative therapeutic method that treats cancer by evoking antitumor immunity, is a promising strategy for treatment of solid tumors and hematologic malignancies [6–8]. Increased expression of PD-1/PD-L1 pathway is linked to T cell exhaustion and poor survival in multiple types of cancers. Blockade of PD-1/PD-L1 interaction with monoclonal antibodies reverses T cell exhaustion and prolongs survival benefit in patients with cancers such as melanoma, hepatocellular carcinoma (HCC), non-small-cell lung cancer (NSCLC), gastric, and urothelial cancers [7,9–11]. Furthermore, many preclinical and clinical studies have demonstrated that the therapeutic efficacy of PD-1/PD-L1 blocking antibodies can be enhanced with immunologic or non-immunologic agents [4].

Interleukin-15 (IL15), the immuno-oncology agent, potentiates antitumor immune via enhancement of CD8$^+$ T and NK cells proliferation and cytotoxic activity. IL15 therapy has been demonstrated to attenuate tumor growth and improve survival rates in murine tumor models. IL15 is also recognized as a potential complementary agent to immunotherapy, effectively increasing anticancer immune response. The first clinical trial of hIL15 was conducted in patients with metastatic renal cell carcinoma and melanoma by daily intravenous administration for 12 consecutive days of recombinant hIL15 expressed by *Escherichia coli* [12–14]. Although some encouraging clinical results were observed, the bioactivity of IL15 is limited due to short in vivo half-life. N-803, formerly ALT-803, composed of N72D IL15 mutant, sushi domain of IL15Rα, and Fc domain of human IgG1, has been demonstrated to have longer serum half-life and more potent stimulatory effect on NK cells and T-lymphocytes than that of WT hIL15 [15,16]. N-803 has been shown to boost antitumor response of anti-PD-L1 antibody in triple negative breast and colon cancers in vivo and its combination with anti-PD-1 monoclonal antibody, Nivolumab, has been verified in safety to treat refractory metastatic non-small cell lung cancer patients with observed tumor responses [17].

We have generated a recombinant fusion protein (hIL15-ABD), which is composed of human IL15, albumin binding domain (ABD), and hexahistidine tag (his6). hIL15-ABD could be expressed by *E. coli* and refolded into active fusion protein, which simultaneously binds to albumin and stimulate CTLL-2 proliferation as well as downstream signaling pathway evidenced by enhanced STAT5 phosphorylation. Fusion of hIL15 with ABD greatly enhanced pharmacokinetic parameters, including half-life, Cmax and area under curve (AUC) as compared with those of hIL15 in experimental mice. hIL15-ABD also displayed significant inhibitory effect on the tumor growths of CT26 murine colon cancer (CC) and B16-F10 murine melanoma models. Moreover, combination of hIL15-ABD with a rat antibody against murine PD-L1 antibody, 10F.9G2, demonstrated greater anti-tumor effect than that of either monotherapy, by enhancing the activity of anti-tumor effector cells associated with both innate and adaptive immunities as well as decreasing the TME's immunosuppressive cells.

2. Materials and Methods

2.1. Reagents and Antibodies

FITC Rat Anti-Mouse CD3 (#561798), PerCP-Cy™5.5 Rat Anti-Mouse CD4 (#561115), FITC Rat Anti-Mouse CD8a (#561966), PE Rat Anti-Mouse CD25 (#561065), FITC Rat Anti-CD11b (#561688), PE Rat Anti-Mouse CD49b (DX5, #561066), PerCP-Cy™5.5 Rat Anti-Mouse CD335 (#560800), Alexa Fluor® 488 Rat anti-Mouse Foxp3 (#560407), PE Rat Anti-Mouse Ly-6G and Ly-6C (Gr-1, #561084), PerCP-Cy™5.5 Mouse Anti-Mouse NK-1.1 (#561111) and Foxp3 Fixation/Permeabilization Buffer Set (#560409) were all purchased from BD Pharmingen™ (BD Biosciences, San Diego, CA, USA). Cleaved-caspase-3 (E-AB-30004, Elabscience Biotechnology Inc, Houston, TX, USA), BAX (#50599-2-lg, Proteintech Inc., Rosemont, IL, USA), Ki-67 (#E-AB-2202, Elabscience Biotechnology Inc.), granzyme B, Indoleamine 2,3-dioxygenase (IDO), Forkhead box protein P3 (FOXP3), CD49b and Interferon-gama (IFN-γ Rat Anti-Mouse PD-L1 (10F.9G2, Bioxcell, Lebann, NH, USA) antibodies were all purchased from different companies as listed.

2.2. In Vitro Characterization for CTLL2 Stimulation and Albumin Binding of hIL15-ABD

CTLL-2 cells in logarithmic phase were harvested, washed and resuspended in RPMI-1640 medium supplemented with 10% FBS at a concentration of 8×10^3 cells per well of 96-well plate. hIL15-ABD in 1 µL of various refolding buffers was incubated with 100 µL of CTLL-2 cell culture of 96-well at 37 °C for 2 days followed by addition of 20 µL MTS reagent (CellTiter 96® AQueous Non-Radioactive Cell Proliferation Assay, Promega, Madison, WI, USA) according to manufacturer's instructions. The viable CTLL-2 cells were measured at 490 nm on a TECAN Sunrise™ multichannel microtiter plate reader. For STAT5 phosphorylation assay, CTLL2 cells were treated with increasing concentrations of either hIL15 or hIL15-ABD for 20 min followed by fixation with formaldehyde (2% v/v final concentration) for 15 min at room temperature. Fixed cells were then spun down ($\times$500 g), and cell pellets were permeabilized with ice cold 100% methanol and incubated on ice for 20 min. Cells were rehydrated by washing twice with 250 µL of PBS in the presence of 0.8% BSA. Cells were incubated with antibody against phosphorylated STAT5 for 16 h at 4°, washed twice with PBS with 0.5% BSA, and lastly treated with FITC-conjugated anti-rabbit IgG for 60 min at room temperature in the dark. The positive events were detected with a BD FACSCalibur flow cytometer and analyzed with CellQuest Pro and Cytexpert software (Beckman Coulter). To examine the ability of hIL15-ABD for binding to albumin, human or murine albumin was diluted in PBS and immobilized on an ELISA plate by incubation at 4 °C overnight. Albumin coated well were blocked with 300 µL 3% milk for 2 h at room temperature, followed by washing the plate three times with wash buffer (0.05% Tween-20 in PBS). Various concentrations of hIL15 or hIL15-ABD were incubated with immobilized human or murine albumin at room temperature for one hour, followed by washing the plate three times with wash buffer. The *in vitro* binding of his6-tagged hIL15 and hIL15-ABD with human albumin was detected using an HRP-tagged, anti-his6 antibody and developed by the addition of the HRP substrate (100 µL/well), 3,3′,5,5′-tetramethylbenzidine (TMB). The peroxidase reaction was stopped 20 min after the addition of 0.5 M H_2SO_4 (50 µL/well), and the absorbance was measured at 450 nm with a multichannel microtiter plate reader.

2.3. Expression, Refolding and Purification of hIL15 and hIL15-ABD

BL21 (DE3) *E. coli* strain transformed with plasmids encoding hIL15 or hIL15-ABD was cultured with terrific broth (TB)/ampicillin (100 µg/mL) and grew at 37 °C in a shaker at 250 rpm. When reaching logarithmic phase with OD_{600} at 0.7, the *E. coli* culture was treated with increasing concentration of IPTG ranging from 0, 0.01, 0.05, 0.1, 0.5, to 1 mM.

Additionally, the cultures were further incubated at either 30 or 37 °C overnight. The induced cultures were centrifuged at 6000 rpm for 15 min, and supernatant was removed followed by re-suspending the pellet in PBS. The pellet was subjected to continuous high-pressure cell disrupter twice at pressure of 28 kpsi followed by centrifugation at 4500 rpm for 15 min for removal of supernatant. The pellets were washed six-time with either 200 mL of H_2O or Tris-HCl buffer in the presence or absence of 1%SDS, and 0.5 M NaCl. Each washing steps were followed by centrifugation at 6000 rpm for 15 min. The inclusion body pellets were solubilized with by 8 M urea buffer containing 25 mM imidazole and loaded onto Ni-NTA resin column. Elution of immobilized protein was conducted by stair-wise increase of imidazole from 75, 300, to 500 mM of imidazole in PBS buffer (pH 7.5). Fractions containing hIL15 or hIL15-ABD were pooled and dilution refolded to a concentration of 0.1 mg/mL by slow dripping into refolding buffer matrix with various combinations of Triton (0.05%), EDTA (2 mM), NaCl (250 mM), GSH (1 mM)/GSSG (0.1 mM), and L-Arginine (0.4 M). The refolding process lasted for 24 h at 4°. After examining activities of hIL15ABD and hIL15 for albumin binding and STAT5 phosphorylation, buffer 26 containing NaCl (250 mM), GSH (1 mM) and GSSG (0.1 mM) in Tris-HCl buffer (50 mM, pH 8.5) was selected for later large-scale protein refolding.

2.4. In Vitro Characterization for CTLL2 Stimulation and Albumin Binding of hIL15-ABD

The ability of hIL15 and hIL15-ABD for T-cell activation was examined for either proliferation of or phosphorylation of STAT5 in CTLL2 cells using MTS assay and flow cytometry, respectively. The detail procedure was described in material and methods section.

2.5. Pharmacokinetics

Female Balb/c mice ($n = 3$, 6 weeks, 20 g) received an intraperitoneal injection of either hIL15 (1 µg) or equimolar hIL15-ABD (1.5 µg) in 300 µL of PBS. For mice injected with hIL15, in time intervals of 15 min, 30 min, 45 min, 1 h, 2 h, 4 h, 6 h, whereas for mice treated with hIL15-ABD, in time intervals of 15 min, 30 min, 45 min, 1 h, 2 h, 4 h, 8 h, 24 h, 36 h, and 52 h, blood samples were withdrawn from the tail and placed on ice. Serum samples were obtained by centrifuging clotted blood at 800 g for 10 min at 4 °C. Serum concentrations of hIL15 and hIL15-ABD were determined by ELISA specific for human hIL15 (DY247-05, R&DSystem, Minneapolis, MN, USA). Pharmacokinetic parameters were determined using the Phoenix® WinNonlin software version 7.0 (Certara USA Inc., Princeton, NJ, USA). Noncompartmental analysis (extravascular input) was used with the log/linear trapezoidal rule. Parameters, including terminal half-life ($T_{1/2\lambda z}$), Tmax, Cmax and area under the curve (AUC) were determined. Pharmacokinetic parameters associated with the terminal phase were calculated using the last four measured time points to estimate the terminal half-life.

2.6. Cell Culture

CT26 mouse colon cancer (BCRC #60447) cell line and B16-F10 (BCRC #60031) mouse melanoma were purchased from the Bioresource Collection and Research Center (Hsinchu, Taiwan). Cells were cultured in Roswell Park Memorial Institute (RPMI) 1640 medium and Dulbecco's Modified Eagle medium (DMEM) supplemented with 10% heat-inactivated FBS, 2mMl-glutamine, 100units/mL penicillin and 100µg/mL streptomycin in a humidity atmosphere containing 5% CO_2 and at 37 °C. Cells culture related reagents and medium were all purchased from Gibco BRL, Grand Island, NY, USA.

2.7. Transfection and Stable Clone Selection

The vector containing CMV-luciferase2 vector (pGL4.50[luc2/CMV]) (Promega, Madison, WI, USA) and transfection reagent (Polyplus transfection, France) were prepared in advance. B16-F10 cells were seeded in 6 cm plate one day before transfection. Cells density is around 70% during transfection procedure. The jetPEI™ reagent (10 µL) dissolved in

250 µL of NaCl buffer was then added into DNA buffer (5 µg plasmids with 250 µL of NaCl buffer), the mixture was then incubated at 25 °C for 25 min. The mixture was finally added to the B16-F10 cells for an incubation period of 1 day. Luc2 expression cells were selected by hygromycin B 200 µg/mL for another two weeks and named as B16-F10/*luc2* cells [18].

2.8. Immune Cells (CD8+T Cells and NK Cells) Validation

The CD8$^+$T and NK cell percentages and functions were used to evaluate immune activation status. CD8$^+$ T cell percentage and function were identified by CD8, IL-2, and IFN-γ markers in tumor-draining lymph node (TDLN) and spleen. Intracellular staining was performed with Fixation/Permeabilization kit following the manufacturer's protocol. In addition, NK cells on two different strains of animal models were also identified by various markers, CD3$^-$/CD49b$^+$/CD335$^+$ on BALB/c and CD3$^-$/CD49b$^+$/NK1.1$^+$ on C57BL/6, respectively [19]. The percentages of these cell types were acquired by NovoExpress® flow cytometry (Agilent, Santa Clara, CA, USA) and data was analyzed by FlowJo software (BD Pharmingen™).

2.9. Immune Suppressive Cells (Treg Cells and MDSCs) Validation

The percentages of regulatory T cells (Tregs) and myeloid-derived suppressor cells (MDSCs) cells were used to evaluate immunosuppressive function. Immune suppressive cells isolated from tumor-draining lymph node (TDLN), spleen [20], and bone marrow (BM), were stained with anti-FOXP3-Alexa Fluor 488/CD4-PerCP-Cy™5.5/CD25-PE antibodies using a Mouse Treg Flow Kit according to manufacturer's protocol. CD11b-FITC/Gr-1-PE antibodies were used to detecting Tregs and MDSCs [21], respectively. The percentages of these cell types were acquired by NovoExpress® flow cytometry and data were analyzed by FlowJo software (BD Pharmingen™).

2.10. Animal Experiments

The animal experiments were performed in accordance with the protocols approved by the Animal Care and Use Committee at China Medical University (approval number: CMU IACUC-2019-208). Six-week-old male BALB/c and C57BL/6 mice were purchased from the National Laboratory Animal Center and housed in a pathogen-free animal facility. The establishment of animal model was described in material and methods section. All experiment was repeated at least twice (n = 6).

2.11. Animal Treatment Procedure

The animals were anaesthetized with 1–2% isofluorane during surgery and imaging. The animals were fed sterilized mouse chow and water. Five million of CT26 or B16F10/luc2 cells were administered to mice (20–25 g) by subcutaneous injection on right thigh. The body weight and tumor volume were measured 3 times per week. Tumor volume was calculated by following formula: volume = length × width^2 × 0.523. The animals were separated into various groups and administered with 100 µL of indicated treatment by i.p.: control (DMSO 0.1%), hIL15 (5 µg/injection), hIL15-ABD (5 µg/injection), or 10F.9G2 alone (anti-PD-L1, 100 µg/injection), co-treatments of hIL15-ABD and 10F.9G2. The drugs for animal treatment were dissolved in 100 µL H$_2$O with 0.1% DMSO.

2.12. Enzyme-Linked Immunosorbent Assay (ELISA)

Secreted IL15, and VEGF were collected from serum and assayed by ELISA. All the procedures followed commercially provided protocol. IL15, and VEGF ELISA kits were all purchased from Elabscience (Houston, TX, USA). ELISA readings were determined by OD scanning at 450 nm using SpectraMax iD3 microplate reader from (Molecular Devices, Downingtown, PA, USA).

2.13. Bioluminescence Imaging (BLI)

Mice bearing B16-F10/*luc2* tumors of each group ($n = 6$) were intraperitoneally injected with 200 μL of 150 mg/kg D-luciferin in PBS before anesthetization with 1–2% isoflurane 10 min prior to imaging. Mice were then set onto the imaging platform and continuously exposed to 1–2% isoflurane throughout the time. The luc2 signal from tumor region was collected by BLI using IVIS50 Imaging System (Xenogen) once per week. The photons emitted from the tumor were assayed using IVIS50 Imaging System with an acquisition time of 1 min. Regions of interest (ROIs) were drawn around the tumor and quantified with the Living Image software as photons/s/cm^2/sr.

2.14. Immunohistochemistry (IHC)

Formalin-fixed and paraffin-embedded tissues from mice were subjected to IHC staining. In brief, sections of paraffin-embedded tumor tissue on slides obtained from each group was deparaffinized in xylene, rehydrated with decreasing concentrations of ethanol (100%, 70%, 30%, 0%), and then incubated in 3% H_2O_2 for 10 min. After washing, the slides were blocked with 5% normal goat serum for 5 min in a tight container, followed by incubation with different primary antibodies in a dilution of 1:100–500 at 4 °C overnight. Finally, slides were counterstained with hematoxylin. At least three slides from each group were studied. Slides were photographed at 200 × magnifications by Nikon ECLIPSE Ti-U microscope and quantified by ImageJ software (National Institutes of Health, Bethesda, MD, USA).

2.15. Statistical Analysis

Statistical analysis was performed utilizing excel 2017 software (Microsoft, Redmond, WA, USA) and GraphPad *Prism* 8.0 (GraphPad Software, Inc., San Diego, CA, USA). Values were expressed as means ±SD. Comparison of means between several groups were performed by one-way analysis of variance (ANOVA) and independent-test was used to compare between two groups. Tukey's test was used to compare all groups as post-hoc test. Values were considered statistically significant at $p \leq 0.05$.

3. Results

3.1. Expression and Purification of Active hIL15-ABD

hIL15-ABD expression by transformed *E. coli*, BL21(DE3), were initiated when OD of culture reached 0.7, in the presence of increasing concentration of IPTG, ranging from 0, 0.05, 0.1, 0.3, 0.5, to 1 mM) at either 30 or 37 °C and 200 rpm. SDS-PAGE (upper panel, Figure 1A) and Western blot (lower panel, Figure 1A) analysis displayed comparable expressions of hIL15-ABD (21.0 kDa) induced by all the IPTG concentrations and two temperatures tested. Large scale protein expression was initiated by addition of 0.1 mM IPTG to 1 L of transformed *E. coli* culture when OD600 value reached 0.7 and kept in rotating shaker at 37 °C and 200 rpm (Figure 1). The majority of expressed hexahistidine-tagged hIL15-ABD was in the inclusion bodies, which were washed and dissolved in an 8 M urea denaturing buffer (Figure 1B) before being loaded onto a Ni-NTA resin column. The immobilized hexahistidine-tagged hIL15-ABD was eluted sequentially using 75, 300, to 500 mM of imidazole in PBS buffer (Figure 1C). The most significant portion of hexahistidine-tagged hIL15-ABD was eluted with 300 mM as displayed by SDS-PAGE (Figure 1D) and Western blotting using anti-hexahistidine antibody (Figure 1E). The purified denatured hIL15-ABD was investigated for optimal refolding condition using buffer matrix as listed in Table 1, and the bioactivities of the refolded hIL15-ABD were examined for binding to human albumin (Figure 1F) as well as stimulation CTLL-2 proliferation (Figure 1G). It turns out that refolding of ABD moiety of hIL15-ABD was quite robust in most of the buffers examined, whereas there is more significant difference in stimulatory effects of hIL15-ABD refolded in individual buffers. We selected buffer 26 composed of NaCl (250 mM), GSH (1 mM) and GSSG (0.1 mM) in Tris-HCl buffer (50 mM, pH 8.5) for later large-scale protein refolding (Figure 2A). We were able to obtain approximately 60 mg of recombinant hIL15-

ABD with purity higher than 90% per liter of TB culture in shake flasks IPTG-induced E. coli. The bioactivities of the refolded hIL15-ABD were examined for binding to either human or murine albumin as well as stimulation of STAT5 phosphorylation in CTLL-2 cells. hIL15-ABD demonstrates similar binding affinity for both human and murine albumin with Kd values of 3.0 and 2.8 nM, respectively (Figure 2B), whereas it displayed comparable stimulatory effect on STAT5 phosphorylation with that of hIL15 positive control with EC_{50} of 0.17 and 0.10 nM, respectively (Figure 2C).

Table 1. Combinations of buffer used for protein refolding.

	Triton (0.05%)		EDTA (2 mM)		NaCl (250 mM)	
	[1] GSH	[2] Arginine	[1] GSH	[2] Arginine	[1] GSH	[2] Arginine
50 mM Tris-HCl pH6.5	1	7	13	19	25	31
50 mM Tris-HCl pH8.5	2	8	14	20	26	32
50 mM Tris-HCl 1 M Urea pH6.5	3	9	15	21	27	33
50 mM Tris-HCl 1 M Urea pH8.5	4	10	16	22	28	34
50 mM Tris-HCl 1 M GdnHCl pH6.5	5	11	17	23	29	35
50 mM Tris-HCl 1 M GdnHCl pH8.5	6	12	18	24	30	36

[1] 1 mM GSH/0.1 mM GSSG; [2] 0.4 M L-Arginine.

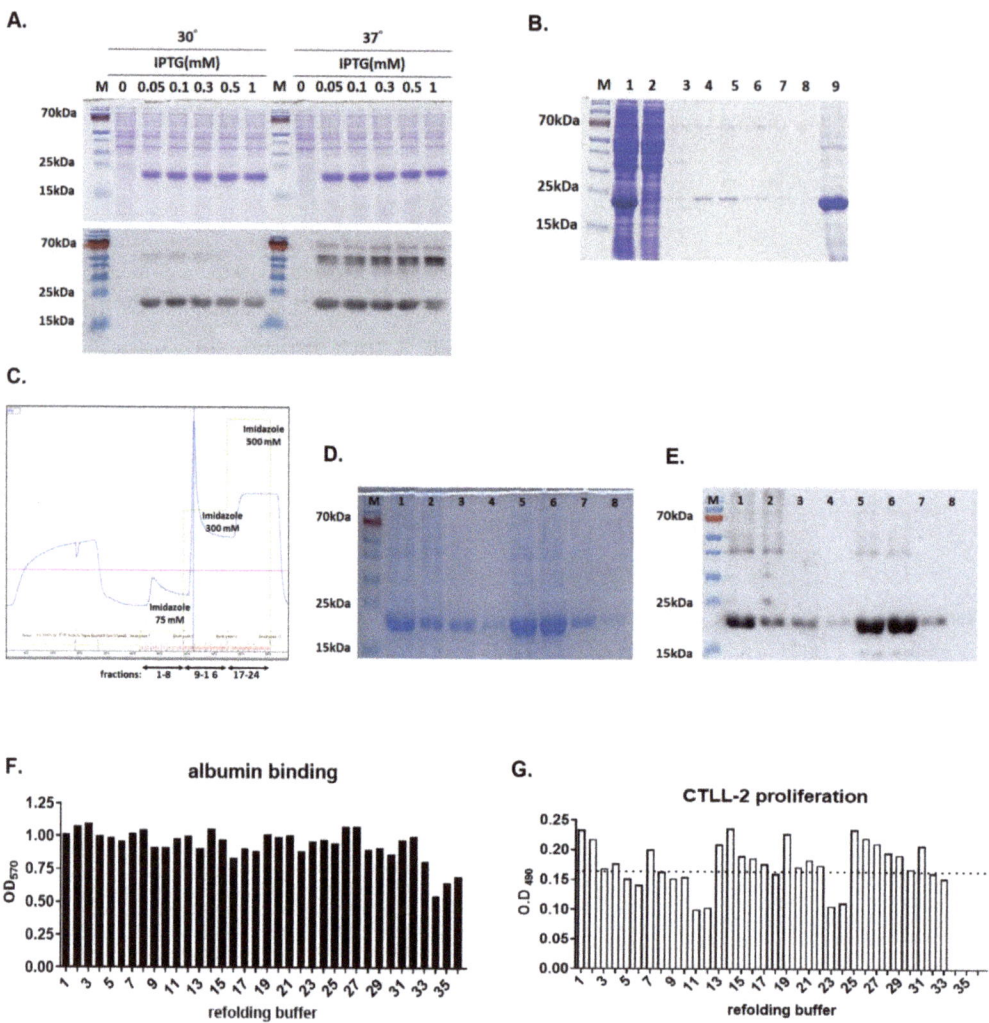

Figure 1. hIL15-ABD expression, refolding and purification. (**A**) SDS-PAGE (upper panel) and Western blot (lower panel) analysis of hIL15-ABD expression induced with increasing concentrations of IPTG, ranging from 0, 0.05, 0.1, 0.3, 0.5, to 1 mM in transformed *E. coli*, BL21(DE3) at either 30 or 37 °C. (**B**) SDS-PAGE analysis of fractions from transformed *E. coli* lysates (lane 1); supernatant of the lysates after centrifugation at 10,000 rpm for 20 min (lane 2); supernatant after washing with H$_2$O (lane 3); supernatant after washing with 20 mM Tris-HCl (lane 4); supernatant after washing with 50 mM Tris-HCl buffer containing 2 mM EDTA and 0.1% SDS (lane 5); supernatant after washing with 50 mM Tris-HCl, 150 mM NaCl and 2 mM EDTA (lane 6); supernatant after washing with H$_2$O (lane 7); supernatant after washing with H$_2$O (lane 8); denatured inclusion body in 8 M urea (lane 9). Lane 1 to 8 each are loaded protein equal to 75 µL and lane 9 equal to 37.5 µL of culture medium. (**C**) Purification of solubilized inclusion body from *E. coli* expressing hIL15-ABD through Ni-column. The blue curve indicates the absorption at 280 nm in mAU, whereas the green line represents the concentration of imidazole. (**D**) SDS-PAGE and (**E**) western blot analysis of elution fractions number 3–6 (lane 1–4), 10, 11, and 12 (lane 5–7), and 21 (lane 9). (**F**, **G**) The purified denatured hIL15-ABD was refolded with buffer matrix listed in Table 1 and resulted L15-ABD is examined for (**F**) human albumin binding and (**G**) stimulation of CTLL-2 proliferation. The dot line indicates the average OD490 values representing the extents of viable CTLL-2 in 96-well plates cultured with refolded hIL5-ABD.

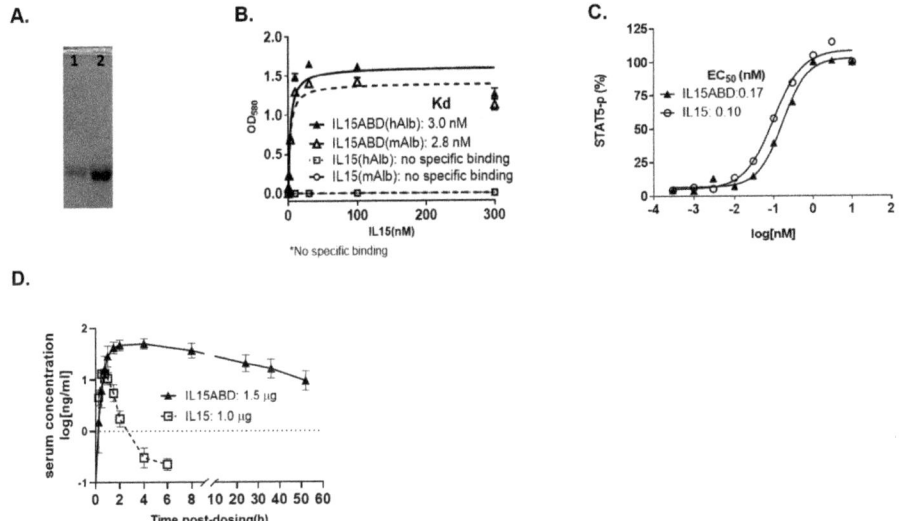

Figure 2. *In vitro* characterizations and pharmacokinetics of hIL15-ABD purified in large scale. (**A**) Lane 1 of the SDS-PAGE indicates the purified hIL1-ABD in refolding buffer number 26 (NaCl (250 mM), GSH (1 mM) and GSSG (0.1 mM) in Tris-HCl buffer (50 mM, pH 8.5)) and lane 2 represents hIL15-ABD after being condensed following the process of refolding. (**B**) Refolded purified hIL-15-ABD displays similar affinity for binding to both human and murine albumin, whereas there is no measurable specific binding to albumin by hIL15. (**C**) hIL15 and IL15-ABD demonstrate comparable EC50 values of stimulation of STAT-5 phosphorylation in CTLL-2 cells, which are 0.10 and 0.17 nM, respectively. (**D**) Serum concentration-time curves of hIL15 in Balb/c mice following single intraperitoneal injection of hIL15-ABD and hIL15 at 1.5 and 1.0 µg/mouse. Data are presented as mean ± SD. N = 3 at each time point.

3.2. Pharmacokinetics Studies

The serum concentration-time curves from derived from Balb/c mice intraperitoneally injected with 1 µg hIL15 and equimolar of 1.5 µg hIL15-ABD are shown in Figure 2D. The pharmacokinetic parameters are summarized in Table 2. The maximum serum concentrations (Cmax) and times to reach Cmax (Tmax) were determined as 13.59 ng/mL at 0.75 h for hIL15 and 51.28 ng/mL at 4 h for hIL15ABD, respectively. The terminal half-lives ($T_{1/2\lambda z}$) of hIL15 and hIL15-ABD were 0.88 h and 23.37 h after injection, respectively. The results showed that $T_{1/2\lambda z}$ of hIL15-ABD was 26-fold longer than that of hIL15 in serum, which confirmed that the long circulation of the hIL15-ABD has been achieved. The AUC (0→∞) of hIL15 and hIL15-ABD were 18.8 ng/mL×h and 1602.4 ng/mL×h, respectively. The AUC (0→∞) of hIL15-ABD in serum was 180-fold larger than that of hIL15.

Table 2. Pharmarcokinetic parameters of hIL-15 and hIL-15-ABD after intraperitoneal injection in BALB/c mice.

Parameter	Unit	hIL-15 Value	hIL-15-ABD Value
$T_{1/2\lambda z}$	h	0.88	23.37
Tmax	h	0.75	4.00
Cmax	ng/mL	13.59	51.28
AUC(0→∞)	ng/mL×h	18.90	1602.4

Calculated with WinNonlin 7.0 for a noncompartmental model.

3.3. hIL15-ABD Showed Superior Tumor Growth Inhibition and Positive Regulation of Immune Response on CC Model

After confirming that hIL15-ABD has more than 20- and 80-fold increase of biological half-life and AUC, respectively, compared to those of hIL15 (Table 2), we further validated the treatment efficacy of both on colon cancer-bearing animal model (Figure 3A). In light of the superior pharmacokinetic profiles of hIL15-ABD and to demonstrate the potent in vivo anticancer effect of hIL15-ABD, we used 5 µg for each injection, which represent 0.36 and 0.24 nanomole of hIL15 and hIL15-ABD, respectively, instead of using equimolar proteins. As shown in Figure 3B and C, hIL15-ABD displayed better tumor growth inhibition ability as compared to hIL15 past day 9 post-treatment. Additionally, hIL15-ABD showed potential to suppress the accumulation of MDSCs, immunosuppressive cells, in spleen and bone marrow (Figure 3D). Percentage of $CD11b^+/Gr-1^+$ cells that are recognized as MDSCs were effectively decreased in hIL15-ABD treated group (Figure 3E,F). Percentage of another group of immunosuppressive cells, regulatory T cells (Tregs), from TDLN and spleen was also identified by flow cytometry after treatment (Figure 3G). Number of $CD4^+/CD25^+/FOXP3^+$ cells was more effectively reduced around one of two by hIL15-ABD as compared to hIL15 (Figure 3H,I). The Treg population was reduced by more than a half in hIL15-ABD-treated group compared to non-treatment group, and also had significantly larder reduction compared to hIL15-treated group. Furthermore, we observed that hIL15-ABD may also increase the population of $CD8^+$ T cells in TDLN and spleen (Figure 3J). Two times more percentage of $CD8^+$ T cells was detected in hIL15-ABD group compared to non-treatment group (Figure 3K,L). Other than induction of adaptive immunity, NK cells, which plays role in innate immunity, was also effectively triggered in hIL15-ABD-treated group (Figure 3M). $CD3^-/CD49b^+$, $CD3^-/CD335^+$ and $CD3^-/CD49b^+/CD335^+$ cells population were all significantly elevated in hIL15-ABD-treated group compared to hIL15-treated group (Figure 3N). Decrease in mice body weight was only observed in hIL15-treated group (Figure 3O), indicating the possibility of toxicity caused by prolonged treatment with unmodified form of hIL15. In sum, hIL15-ABD not only demonstrates better tumor inhibition, but also provides a positive microenvironment for cells involved in innate and adaptive immunities to function.

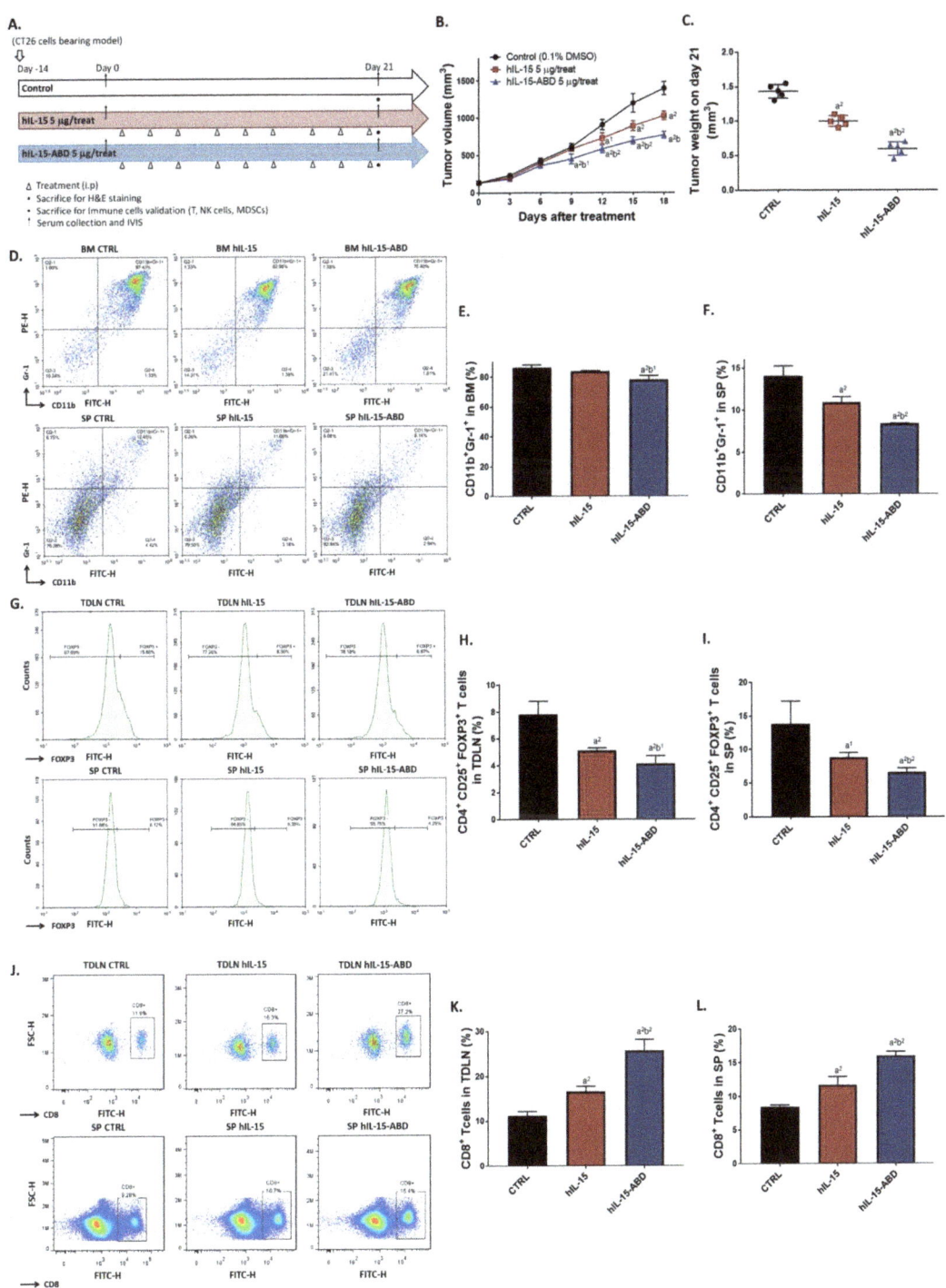

Figure 3. *Cont.*

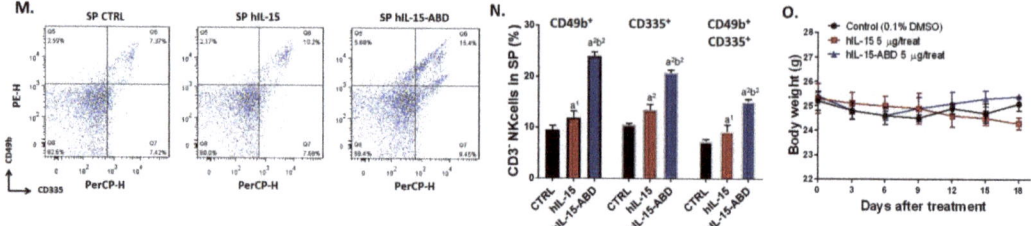

Figure 3. hIL15-ABD induced the accumulation of CD8+ T cells and NK cells, but diminished Tregs and MDSCs, resulting in colon cancer growth inhibition. (**A**) Animal flow chart of different treatment materials is displayed. (**B**) Tumor volume is recorded every 3 days and (**C**) tumor weight is weighed after isolation from mice on day 21. (**D**) Flow cytometry pattern of CD11b+/Gr-1+ MDSCs isolated from BM and SP. Percentage of CD11b+/Gr-1+ MDSCs from (**E**) BM and (**F**) SP are gated and quantified by FlowJo software. (**G**) Flow cytometry pattern of CD4+/CD25+/FOXP+ Tregs isolated from TDLN and SP. Percentage of CD4+/CD25+/FOXP3+ Tregs from (**H**) TDLN and (**I**) SP. (**J**) Flow cytometry pattern of CD8+ T cells isolated from TDLN and SP. Percentage of CD8+ T cells from (**K**) TDLN and (**L**) SP. (**M**) Flow cytometry pattern of CD3−/CD49b+/CD335+ NK cells isolated from TDLN and SP. Percentage of CD3−/CD49b+/CD335+ NK cells from (**N**) SP. (**O**) Mice body weight are measured 3 time per week. [BM = bone marrow, TDLN = tumor-draining lymph node and SP = spleen] (a^1 $p < 0.05$, a^2 $p < 0.01$ vs. CTRL; b^1 $p < 0.05$, b^2 $p < 0.01$ vs. hIL15).

3.4. hIL15-ABD Enhanced Tumor Inhibition Capacity and Triggered Apoptosis Effect of Anti-PD-L1 Therapy on Both CC and Melanoma Models

Though hIL15-ABD monotherapy demonstrated tumor inhibition potential, the inhibition ability remained limited. Therefore, we further validated whether hIL15-ABD may positively augment the function of checkpoint inhibitor-related therapy. In Figure 4A, we show the effects of hIL15-ABD and anti-PD-L1 anti-body (10F.9G2) monotherapies as well as their combined therapy effect. Combined therapy not only showed superior tumor growth inhibition in colon cancer (CC) (Figure 4B), but also melanoma bearing animal model (Figure 4D). Tumors isolated from combined therapy groups in CC and melanoma models on day 21 displayed significant tumor shrinking effect as compared to those isolated from monotherapy groups (Figure 4C,E). In addition, tumor weight of combined therapy groups also showed more significant decreases compared to either hIL15-ABD or anti-PD-L1 anti-body monotherapy groups (Figure 4F,G). Luc2 signal emitted from melanoma (B16-F10/luc2) was recognized as amount of living cells within tumor region that also presented the minimal signal intensity in combined therapy group (Figure 4H). Quantification result from BLI (Figure 4I) was corresponded to tumor volume, and the combined therapy group was found to exhibit superior tumor growth inhibition compared to the monotherapy groups. No obvious body weight loss of each treatment procedure was found in both CC and melanoma models (Figure 4J,K). Ki-67, a cell proliferation marker, was showed to be effactually suppressed by combination therapy (Figure 4L,M). Finally, we measured BAX and cleaved caspase-3 protein expression levels in CC and melanoma (Figure 4N). BAX and cleaved caspase-3 stain signals were markedly increased in combination therapy group (Figure 4O,P). Taken together, these results demonstrate that hIL15-ABD can successfully enhance tumor inhibition ability of anti-PD-L1 by disrupting proliferation effect and induction of apoptosis signaling.

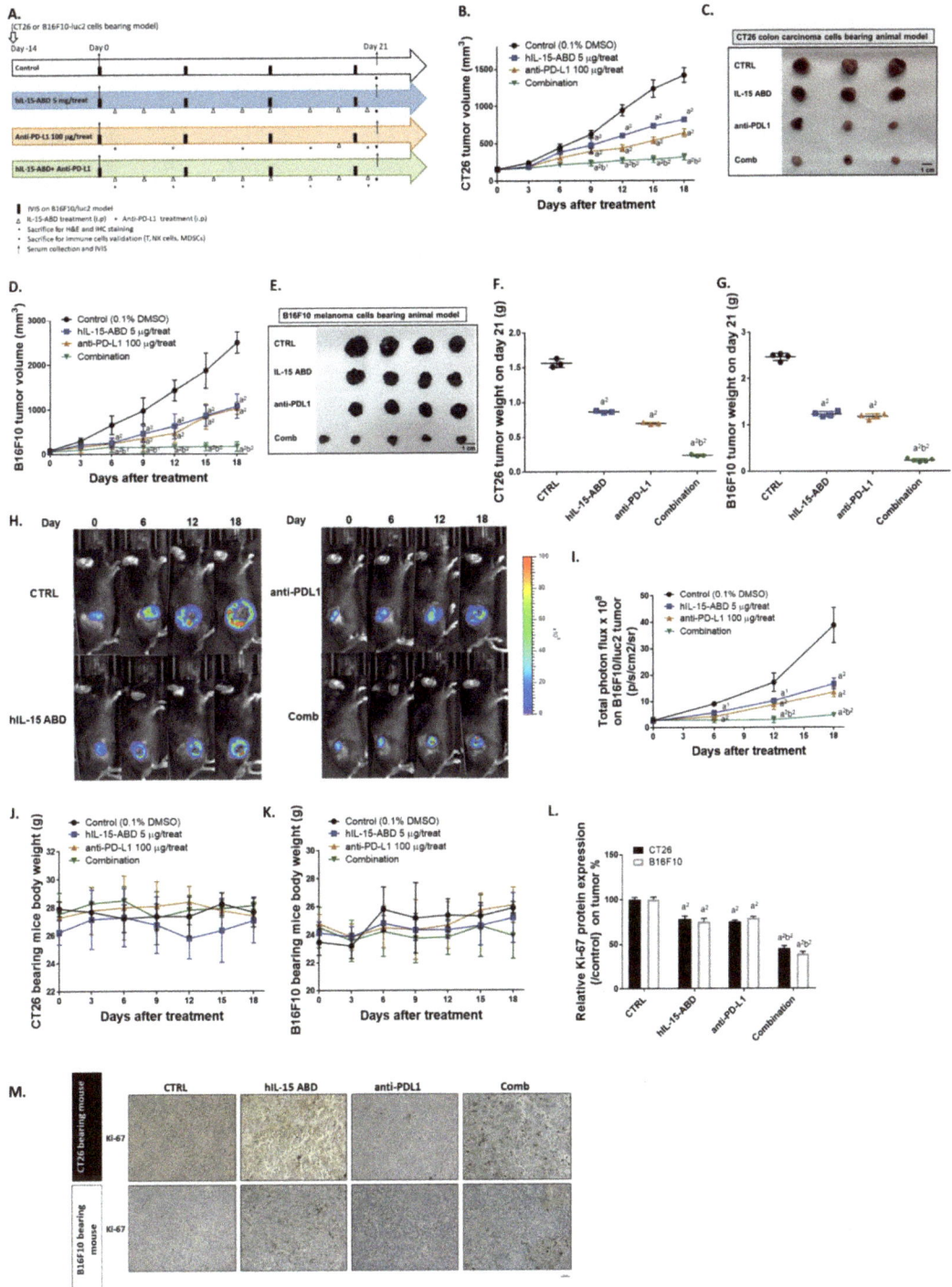

Figure 4. *Cont.*

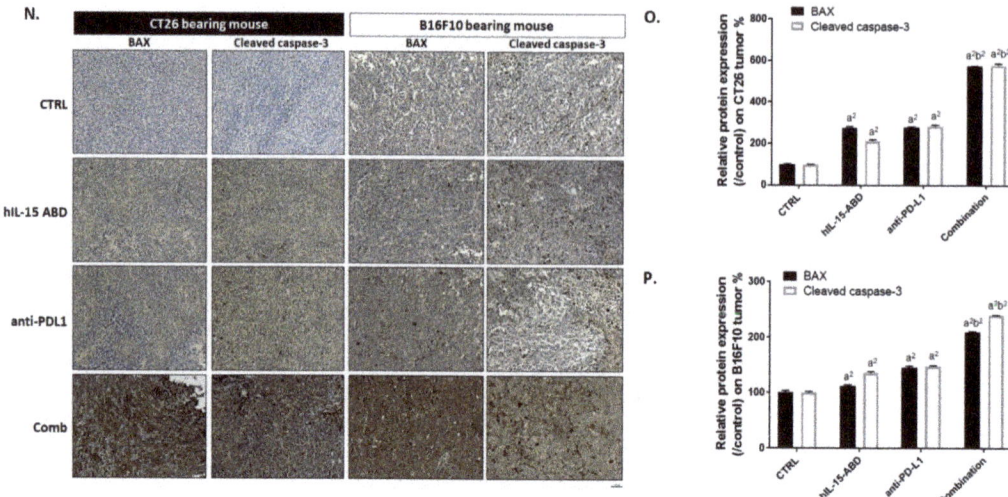

Figure 4. hIL15-ABD facilitated anti-tumor efficacy of anti-PD-L1 antibody via enhancing apoptosis mechanism. (**A**) Animal flow chart of hIL15-ABD, anti-PD-L1 and combination treatment is presented. (**B,C**) Colon cancer (CC) tumor growth from day 0–18 and tumor photographed on day 21 are displayed. (**D,E**) Melanoma tumor growth from day 0–18 and tumor photographed on day 21 are displayed. Tumor weight from (**F**) CC and (**G**) melanoma on day 18 are summarized. (**H**) BLI and (**I**) quantification results from B16-F10/luc2 bearing mice are presented. Mice body weight from (**J**) CC (**K**) melanoma model is recorded every 3 days during therapy. (**M,N**) IHC staining images and (**L,O,P**) relative proteins quantification level on CC and melanoma are presented. (a^1 $p < 0.05$, a^2 $p < 0.01$ vs. CTRL; b^1 $p < 0.05$, b^2 $p < 0.01$ vs. hIL15-ABD and anti-PD-L1; scale bar = 100 µm).

3.5. hIL15-ABD Strengthened Anti-PD-L1-Induced Function of CD8+ T Cells on Both CC and Melanoma Models

To further investigate the effect of hIL15-ABD and anti-PD-L1 combination on tumor microenvironment, we determined the function of CD8+ T cells by observing activation of intracellular IFN-γ and IL-2. As shown in Figure 5A, CD8+ cells with the expression of IFN-γ were increased to 50% in combination treatment in TDLN. Both CC and melanoma displayed an increasing percentage of IFN-γ in CD8+ cells from TDLN after combination therapy (Figure 5B). The expression level of IFN-γ in CD8+ Tcells from SP has showed similar elevations, especially in combination therapy group (Figure 5C,D). Furthermore, we found that IL-2 activation in CD8+ T cells from TDLN (Figure 5E,F) and SP (Figure 5G,H) were both effectually increased in the hIL15-ABD + anti-PD-L1 combined therapy group as compared to monotherapy groups. Lastly, we performed IHC staining on CC and melanoma tumor to validate granzyme B and CD8 protein expression after therapy (Figure 5I). The results show that not only CD8, but also granzyme B, key indicators of cytotoxic T cells (CD8+ T), were raised by combination therapy (Figure 5J,K). Higher activation levels of IFN-γ and IL-2 in CD8+ T cells from TDLN and SP in CC and melanoma models were also found in the combination treatment group relative to those of the monotherapy groups.

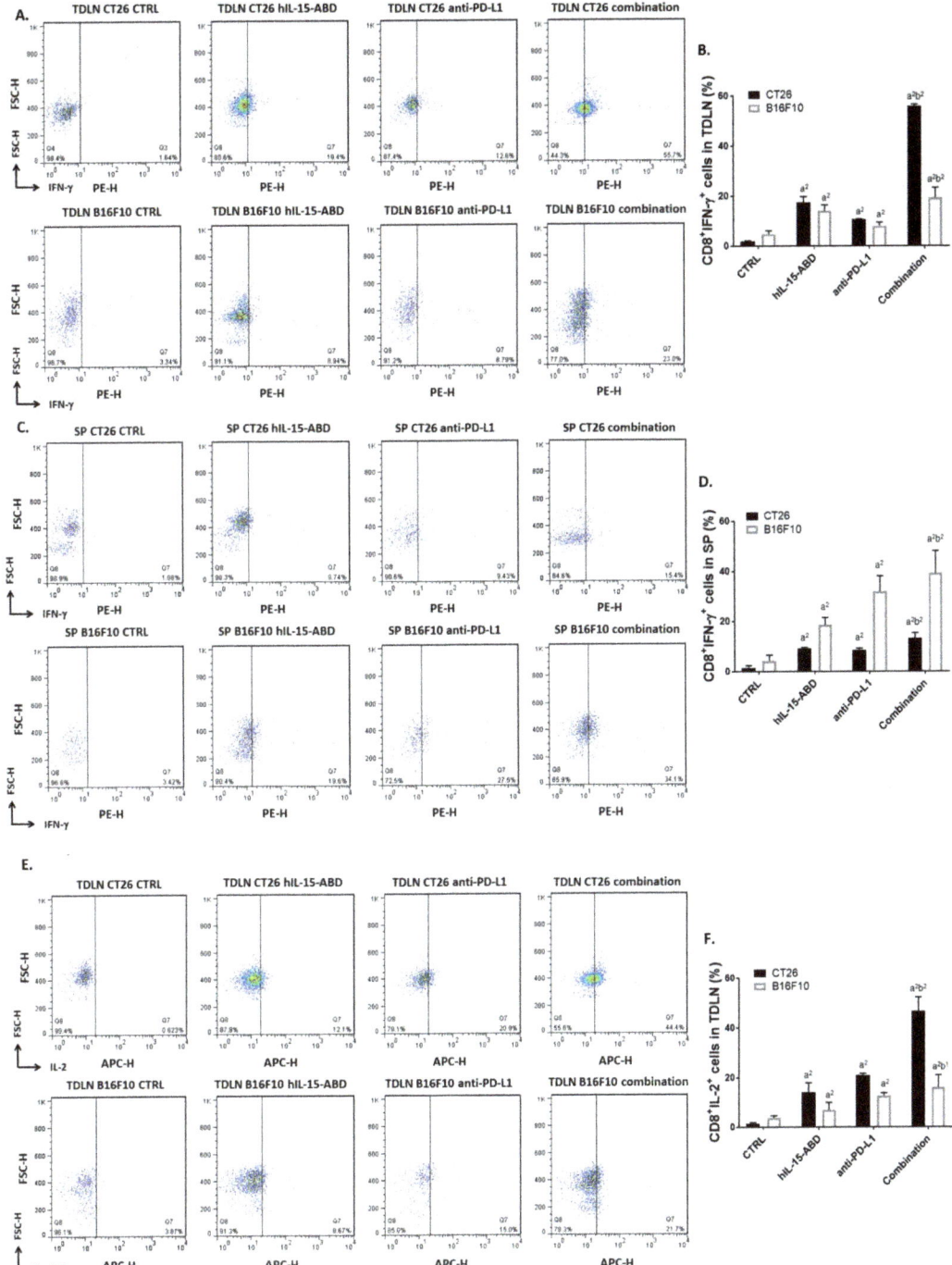

Figure 5. Cont.

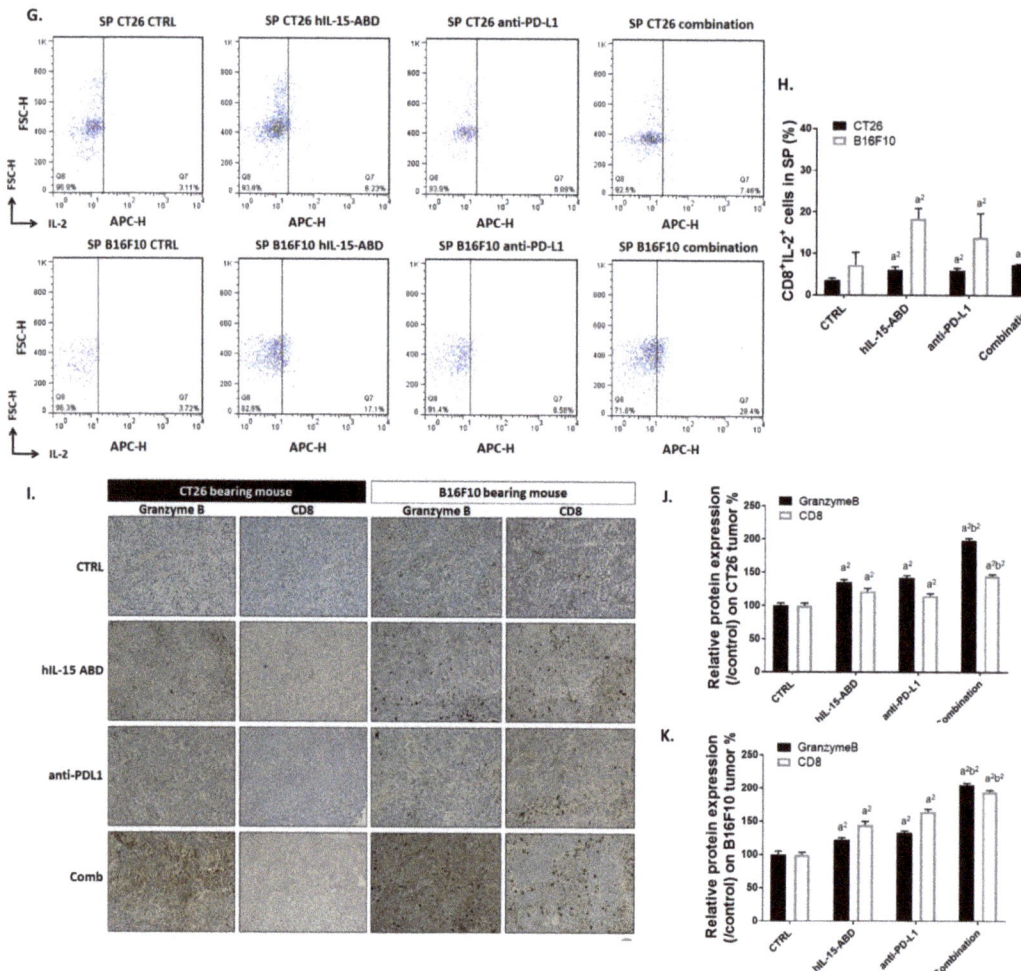

Figure 5. hIP-15-ABD offer a reinforce role of increasing anti-PD-L1 antibody induced CD8+ T cells activation. (**A**) Flow cytometry pattern and (**B**) quantification results of CD8+/IFN-γ^+ cells from TDLN. (**C**) Flow cytometry pattern and (**D**) quantification results of CD8+/IFN-γ^+ cells from SP. (**E**) Flow cytometry pattern and (**F**) quantification results of CD8+/IL-2+ cells from TDLN. (**G**) Flow cytometry pattern and (**H**) quantification results of CD8+/IL-2+ cells from SP. (**I**) Granzyme B and CD8 immunohistochemistry (IHC) staining images and relative proteins quantification level of (**J**) CC and (**K**) melanoma are displayed. (a^1 $p < 0.05$, a^2 $p < 0.01$ vs. CTRL; b^1 $p < 0.05$, b^2 $p < 0.01$ vs. hIL15-ABD and anti-PD-L1; scale bar = 100 μm).

3.6. hIL15-ABD Increased Anti-PD-L1 Antibody Induced Accumulation and Activation of NK Cells in Both CC and Melanoma Models

To identify whether combining hIL15-ABD with anti-PD-L1 promote the function of NK cells, we measured NK cell population and activity in the spleen after treatment. Results from Figure 6A indicate that CD3$^-$/CD49b$^+$, CD3$^-$/CD335$^+$, CD3$^-$/NK1.1$^+$, CD3$^-$/NK1.1$^+$/CD335$^+$ and CD3$^-$/CD49b$^+$/CD335$^+$ cells were all dramatically increased after combination therapy. Based on different species of animal, we separated NK cells according to their specific markers as indicated in Figure 6B,C. Highest amount of CD3$^-$/NK1.1$^+$/CD335$^+$ and CD3$^-$/CD49b$^+$/CD335$^+$ triple positive cells were found in the combined therapy group. Next, we identified whether these NK cells possessed function by detecting intracellular IFN-γ. The activation of IFN-γ was observably increased in CD3⁻/NK1.1⁺ and CD3⁻/CD49b⁺ cells from the combined therapy group (Figure 6D,E). Furthermore, we also investigated the expression levels of CD49b and IFN-γ proteins expression on tumor tissue from CC and melanoma models by IHC staining (Figure 6F). As shown in Figure 6G,H, the protein expression levels of both CD49b and IFN-γ were increased in treated groups. Finally, we checked VEGF (Figure 6I) secretion level in mouse serum to demonstrate the decreasing of immunosuppressive factor after combination therapy. Most importantly, the level of IL15 secretion was also effectively triggered by hIL15-ABD combined with anti-PD-L1 (Figure 6J). These results support the hypothesis that hIL15-ABD combined with anti-PD-L1 may develop a positive regulation of immune response for defending against tumor.

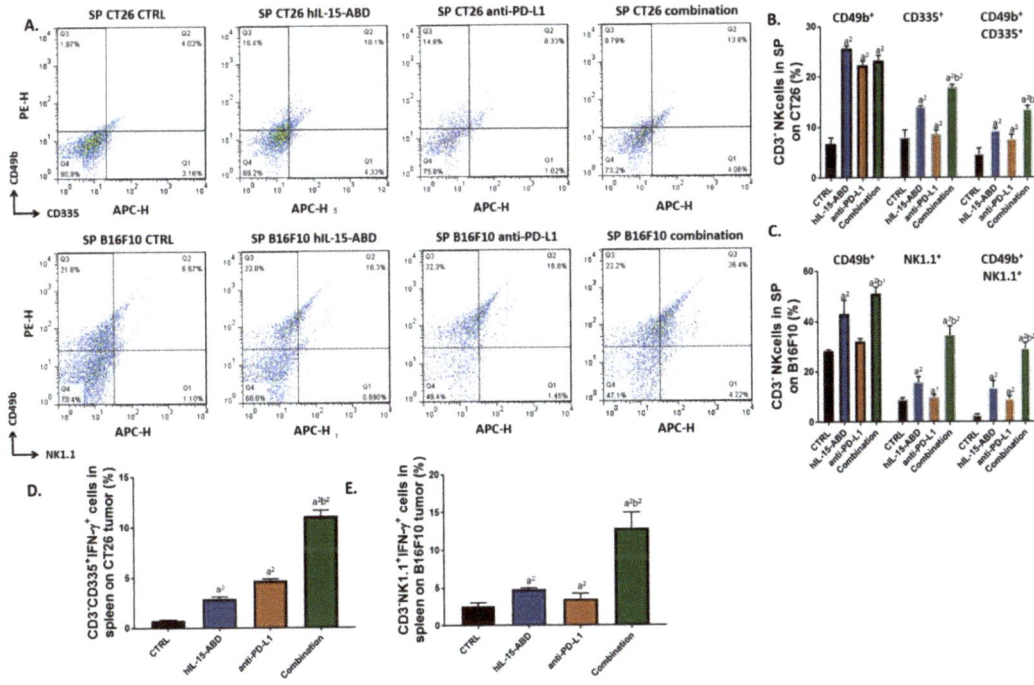

Figure 6. Cont.

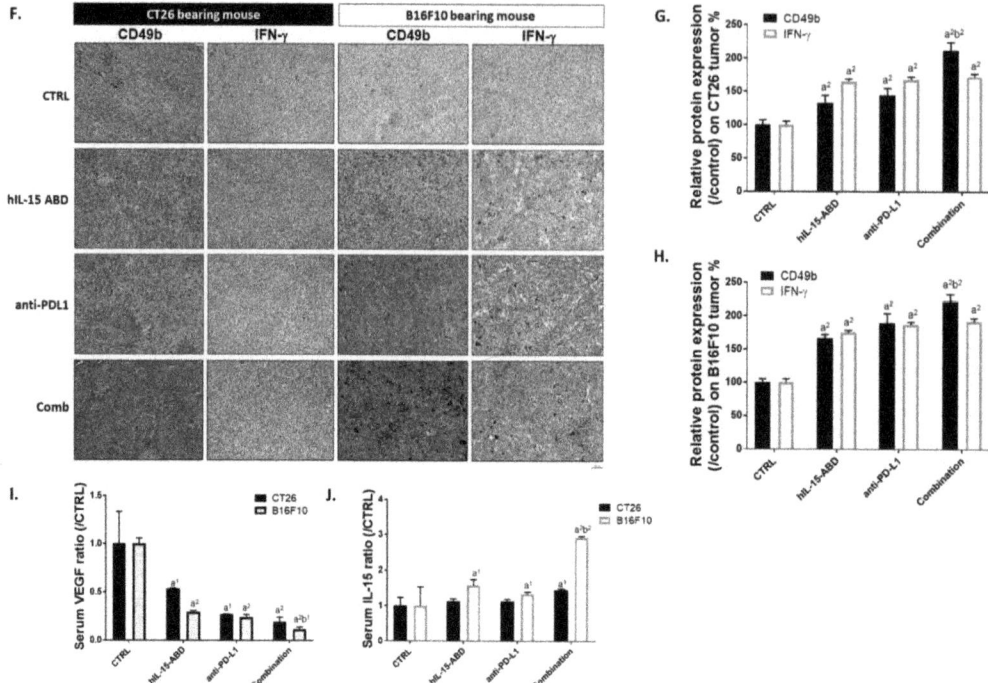

Figure 6. hIP-15-ABD combined anti-PD-L1 antibody effectively trigger the accumulation and function of NK cells. (**A**) Flow cytometry pattern and (**B**) quantification results of CD3$^-$/CD49b$^+$, CD3$^-$/CD335$^+$ and CD3$^-$/CD49b$^+$/CD335$^+$ NK cells from SP on CC bearing BALB/c animal model. (**C**) Quantification results of CD3$^-$/CD49b$^+$, CD3$^-$/NK1.1$^+$ and CD3$^-$/CD49b$^+$/NK1.1$^+$ NK cells from SP on melanoma bearing C57BL/6 animal model. (**D,E**) CD3$^-$/CD335$^+$/IFN-γ$^+$ and CD3-/NK1.1$^+$/IFN-γ$^+$ NK cells from SP on CC and melanoma model is displayed. (**F**) CD49b and IFN-γ IHC staining images and relative proteins quantification level on (**G**) CC and (**H**) melanoma are shown. Expression level of secreted (**I**) VEGF and (**J**) IL15 are shown as quantification results. (a^1 $p < 0.05$, a^2 $p < 0.01$ vs. CTRL; b^1 $p < 0.05$, b^2 $p < 0.01$ vs. hIL15-ABD and anti-PD-L1; scale bar = 100 μm).

3.7. hIL15-ABD Combined Anti-PD-L1 Antibody Diminished the Accumulation of Immunosuppressive Cells in Both CC and Melanoma Models

Next, we determined whether hIL15-ABD promotes anti-tumor capacity of anti-PD-L1 by reducing accumulation of Tregs and MDSCs. Flow cytometry from mice TDLN showed that CD4$^+$/CD25$^+$/FOXP3$^+$ triple positive cells amount was significantly reduced in combination therapy group (Figure 7A). The amount of Tregs was decreased by around 5–10 fold as compared to non-treated control (Figure 7B). At the same time, the percentage of Tregs decreased the most in the combination treatment group (Figure 7C,D). Moreover, we also detected the population of MDSCs within BM and SP of CC and melanoma mice by flow cytometry. The obtained results of flow cytometry from BM indicated the effective diminishment of CD11b$^+$/Gr-1$^+$ MDSCs in combination treatment group (Figure 7E,F). The percentage of CD11b$^+$/Gr-1$^+$ MDSCs within CC and melanoma mice SP was also decreased after combination therapy (Figure 7G,H). Subsequently, we validated the protein expression level of FOXP3 and IDO in mice tumor by IHC staining (Figure 7I). FOXP3 and IDO are known to be important immunosuppressive factors that allow the tumor to escape immunosurveillance. As indicated in Figure 7J,K, proteins expression levels of FOXP3 and IDO in combination therapy group were decreased to 10–30% of that in non-treated control. Our results illustrate that combination of hIL15-ABD and anti-PD-L1 may develop an environment that inhibits the tumor's ability to escape from immune surveillance.

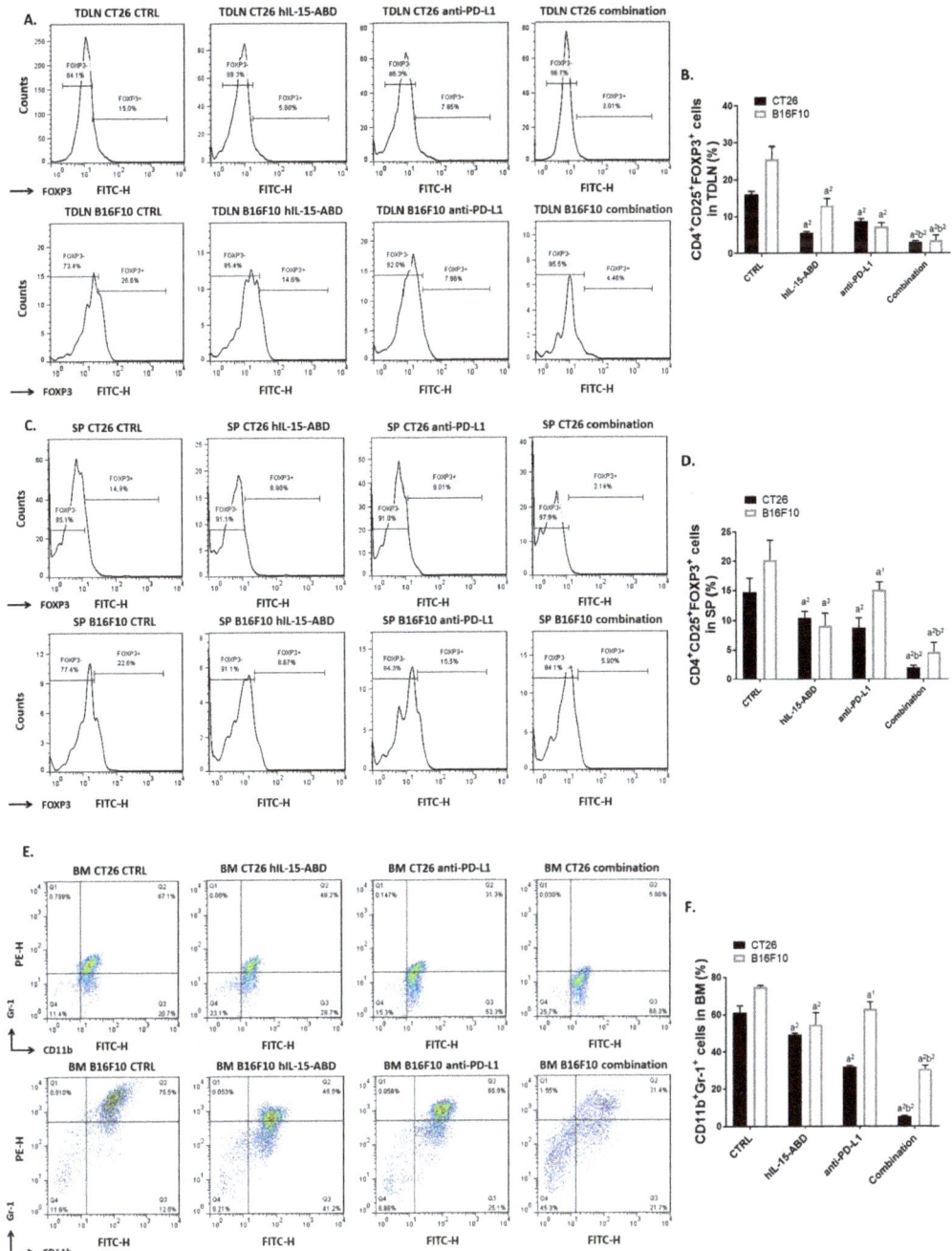

Figure 7. Cont.

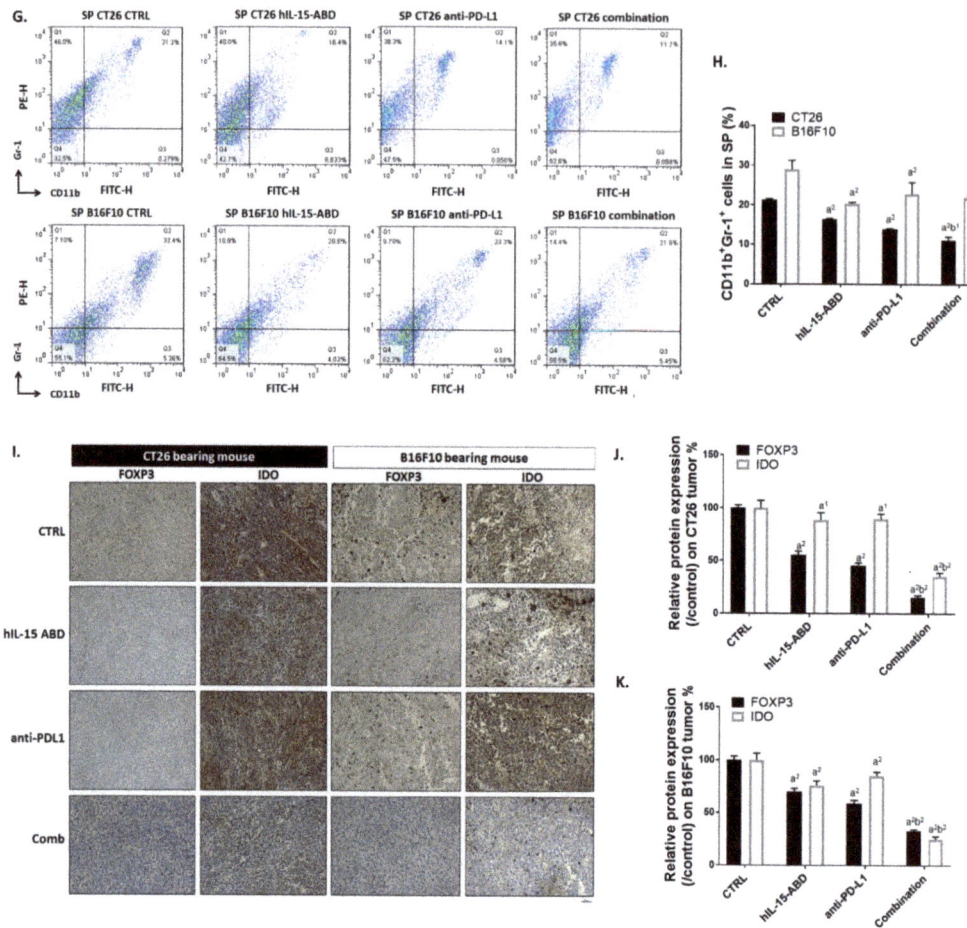

Figure 7. hIP-15-ABD combined anti-PD-L1 antibody successfully suppress the accumulation of immunosuppressive cells. (**A,C**) Flow cytometry pattern and (**B,D**) quantification results of CD3$^-$/CD49b$^+$, CD4$^+$/CD25$^+$/FOXP3$^+$ Tregs from TDLN and SP, respectively. (**E,G**) Flow cytometry pattern and (**F,H**) quantification results of CD11b$^+$/Gr-1$^+$ MDSCs from BM and SP, respectively. (**I**) FOXP3 and IDO IHC staining images and (**J,K**) relative quantification of CC and melanoma are presented. (a^1 $p < 0.05$, a^2 $p < 0.01$ vs. CTRL; b^1 $p < 0.05$, b^2 $p < 0.01$ vs. hIL15-ABD and anti-PD-L1; scale bar = 100 μm).

4. Discussion

In the first human clinical trial, hIL15, as a wild-type (WT) recombinant protein was administrated for 12 consecutive days to patients with metastatic melanoma and renal cell carcinoma [12]. Dose-limiting toxicities of WT hIL15, included grade 3 hypotension, thrombocytopenia, and elevated values of ALT and AST and 0.3 μg/kg per day was determined as the maximum tolerable dose. Although greatly altered homeostasis of lymphocyte subsets, such as NK cells and memory CD8 T cells, as well as anticancer efficacy observed in the first in-human trial of recombinant WT hIL15, it becomes evident that alternative dosing strategies is needed to enhance efficacy while reducing toxicity. Non-human primate pharmacokinetic study verified that constant administration regimens of recombinant IL15 through either continuous intravenous infusion or subcutaneous injection achieve remarkable immune stimulation in the absence of obvious toxicity, indicating potentially better clinical result than the previous bolus intravenous regimen [22]. Clinical trial of recombinant hIL15 administrated subcutaneously daily (Monday through Friday) for two weeks was conducted in patients with refractory solid tumor cancers. This dosing regimen resulted in markedly enhanced circulating $CD56^{bright}$ NK and $CD8^+$ T cells as well as an encouraging safety profile [23].

Although hIL15 displayed encouraging results in early clinical trials, its short half-life suggests potential improvement in anticancer efficacy through engineering hIL15 with prolonged half-life. N-803, the novel hIL15 superagonist complex, comprises N72D mutant IL15 and IL15Rα-IgG Fc fusion protein and displays enhanced affinity for IL-2Rβ and prolonged half-life. It is under multiple clinical trials, including advanced melanoma, renal cell, non-small cell lung, head and neck, hematologic malignancies who relapse after allogeneic hematopoietic cell transplantation and showing encouraging results [16,17,24]. N-803 has been shown to exhibit greater anti-CC activity compared to hIL15 in CT26 bearing model [25]. In this study, CT26 bearing model was also used to evaluate differences in therapeutic efficacy and anticancer immune response between hIL15 and hIL15-ABD treatments. Our results demonstrate that hIL15-ABD group has higher tumor growth inhibition capability and anticancer immunity than hIL15 group (Figure 3). Immunosuppressive cells such as Tregs and MDSCs restrain antitumor immunity through the downregulation of effector T cells and NK cells [26–28]. The increased abundance of Treg or MDSCs in peripheral blood and tumor are associated with poor prognosis in different types of cancer [29–31]. Although the relationship between Treg population and prognosis in patients with colorectal cancer remains uncertain [32,33], depletion of Tregs and MDSCs has been indicated to promote anticancer immunity in colorectal cancer [34,35]. In our results, we present that hIL15-ABD not only significantly increased percentage of $CD8^+$ T and NK cells (Figure 3J–O), but also effectively reduced population of Tregs and MDSCs compared to hIL-15 treatment (Figure 3D–I).

Anti-PD-L1 therapies have been shown promising results as a member of an increasing number of immunotherapies against cancer [36]. However, despite its potential, anti-PD-L1 antibody has failed to elicit objective response in a majority of patients treated [37]. A common explanation for the lack of response is the lack of anti-tumor effector cells in the TME. Both tumor cells and MDSCs express PD-L1, which binds to PD-1 on T cells and causes T cell exhaustion as well as conversion of T helper type 1 (Th1) cells to Tregs. The combination of N-803 and anti-PD-L1 therapy reduced numbers of Tregs and MDSCs in lung [38,39]. Having verified that hIL15-ABD is superior to hIL15 in inhibiting tumor growth and regulating anti-cancer immunity, we investigated the anticancer efficacy and immune response induction of hIL15-ABD combined with anti-PD-L1 in both CC and melanoma models. Our results indicate an obvious enhancement of tumor growth inhibition in CT26 or B16-F10 bearing mice after hIL15-ABD and anti-PD-L1 combined therapy (Figure 4B–G). Furthermore, the combination group had significantly smaller population of Tregs (within TDLN and SP, Figure 7A–D) and MDSCs (within BM and SP, Figure 7E–H) compared to hIL15-ABD or anti-PD-L1 therapy monotherapies.

Both $CD8^+$ T and NK cells are critical executors that mediate tumor cell apoptosis through secretion of granzyme-B and IFN-γ in immunotherapy modulating tumor regression. In addition to hIL15, anti-PD-1/-L1 therapy has also been indicated to enhance anti-tumor efficacy of $CD8^+$ T and NK cells [4,9,40–42]. The combination of N-803 and anti-PD-L1 therapy significantly induced the activated $CD8^+$ T cell phenotype compared to N-803 or anti-PD-L1 monotherapies in murine breast cancer models [43]. The increased number and function of $CD8^+$ T or NK cells were linked to favorable prognosis in patients with colorectal cancer or melanoma [44–46]. Therefore, it is worthwhile to investigate whether hIL15-ABD promotes anti-PD-L1 therapy-elicited activity and percentage of $CD8^+$ T and NK cells in CT26 or B16-F10 bearing mice. In our results, the combination of hIL15-ABD and anti-PD-L1 monotherapy effectively increased percentage of $CD8^+IFN-γ$ or $CD8^+IL-2$ cells in spleen and TDLN compared to hIL15-ABD or anti-PD-L1 therapy (Figure 5A–H). Human IL15-ABD also significantly promoted anti-PD-L1 therapy-induced accumulation and function of NK cells in spleen and TDLN (Figure 6A–H).

Granzyme-B, the granule protease secreted by NK and CD8[+] T cells, induces apoptosis through BAX/BAK-mediated mitochondrial apoptotic pathway [47]. The increased level of granzyme-B in serum or tumor was correlated with favorable outcomes in patients with colorectal cancer or NSCLC [48,49]. In our results indicated the combination group presented significantly higher expression of granzyme-B and apoptotic proteins (BAX and cleaved-caspase-3) in CT-26 or B16-F10 tumor tissues compared to hIL15-ABD or anti-PD-L1 therapy (Figure 5I–K and Figure 4N–P). VEGF, the major angiogenic mediator, contributes to tumor growth and metastasis through promoting new vessel formation. VEGF participates in regulation of Tregs and MDSCs leading to restriction of anti-tumor immunity. The high level of serum VEGF was correlated with poor overall survival of melanoma patients treated with the immune checkpoint inhibitor [50]. Indoleamine 2,3-dioxygenase (IDO), immunosuppressive protein, attenuates anti-tumor function of T cells by regulating the conversion of tryptophan to kynurenine [51]. The decreased expression of IDO was associated with better prognosis in patients with colorectal cancer or melanoma [52]. Our results demonstrated expression of IDO and VEGF was obviously reduced by hIL15-ABD, anti-PD-L1, or combination therapy (Figures 6I and 7I–K). The combination group had lower expression of IDO or VEGF compared to hIL15-ABD or anti-PD-L1 monotherapy in CT26 or B16-F10 bearing mice.

In light of the great potential of IL15 as one of the critical weapons in the arsenal of anticancer immunotherapy, many related therapeutics, ranging from the wild-type IL15, IL15 superagonist (N-803) to PD-L1–targeting IL15 (KD033 [53] and N-809 [54]), are actively being developed in either preclinical or clinical settings. In our study, hIL15-ABD displays a much-extended half-life and superior inhibitory effect on CT26 and B16 growth in experimental mice than WT hIL15, which has shown encouraging results in early human trial [12]. hIL15-ABD could be easily purified and refolded into active form with yields of approximately 60 and 300 mg/L of transformed E. coli TB cultures in shake flask and fermenter, respectively, indicating a relatively lower production cost than those of modified hIL15, such as N-803, KD033 and N-809, expressed by mammalian cells. Intriguingly, while displaying anticancer effect as a monotherapy (Figure 3B), hIL15 treated CT26-bearing mice showed statistically significant body weight loss as compared with those treated with vehicle and hIL15-ABD (Figure 3O), suggesting a better therapeutic window of hIL15-ABD comparing with hIL15. Given that albumin-based carriers for anticancer therapeutics has shown promising results in both preclinical and clinical studies not only through half-life prolongation but also enhanced tumor localization [55], it is of great interest to investigate whether albumin associated hIL15-ABD will obtain a more favorable biodistribution profile, thereby increasing anticancer effects while reducing toxicity to normal organs.

5. Conclusions

In conclusion, for the first time, we presented the hIL15-ABD, the novel recombinant IL15 protein, was superior to in induction of tumor regression and antitumor immunity. hIL15-ABD may suppress the accumulation of MDSCs and Treg at the site of the tumor. In addition, hIL15-ABD can also promote the activity of IL-2 and IFN-γ in CD8[+] T cells or NK cells, supporting more effective anti-tumor activity by effector cells. Importantly, hIL15-ABD can trigger innate immunity by enhancement of NK cells toxicity effect. Furthermore, the combination of hIL15-ABD and anti-PD-L1 therapy significantly inhibited tumor growth and promoted anti-tumor immune response compared to either monotherapy in mouse models of CC or melanoma. We demonstrated enhancement of CD8[+] T and NK cells accumulation and cytotoxic function and reduction of Tregs and MDSCs population are associated with antitumor properties of hIL15-ABD combined with anti-PD-L1 therapy in CC or melanoma. We suggested the combination of hIL15-ABD and anti-PD-L1 therapy as potential immune therapy may offers therapeutic activity for treatment of CC or melanoma.

Author Contributions: Conceptualization, F.-T.H. and K.-L.L.; data curation, F.-T.H., Y.-C.L., C.-L.T. and P.-F.Y.; funding acquisition, F.-T.H.; investigation, P.-F.Y.; methodology, F.-T.H.; project administration, F.-T.H.; supervision, F.-T.H.; validation, F.-T.H. and C.-L.T.; visualization, F.-T.H. and K.-L.L.;

writing—original draft, F.-T.H. and Y.-C.L.; writing—review and editing, F.-T.H., Y.-C.L., C.-H.C. and K.-L.L. All authors have read and agreed to the published version of the manuscript.

Funding: This study was financially supported by a grant from the Ministry of Science and Technology (Taipei), (grant number: MOST 108-2314-B-039-007-MY3, 109-2623-E-010-002-NU and MOST 109-2314-B-758-001). This work was also financially supported by the "Drug Development Center, China Medical University" from The Featured Areas Research Center Program within the framework of the Higher Education Sprout Project by the Ministry of Education (MOE) in Taiwan.

Institutional Review Board Statement: The animal experiments were performed in accordance with the protocols approved by the Animal Care and Use Committee at China Medical University (approval number: CMU IACUC-2019-208).

Informed Consent Statement: Not applicable.

Data Availability Statement: The data generated and analyzed will be made available from the corresponding author on reasonable request.

Acknowledgments: Experiments and data analysis were performed in part through the use of the Medical Research Core Facilities Center, Office of Research & Development at China Medical University, Taichung, Taiwan.

Conflicts of Interest: The authors declare no conflict of interest.

References

1. Munhoz, R.R.; Postow, M.A. Recent advances in understanding antitumor immunity. *F1000Research* **2016**, *5*, 2545. [CrossRef] [PubMed]
2. Zhang, H.; Chen, J. Current status and future directions of cancer immunotherapy. *J. Cancer* **2018**, *9*, 1773–1781. [CrossRef]
3. Jiang, X.; Wang, J.; Deng, X.; Xiong, F.; Ge, J.; Xiang, B.; Wu, X.; Ma, J.; Zhou, M.; Li, X.; et al. Role of the tumor microenvironment in PD-L1/PD-1-mediated tumor immune escape. *Mol. Cancer* **2019**, *18*, 10. [CrossRef]
4. Sun, H.; Sun, C. The Rise of NK Cell Checkpoints as Promising Therapeutic Targets in Cancer Immunotherapy. *Front. Immunol.* **2019**, *10*. [CrossRef]
5. Yu, Y.R.; Ho, P.C. Sculpting tumor microenvironment with immune system: From immunometabolism to immunoediting. *Clin. Exp. Immunol.* **2019**, *197*, 153–160. [CrossRef]
6. Wu, Y.; Chen, W.; Xu, Z.P.; Gu, W. PD-L1 Distribution and Perspective for Cancer Immunotherapy—Blockade, Knockdown, or Inhibition. *Front. Immunol.* **2019**, *10*. [CrossRef] [PubMed]
7. Nixon, N.A.; Blais, N.; Ernst, S.; Kollmannsberger, C.; Bebb, G.; Butler, M.; Smylie, M.; Verma, S. Current landscape of immunotherapy in the treatment of solid tumours, with future opportunities and challenges. *Curr. Oncol.* **2018**, *25*, e373–e384. [CrossRef]
8. Dong, S.; Ghobrial, I.M. Immunotherapy for hematological malignancies. *J. Life Sci. (Westlake VillageCalif.)* **2019**, *1*, 46–52. [CrossRef]
9. Pauken, K.E.; Wherry, E.J. Overcoming T cell exhaustion in infection and cancer. *Trends Immunol.* **2015**, *36*, 265–276. [CrossRef]
10. Xue, S.; Song, G.; Yu, J. The prognostic significance of PD-L1 expression in patients with glioma: A meta-analysis. *Sci. Rep.* **2017**, *7*, 4231. [CrossRef]
11. Yau, T.; Hsu, C.; Kim, T.-Y.; Choo, S.-P.; Kang, Y.-K.; Hou, M.-M.; Numata, K.; Yeo, W.; Chopra, A.; Ikeda, M.; et al. Nivolumab in advanced hepatocellular carcinoma: Sorafenib-experienced Asian cohort analysis. *J. Hepatol.* **2019**, *71*, 543–552. [CrossRef]
12. Conlon, K.C.; Lugli, E.; Welles, H.C.; Rosenberg, S.A.; Fojo, A.T.; Morris, J.C.; Fleisher, T.A.; Dubois, S.P.; Perera, L.P.; Stewart, D.M.; et al. Redistribution, Hyperproliferation, Activation of Natural Killer Cells and CD8 T Cells, and Cytokine Production During First-in-Human Clinical Trial of Recombinant Human Interleukin-15 in Patients With Cancer. *J. Clin. Oncol. Off. J. Am. Soc. Clin. Oncol.* **2015**, *33*, 74–82. [CrossRef] [PubMed]
13. Yu, P.; Steel, J.C.; Zhang, M.; Morris, J.C.; Waldmann, T.A. Simultaneous blockade of multiple immune system inhibitory checkpoints enhances antitumor activity mediated by interleukin-15 in a murine metastatic colon carcinoma model. *Clin. Cancer Res.* **2010**, *16*, 6019–6028. [CrossRef]
14. Robinson, T.O.; Schluns, K.S. The potential and promise of IL-15 in immuno-oncogenic therapies. *Immunol. Lett.* **2017**, *190*, 159–168. [CrossRef]
15. Hu, Q.; Ye, X.; Qu, X.; Cui, D.; Zhang, L.; Xu, Z.; Wan, H.; Zhang, L.; Tao, W. Discovery of a novel IL-15 based protein with improved developability and efficacy for cancer immunotherapy. *Sci. Rep.* **2018**, *8*, 7675. [CrossRef]
16. Margolin, K.; Morishima, C.; Velcheti, V.; Miller, J.S.; Lee, S.M.; Silk, A.W.; Holtan, S.G.; Lacroix, A.M.; Fling, S.P.; Kaiser, J.C.; et al. Phase I Trial of ALT-803, A Novel Recombinant IL15 Complex, in Patients with Advanced Solid Tumors. *Clin. Cancer Res.* **2018**, *24*, 5552–5561. [CrossRef]

17. Wrangle, J.M.; Velcheti, V.; Patel, M.R.; Garrett-Mayer, E.; Hill, E.G.; Ravenel, J.G.; Miller, J.S.; Farhad, M.; Anderton, K.; Lindsey, K.; et al. ALT-803, an IL-15 superagonist, in combination with nivolumab in patients with metastatic non-small cell lung cancer: A non-randomised, open-label, phase 1b trial. *Lancet Oncol.* **2018**, *19*, 694–704. [CrossRef]
18. Weng, M.C.; Wang, M.H.; Tsai, J.J.; Kuo, Y.C.; Liu, Y.C.; Hsu, F.T.; Wang, H.E. Regorafenib inhibits tumor progression through suppression of ERK/NF-κB activation in hepatocellular carcinoma bearing mice. *Biosci. Rep.* **2018**, *38*. [CrossRef]
19. Ghanekar, S.A.; Nomura, L.E.; Suni, M.A.; Picker, L.J.; Maecker, H.T.; Maino, V.C. Gamma interferon expression in CD8(+) T cells is a marker for circulating cytotoxic T lymphocytes that recognize an HLA A2-restricted epitope of human cytomegalovirus phosphoprotein pp65. *Clin. Diagn. Lab. Immunol.* **2001**, *8*, 628–631. [CrossRef]
20. Werner, J.M.; Busl, E.; Farkas, S.A.; Schlitt, H.J.; Geissler, E.K.; Hornung, M. DX5+NKT cells display phenotypical and functional differences between spleen and liver as well as NK1.1-Balb/c and NK1.1+ C57Bl/6 mice. *BMC Immunol.* **2011**, *12*, 26. [CrossRef]
21. Hsu, F.T.; Chen, T.C.; Chuang, H.Y.; Chang, Y.F.; Hwang, J.J. Enhancement of adoptive T cell transfer with single low dose pretreatment of doxorubicin or paclitaxel in mice. *Oncotarget* **2015**, *6*, 44134–44150. [CrossRef]
22. Sneller, M.C.; Kopp, W.C.; Engelke, K.J.; Yovandich, J.L.; Creekmore, S.P.; Waldmann, T.A.; Lane, H.C. IL-15 administered by continuous infusion to rhesus macaques induces massive expansion of CD8+ T effector memory population in peripheral blood. *Blood* **2011**, *118*, 6845–6848. [CrossRef]
23. Miller, J.S.; Morishima, C.; McNeel, D.G.; Patel, M.R.; Kohrt, H.E.K.; Thompson, J.A.; Sondel, P.M.; Wakelee, H.A.; Disis, M.L.; Kaiser, J.C.; et al. A First-in-Human Phase I Study of Subcutaneous Outpatient Recombinant Human IL15 (rhIL15) in Adults with Advanced Solid Tumors. *Clin. Cancer Res.* **2017**. [CrossRef]
24. Romee, R.; Cooley, S.; Berrien-Elliott, M.M.; Westervelt, P.; Verneris, M.R.; Wagner, J.E.; Weisdorf, D.J.; Blazar, B.R.; Ustun, C.; DeFor, T.E.; et al. First-in-human phase 1 clinical study of the IL-15 superagonist complex ALT-803 to treat relapse after transplantation. *Blood* **2018**, *131*, 2515–2527. [CrossRef]
25. Rhode, P.R.; Egan, J.O.; Xu, W.; Hong, H.; Webb, G.M.; Chen, X.; Liu, B.; Zhu, X.; Wen, J.; You, L.; et al. Comparison of the Superagonist Complex, ALT-803, to IL15 as Cancer Immunotherapeutics in Animal Models. *Cancer Immunol. Res.* **2016**, *4*, 49–60. [CrossRef]
26. Wang, Y.; Ma, Y.; Fang, Y.; Wu, S.; Liu, L.; Fu, D.; Shen, X. Regulatory T cell: A protection for tumour cells. *J. Cell Mol. Med.* **2012**, *16*, 425–436. [CrossRef]
27. Bruno, A.; Mortara, L.; Baci, D.; Noonan, D.M.; Albini, A. Myeloid Derived Suppressor Cells Interactions With Natural Killer Cells and Pro-angiogenic Activities: Roles in Tumor Progression. *Front. Immunol.* **2019**, *10*, 771. [CrossRef]
28. Özkan, B.; Lim, H.; Park, S.G. Immunomodulatory Function of Myeloid-Derived Suppressor Cells during B Cell-Mediated Immune Responses. *Int. J. Mol. Sci.* **2018**, *19*, 1468. [CrossRef] [PubMed]
29. Togashi, Y.; Shitara, K.; Nishikawa, H. Regulatory T cells in cancer immunosuppression - implications for anticancer therapy. *Nat. Rev. Clin. Oncol.* **2019**, *16*, 356–371. [CrossRef]
30. Jordan, K.R.; Amaria, R.N.; Ramirez, O.; Callihan, E.B.; Gao, D.; Borakove, M.; Manthey, E.; Borges, V.F.; McCarter, M.D. Myeloid-derived suppressor cells are associated with disease progression and decreased overall survival in advanced-stage melanoma patients. *Cancer Immunol. Immunother.* **2013**, *62*, 1711–1722. [CrossRef] [PubMed]
31. Xu, J.; Peng, Y.; Yang, M.; Guo, N.; Liu, H.; Gao, H.; Niu, F.; Wang, R.; Wang, C.; Yu, K. Increased levels of myeloid-derived suppressor cells in esophageal cancer patients is associated with the complication of sepsis. *Biomed. Pharmacother. Biomed. Pharmacother.* **2020**, *125*, 109864. [CrossRef]
32. Saito, T.; Nishikawa, H.; Wada, H.; Nagano, Y.; Sugiyama, D.; Atarashi, K.; Maeda, Y.; Hamaguchi, M.; Ohkura, N.; Sato, E.; et al. Two FOXP3(+)CD4(+) T cell subpopulations distinctly control the prognosis of colorectal cancers. *Nat. Med.* **2016**, *22*, 679–684. [CrossRef]
33. Mougiakakos, D. Regulatory T cells in colorectal cancer: From biology to prognostic relevance. *Cancers* **2011**, *3*, 1708–1731. [CrossRef] [PubMed]
34. Terme, M.; Pernot, S.; Marcheteau, E.; Sandoval, F.; Benhamouda, N.; Colussi, O.; Dubreuil, O.; Carpentier, A.F.; Tartour, E.; Taieb, J. VEGFA-VEGFR pathway blockade inhibits tumor-induced regulatory T-cell proliferation in colorectal cancer. *Cancer Res.* **2013**, *73*, 539–549. [CrossRef]
35. Kim, H.S.; Park, H.M.; Park, J.S.; Sohn, H.J.; Kim, S.G.; Kim, H.J.; Oh, S.T.; Kim, T.G. Dendritic cell vaccine in addition to FOLFIRI regimen improve antitumor effects through the inhibition of immunosuppressive cells in murine colorectal cancer model. *Vaccine* **2010**, *28*, 7787–7796. [CrossRef]
36. Brahmer, J.R.; Drake, C.G.; Wollner, I.; Powderly, J.D.; Picus, J.; Sharfman, W.H.; Stankevich, E.; Pons, A.; Salay, T.M.; McMiller, T.L.; et al. Phase I study of single-agent anti-programmed death-1 (MDX-1106) in refractory solid tumors: Safety, clinical activity, pharmacodynamics, and immunologic correlates. *J. Clin. Oncol. Off. J. Am. Soc. Clin. Oncol.* **2010**, *28*, 3167–3175. [CrossRef]
37. Apolo, A.B.; Infante, J.R.; Balmanoukian, A.; Patel, M.R.; Wang, D.; Kelly, K.; Mega, A.E.; Britten, C.D.; Ravaud, A.; Mita, A.C.; et al. Avelumab, an Anti-Programmed Death-Ligand 1 Antibody, In Patients With Refractory Metastatic Urothelial Carcinoma: Results From a Multicenter, Phase Ib Study. *J. Clin. Oncol. Off. J. Am. Soc. Clin. Oncol.* **2017**, *35*, 2117–2124. [CrossRef]
38. Amarnath, S.; Mangus, C.W.; Wang, J.C.; Wei, F.; He, A.; Kapoor, V.; Foley, J.E.; Massey, P.R.; Felizardo, T.C.; Riley, J.L.; et al. The PDL1-PD1 axis converts human TH1 cells into regulatory T cells. *Sci. Transl. Med.* **2011**, *3*, 111ra120. [CrossRef]
39. Beldi-Ferchiou, A.; Caillat-Zucman, S. Control of NK Cell Activation by Immune Checkpoint Molecules. *Int. J. Mol. Sci.* **2017**, 2129. [CrossRef]

40. Cullen, S.P.; Brunet, M.; Martin, S.J. Granzymes in cancer and immunity. *Cell Death Differ.* **2010**, *17*, 616–623. [CrossRef]
41. Wall, L.; Burke, F.; Barton, C.; Smyth, J.; Balkwill, F. IFN-gamma induces apoptosis in ovarian cancer cells in vivo and *in vitro*. *Clin. Cancer Res. Off. J. Am. Assoc. Cancer Res.* **2003**, *9*, 2487–2496.
42. Calik, I.; Calik, M.; Turken, G.; Ozercan, I.H.; Dagli, A.F.; Artas, G.; Sarikaya, B. Intratumoral Cytotoxic T-Lymphocyte Density and PD-L1 Expression Are Prognostic Biomarkers for Patients with Colorectal Cancer. *Medicina* **2019**, *55*, 723. [CrossRef]
43. Knudson, K.M.; Hicks, K.C.; Alter, S.; Schlom, J.; Gameiro, S.R. Mechanisms involved in IL-15 superagonist enhancement of anti-PD-L1 therapy. *J. Immunother. Cancer* **2019**, *7*, 82. [CrossRef]
44. Jung, Y.S.; Kwon, M.J.; Park, D.I.; Sohn, C.I.; Park, J.H. Association between natural killer cell activity and the risk of colorectal neoplasia. *J. Gastroenterol. Hepatol.* **2018**, *33*, 831–836. [CrossRef]
45. Cursons, J.; Souza-Fonseca-Guimaraes, F.; Foroutan, M.; Anderson, A.; Hollande, F.; Hediyeh-Zadeh, S.; Behren, A.; Huntington, N.D.; Davis, M.J. A Gene Signature Predicting Natural Killer Cell Infiltration and Improved Survival in Melanoma Patients. *Cancer Immunol. Res.* **2019**, *7*, 1162. [CrossRef]
46. Metkar, S.S.; Wang, B.; Ebbs, M.L.; Kim, J.H.; Lee, Y.J.; Raja, S.M.; Froelich, C.J. Granzyme B activates procaspase-3 which signals a mitochondrial amplification loop for maximal apoptosis. *J. Cell Biol.* **2003**, *160*, 875–885. [CrossRef]
47. Prizment, A.E.; Vierkant, R.A.; Smyrk, T.C.; Tillmans, L.S.; Nelson, H.H.; Lynch, C.F.; Pengo, T.; Thibodeau, S.N.; Church, T.R.; Cerhan, J.R.; et al. Cytotoxic T Cells and Granzyme B Associated with Improved Colorectal Cancer Survival in a Prospective Cohort of Older Women. *Cancer Epidemiol. Prev. Biomark.* **2017**, *26*, 622–631. [CrossRef]
48. Hurkmans, D.P.; Basak, E.A.; Schepers, N.; Oomen-De Hoop, E.; Van der Leest, C.H.; El Bouazzaoui, S.; Bins, S.; Koolen, S.L.W.; Sleijfer, S.; Van der Veldt, A.A.M.; et al. Granzyme B is correlated with clinical outcome after PD-1 blockade in patients with stage IV non-small-cell lung cancer. *J. Immunother. Cancer* **2020**, *8*, e000586. [CrossRef]
49. Ott, P.A.; Hodi, F.S.; Buchbinder, E.I. Inhibition of Immune Checkpoints and Vascular Endothelial Growth Factor as Combination Therapy for Metastatic Melanoma: An Overview of Rationale, Preclinical Evidence, and Initial Clinical Data. *Front. Oncol.* **2015**, *5*, 202. [CrossRef]
50. Terai, M.; Londin, E.; Rochani, A.; Link, E.; Lam, B.; Kaushal, G.; Bhushan, A.; Orloff, M.; Sato, T. Expression of Tryptophan 2,3-Dioxygenase in Metastatic Uveal Melanoma. *Cancers* **2020**, *12*, 405. [CrossRef]
51. Brandacher, G.; Perathoner, A.; Ladurner, R.; Schneeberger, S.; Obrist, P.; Winkler, C.; Werner, E.R.; Werner-Felmayer, G.; Weiss, H.G.; Göbel, G.; et al. Prognostic value of indoleamine 2,3-dioxygenase expression in colorectal cancer: Effect on tumor-infiltrating T cells. *Clin. Cancer Res. Off. J. Am. Assoc. Cancer Res.* **2006**, *12*, 1144–1151. [CrossRef]
52. Rubel, F.; Kern, J.S.; Technau-Hafsi, K.; Uhrich, S.; Thoma, K.; Häcker, G.; von Bubnoff, N.; Meiss, F.; von Bubnoff, D. Indoleamine 2,3-Dioxygenase Expression in Primary Cutaneous Melanoma Correlates with Breslow Thickness and Is of Significant Prognostic Value for Progression-Free Survival. *J. Investig. Dermatol.* **2018**, *138*, 679–687. [CrossRef]
53. Martomo, S.A.; Lu, D.; Polonskaya, Z.; Luna, X.; Zhang, Z.; Feldstein, S.; Lumban-Tobing, R.; Almstead, D.K.; Miyara, F.; Patel, J. Single-Dose Anti-PD-L1/IL-15 Fusion Protein KD033 Generates Synergistic Antitumor Immunity with Robust Tumor-Immune Gene Signatures and Memory Responses. *Mol. Cancer* **2021**, *20*, 347–356. [CrossRef]
54. Jochems, C.; Tritsch, S.R.; Knudson, K.M.; Gameiro, S.R.; Rumfield, C.S.; Pellom, S.T.; Morillon, Y.M.; Newman, R.; Marcus, W.; Szeto, C.; et al. The multi-functionality of N-809, a novel fusion protein encompassing anti-PD-L1 and the IL-15 superagonist fusion complex. *Oncoimmunology* **2019**, *8*, e1532764. [CrossRef]
55. Hoogenboezem, E.N.; Duvall, C.L. Harnessing albumin as a carrier for cancer therapies. *Adv. Drug Deliv. Rev.* **2018**, *130*, 73–89. [CrossRef]

Article

Vaccine Increases the Diversity and Activation of Intratumoral T Cells in the Context of Combination Immunotherapy

Lucas A. Horn [1], Kristen Fousek [1], Duane H. Hamilton [1], James W. Hodge [1], John A. Zebala [2], Dean Y. Maeda [2], Jeffrey Schlom [1] and Claudia Palena [1,*]

1 Laboratory of Tumor Immunology and Biology, Center for Cancer Research, National Cancer Institute, National Institutes of Health, Bethesda, MD 20892, USA; lucas.horn@nih.gov (L.A.H.); kristen.fousek@nih.gov (K.F.); duane.hamilton@nih.gov (D.H.H.); hodgej@mail.nih.gov (J.W.H.); schlomj@mail.nih.gov (J.S.)
2 Syntrix Pharmaceuticals, Auburn, WA 98001, USA; jzebala@syntrixbio.com (J.A.Z.); dmaeda@syntrixbio.com (D.Y.M.)
* Correspondence: palenac@mail.nih.gov; Tel.: +1-240-858-3475; Fax: +1-240-541-4558

Simple Summary: Innovative strategies to reduce immune suppression and activate tumor-specific immunity are needed to help patients who do not respond or become resistant to immune checkpoint blockade therapies. In this study, we demonstrate that the addition of a cancer vaccine targeting a tumor-associated antigen to a checkpoint inhibitor-based immunotherapy induces greater numbers of proliferative, activated, and cytotoxic tumor-infiltrating T cells, leading to improved antitumor activity in tumors otherwise resistant to immunotherapy. Our results provide the rationale for the addition of cancer vaccines in combination immunotherapy approaches being evaluated in the clinic.

Abstract: Resistance to immune checkpoint blockade therapy has spurred the development of novel combinations of drugs tailored to specific cancer types, including non-inflamed tumors with low T-cell infiltration. Cancer vaccines can potentially be utilized as part of these combination immunotherapies to enhance antitumor efficacy through the expansion of tumor-reactive T cells. Utilizing murine models of colon and mammary carcinoma, here we investigated the effect of adding a recombinant adenovirus-based vaccine targeting tumor-associated antigens with an IL-15 super agonist adjuvant to a multimodal regimen consisting of a bifunctional anti-PD-L1/TGF-βRII agent along with a CXCR1/2 inhibitor. We demonstrate that the addition of vaccine induced a greater tumor infiltration with T cells highly positive for markers of proliferation and cytotoxicity. In addition to this enhancement of cytotoxic T cells, combination therapy showed a restructured tumor microenvironment with reduced T_{regs} and $CD11b^+Ly6G^+$ myeloid cells. Tumor-infiltrating immune cells exhibited an upregulation of gene signatures characteristic of a Th1 response and presented with a more diverse T-cell receptor (TCR) repertoire. These results provide the rationale for the addition of vaccine-to-immune checkpoint blockade-based therapies being tested in the clinic.

Keywords: cancer vaccine; combination immunotherapy; TCR diversity

1. Introduction

Immune checkpoint blockade therapies have led to successful and durable responses in patients with various tumor types [1,2]. Despite this great success, only a small percentage of patients with solid malignancies experience complete responses with antibodies directed against programmed cell death protein 1 (PD-1), programmed death ligand 1 (PD-L1), or cytotoxic T-lymphocyte associated protein 4 (CTLA-4) as monotherapies [3]. Expanding knowledge of the mechanisms of immunoregulation and resistance to immune checkpoint blockade therapy has allowed researchers to better formulate combinations of drugs aimed at simultaneously targeting the numerous inhibitory factors and cell types responsible for tumor-induced immune suppression and treatment failure [4,5].

Immunologically "cold" or non-inflamed tumors present with a series of unique problems that cannot be overcome by immune checkpoint blockade or modification of the tumor microenvironment (TME) [6,7], including deficiencies in T-cell recognition of tumor antigens, dendritic cell priming, and lymphocyte homing to the tumor tissue. One approach being investigated to potentially address these additional problems is the incorporation of a therapeutic cancer vaccine to other immunotherapeutic regimens. Studies in murine models have demonstrated that checkpoint blockade antibodies are more effective when combined with cancer vaccines than checkpoint blockade alone, even in tumors that are refractory to checkpoint blockade monotherapy [8,9]. Other studies have shown that addition of a cancer vaccine can promote epitope spreading and antigen cascade [10]; this increase in T-cell receptor (TCR) diversity has been shown to drive more potent antitumor immunity and tumor clearance [11]. Furthermore, cancer vaccines targeted to cancer-associated antigens or neoantigens have had success in the clinic and have been shown to be safe and well tolerated by patients [12–14].

Bintrafusp alfa is a first-in-class bifunctional fusion protein composed of the extra-cellular domain of the human transforming growth factor β receptor II (TGF-βRII) fused to the C-terminus of each heavy chain of an IgG1 antibody blocking PD-L1. This agent is currently being evaluated in multiple clinical studies, showing clinical activity with a confirmed objective response rate of 30.5% in patients with human papillomavirus-associated malignancies [15,16]. In a previous study, we showed that the combination of bintrafusp alfa with SX-682, a small molecule inhibitor of the chemokine receptors CXCR1 and CXCR2 that blocks signaling initiated by IL-8 and other chemokines of the CXCL family, synergizes to mediate antitumor activity in murine models of breast and lung cancer [17]. To test our hypothesis that a vaccine could help overcome some of the challenges presented by tumors that are refractory to checkpoint blockade, in the present study we investigated the effect of adding a vaccine consisting of a recombinant adenovirus serotype-5 (Ad5) vector encoding a tumor-associated antigen in combination with N-803 as an adjuvant [18] to the bintrafusp alfa/SX-682 combination. N-803 is an IL-15 super agonist that helps activate antigen-specific T cells and has shown clinical activity in combination with checkpoint blockade in non-small cell lung cancer [19,20].

Using murine models of colon and breast cancer, we demonstrate that the addition of vaccine to bintrafusp alfa/SX-682 significantly increases tumor infiltration with T cells, enhances T-cell activation and TCR diversity at the tumor site, and diversifies the number of tumor antigens being recognized by TCRs through the phenomenon of antigen cascade or epitope spreading. These results provide the rationale for the addition of cancer vaccines as integral components in combination immunotherapy approaches being evaluated in the clinic.

2. Materials and Methods

2.1. Cell Lines

BALB/c-derived 4T1 mammary carcinoma cells were obtained and cultured as recommended by the American Type Culture Collection (ATCC, Manassas, VA, USA). MC38-CEA cells were previously obtained by retroviral transduction of C57BL/6-derived MC38 colon cancer cells to overexpress human carcinoembryonic antigen (CEA) [21]. Cell lines were tested to be mycoplasma free using a MycoAlert Mycoplasma Detection Kit (Lonza, Basel, Switzerland) and used at low passage number.

2.2. Mice

Female BALB/c mice were obtained from the NCI Frederick Cancer Research Facility. Mice expressing human CEA on a C57BL/6 background (CEA.Tg) were generously provided by Dr. John Shively (City of Hope, Duarte, CA, USA). Mice were approximately 4 to 6 weeks old at start of experiments and were maintained under pathogen-free conditions in accordance with the Association for Assessment and Accreditation of Laboratory Animal

Care guidelines. All animal studies were approved by the NIH Intramural Animal Care and Use Committee (LTIB-038) on 9 January 2018.

2.3. Tumor Inoculation, Treatment Schedule, and Metastasis Assay

BALB/c mice were injected in the abdominal mammary fat pad with 3×10^4 4T1 cells. CEA transgenic mice (CEA.Tg) were injected subcutaneously (s.c.) in the flank with 3×10^5 MC38-CEA cells. Control diet feed or SX-682-containing feed (1428.5 mg/kg, equivalent to a dose of 200 mg/kg body weight/day; Research Diets, New Brunswick, NJ, USA) were administered to mice starting on day 7. SX-682 was provided by Syntrix Pharmaceuticals under a Cooperative Research and Development Agreement (CRADA) with the NCI. In tumor volume experiments, intraperitoneal injections (i.p.) of bintrafusp alfa (kindly provided by EMD Serono under a CRADA) were given at a dose of 200 µg per mouse starting on day 14 and every 7 days thereafter, as noted. The vaccine utilized in this study consisted of a recombinant Ad5 encoding either the tumor antigen murine Twist1, a transcription factor that is overexpressed in 4T1 tumors [22], or human CEA, which is over-expressed in MC38-CEA tumors. The Ad-vector was combined with the IL-15 super agonist N-803 as an adjuvant. The antitumor efficacy of this vaccine formulation was previously described [18], and its optimized performance was confirmed here in terms of induction of higher levels of the Th1 cytokine, TNFα, in the serum of animals in the combined Ad-vector + N-803 group versus each single agent (Figure S1). Adenovirus vaccine was administered s.c. (1×10^{10} viral particles) on day 7 (prime) followed by s.c. adenovirus vaccine (1×10^{10} viral particles) plus N-803 (1 µg, s.c.) every 7 days as noted (boosts).

Metastasis assays were performed as previously described with some modifications [17]. Lungs were harvested from 4T1 tumor-bearing mice under sterile conditions, rinsed in phosphate buffer saline (PBS), transferred to gentleMACS C tubes (Miltenyi Biotec, Waltham, MA, USA) in RPMI-1640 medium containing 5% fetal bovine serum (FBS), 5 mg/mL collagenases IV and I (Gibco, Gaithersburg, MD, USA), and 40 U/mL DNase, and dissociated using a gentleMACS tissue dissociator (Miltenyi Biotec), following the manufacturer's recommended procedure. Cells were passed through a 70 µm filter, pelleted and washed with PBS, and resuspended in 10 mL RPMI-1640 medium supplemented with 10% FBS, 1% Na pyruvate, 1% Hepes, 1× glutamine, 1× gentamicin, and 1× penicillin-streptomycin. A 250 µL aliquot of this suspension, representing 1/40 of the total lung, was cultured in the same medium containing 60 µM 6-thioguanine for 14 days. Colonies were fixed with methanol, stained with 0.05% (w/v) methylene blue, air-dried, and counted. The number of metastases per lung was calculated as the number of colonies counted per flask ×40.

In mouse experiments quantifying TCR diversity, control or SX-682-containing feed were administered to mice starting on day 7 with i.p. injections of bintrafusp alfa given at a dose of 492 µg per mouse on days 9 and 11. The vaccine was administered s.c. (1×10^{10} viral particles) plus s.c. N-803 (1 µg) on day 9. Tumors were collected on day 17 post-tumor injection for subsequent TCR sequence analysis, as indicated below. Adenovirus vaccines and N-803 were kindly provided by ImmunityBio under a CRADA. In all experiments, tumors were measured every 2–3 days in two perpendicular diameters. Tumor volume = (short diameter^2 × long diameter)/2.

2.4. Depletion Studies

To deplete CD8$^+$ T cells from MC38-CEA tumor-bearing mice, 100 µg of anti-CD8 (clone 2.43, BioXcell, Lebanon, NH, USA) depletion antibodies were administered i.p. starting on days 5, 6, and 7 post-tumor implantation and then once per week for the duration of the experiment. Blood was obtained from all animals upon termination of the experiment to determine immune cell population depletion efficiency by flow cytometry.

2.5. Flow Cytometry

Prior to staining, tumors were weighed, mechanically dissociated, incubated in a shaker at 37 °C for 30 min at a speed of 300 rpm in RPMI-1640 medium containing 5% FBS, 5 mg/mL collagenases IV and I (Gibco), and 40 U/mL DNase, and then passed through a 70 μm filter as a single-cell suspension. Spleens were crushed through a 70 μm filter and red cell lysis was performed with ammonium-chloride-potassium (ACK) buffer (Gibco). All antibodies used for flow cytometry were purchased from Thermo Fisher Scientific (Waltham, MA, USA), BioLegend (San Diego, CA, USA), or BD Biosciences (San Jose, CA, USA). Cells were stained for cell surface expression in flat-bottom 96-well plates on ice in phosphate buffered saline with 2% FBS. Intracellular markers were stained using the eBioscience Foxp3/Transcription Factor Staining Buffer Set according to the manufacturer's instructions. Fluorescently conjugated antibodies for CD45 (30-F11), CD3 (500A2), CD4 (RM4-5), CD8 (53-6.7), CD44 (IM7), CD62L (MEL14), Foxp3 (150D), Ki67 (16A8), GzmB (QA18A28), Ly6G (1A8), Ly6C (HK1.4), CD11b (M1/70), F4/80 (BM8), and CD11c (N418) were used as per the manufacturers' instructions. LIVE/DEAD Fixable Aqua Dead Cell Stain Kit (Thermo Fisher Scientific) was used to gate on live cells. Data were acquired on an Attune NxT Flow Cytometer (Thermo Fisher Scientific) and analyzed via FlowJo (FlowJo, Ashland, OR). Immune cell subsets were defined as: CD4 = $CD3^+CD4^+$; CD8 = $CD3^+CD8^+$; T_{CM} = $CD3^+CD44^+CD62L^+$; $T_{Eff\&EM}$ = $CD3^+CD44^+CD62L^-$; T_{regs} = $CD4^+Foxp3^+$.

2.6. ELISPOT Assays

CEA.Tg mice bearing MC38-CEA tumors were fed an SX-682-containing diet starting on day 7; on days 14 and 21, mice received i.p. injections of bintrafusp alfa, with a priming vaccine dose of s.c. Ad-CEA administered on day 7 and boosting doses of Ad-CEA/N-803 vaccine on days 14 and 21. Control mice were left untreated and fed a base diet without SX-682. Splenocytes were harvested from control versus treated mice and assayed ex vivo on day 24 for antigen-dependent cytokine secretion using an IFNγ ELISPOT assay (BD Biosciences), according to the manufacturer's instructions. Briefly, 0.5×10^6 splenocytes were incubated overnight with 10 μg/mL of $CEA_{526-533}$, $p15e_{604-611}$, the MC38 neoepitope PTGFR, or a negative control peptide [10]. Spot-forming cells were quantified using an ImmunoSpot analyzer (Cellular Technology, Ltd, Shaker Heights, OH, USA). The amount of $CD8^+$ T cells added per well was calculated by flow cytometry analysis. Data were adjusted to the number of spots/0.5×10^5 $CD8^+$ T cells present in the assay, subtracting the number of spots in paired wells containing the control peptide.

2.7. Real-Time PCR, Nanostring and TCR Analysis

Total RNA from flash-frozen tumor sections was prepared using the RNeasy Mini Kit (Qiagen, Hilden, Germany). For some experiments, RNA was then reverse-transcribed using SMARTer® PCR cDNA Synthesis Kit (Takara Bio Inc, Mountain View, CA, USA) or the High-Capacity cDNA Reverse Transcription Kit (ThermoFisher Scientific) as per the manufacturer's instructions. cDNA was amplified in triplicate using TaqMan Master Mix in an Applied Biosystems 7500 Real-Time PCR System (ThermoFisher Scientific). The following Taqman gene expression assays were used (ThermoFisher Scientific): Cd247 (Mm00446171_m1), Gzmk (Mm00492530_m1), CD8a (Mm01182107_g1), Prf1 (Mm00812512_m1), Gzmb (Mm00442837_m1), Cd3e (Mm01179194_m1), Pdcd1 (Mm0043494 6_m1), Tbx21 (Mm00450960_m1). NanoString analysis was performed on purified RNA samples from indicated tumors by using the PanCancer Immune Profiling Gene Expression Panel. The nSolver analysis software was used for data normalization (NanoString Technologies, Seattle, WA, USA). Further clustering and pathway analyses were performed using Ingenuity Pathway Analysis (Qiagen). To assess TCR diversity, genomic DNA was purified from whole tumor using the QIAamp DNA Mini Kit (Qiagen). TCRβ chain sequencing was then performed by Adaptive Biotechnologies and analyzed using the Immunoseq analyzer. Simpson clonality (square root of sum over all observed rearrangements of the square fractional abundances of each rearrangement) was calculated as a measurement of the observed TCRβ repertoire. The number of clones representing the top 25% of TCR sequences was used as a metric of the relative diversity of the immune response.

2.8. OPAL Immunofluorescence

Tumor tissue was fixed in Z-fix (Anatech, Battle Creek, MI, USA), embedded in paraffin, and sectioned onto glass slides (American HistoLabs, Gaithersburg, MD, USA). Slides were stained using the Opal 4-Color Manual IHC Kit (PerkinElmer, Waltham, MA, USA). Antigen retrieval was performed with Rodent Decloaker (BioCare Medial, Pacheco, CA, USA) antigen retrieval solution and blocked with BLOXALL Blocking Solution (Vector Laboratories, Burlingame, CA, USA). All other steps, including staining with primary and secondary antibodies and OPAL fluorophore working solution, were conducted following the manufacturer's instructions. Antibodies used included anti-CD4 (4SM95, Invitrogen, Carlsbad, CA) and anti-CD8a (4SM16, Invitrogen). Slide scanning was performed on an Axio Scan.Z1 and Zen software (Zeiss, Oberkochen, Germany).

2.9. Statistical Methods

All statistical analyses were performed using GraphPad Prism V.7 for Windows (GraphPad Software, La Jolla, CA, USA). Analysis of tumor growth curves was conducted using two-way analysis of variance (ANOVA). Statistical differences between two sets of data were determined through a two-tailed Student's t-test, while one-way ANOVA with Tukey's post hoc test was used to determine statistical differences among three or more sets of data. Statistical differences between survival plots were determined using Log-rank (Mantel-Cox) test. Error bars represent SEM where noted. Asterisks indicate that the experimental p value is statistically significantly different from the associated controls at * $p \leq 0.05$; ** $p \leq 0.01$; *** $p \leq 0.001$, **** $p \leq 0.0001$.

3. Results

3.1. Addition of Vaccine to Checkpoint Blockade-Based Therapy Enhances Immune T-Cell Infiltration and Promotes a Th1 Tumor-Infiltrating Lymphocyte (TIL) Phenotype

The effect of adding a cancer vaccine to the combination bintrafusp alfa/SX-682 was first evaluated with CEA.Tg mice, where CEA is a self-antigen [23,24], bearing subcutaneous MC38-CEA tumors. To model a scenario where tumors do not respond to checkpoint-based immunotherapy, control feed or SX-682-containing feed were administered to mice starting on day 7, while administration of bintrafusp alfa at a low dose was delayed until day 14 to ensure response failure. In the vaccine treatment groups, mice were administered a priming vaccine dose of Ad-CEA on day 7 and a boosting dose of Ad-CEA/N-803 given on day 14 (hereafter designated "Vaccine"). As expected, the modified schedule of bintrafusp alfa plus SX-682 (Bintrafusp/SX) was unable to exert tumor control (Figure 1A). The use of vaccine as a monotherapy also failed to control tumors; the average tumor growth in the Vaccine group was statistically not different from that of the Control group (Figure 1A). Although the addition of vaccine to the Bintrafusp/SX therapy was able to induce a significant albeit modest delay in primary tumor growth in this experiment, the triple combination Vaccine/Bintrafusp/SX resulted in significant changes in the composition of the tumor immune infiltrate when compared with the other groups (Figure 1B). Overall, Vaccine/Bintrafusp/SX showed a significant enhancement of CD4$^+$ and CD8$^+$ T cells characterized by an effector and effector-memory phenotype (CD4$_{Eff\&Em}$ and CD8$_{Eff\&Em}$ TIL) above the levels achieved in the Vaccine monotherapy, Bintrafusp/SX, and Control groups (Figure 1B). Also remarkable was the ability of vaccine to decrease the percentage of regulatory T cells (T$_{regs}$) in the CD4$^+$ TIL population, compared to the Control and Bintrafusp/SX groups (Figure 1B). Previously, we demonstrated that Bintrafusp/SX therapy can significantly reduce tumor infiltration with suppressive granulocytic myeloid-derived suppressor cells (G-MDSC), defined as CD11b$^+$F4/80$^-$Ly6C^{lo}Ly6G$^+$, an effect attributed to the ability of SX-682 to block the CXCR1/2-mediated migration of G-MDSC into the tumor. The effect was not observed with monocytic MDSC, defined as

$CD11b^+F4/80^-Ly6G^-Ly6C^+$. Here, $CD11b^+F4/80^-Ly6C^{lo}Ly6G^+$ cells were significantly reduced in the tumors of mice treated with both Bintrafusp/SX and Vaccine/Bintrafusp/SX, an effect that was not observed with $CD11b^+F4/80^-Ly6G^-Ly6C^+$ fractions (Figure 1C). Neither fraction of myeloid cells was altered in the spleen of mice in any of the treatment groups (Figure 1D). As shown in Figure 1E, only Vaccine/Bintrafusp/SX treatment induced a significant increase in the ratio of $CD8^+$ TIL to both T_{regs} and $CD11b^+F4/80^-Ly6C^{lo}Ly6G^+$ cells in the TME compared to Control mice.

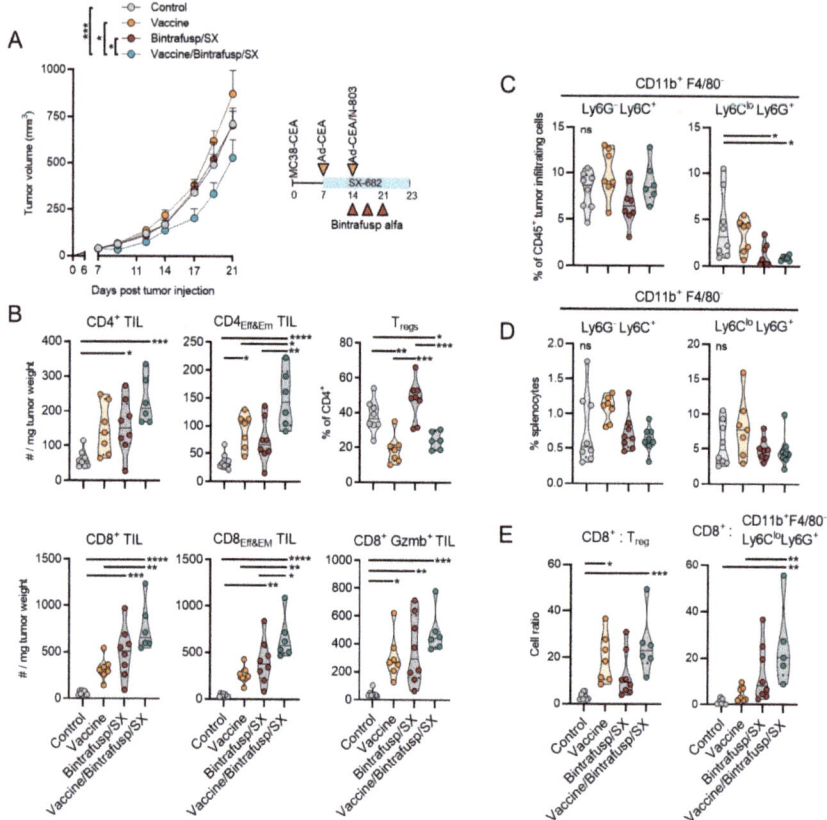

Figure 1. Vaccine synergizes with Bintrafusp alfa/SX-682 and increases TIL in MC38-CEA tumors. (**A**) CEA.Tg mice were injected s.c. with 3×10^5 MC38-CEA in the flank. On day 7, mice were started on a control or SX-682 diet (200 mg/kg body weight/day), and on days 14, 17, and 21 mice received i.p. injections of 200 µg bintrafusp alfa. Priming vaccine dose of s.c. Ad-CEA (1×10^{10} viral particles) was administered on day 7 with a boosting dose of Ad-CEA/N-803 (1×10^{10} viral particles, N-803, 1 µg, s.c.) on day 14. Graph shows average tumor growth and error bars indicate SEM of biological replicates; n = 8 mice/group. * $p \leq 0.05$; *** $p \leq 0.001$ for two-way ANOVA in (**A**). Control indicates mice that were left untreated and fed a base diet without SX-682. Tumors (**B,C**) and spleens (**D**) were harvested and analyzed by flow cytometry on day 23 for lymphocytes (**B**) and myeloid cells (**C,D**). (**E**) Cell ratios comparing the number of cells per mg tumor weight were also calculated. Individual points represent data from one tumor. ns, not significant; * $p \leq 0.05$; ** $p \leq 0.01$; *** $p \leq 0.001$; **** $p \leq 0.0001$ for one-way ANOVA followed by Tukey's post hoc test in (**B–E**). i.p. = intraperitoneal. s.c. = subcutaneous. TIL = tumor-infiltrating lymphocyte. Tregs = regulatory T cells.

To understand whether both bintrafusp alfa and SX-682 were needed for the antitumor efficacy of the combination Vaccine/Bintrafusp/SX, in the next study we also evaluated the addition of vaccine to SX-682 (Vaccine/SX) or bintrafusp alfa alone (Vaccine/Bintrafusp). In this experiment, an additional boosting dose of vaccine was administered on day 21. While the growth of MC38-CEA tumors was not delayed with Vaccine/SX or Vaccine/Bintrafusp combinations, there was a significant delay in tumor growth in the Vaccine/Bintrafusp/SX group (Figure 2A). Interestingly, some tumors began to completely regress in the Vaccine/Bintrafusp/SX group immediately after the final dose of vaccine plus bintrafusp alfa administered on day 21. Sections of tumor tissue stained by immunofluorescence revealed high levels of infiltrating CD4+ and CD8+ T cells in the Vaccine/Bintrafusp/SX group that were distributed uniformly throughout the tumors, compared to the other groups (Figure 2B).

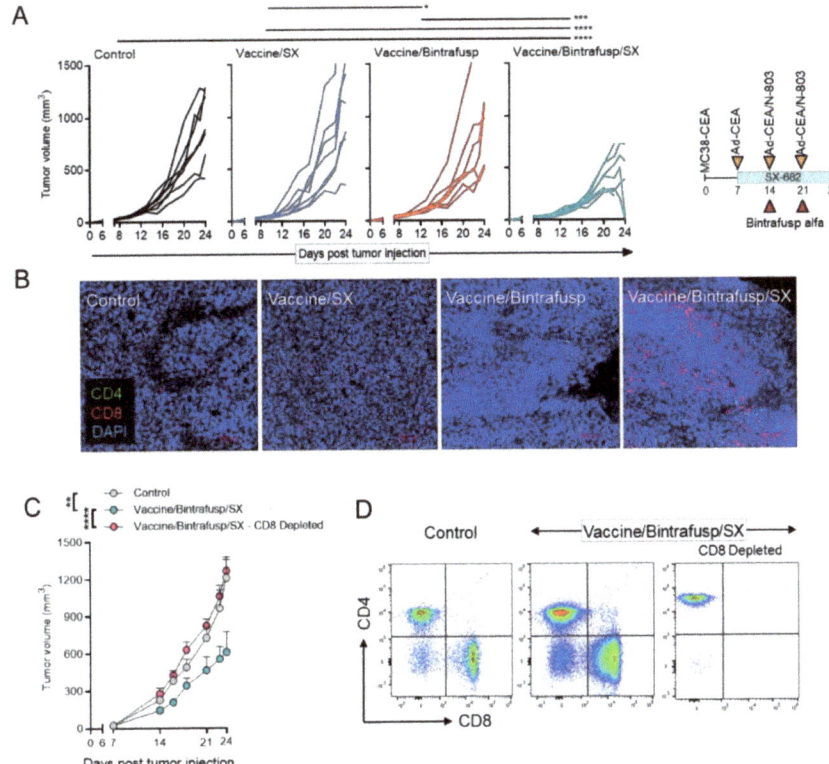

Figure 2. Vaccine combination immunotherapy is dependent on CD8+ TIL. (**A**) CEA.Tg mice were injected s.c. with 3×10^5 MC38-CEA in the flank. On day 7, mice were started on a control or SX-682 diet (200 mg/kg body weight/day). On days 14 and 21, mice received i.p. injections of 200 µg bintrafusp alfa. A priming vaccine dose of s.c. Ad-CEA (1×10^{10} viral particles) was administered on day 7 with a boosting dose of Ad-CEA/N-803 vaccine on days 14 and 21 (1×10^{10} viral particles, N-803, 1 µg, s.c.). Shown are the individual tumor growths for mice in the Control, Vaccine/SX, Vaccine/Bintrafusp, and Vaccine/Bintrafusp/SX groups; $n = 7$ mice/group. Control indicates mice that were left untreated and fed a base diet without SX-682. (**B**) Representative images of indicated tumors stained for CD4+ (green) and CD8+ (red) T cells and DAPI (blue) by immunofluorescence. (**C**) MC38-CEA tumor-bearing CEA.Tg mice received Vaccine/Bintrafusp/SX as in (**A**). Additionally, mice receiving Vaccine/Bintrafusp/SX also received depleting antibodies for CD8+ cells starting on day 5; $n = 7$ (Control and Vaccine/Bintrafusp/SX − CD8 Depleted) or 5 (Vaccine/Bintrafusp/SX) mice/group. (**D**) Flow profiles confirming efficacy of CD8 depletion antibodies from (**C**). Error bars indicate SEM of biological replicates. * $p \leq 0.05$; ** $p \leq 0.01$; *** $p \leq 0.001$; **** $p \leq 0.0001$ for two-way ANOVA in (**A**,**C**). i.p. = intraperitoneal. s.c. = subcutaneous. TIL = tumor-infiltrating lymphocyte.

The importance of the CD8+ T-cell fraction for the effectiveness of the multimodal therapy was evaluated with CEA.Tg mice bearing MC38-CEA tumors that were either left untreated and fed a base diet without SX-682 (Control group), treated with Vaccine/Bintrafusp/SX multimodal therapy, or treated with multimodal therapy with simultaneous depletion of CD8+ T cells (Vaccine/Bintrafusp/SX – CD8 Depleted group, Figure 2C,D). As shown in Figure 2C, depletion of CD8+ T cells completely abrogated the antitumor efficacy of Vaccine/Bintrafusp/SX treatment. The triple combination also had a modest yet significant effect on the survival of MC38-CEA tumor-bearing mice over that of Bintrafusp/SX-treated or Control mice (Figure S2).

It has been previously reported that combination therapy consisting of vaccine and various immune modulatory agents, including immune checkpoint blockade, can enhance antitumor immunity by diversifying the number of tumor antigens being recognized by TCRs through the phenomenon of antigen cascade or epitope spreading [10]. In this study, splenocytes from Control and Vaccine/Bintrafusp/SX-treated mice were evaluated for potential epitope spreading by quantifying on an ELISPOT assay the number of CD8+ T cells specific for CEA, the MC38-neoantigen PTGFR [10], or P15e, compared to a negative control peptide (Figure 3A). While there was a modest enhancement of the number of T cells specific for CEA in the spleens of vaccinated mice (~2-fold increase), high numbers of both PTGFR-specific and P15e-specific T cells were observed in the Vaccine/Bintrafusp/SX-treated mice, compared to the Control group (2.9-fold and 3.6-fold, respectively) (Figure 3A).

To understand how the combination of these agents restructures the immune profile of the TME in Vaccine/Bintrafusp/SX-treated tumors, NanoString gene expression analysis was performed on whole tumor tissue-derived RNA. Table 1 lists genes that were found to be up- or down-regulated more than 2.0-fold in Vaccine/Bintrafusp/SX-treated mice compared to Control tumors. Ingenuity Pathway Analysis demonstrated an upregulation of many immune-specific canonical pathways, with Th1 and Th2 being the two most significantly upregulated pathways (Figure 3B) in Vaccine/Bintrafusp/SX versus Control tumors. In addition, strong upregulation of inducible T-cell costimulator (ICOS) signaling, nuclear factor of activated T cells (NFAT) regulation, CTL-mediated apoptosis of target cells, and CD28 signaling were observed in tumors treated with the multimodal therapy Vaccine/Bintrafusp/SX versus Control. Figure 3C shows genes that were up- or down-regulated >2.5-fold in the triple combination group, with some of them being confirmed by PCR analysis in tumors of mice treated with Vaccine/Bintrafusp/SX versus Control (Figure 3D). There was a significant upregulation of Cd3e, Cd8a, Tbx21, Pdcd1, Cd247, and genes encoding for the effector molecules, Prf1, Gzmb, and Gzmk, suggesting a highly cytotoxic phenotype in TIL isolated from Vaccine/Bintrafusp/SX-treated tumors. Additional PCR analysis of expression of CD8a, Tbx21, Gzmk, and Prf1 mRNA was conducted in individual tumors from the Control, Vaccine, Bintrafusp/SX and Vaccine/Bintrafusp/SX groups. While vaccine used as monotherapy induced only a modest upregulation of these genes in some of the tumors compared with Control tumors, a stronger upregulation was observed in the Bintrafusp/SX group, though the level of upregulation was variable among genes and across tumor samples (Figure 3E). Supporting the benefit of adding all agents together, tumors in the Vaccine/Bintrafusp/SX group exhibited a more robust upregulation of all four genes in the majority of samples evaluated (Figure 3E). These data indicated that addition of vaccine can further enhance immune infiltration and activation above the induction mediated by blockade of PD-L1, TGF-β and CXCR1/2.

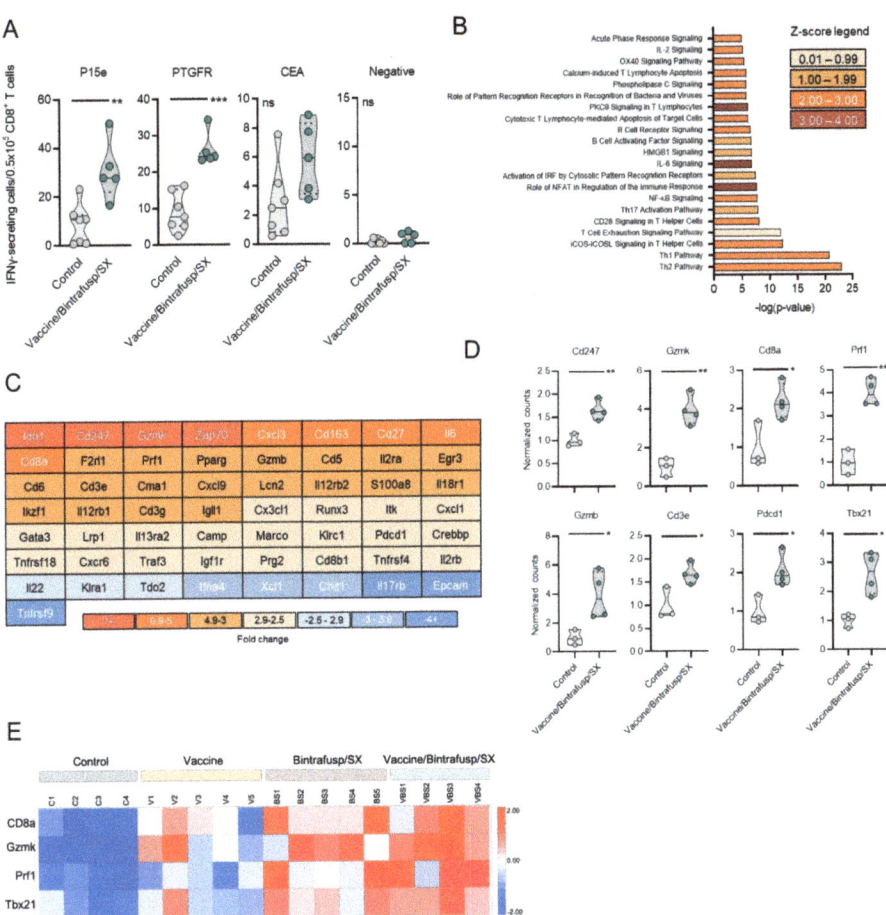

Figure 3. Immune activation signature observed in MC38-CEA tumors treated with Vaccine/Bintrafusp/SX combination. CEA.Tg mice were injected s.c. with 3×10^5 MC38-CEA in the flank. On day 7, mice were started on a control or SX-682 diet (200 mg/kg body weight/day). On days 14 and 21, mice received i.p. injections of 200 µg bintrafusp alfa. A priming vaccine dose of s.c. Ad-CEA was administered on day 7 (1×10^{10} viral particles) with a boosting dose of Ad-CEA/N-803 vaccine on days 14 and 21 (1×10^{10} viral particles, N-803, 1 µg, s.c.). (**A**) IFNγ ELISPOT analysis of spleens collected on day 24 from Control and Vaccine/Bintrafusp/SX-treated mice against MC38-CEA tumor antigens. Control indicates mice that were left untreated and fed a base diet without SX-682; n = 7 (Control) or 5 (Vaccine/Bintrafusp/SX) mice/group. Tumors collected on day 24 were used for RNA preparation and NanoString analysis as described in the Materials and Methods. Shown in (**B**) is an Ingenuity Pathway Analysis performed on genes that were found to be up- or down-regulated more than 2-fold in Vaccine/Bintrafusp/SX-treated tumors compared to Control tumors; n = 3 mice/group. (**C**) Heat map of genes differentially expressed >2.5-fold in Vaccine/Bintrafusp/SX-treated tumors compared to Control tumors; n = 3 mice/group. (**D**) Real-time PCR analysis confirming selected genes upregulated in Vaccine/Bintrafusp/SX-treated tumors compared to Control tumors; n = 3 (Control) or 4 (Vaccine/Bintrafusp/SX) mice/group. Individual points represent data from one tumor. ns, not significant; * $p \leq 0.05$; ** $p \leq 0.01$; *** $p \leq 0.001$ for two-tailed Student's t-test in (**A**,**D**). (**E**) Heat map expression of indicated genes in MC38-CEA tumors treated as per the schedule of administration in Figure 1. Tumor RNA was prepared at day 23; RNA expression of indicated genes was evaluated by real-time PCR as described in the Materials and Methods.

Table 1. Genes that were found to be up- or down-regulated more than 2.0-fold in Vaccine/Bintrafusp/SX-treated mice compared to Control tumors.

Gene	Fold Change	Gene	Fold Change	Gene	Fold Change
Ido1	10.13	Tnfrsf18	2.59	Csf1	2.13
Cd247	8.73	Cxcr6	2.53	Pou2f2	2.13
Gzmk	8.56	Traf3	2.53	Igf2r	2.12
Zap70	7.41	Igf1r	2.52	Itgal	2.11
Cxcl3	6.93	Prg2	2.52	Notch1	2.11
Cd163	6.77	Cd8b1	2.51	Pnma1	2.11
Cd27	6.75	Tnfrsf4	2.51	Hc	2.1
Il6	6.75	Il2rb	2.5	Cmah	2.09
Cd8a	6.7	Nfatc2	2.49	Inpp5d	2.09
F2rl1	5.76	Dmbt1	2.47	Cxcl2	2.08
Prf1	5.37	CD209e	2.46	Smad3	2.07
Pparg	5.35	Cxcl5	2.46	Angpt1	2.06
Gzmb	5.21	Ccl3	2.45	Tfe3	2.05
Cd5	5.16	Itga4	2.43	Fcer1a	2.04
Il2ra	4.85	Polr2a	2.43	Masp1	2.04
Egr3	4.33	Egr1	2.42	Bst1	2.02
Cd6	4.01	Gbp5	2.42	Erbb2	2.02
Cd3e	3.8	Sap130	2.39	Rel	2.02
Cma1	3.73	Tlr9	2.36	Tapbp	2.02
Cxcl9	3.59	Nlrc5	2.35	Tirap	2.01
Lcn2	3.56	Il25	2.33	Sdha	2.01
Il12rb2	3.42	Pin1	2.33	Cr2	2
S100a8	3.42	C8b	2.3	Cd7	−2.01
Il18r1	3.34	Icos	2.28	Il17b	−2.03
Ikzf1	3.18	Lyve1	2.28	Aire	−2.08
Il12rb1	3.18	Elk1	2.27	Tnfrsf17	−2.15
Cd3g	3.17	Ep300	2.27	Ms4a1	−2.21
Igll1	3.16	Gbp2b	2.23	Cfd	−2.43
Cx3cl1	2.86	C4b	2.22	Il12a	−2.48
Runx3	2.84	Crp	2.22	Il22	−2.5
Itk	2.83	Nfatc3	2.22	Klra1	−2.54
Cxcl1	2.82	Cxcl13	2.21	Tdo2	−2.82
Gata3	2.8	Atm	2.2	Ifna4	−3.04
Lrp1	2.79	Il6ra	2.2	Xcl1	−3.17
Il13ra2	2.76	Tnfrsf11b	2.2	Chit1	−3.7
Camp	2.67	Fasl	2.19	Il17rb	−3.94
Marco	2.67	Jun	2.19	Epcam	−4.85
Klrc1	2.65	Ddx58	2.18	Tnfrsf9	−5.48
Pdcd1	2.63	Il18rap	2.15		
Crebbp	2.62	Tigit	2.14		

3.2. Addition of Vaccine to Checkpoint Blockade-Based Therapy Enhances Immune T-Cell Activation and TCR Diversity

To corroborate the results in a different tumor model, a single dose of bintrafusp alfa in combination with SX-682 was given to 4T1 tumor-bearing mice which, as expected, failed to control tumor growth (Bintrafusp/SX, Figure 4A). In this mammary carcinoma model, vaccine was administered as a priming dose of Ad-Twist on day 7 with a boosting vaccine on day 14 consisting of Ad-Twist plus N-803. Twist1, a transcription factor that drives metastasis, was identified and characterized as a targetable "self" tumor-associated antigen in 4T1 tumor cells [22]. Addition of vaccine to Bintrafusp/SX therapy induced only a modest delay in primary tumor growth (Vaccine/Bintrafusp/SX, Figure 4A), and a trend towards reduced number of lung metastases (Figure 4B), with a 76% reduction of metastases in the Vaccine/Bintrafusp/SX group compared with the Control (Figure 4C). Two caveats with these results, however, are the low number of mice evaluated in each group, and the reduction of primary tumor volume in the Vaccine/Bintrafusp/SX group that could directly impact the number of disseminated cells.

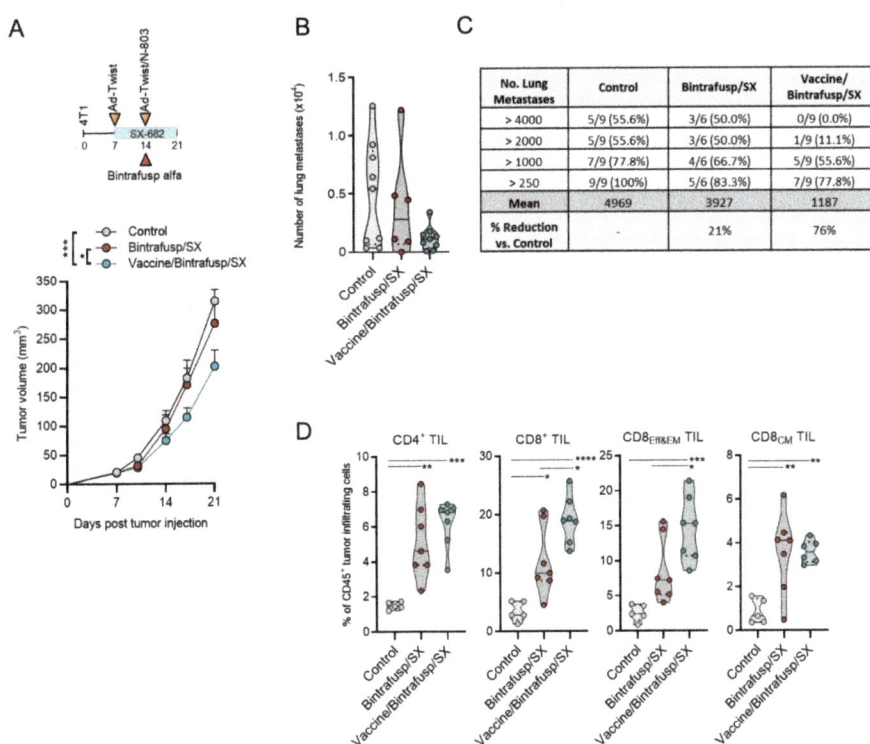

Figure 4. Vaccine synergizes with Bintrafusp alfa and SX-682 and increases TIL in 4T1 tumors. (**A**) BALB/c mice bearing 4T1 tumors in the mammary fat pad received control or SX-682 diet on day 7 (200 mg/kg body weight/day), with a priming vaccine dose of s.c. Ad-Twist (1 × 10^{10} viral particles). On day 14, mice received an i.p. injection of 200 μg bintrafusp alfa with a boosting vaccine dose of Ad-Twist/N-803 (1 × 10^{10} viral particles, N-803, 1 μg, s.c.). Graph shows average tumor growth and error bars indicate SEM of biological replicates; n = 6 (Control) or 7 (Bintrafusp/SX, Vaccine/Bintrafusp/SX) mice/group. Control indicates mice that were left untreated and fed a base diet without SX-682. * $p \leq 0.05$; *** $p \leq 0.001$ for two-way ANOVA. (**B**) Number of metastases quantified in the lungs of 4T1 tumor-bearing mice on day 21; individual points represent data from one mouse. (**C**) Table depicting the number and percentage of mice with the indicated range of lung metastases in each group, the mean number of metastases in each group, and the % reduction of the mean in each group vs. the Control group. Data are pooled from 2 independent experiments. (**D**) Tumors were harvested and analyzed by flow cytometry on day 21. Individual points represent data from one tumor. * $p \leq 0.05$; ** $p \leq 0.01$; *** $p \leq 0.001$; **** $p \leq 0.0001$ for one-way ANOVA followed by Tukey's post hoc test. i.p. = intraperitoneal. s.c. = subcutaneous. TIL = tumor-infiltrating lymphocyte.

Similar to the results observed with MC38-CEA tumors, addition of vaccine had a marked impact on the composition of 4T1 primary tumor T-cell infiltrates. As shown in Figure 4D, flow cytometry analysis of tumors collected at 1 week post-bintrafusp alfa ± vaccine administration (day 21 post-tumor injection) revealed significantly higher frequencies of CD8+ T cells characterized by an effector and effector-memory phenotype ($CD8_{Eff\&EM}$) in the Vaccine/Bintrafusp/SX group compared with the Bintrafusp/SX group or Control tumors. In contrast, the frequency of CD4+ T cells and central memory CD8+ T cells ($CD8_{CM}$) were similar among the two treatment groups, irrelevant of vaccine. In agreement with the flow cytometry data, immunofluorescence-based analysis of TIL in sections of Formalin-Fixed Paraffin-Embedded (FFPE) tumor tissues (Figure S3A) showed large clusters of CD4$^+$ and CD8$^+$ T cells homogenously distributed throughout the tumor in Vaccine/Bintrafusp/SX-treated tumors and not solely contained to the tumor boundaries. Consistent with previous findings, immune subset profiling of Vaccine/Bintrafusp/SX-treated tumors also revealed a significant decrease in the frequency of tumor-infiltrating CD11b$^+$F4/80$^-$Ly6G$^+$Ly6C^{lo} myeloid cells and CD11b$^+$F4/80^{hi} macrophages, together with a marked increase of CD4$^+$ and CD8$^+$ T cells (Figure S3B). Additionally, no adverse events or toxicity were observed with the total combination of therapeutics. These results suggested that addition of a prime-boost vaccine to a checkpoint blockade-based immunotherapy can further enhance frequency of effector T lymphocytes in the TME.

The quality of the T-cell infiltrates in 4T1 tumors of Bintrafusp/SX ± vaccine-treated mice was further evaluated. Intracellular flow cytometry-based analysis of tumor-infiltrating T cells from Vaccine/Bintrafusp/SX-treated mice revealed significantly higher frequencies of proliferative (CD8$^+$ Ki67$^+$) and cytotoxic (CD8$^+$ Granzyme B$^+$) TIL compared to tumors in the Bintrafusp/SX and Control groups (Figure 5A). TCRβ sequencing analysis was also performed on whole tumor lysates from 3 individual tumors per group; addition of vaccine to Bintrafusp/SX resulted in reduced clonality (Figure 5B) and expanded the T-cell repertoire compared with Control and Bintrafusp/SX-treated tumors, with an average of 481 ± 240, 907 ± 372, and 1897 ± 1469 productive TCRβ rearrangements in the Control, Bintrafusp/SX and Vaccine/Bintrafusp/SX groups, respectively (Figure 5C).

In addition, analysis of sequence similarities revealed a higher number of TCRβ sequences shared among tumors in the Vaccine/Bintrafusp/SX > Bintrafusp/SX > Control group, as shown by the numbers in the regions of intersection. Analysis of the top 25% of TCRβ sequences present in tumors from 3 mice in each group revealed a more diversified TCR repertoire in the Vaccine/Bintrafusp/SX-treated mice (Figure 5D) comprising 21, 17, and 13 clones per individual, while tumors from Control and Bintrafusp/SX-treated mice contained 5, 7, 6 and 6, 18, and 3 different TCRβ clones, respectively. These data indicated that the addition of a vaccine consisting of Ad-vector plus N-803 adjuvant to bintrafusp alfa plus SX-682 therapy has the potential to increase the proliferation and cytotoxic functionality of tumor-infiltrating CD8$^+$ T cells, while promoting a more diversified TCR repertoire in the tumor (Figure 6).

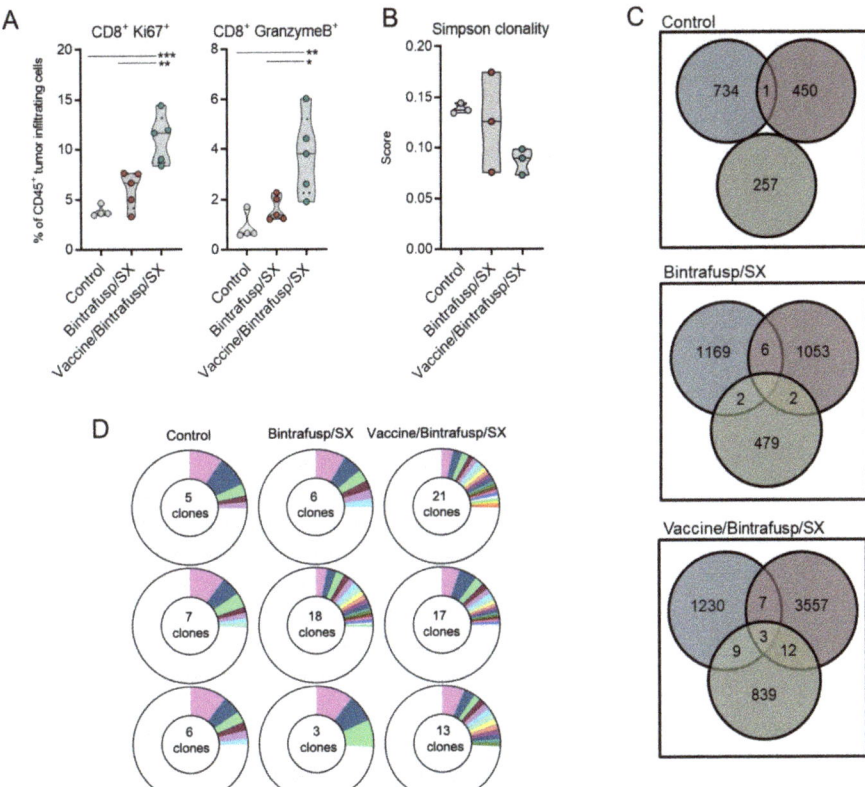

Figure 5. Vaccine enhances activation and TCR diversity of TIL when incorporated into combination immunotherapy in the 4T1 carcinoma model. (**A**) BALB/c mice bearing 4T1 tumors in the mammary fat pad received control or SX-682 diet on day 7 (200 mg/kg body weight/day), with a priming vaccine dose of s.c. Ad-Twist (1 × 10^{10} viral particles). On day 14, mice received an i.p. injection of 200 µg bintrafusp alfa with a boosting vaccine dose of Ad-Twist/N-803 (1 × 10^{10} viral particles, N-803, 1 µg, s.c.). Graphs show immune subsets determined by flow cytometry analysis of tumors at day 21. Individual points represent data from one tumor. * $p \leq 0.05$; ** $p \leq 0.01$; *** $p \leq 0.001$ for one-way ANOVA followed by Tukey's post hoc test. (**B**) Simpson clonality score for individual tumor samples in each indicated group determined as indicated in the Materials and Methods. (**C**) Number of productive TCRβ rearrangements per individual tumor in the indicated groups, showing the number of overlapping TCRβ sequences among individuals. (**D**) The number of TCRβ clones comprising the top 25% of detected sequences. $n = 3$ mice/group. i.p. = intraperitoneal. s.c. = subcutaneous. TCR = T-cell receptor. TIL = tumor-infiltrating lymphocyte.

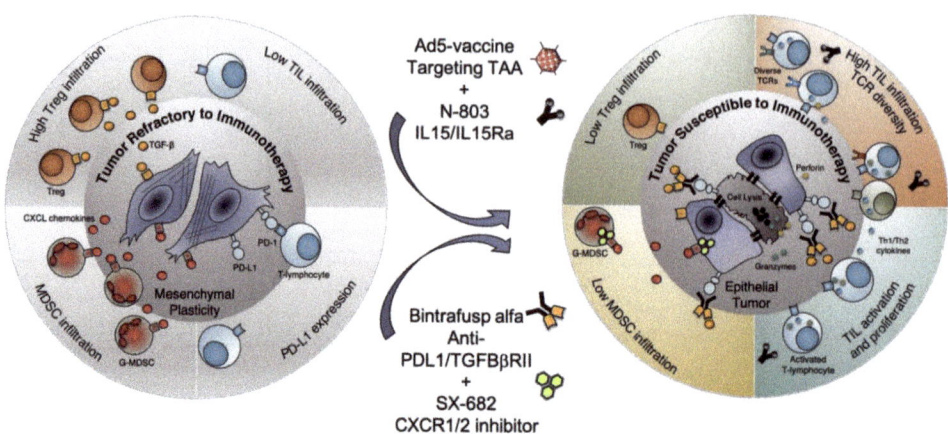

Figure 6. Schematic representation of the mechanism of action of the combination Ad5-vaccine, N-803, Bintrafusp alfa and SX-682. G-MDSC = granulocytic myeloid-derived suppressor cells. TCR = T-cell receptor. TIL = tumor-infiltrating lymphocyte. Tregs = regulatory T cells.

4. Discussion

In this study, we demonstrate the effect of adding a cancer vaccine to immune checkpoint blockade therapy. Our data show that a vaccine consisting of a recombinant adenovirus with a target antigen transgene coupled with an IL-15 super agonist adjuvant is able to contribute to checkpoint-based immunotherapy by increasing T-cell migration to the tumor, enhancing T-cell activation and cytotoxicity, and promoting TCR diversity and antigen cascade.

The mechanism of action and immunological benefits of both bintrafusp alfa and SX-682 have been extensively studied as monotherapies and in combination by our group and others. Bintrafusp alfa, designed as a checkpoint inhibitor and to "trap TGF-β" in the TME, has been shown to promote T- and NK-cell killing of tumor cells, promote antibody-dependent cell cytotoxicity, revert TGF-β-induced epithelial-mesenchymal phenotypic changes in cancer cells (tumor cell plasticity), and delay tumor growth in numerous mouse models of cancer [15,25–27]. There are numerous ongoing clinical studies of bintrafusp alfa in patients with a variety of cancer types, with several of these studies investigating its use in combination with other immunotherapies, chemotherapy or radiation [15]. SX-682 is a small molecule inhibitor that allosterically binds to the CXCR1 and CXCR2 receptors to irreversibly inhibit downstream signaling from CXC family ligands CXCL1-3 and CXCL5-8. One of the most notable CXCR1/2 ligands, IL-8 (CXCL8), is a known inducer of tumor cell plasticity, attractant of suppressive myeloid-derived suppressor cells to the tumor, and correlates with failure of treatment in numerous cancer types, including failure to checkpoint inhibitor therapy [28–31]. SX-682 has been shown to inhibit tumor growth, block migration of G-MDSC to tumors in vivo, and decrease markers of tumor cell plasticity in human xenografts and murine tumors [17,32,33], and is currently undergoing clinical evaluation in several clinical trials [29]. In a previous study, we demonstrated that the combination of bintrafusp alfa and SX-682 reduces mesenchymal tumor features and increases epithelial protein expression in murine models of breast and lung cancer, reduces tumor infiltration with G-MDSC, and enhances T-cell infiltration and activation in tumors [17].

Tumor immunologists have been attempting to develop highly specific yet off-the-shelf immune activating vaccines for the treatment of cancer patients prior to the immune checkpoint blockade revolution. These vaccines often targeted tumor-associated antigens and were combined with immune-activating adjuvants or costimulatory molecules to promote T-cell infiltration into tumors and kick-start antitumor immunity [8,34,35]. More recent studies have also found efficacy with the use of neoantigen-based vaccines and irradiated cancer cell vaccines. However, the subsequently activated T-cell population can still be rapidly inhibited by immune checkpoint pathways or immune suppressive cells once arriving to the tumor. Additionally, many tumor types with low degree of T-cell infiltration which respond poorly to immunotherapy such as pancreatic, colon, and prostate cancers upregulate additional immune suppressive mechanisms including TGF-β, MDSC, and mesenchymal features [36–39]. In this study, we lowered the dose and delayed the administration of bintrafusp alfa in combination with SX-682 with the idea of preventing antitumor activity to mimic the situation of non-responsive tumors. We were able to demonstrate that the addition of vaccine in this context promoted further T-cell infiltration and activation, and enhanced TCR diversity in the tumor above what was induced by bintrafusp alfa/SX-682 treatment (Figure 6). We also showed here that addition of vaccine further enhanced the expression of genes indicative of immune activation and T-cell infiltration in the TME (CD8a, Tbx21, Gmzk, Prf1). These data are in agreement with the flow cytometric analysis of MC38-CEA tumors, which demonstrated an increased number of infiltrating $CD4^+$ effector/effector memory T cells as well as $CD8^+$ effector/effector-memory T cells in Vaccine/Bintrafusp/SX-treated tumors versus tumor in the Bintrafusp/SX group. Similarly, infiltration with $CD8^+$ effector/effector-memory T cells was significantly enhanced in 4T1 tumors treated with Vaccine/Bintrafusp/SX versus Bintrafusp/SX treatment. Additionally, increased proliferation and cytolytic effect of T cells was observed in the TME of Vaccine/Bintrafusp/SX-treated 4T1 tumors, denoted by a higher percentage of $CD8^+$ T cells positive for Ki67 or Granzyme B, compared with tumors in the Bintrafusp/SX group.

Analysis of splenocytes via ELISPOT assay also revealed epitope spreading in the Vaccine/Bintrafusp/SX-treated mice, with an increase in the number of T cells specific for antigens found in the tumor but not in the vaccine (PTGFR and P15e), compared with the Control group. One could hypothesize that these activated, tumor-specific T cells from spleens of Vaccine/Bintrafusp/SX-treated mice could mediate some degree of tumor control if adoptively transferred into MC38-CEA tumor-bearing mice; however, such experiments would not be able to reveal the full potential of this combination immunotherapy, which relies on tumor-localized effects mediated by SX-682 and bintrafusp alfa. As we have previously shown, inhibition of CXCR1/2 via SX-682 significantly reduces the migration of suppressive $CXCR2^+$ G-MDSC into tumors. At the same time, SX-682 directly affects the phenotype of the tumor cells resulting in reduced mesenchymal features which, in turn, improves tumor susceptibility to immune-mediated lysis [17]. Similarly, bintrafusp alfa is able to mediate neutralization of PD-L1 and TGF-β in the TME, leading to alleviation of local tumor immunosuppression mediated by both pathways, including the reversion of tumor mesenchymal features for improved susceptibility to immune attack [15,17].

Despite increased infiltration of tumors with activated T cells and increased numbers of tumor-specific T cells in the Vaccine/Bintrafusp/SX group, the treatment schedules investigated here did not result in a significant number of tumor cures. We hypothesize that this could have been due to various factors, including the limited therapeutic window in which the human drugs employed here could be administered to immune competent mice without production of anti-drug antibodies. Another possibility is the very rapid tumor growth characteristic of the two murine models utilized in this study, combined with a delayed initiation of therapy, which limited time for treatment. Notably, in the clinical setting, multiple agents can be administered continuously with optimal dosing over an extended period of time for maximum benefit, as in the case of the combination of Adenoviral-based vaccines, N-803, and bintrafusp alfa currently being tested in the clinic [40]. Alternatively, other mechanisms of immune suppression may have limited tumor control in the combination group, even in the presence of activated, infiltrating T cells. Interestingly, one of the genes most upregulated in MC38-CEA tumors treated with Vaccine/Bintrafusp/SX was Ido1, suggestive of the possibility of adding an IDO inhibitor to this therapeutic regimen. Overall, the combination Vaccine/Bintrafusp/SX therapy was more effective at controlling MC38 compared with 4T1 tumor growth, an effect that could be related to the higher mutational burden and neoepitope expression in MC38 versus 4T1 tumors.

In conclusion, this study highlights the mechanistic synergy between vaccine and combination checkpoint immunotherapy and provides rationale for an ongoing clinical trial combining a cancer vaccine with bintrafusp alfa plus SX-682 therapy in patients with advanced solid tumors (NCT04574583).

Supplementary Materials: The following are available online at https://www.mdpi.com/2072-6694/13/5/968/s1, Figure S1: Optimization of the combination Ad-CEA plus N-803, Figure S2: Survival of CEA.Tg mice bearing MC38-CEA tumors in response to indicated treatments, Figure S3: Multimodal therapy effect on 4T1 tumor immune cell infiltration.

Author Contributions: L.A.H., D.H.H., J.W.H., C.P. and J.S. conceived various aspects of the project, designed experiments, and interpreted the results. L.A.H., K.F. and D.H.H. were responsible for performing experiments, data compilation and analysis. J.A.Z. and D.Y.M. provided reagents and in vivo dosing and formulation recommendations. L.A.H., C.P. and J.S. were responsible for manuscript writing and coordination. All authors have read and agreed to the published version of the manuscript.

Funding: This work was supported by the Intramural Research Program of the Center for Cancer Research, National Cancer Institute (NCI), National Institutes of Health (NIH), as well as through Cooperative Research and Development Agreements (CRADA) between the NCI/NIH and Syntrix Pharmaceuticals, the NCI/NIH and EMD Serono, and the NCI/NIH and ImmunityBio.

Institutional Review Board Statement: Mice were maintained under pathogen-free conditions in accordance with the Association for Assessment and Accreditation of Laboratory Animal Care guidelines. All animal studies were approved by the NIH Intramural Animal Care and Use Committee (ACUC); protocol LTIB-038.

Informed Consent Statement: Not applicable.

Data Availability Statement: The data presented in this study will be provided upon reasonable request.

Acknowledgments: The authors thank Haiyan Qin for her technical assistance with animal studies, Masafumi Iida for help with tumor collections, and Debra Weingarten for editorial assistance in the preparation of this manuscript.

Conflicts of Interest: The NCI/NIH authors do not have any competing interests to disclose. J.A.Z. and D.Y.M. are paid employees of Syntrix Pharmaceuticals. The NCI/NIH has ongoing Collaborative Research and Development Agreements (CRADA) with Syntrix Pharmaceuticals, EMD Serono, and ImmunityBio.

References

1. Hodi, F.S.; O'Day, S.J.; McDermott, D.F.; Weber, R.W.; Sosman, J.A.; Haanen, J.B.; Gonzalez, R.; Robert, C.; Schadendorf, D.; Hassel, J.C.; et al. Improved Survival with Ipilimumab in Patients with Metastatic Melanoma. *N. Engl. J. Med.* **2010**, *363*, 711–723. [CrossRef] [PubMed]
2. Brahmer, J.; Reckamp, K.L.; Baas, P.; Crinò, L.; Eberhardt, W.E.; Poddubskaya, E.; Antonia, S.; Pluzanski, A.; Vokes, E.E.; Holgado, E.; et al. Nivolumab versus Docetaxel in Advanced Squamous-Cell Non–Small-Cell Lung Cancer. *N. Engl. J. Med.* **2015**, *373*, 123–135. [CrossRef] [PubMed]
3. Fares, C.M.; Van Allen, E.M.; Drake, C.G.; Allison, J.P.; Hu-Lieskovan, S. Mechanisms of Resistance to Immune Checkpoint Blockade: Why Does Checkpoint Inhibitor Immunotherapy Not Work for All Patients? *Am. Soc. Clin. Oncol. Educ. Book* **2019**, *39*, 147–164. [CrossRef]
4. Horn, L.A.; Fousek, K.; Palena, C. Tumor Plasticity and Resistance to Immunotherapy. *Trends Cancer* **2020**, *6*, 432–441. [CrossRef]
5. Drake, C.G. Combination immunotherapy approaches. *Ann. Oncol.* **2012**, *23*, viii41–viii46. [CrossRef]
6. Li, J.; Byrne, K.T.; Yan, F.; Yamazoe, T.; Chen, Z.; Baslan, T.; Richman, L.P.; Lin, J.H.; Sun, Y.H.; Rech, A.J.; et al. Tumor Cell-Intrinsic Factors Underlie Heterogeneity of Immune Cell Infiltration and Response to Immunotherapy. *Immunity* **2018**, *49*, 178–193e7. [CrossRef]
7. Gajewski, T.F.; Corrales, L.; Williams, J.; Horton, B.; Sivan, A.; Spranger, S. Cancer Immunotherapy Targets Based on Understanding the T Cell-Inflamed Versus Non-T Cell-Inflamed Tumor Microenvironment. *Adv. Exp. Med. Biol.* **2017**, *1036*, 19–31. [CrossRef] [PubMed]

8. Fu, J.; Kanne, D.B.; Leong, M.; Glickman, L.H.; McWhirter, S.M.; Lemmens, E.; Mechette, K.; Leong, J.J.; Lauer, P.; Liu, W.; et al. STING agonist formulated cancer vaccines can cure established tumors resistant to PD-1 blockade. *Sci. Transl. Med.* **2015**, *7*, 283ra52. [CrossRef]
9. Ali, O.A.; Lewin, S.A.; Dranoff, G.; Mooney, D.J. Vaccines Combined with Immune Checkpoint Antibodies Promote Cytotoxic T-cell Activity and Tumor Eradication. *Cancer Immunol. Res.* **2016**, *4*, 95–100. [CrossRef] [PubMed]
10. Lee, K.L.; Benz, S.C.; Hicks, K.C.; Nguyen, A.; Gameiro, S.R.; Palena, C.; Sanborn, J.Z.; Su, Z.; Ordentlich, P.; Rohlin, L.; et al. Efficient Tumor Clearance and Diversified Immunity through Neoepitope Vaccines and Combinatorial Immunotherapy. *Cancer Immunol. Res.* **2019**, *7*, 1359–1370. [CrossRef] [PubMed]
11. Rudqvist, N.-P.; Pilones, K.A.; Lhuillier, C.; Wennerberg, E.; Sidhom, J.-W.; Emerson, R.O.; Robins, H.S.; Schneck, J.; Formenti, S.C.; DeMaria, S. Radiotherapy and CTLA-4 Blockade Shape the TCR Repertoire of Tumor-Infiltrating T Cells. *Cancer Immunol. Res.* **2018**, *6*, 139–150. [CrossRef]
12. Collins, J.M.; Donahue, R.N.; Tsai, Y.; Manu, M.; Palena, C.; Gatti-Mays, M.E.; Marté, J.L.; Madan, R.A.; Karzai, F.; Heery, C.R.; et al. Phase I Trial of a Modified Vaccinia Ankara Priming Vaccine Followed by a Fowlpox Virus Boosting Vaccine Modified to Express Brachyury and Costimulatory Molecules in Advanced Solid Tumors. *Oncology* **2019**, *25*, 560–e1006. [CrossRef]
13. Gatti-Mays, M.E.; Redman, J.M.; Donahue, R.N.; Palena, C.; Madan, R.A.; Karzai, F.; Bilusic, M.; Sater, H.A.; Marté, J.L.; Cordes, L.M.; et al. A Phase I Trial Using a Multitargeted Recombinant Adenovirus 5 (CEA/MUC1/Brachyury)-Based Immunotherapy Vaccine Regimen in Patients with Advanced Cancer. *Oncology* **2019**, *25*, 479. [CrossRef] [PubMed]
14. Gatti-Mays, M.E.; Strauss, J.; Donahue, R.N.; Palena, C.; Del Rivero, J.; Redman, J.M.; Madan, R.A.; Marté, J.L.; Cordes, L.M.; Lamping, E.; et al. A Phase I Dose-Escalation Trial of BN-CV301, a Recombinant Poxviral Vaccine Targeting MUC1 and CEA with Costimulatory Molecules. *Clin. Cancer Res.* **2019**, *25*, 4933–4944. [CrossRef] [PubMed]
15. Lind, H.; Gameiro, S.R.; Jochems, C.; Donahue, R.N.; Strauss, J.; Gulley, J.L.; Palena, C.; Schlom, J. Dual targeting of TGF-β and PD-L1 via a bifunctional anti-PD-L1/TGF-βRII agent: Status of preclinical and clinical advances. *J. Immunother. Cancer* **2020**, *8*, e000433. [CrossRef] [PubMed]
16. Strauss, J.; E Gatti-Mays, M.; Cho, B.C.; Hill, A.; Salas, S.; McClay, E.; Redman, J.M.; A Sater, H.; Donahue, R.N.; Jochems, C.; et al. Bintrafusp alfa, a bifunctional fusion protein targeting TGF-β and PD-L1, in patients with human papillomavirus-associated malignancies. *J. Immunother. Cancer* **2020**, *8*, e001395. [CrossRef] [PubMed]
17. A Horn, L.; Riskin, J.; A Hempel, H.; Fousek, K.; Lind, H.; Hamilton, D.H.; McCampbell, K.K.; Maeda, D.Y.; A Zebala, J.; Su, Z.; et al. Simultaneous inhibition of CXCR1/2, TGF-β, and PD-L1 remodels the tumor and its microenvironment to drive antitumor immunity. *J. Immunother. Cancer* **2019**, *8*, e000326. [CrossRef] [PubMed]
18. Fabian, K.P.; Malamas, A.S.; Padget, M.R.; Solocinski, K.; Wolfson, B.; Fujii, R.; Sater, H.A.; Schlom, J.; Hodge, J.W. Therapy of Established Tumors with Rationally Designed Multiple Agents Targeting Diverse Immune–Tumor Interactions: Engage, Expand, Enable. *Cancer Immunol. Res.* **2021**, *9*, 239–252. [CrossRef] [PubMed]
19. Kim, P.S.; Kwilas, A.R.; Xu, W.; Alter, S.; Jeng, E.K.; Wong, H.C.; Schlom, J.; Hodge, J.W. IL-15 superagonist/IL-15RαSushi-Fc fusion complex (IL-15SA/IL-15RαSu-Fc; ALT-803) markedly enhances specific subpopulations of NK and memory CD8+ T cells, and mediates potent anti-tumor activity against murine breast and colon carcinomas. *Oncotarget* **2016**, *7*, 16130–16145. [CrossRef]
20. Wrangle, J.M.; Velcheti, V.; Patel, M.R.; Garrett-Mayer, E.; Hill, E.G.; Ravenel, J.G.; Miller, J.S.; Farhad, M.; Anderton, K.; Lindsey, K.; et al. ALT-803, an IL-15 superagonist, in combination with nivolumab in patients with metastatic non-small cell lung cancer: A non-randomised, open-label, phase 1b trial. *Lancet Oncol.* **2018**, *19*, 694–704. [CrossRef]
21. Robbins, P.F.; A Kantor, J.; Salgaller, M.; Hand, P.H.; Fernsten, P.D.; Schlom, J. Transduction and expression of the human carcinoembryonic antigen gene in a murine colon carcinoma cell line. *Cancer Res.* **1991**, *51*, 3657–3662.
22. Ardiani, A.; Gameiro, S.R.; Palena, C.; Hamilton, D.H.; Kwilas, A.; King, T.H.; Schlom, J.; Hodge, J.W. Vaccine-Mediated Immunotherapy Directed against a Transcription Factor Driving the Metastatic Process. *Cancer Res.* **2014**, *74*, 1945–1957. [CrossRef]
23. Clarke, P.; Mann, J.; Simpson, J.F.; Rickard-Dickson, K.; Primus, F.J. Mice transgenic for human carcinoembryonic antigen as a model for immunotherapy. *Cancer Res.* **1998**, *58*, 1469–1477. [PubMed]
24. Hance, K.W.; Zeytin, H.E.; Greiner, J.W. Mouse models expressing human carcinoembryonic antigen (CEA) as a transgene: Evaluation of CEA-based cancer vaccines. *Mutat. Res. Mol. Mech. Mutagen.* **2005**, *576*, 132–154. [CrossRef] [PubMed]
25. Knudson, K.M.; Hicks, K.C.; Luo, X.; Chen, J.-Q.; Schlom, J.; Gameiro, S.R. M7824, a novel bifunctional anti-PD-L1/TGFβ Trap fusion protein, promotes anti-tumor efficacy as monotherapy and in combination with vaccine. *OncoImmunology* **2018**, *7*, e1426519. [CrossRef]
26. Lan, Y.; Zhang, D.; Xu, C.; Hance, K.W.; Marelli, B.; Qi, J.; Yu, H.; Qin, G.; Sircar, A.; Hernández, V.M.; et al. Enhanced preclinical antitumor activity of M7824, a bifunctional fusion protein simultaneously targeting PD-L1 and TGF-β. *Sci. Transl. Med.* **2018**, *10*, eaan5488. [CrossRef]
27. David, J.M.; Dominguez, C.; McCampbell, K.K.; Gulley, J.L.; Schlom, J.; Palena, C. A novel bifunctional anti-PD-L1/TGF-β Trap fusion protein (M7824) efficiently reverts mesenchymalization of human lung cancer cells. *OncoImmunology* **2017**, *6*, e1349589. [CrossRef] [PubMed]
28. Fernando, R.I.; Castillo, M.D.; Litzinger, M.; Hamilton, D.H.; Palena, C. IL-8 Signaling Plays a Critical Role in the Epithelial–Mesenchymal Transition of Human Carcinoma Cells. *Cancer Res.* **2011**, *71*, 5296–5306. [CrossRef] [PubMed]

29. Fousek, K.; Horn, L.A.; Palena, C. Interleukin-8: A chemokine at the intersection of cancer plasticity, angiogenesis, and immune suppression. *Pharmacol. Ther.* **2021**, *219*, 107692. [CrossRef]
30. Yuen, K.C.; Liu, L.-F.; Gupta, V.; Madireddi, S.; Keerthivasan, S.; Li, C.; Rishipathak, D.; Williams, P.; Kadel, E.E.; Koeppen, H.; et al. High systemic and tumor-associated IL-8 correlates with reduced clinical benefit of PD-L1 blockade. *Nat. Med.* **2020**, *26*, 693–698. [CrossRef]
31. Schalper, K.A.; Carleton, M.; Zhou, M.; Chen, T.; Feng, Y.; Huang, S.-P.; Walsh, A.M.; Baxi, V.; Pandya, D.; Baradet, T.; et al. Elevated serum interleukin-8 is associated with enhanced intratumor neutrophils and reduced clinical benefit of immune-checkpoint inhibitors. *Nat. Med.* **2020**, *26*, 688–692. [CrossRef]
32. Sun, L.; Clavijo, P.E.; Robbins, Y.; Patel, P.; Friedman, J.; Greene, S.; Das, R.; Silvin, C.; Van Waes, C.; Horn, L.A.; et al. Inhibiting myeloid-derived suppressor cell trafficking enhances T cell immunotherapy. *JCI Insight* **2019**, *4*, 4. [CrossRef]
33. Greene, S.; Robbins, Y.; Mydlarz, W.K.; Huynh, A.P.; Schmitt, N.C.; Friedman, J.; Horn, L.A.; Palena, C.; Schlom, J.; Maeda, D.Y.; et al. Inhibition of MDSC Trafficking with SX-682, a CXCR1/2 Inhibitor, Enhances NK-Cell Immunotherapy in Head and Neck Cancer Models. *Clin. Cancer Res.* **2020**, *26*, 1420–1431. [CrossRef] [PubMed]
34. Heery, C.R.; Palena, C.; McMahon, S.; Donahue, R.N.; Lepone, L.M.; Grenga, I.; Dirmeier, U.; Cordes, L.; Marté, J.; Dahut, W.; et al. Phase I Study of a Poxviral TRICOM-Based Vaccine Directed Against the Transcription Factor Brachyury. *Clin. Cancer Res.* **2017**, *23*, 6833–6845. [CrossRef] [PubMed]
35. Duraiswamy, J.; Freeman, G.J.; Coukos, G. Therapeutic PD-1 Pathway Blockade Augments with Other Modalities of Immunotherapy T-Cell Function to Prevent Immune Decline in Ovarian Cancer. *Cancer Res.* **2013**, *73*, 6900–6912. [CrossRef]
36. Otsuru, T.; Kobayashi, S.; Wada, H.; Takahashi, T.; Gotoh, K.; Iwagami, Y.; Yamada, D.; Noda, T.; Asaoka, T.; Serada, S.; et al. Epithelial-mesenchymal transition via transforming growth factor betain pancreatic cancer is potentiated by the inflammatory glycoproteinleucine-rich alpha-2 glycoprotein. *Cancer Sci.* **2018**, *110*, 985–996. [CrossRef]
37. Wang, Y.; Yin, K.; Tian, J.; Xia, X.; Ma, J.; Tang, X.; Xu, H.; Wang, S. Granulocytic Myeloid-Derived Suppressor Cells Promote the Stemness of Colorectal Cancer Cells through Exosomal S100A9. *Adv. Sci.* **2019**, *6*, 1901278. [CrossRef]
38. Thorsson, V.; Gibbs, D.L.; Brown, S.; Wolf, D.; Bortone, D.S.; Ouyang, T.-H.; Porta-Pardo, E.; Gao, G.F.; Plaisier, C.L.; Eddy, J.A.; et al. The Immune Landscape of Cancer. *Immunity* **2018**, *48*, 812–830e14. [CrossRef] [PubMed]
39. Villalba, M.; Evans, S.R.; Vidal-Vanaclocha, F.; Calvo, A. Role of TGF-β in metastatic colon cancer: It is finally time for targeted therapy. *Cell Tissue Res.* **2017**, *370*, 29–39. [CrossRef] [PubMed]
40. Redman, J.M.; Steinberg, S.M.; Gulley, J.L. Quick efficacy seeking trial (QuEST1): A novel combination immunotherapy study designed for rapid clinical signal assessment metastatic castration-resistant prostate cancer. *J. Immunother. Cancer* **2018**, *6*, 91. [CrossRef]

Article

Facile Generation of Potent Bispecific Fab via Sortase A and Click Chemistry for Cancer Immunotherapy

Xuefei Bai [1], Wenhui Liu [1,2], Shijie Jin [1], Wenbin Zhao [1], Yingchun Xu [1], Zhan Zhou [1,3], Shuqing Chen [1,3,4,*] and Liqiang Pan [1,5,*]

1. Institute of Drug Metabolism and Pharmaceutical Analysis, College of Pharmaceutical Sciences, Zhejiang University, Hangzhou 310058, China; xfbai@zju.edu.cn (X.B.); liuwenhui@bangshunpharm.com (W.L.); jinsj08@163.com (S.J.); pharmacy_zwb@zju.edu.cn (W.Z.); ycxu66@163.com (Y.X.); zhou@zju.edu.cn (Z.Z.)
2. Hangzhou Biosun Pharmaceutical Co., Ltd., Liangzhu International Life Science Town, 268 Tongyun Street, Yuhang District, Hangzhou 310015, China
3. Collaborative Innovation Center of Artificial Intelligence by MOE and Zhejiang Provincial Govement (ZJU), Hangzhou 310058, China
4. Department of Precision Medicine on Tumor Therapeutics, ZJU-Hangzhou Global Scientific and Technological Innovation Center, Hangzhou 311200, China
5. Key Laboratory of Pancreatic Disease of Zhejiang Province, The First Affiliated Hospital, School of Medicine, Zhejiang University, Hangzhou 310003, China
* Correspondence: chenshuqing@zju.edu.cn (S.C.); panliqiang@zju.edu.cn (L.P.)

Citation: Bai, X.; Liu, W.; Jin, S.; Zhao, W.; Xu, Y.; Zhou, Z.; Chen, S.; Pan, L. Facile Generation of Potent Bispecific Fab via Sortase A and Click Chemistry for Cancer Immunotherapy. *Cancers* 2021, *13*, 4540. https://doi.org/10.3390/cancers13184540

Academic Editors: Subree Subramanian and Xianda Zhao

Received: 14 August 2021
Accepted: 6 September 2021
Published: 10 September 2021

Publisher's Note: MDPI stays neutral with regard to jurisdictional claims in published maps and institutional affiliations.

Copyright: © 2021 by the authors. Licensee MDPI, Basel, Switzerland. This article is an open access article distributed under the terms and conditions of the Creative Commons Attribution (CC BY) license (https://creativecommons.org/licenses/by/4.0/).

Simple Summary: The formats of bispecific antibody have been investigated for many years to enhance the stability of the structure and anti-tumor efficacy. One of the formats combining two Fabs at their C termini provides unmodified variable region and comparable activity to other fragment-based bispecific antibodies that are usually combined in a head-to-tail manner. However, the current strategy to produce the BiFab molecule is limited to a semisynthetic method that introduces unnatural amino acid to antibodies' sequences during production. To improve the application of BiFab format in investigational biodrugs, we have applied sortase A-mediated "bio-click" chemistry to generate BiFab, for facile assembly of Fab molecules that have been expressed and stored as BiFab module candidates. The BiFabs made by our method stimulate T cell proliferation and activation with favorable in vitro and in vivo anti-tumor activit. Our results indicate that BiFab made by sortase A-mediated click chemistry could be used to efficiently generate various BiFabs with high potency, which further supports personalized tumor immunotherapy in the future.

Abstract: Bispecific antibodies (BsAbs) for T cell engagement have shown great promise in cancer immunotherapy, and their clinical applications have been proven in treating hematological malignance. Bispecific antibody binding fragment (BiFab) represents a promising platform for generating non-Fc bispecific antibodies. However, the generation of BiFab is still challenging, especially by means of chemical conjugation. More conjugation strategies, e.g., enzymatic conjugation and modular BiFab preparation, are needed to improve the robustness and flexibility of BiFab preparation. We successfully used chemo-enzymatic conjugation approach to generate bispecific antibody (i.e., BiFab) with Fabs from full-length antibodies. Paired click handles (e.g., N_3 and DBCO) was introduced to the C-terminal LPETG tag of Fabs via sortase A mediated transpeptidation, followed by site-specific conjugation between two click handle-modified Fabs for BiFab generation. Both BiFab$^{CD20/CD3}$ (EC_{50} = 0.26 ng/mL) and BiFab$^{Her2/CD3}$ exhibited superior efficacy in mediating T cells, from either PBMC or ATC, to kill target tumor cell lines while spared antigen-negative tumor cells in vitro. The BiFab$^{CD20/CD3}$ also efficiently inhibited CD20-positive tumor growth in mouse xenograft model. We have established a facile sortase A-mediated click handle installation to generate homogeneous and functional BiFabs. The exemplary BiFabs against different targets showed superior efficacy in redirecting and activating T cells to specifically kill target tumor cells, demonstrating the robustness of sortase A-mediated "bio-click" chemistry in generating various potent BiFabs. This approach also holds promise for further efficient construction of a Fab derivative library for personalized tumor immunotherapy in the future.

Keywords: bispecific antibody; sortase A; chemo-enzymatic approach; anti-CD20 antibody; Fab; BiFab

1. Introduction

Immunotherapies, such as chimeric antigen receptor T cells (CAR-Ts) and T-cell-engaging bispecific antibodies (T-BsAbs), have revolutionized cancer treatments by leveraging the immune system [1,2]. T-BsAbs usually refer to bifunctional antibodies with one arm targeting T cell receptors (e.g., CD3) to engage T cells and another arm targeting antigen on tumor cells, for the purpose of bridging and redirecting T cells to tumor cells. Compared with CAR-T cells, which are autologous T lymphocytes that are genetically engineered to express chimeric antigen receptor for specific tumor cell targeting [3], bispecific antibody can be produced relatively easier and provide off-the-shelf treatment [4,5]. This strategy has generated great interest with more than 50 T-BsAb candidates in clinical trials for a range of indicators nowadays [6].

One representative T-BsAb is Blinatumomab, a bi-specific T cell engager (BiTE) targeting CD19 and CD3 that was approved by FDA in 2014 for the treatment of Acute Lymphoblastic Leukemia (ALL). The flexible tandem arrangement of this single chain bispecific antibody accounts for its superior efficacy in inducing lytic synapse and thereby high T cell activity in comparison with its IgG-based and Fab-based format [7,8]. Despite the high efficacy, BiTE molecule has a very short half-life of ~2 h in blood circulation in the absence of Fc domain [9]. In order to increase stability and activity of fragment-based T-BsAbs, Dual-Affinity Re-Targeting (DART®) protein and tandem diabody (TandAb) were designed to further improve the half-life and stability in vivo [10–13]. However, the variable region spanning engineered constant scaffold might result in the loss of affinity and stability for Fc-free T-BsAbs, such as single-chain variable fragment (scfv) molecules [14,15]. For example, variable regions assembled to a format that deviate significantly from its cognate high stable IgG might compromise its affinity, especially when the N-terminus of Fvs have additional polypeptide chains that function as linkers [14,16,17].

Bispecific antibody binding fragment (BiFab) represents another promising platform for generating bispecific antibodies. Two Fab fragments providing different binding specificities are usually chemically linked in a tail-to-tail manner to generate BiFab. The intact structure of Fab fragments is parallelly grafted into the BiFab format, which maintains a natural association of four domains (VL, CL, VH and CH1) and thus ensures stability [14,18,19]. The BiFabs could also avoid Fc-related side effects since they lack a Fc region. However, the site-specific conjugation of two Fab molecules remains challenging during BiFab preparation [20]. One of the well-known chemical approaches for BiFab generation is the application of click chemistry, in which the click handle is installed through the introduced noncanonical amino acid (ncAA) on Fabs, to realize site-specific conjugation of two Fabs [21–23].

To achieve site-specific click handle installation, an alternative approach is sortase A-mediated transpeptidation. Sortase A is a bacterial enzyme that recognizes C-terminal LPXTG motif (X represents any amino acid) of proteins or peptides, which is used to anchor building blocks of cell walls of Gram-positive bacteria. The enzyme cleavages between Thr and Gly residues and then yields an acyl-enzyme intermediate. Subsequently, the nucleophilic primary amine of oligo-glycine modified substrates resolved the intermediate and then form a covalent bond between oligo-glycine modified substrates and LPETG-tagged protein [24–27]. Therefore, the paired click handles could be modified with oligo-glycines, such as GGG, before installation to the C-terminus of the target protein (e.g., Fab). Herein, we applied sortase A-mediated two-step chemo-enzymatic conjugation to generate BiFabs. The paired click handles that comprising azide and dibenzocyclooctyne function groups was firstly attached to the Fabs by sortase A mediated transpeptidation between LPETG-tagged Fab and click chemistry-functionalized GGG, and subsequently the Fab-linkers are conjugated via click chemistry to form BiFabs. Using this strategy, we successfully constructed homologous BiFab$^{CD20/CD3}$ and BiFab$^{Her2/CD3}$. We have demonstrated the potent in vitro and in vivo efficacy of BiFab$^{CD20/CD3}$, and its ability to stimulate resting PBMC to proliferate and degranulate. In addition, functional BiFab$^{Her2/CD3}$ was generated by simply replacing Fab^{CD20} arm with Fab^{Her2}, further suggesting the potential of this chemo-enzymatic approach on preparing various BiFabs based on prestored Fab derivative library.

2. Materials and Methods

2.1. Reagents and Cell Lines

The human CD20-positive cell lines Ramos, Raji, Daudi and the human CD20-negative cell line K562 were purchased from the American Type Culture Collection (ATCC, San Francisco, CA, USA), and were cultured in 1640 medium (Gibco) with 10% fetal bovine serum (FBS, Gibco). The human HER2-positive cell line SK-OV-3 and HER2-negtive cell line MDA-MB-468 were purchased from ATCC and were cultured in McCoy's 5A or DMEM (Gibco) with 10% FBS, respectively. The expression plasmids of the full-length anti-CD20 antibody Ofatumumab and sortase A enzyme were constructed in our laboratory [28]. The HEK-293F cell line was from Qilin Zhang's laboratory in Tsinghua University. The HEK293F cells were grown in 250 mL SMM-293-TI medium (Sinobiological, Beijing, China) supplemented with 100 U/mL ampicillin, 100 μg/mL streptomycin (Sorlabio), and 1% FBS and the cells were shaking cultured at 37 °C and 210 rpm (Eppendorf). Anti-CD3 Fab sequence was derived from the humanized OKT3 antibody [29]. Anti-Her2 Fab sequence was derived from the Trastuzumab [30].

Triple glycine-modified linker Gly$_3$-(PEG)$_3$-N$_3$ (GPN) were synthesized by Concortis (San Diego, CA, USA). Triple glycine-modified linker Gly$_3$-(PEG)$_4$-dibenzocyclooctyne (DBCO) (GPD) was purchased from Lumiprobe (Hunt Valley, MD, USA).

2.2. Sortase A-Mediated Click Handle Installation

We previously showed that sortase A was used to specifically conjugate LPETG tagged IgG with GGG modified toxins [24], and the enzyme was kept by our lab. Briefly, we used a sortase A mutant ($\triangle$N59) derived from *Staphylococcus aureus*, which is subcloned into pET28a(+) before a six Histidine polypeptide (His$_6$). The expression vector of sortase A was then transfected into BL21 (DE3) Competent Cells (Sangon, Shanghai, China) and the expression is induced by 0.5 M IPTG for 16 h. After incubation, cells were harvested and disrupted by French Press (ThermoFisher Scientific Inc., Shanghai, China). The soluble fraction was collected and purified by Ni-NTA (HiTrap Ni-NTA column, GE) with instruction of the manufacturer's protocol. The purified sortase A protein was buffer exchanged to 50×10^{-3} M Tris–HCl (pH 7.5), 150×10^{-3} M NaCl by ultrafiltration (Amicon Ultra-10k, Millipore, MA, USA), sterile filtered and stored at −80 °C. Sequences of light chain and heavy chain of antibody fragments (Fabs) were, respectively, inserted into pMH3 expression vector behind human signal peptide sequence, and Fabs of heavy chain were C-terminally tagged with nucleotide sequence that express polypeptide GGGGSGGGGSGGGGS-LPETG-6 × His ((G$_4$S)$_3$-LPETG-His$_6$). G$_4$S linker was used to facilitate sortase A mediated transpeptidation. The expression vector of Fabs was transiently expressed in HEK293F cells for 3–4 days.

To optimize the reaction conditions of the sortase A-mediated conjugation, the reaction molar ratio of antibody fragments to glycine modified linkers (e.g., GPD and GPN) was explored. The reaction molar ratios (1:25 and 1:50) and different reaction time (6 h, 12 h or 24 h) at 37 °C were investigated in reaction buffer (50 mM Tris-HCl, 150 mM NaCl, 5 mM CaCl$_2$, pH 7.4) solution in the presence of 50 μM sortase A enzyme (the molar ratio of sortase A/Fab was 1:8.3). To evaluate the conjugation efficiency, the reverse-phase high pressure liquid chromatography (RP-HPLC) with a Varian PLRP-S 100 Å column was used as previously described [28,31]. The conjugation reaction was scaled up under optimal reaction condition. Since the His tag was cut off by sortase A during transpeptidation, the flow-through fluid containing modified Fabs (e.g., Fab^{CD3}-DBCO and Fab^{CD20}-N$_3$) was collected during HiTrap Ni-NTA affinity chromatography. All modified Fabs were buffer exchanged to PBS (pH 7.4) by ultracentrifugation (Millipore Amicon Ultra Filters, 10 kDa cut-off).

2.3. Click Chemistry Mediated Generation of Bispecific Fab (BiFab)

The copper-free click reaction between Fab-GPN and Fab-GPD was reacted in a buffered solution contained 50 mM Tris-HCl, 150 mM NaCl (pH 7.4). Fab^{CD3}-DBCO was reacted with Fab^{CD20}-N_3 or Fab^{Her2}-N_3 at a molar ratio of 1:1 at 4 °C for 12 h. After reaction, BiFabs were purified from free Fab by size exclusion chromatography (SEC) (Superdex 200 increase 10/300 GL, GE) on AKTA purifier (Amersham Biosciences, MA, USA). Sample from each peak was analyzed by SDS-PAGE under reducing condition and non-reducing condition. The purified protein from SEC was also analyzed by RP-HPLC with the following condition, a linear gradient elution starting from 75% buffer A (1.5 M $(NH_4)_2SO_4$, 25 mM Na_3PO_4, pH 7.0), 25% buffer B (25 mM Na_3PO_4, pH 7.0) and 0% isopropanol, to 0% buffer A (1.5 M $(NH_4)_2SO_4$, 75 mM Na_3PO_4, pH 7.0), 75% buffer B (25 mM Na_3PO_4, pH 7.0) and 25% isopropanol.

2.4. Flow Cytometry

All flow cytometry studies were conducted on ACEA NovoCyteTM (ACEA Biosciences Inc., San Diego, CA, USA). Data were processed with FlowJo 10.1 (FlowJo, LLC, Ashland, OR, USA) and Prism 8.0.1 (GraphPad Software Inc., San Diego, CA, USA).

To evaluate the binding ability of $BiFab^{CD20/CD3}$, 1×10^6 CD20-positive cells or 1×10^6 CD3-positive Jurkat cells were incubated with serial concentrations of Fab^{CD20}, Fab^{CD3} and $BiFab^{CD20/CD3}$ in ice-cold PBS (pH 7.4) for 30 min, followed by incubation with the primary anti-human IgG-Fab fragment (Abcam, Cambridge, UK) for 30 min. After washing three times with cold PBS (pH 7.4), cells were incubated with secondary goat anti-mouse IgG-FITC (Beyotime, Shanghai, China) for 30 min. After washing step, immune-stained cells were analyzed by flow cytometry.

2.5. Preparation of Active T Cells (ATC) from Peripheral Blood Mononuclear Cells (PBMC)

Human blood samples were obtained from healthy volunteers. PBMC were extracted from fresh blood samples by density centrifugation (Ficoll-Paque) following manufacturer's instruction.

PBMC were stimulated with Dynabeads™ Human T-Activator CD3/CD28 (Thermo Fisher) for T cell expansion and activation to generate active T cells (ATC). Briefly, PBMC were mixed with dynabeads at a cell-to-bead ratio of 1:1, and co-incubated for 4 days in the presence of 30 U/mL recombinant IL-2.

2.6. Cell Apoptosis

PBMCs were used as effector cells in all experiments. For LDH releasing assay, 96-well plates were seeded with 3×10^4 tumor cells (e.g., Ramos or Daudi cells) and 6×10^4 ATC per well, and then added with serial concentrations of BiFabs for a 24 h incubation at 37 °C. After incubation, the release of the intracellular enzyme lactate dehydrogenase (LDH) was determined by LDH cytotoxicity assay kit (Beyotime, Shanghai, China) to measure cell death. The percentage of necrotic cells was calculated according to the absorbance of each well at 450 nm.

For flow cytometry studies on cell apoptosis, ATC were prestained with Carboxyfluorescein succinimidyl ester (CFSE), and then co-cultured with 2×10^5 tumor cells at an effector: target (E:T) ratio of 2:1 for 24 h. When PBMC were used as effector cells, the E:T ratio was 5:1. Cells were then stained with Annexin-Cy5/Propidium Iodide (PI), and the percentage of apoptotic and necrotic cells were determined by flow cytometry.

2.7. T Cell Activation

CD69 and CD25 are early and late activation markers for T cells, respectively. We therefore used flow cytometry to evaluate T cell activation via measuring cell surface CD69 and CD25 expression. Fresh PBMC were mixed with target tumor cells (e.g., Ramous, Raji, Daudi and K562 cells) at E: T ratio of 5:1 before adding serial concentrations of BiFabs (BiFab$^{Her2/CD3}$ or BiFab$^{CD20/CD3}$) to initiate specific killing, and the co-incubation lasted 48 h. Naïve T cells were labeled with FITC-αCD4 and FITC-αCD8, and active T cells were further labelled with APC-αCD25 and PE-αCD69 (BD Biosciences). When CD20 positive tumor cells were used as target cells, fresh PBMC were pre-treated with anti-CD20 antibody-coated magnetic beads to deplete CD20-positive B cells.

Enzyme-linked immunosorbent assay (ELISA) was used to detect interferon gamma (IFN-γ) that was secreted from the activated T cells. Fresh PBMC were co-incubated with target tumor cells (e.g., K562 cells) at an E: T ratio of 5:1, and then treated with BiFabs or IgG format bispecific antibodies for 48 h. Supernatants were collected for IFNγ detection through Human IFN-γ CytoSetTM KIT (Invitrogen, Shanghai, China). The absorbance at 450 nm was measured by 680 Microplate reader (Bio-Rad, Hercules, CA, USA).

To evaluate T cell proliferation after activation, fresh PBMC were pre-labelled with CFSE and then mixed with target tumor cells at an E: T ratio of 5:1. The cell mixtures were treated with different concentrations of BiFabs for 48 h. T cell proliferation was further determined by flow cytometry. Ramos cells were used in experimental groups as CD20-positive cells, while K562 cells served as CD20-negative cell control.

2.8. In Vivo Antitumor Activity of BiFab in Mouse Xenograft Model

Eight-week-old female SCID Beige mice were inoculated subcutaneously with 2.5×10^6 Ramos cells and 1×10^7 PBMC (E:T = 4:1) into the right flank of the nude mice. Inoculated mice were randomly divided into 4 groups: vehicle group, Fab^{CD3} group, Fab$^{CD20/CD3}$ (3 mg/kg) group and BiFab$^{CD20/CD3}$ (1 mg/kg) group. Mice in the experimental groups received BiFab$^{CD20/CD3}$ (1 mg/kg or 3 mg/kg) by intravenous (i.v.) injection into tail vein. Mice in the control groups received Fab^{CD3} (1 mg/kg) or saline. Each treatment was given four times at 2-day intervals (q2d × 4). The mean tumor volume and mouse body weight were measured using calipers and an electronic balance, respectively. The mean tumor volume was calculated using the formula: tumor volume (mm^3) = tumor length × tumor width × tumor width/2.

2.9. Statistical Analysis

Statistical analysis was performed by using GraphPad Prism 6.01 software. Student's *t*-test was used when two independent groups are compared, while Dunnett's multiple comparison test was used for comparison of multiple groups. Statistical significance was determined by the *p* value (* $p < 0.05$, ** $p < 0.01$, and *** $p < 0.001$).

3. Results

3.1. Generation of Bispecific Fab via Sortase-Mediated Transpeptidation and Click Chemistry

The whole procedure to generate BiFabs was summarized in Figure 1a. Fabs targeting CD20, CD3 or HER2 were first expressed with LPETG-His$_6$ tail at C terminus of heavy chains (Figure 1b) and stored for future assembly after purification. GGG-PEG$_3$-N$_3$ or GGG-PEG$_4$-DBCO was linked onto Fabs via sortase A transpeptidation, and His-tag was released from Fabs, which spared linker-Fab components from the reaction mixture when purified by Ni-NTA affinity chromatography. Before click reaction, the optimal molar ratio and reaction time for sortase A-catalyzed reaction was investigated. According to peak shifting of H-DBCO, the optimal reaction condition is 1:25 of Fab^{CD3} and GPD and reacted for 12 h (Figure 1c), in which there is much less unconjugated heavy chain (peak "H") compared to other reaction conditions. Click reaction between Fab^{CD3}-DBCO and Fab^{CD20}-N$_3$ at a molar ratio of 1:1 efficiently generated BiFab$^{CD20/CD3}$. After click reaction, homogenous BiFab$^{CD20/CD3}$ was obtained by size exclusion chromatography purification and further confirmed by SDS-PAGE (Supplementary Figure S1). The assembly of Fab^{Her2} and Fab^{CD3} was conducted in the same way to generate homologous BiFab$^{Her2/CD3}$ (Figure 1d). The purity of BiFab$^{CD20/CD3}$ was further confirmed by RP-HPLC analysis (Figure 1e). According to the peak area, the content of BiFab$^{CD20/CD3}$ in the final buffered solution is about 95% after SEC purification and ultraconcentration.

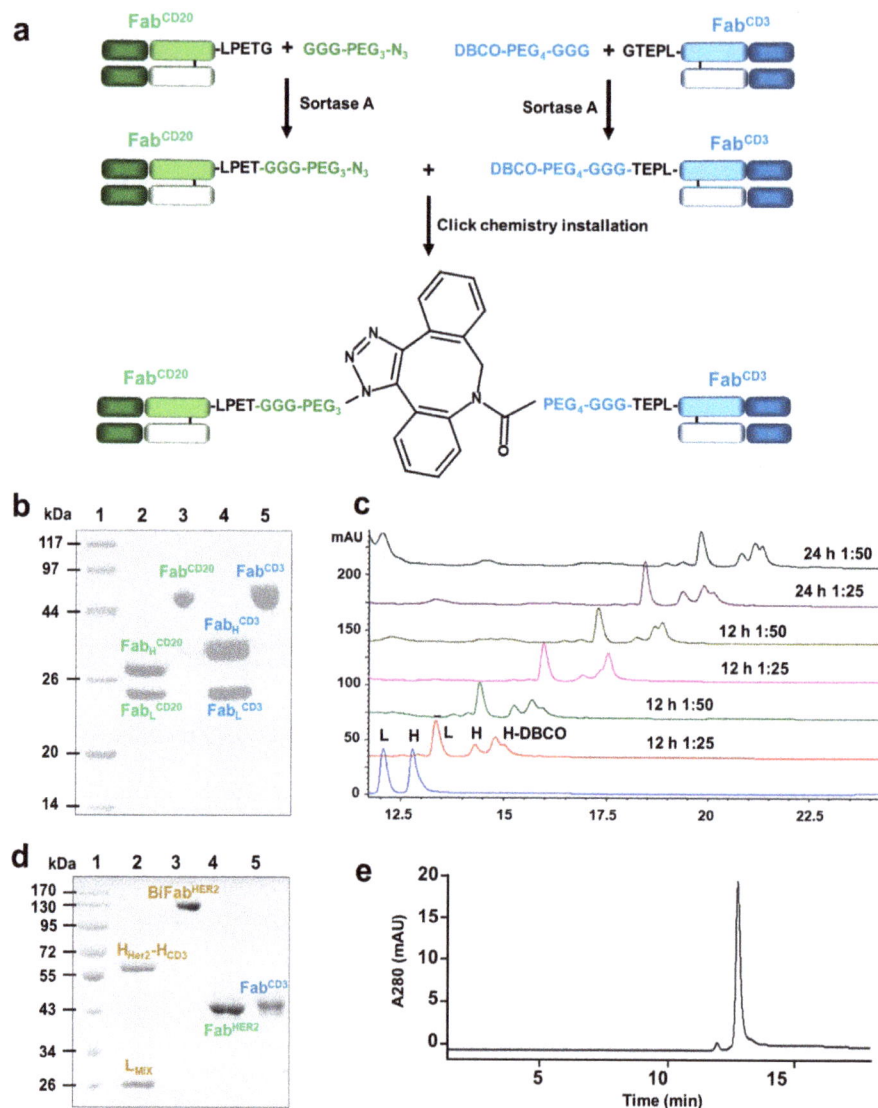

Figure 1. Generation and characterization of BiFabs. (a) Schematic diagram of sortase A-mediated click chemistry installation for BiFab preparation. (b) Characterization of the purified Fabs by SDS-PAGE. Lane 1, high molecular weight protein marker; Lane 2, the reduced Fab^{CD20}; Lane 3, the intact Fab^{CD20}; Lane 3, the reduced Fab^{CD3}; Lane 4, the intact Fab^{CD3}. (c) Reverse-phase HPLC analysis of Fab-click handle conjugation through sortase A-mediated transpeptidation, under different reaction conditions. (d) Characterization of BiFabs by SDS-PAGE. Lane 1, high molecular weight protein marker; Lane 2, the reduced BiFab$^{Her2/CD3}$; Lane 3, the intact BiFab$^{Her2/CD3}$; Lane 4, the intact Fab^{Her2}; Lane 5, the intact Fab^{CD3}. (e) Reverse phase high-performance liquid chromatography (RP-HPLC) analysis of the purity of BiFab$^{CD20/CD3}$.

3.2. The Binding Ability of BiFabs with Target and Effector Cells

To confirm whether BiFab$^{CD20/CD3}$ maintained the binding ability of two Fabs, we used Jurkat cells (CD3 positive) and Ramos cells (CD20 positive) for flow cytometric analysis of BiFab$^{CD20/CD3}$. The BiFab$^{CD20/CD3}$ showed concentration-dependent binding with CD20-positive Ramos cells and CD3-positive Jurkat cells (Figure 2a). Interestingly, BiFab$^{CD20/CD3}$ had a higher binding affinity compared to that of Fab^{CD20} or Fab^{CD3} monomers to target cells (Figure 2a). Upon binding with CD3 on T cells and CD20 on tumor cells, the BiFab$^{CD20/CD3}$ could efficiently activate T cells according to the measurement of cell-surfaceCD69 and CD25, which represent early and late activation markers on T cells, respectively (Figure 2b). At the same concentrations of BiFab$^{CD20/CD3}$, the expression level of CD69 was much higher than that of CD25, which exhibited a quicker response curve of CD69 comparing to CD25. Similarly, the BiFab$^{Her2/CD3}$ generated by replacing Fab^{CD20} could also bind to HER2-positive SK-OV-3 cells and CD3-positive Jurkat cells (Figure 2c).

3.3. BiFab Efficiently and Specifically Induced Cytokine Release and Proliferation of T Cells

The release of interferon-γ (INF-γ) was evaluated as this cytokine is essential for mediating the antitumor activity. We measured INF-γ release by ELISA kit with CD20+ Daudi and Raji cells as target cells and unstimulated PBMC as effector cells. In both types of target cells, high level of IFN-γ release was detected in the culture supernatants in the presence of BiFab$^{CD20/CD3}$ (400 and 2000 ng/mL) (Figure 2d,e). We also noticed that BiFab$^{CD20/CD3}$ induced stronger T cell activation at a concentration of 80 ng/mL when the target cells were Daudi cells in comparison to Raji cells. Almost no T cell activation was observed in the absence of BiFab$^{CD20/CD3}$, suggesting the specific mode of action underlying BiFab mediated T cell engaging.

For the analysis of T cell proliferation after BiFab stimulation, fresh PBMC were pre-stained with CFSE, a cell permeant green fluorescent molecule whose succinimidyl ester group reacts indiscriminately and covalently with primary amines of intracellular proteins, to facilitate fluorescent labeling of T cell population. Upon incubating with BiFab$^{CD20/CD3}$, PBMC significantly proliferated after 48 h in the presence of target cells (i.e., CD20-positive Ramos Cells) (Figure 2e). No obvious T cell proliferation was observed in negative control group (CD20-negative K562 cells). We further studied the proliferation rate with various concentrations of BiFab$^{CD20/CD3}$ measured by flow cytometry. BiFab$^{CD20/CD3}$ triggered T cells proliferation in a concentration-dependent manner (Figure 2f).

3.4. BiFabs Redirected T Cells to Kill Target Tumor Cells

In vitro cytotoxicity of BiFab$^{CD20/CD3}$ was measured by LDH releasing assay on Daudi and Ramos cell lines. BiFab$^{CD20/CD3}$ efficiently induced tumor cell apoptosis at an E:T ratio of 2:1, achieving half maximal-apoptosis rate at a concentration of 0.262 ng/mL (2.62 pM) on Daudi cells and 0.275 ng/mL (2.75 pM) on Ramos cells (Figure 3a). The apoptosis-inducing efficacy of BiFab$^{CD20/CD3}$ was further assessed by FITC-Annexin V/PI staining assay on Daudi and Ramos cell lines. BiFab$^{CD20/CD3}$ could induce maximal apoptosis, including early (Annexin V+/PI−) and late (Annexin V+/PI+) apoptotic cells, on Daudi cells at various concentrations (10 ng/mL–10 ug/mL) (Figure 3b). For Ramos cells, the apoptosis rate, ranging from 70–100%, was concentration-dependent and lower than that of Daudi cells (Figure 3b). Fc-mediated nonspecific activation through binding to Fc receptors on immune cells could probably cause toxicity [32]. Comparing to IgG$^{CD20/CD3}$ which could elicit Fc-mediated non-specific killing, the BiFab$^{CD20/CD3}$ was demonstrated to have minimal killing towards CD20 negative K562 cells, suggesting the advantage of Fc truncation in eliminating Fc-mediated side effects (Figure 3c). Similar to BiFab$^{CD20/CD3}$, the BiFab$^{Her2/CD3}$ exhibited remarkable killing efficacy on HER2-positive SK-OV-3 cells while spared HER2 negative MDA-MB-468 cells with marginal cell killing (Figure 3d). The results showed here suggested that sortase A-mediated chemo-enzymatic approach was successfully applied to the generation of other BiFabs.

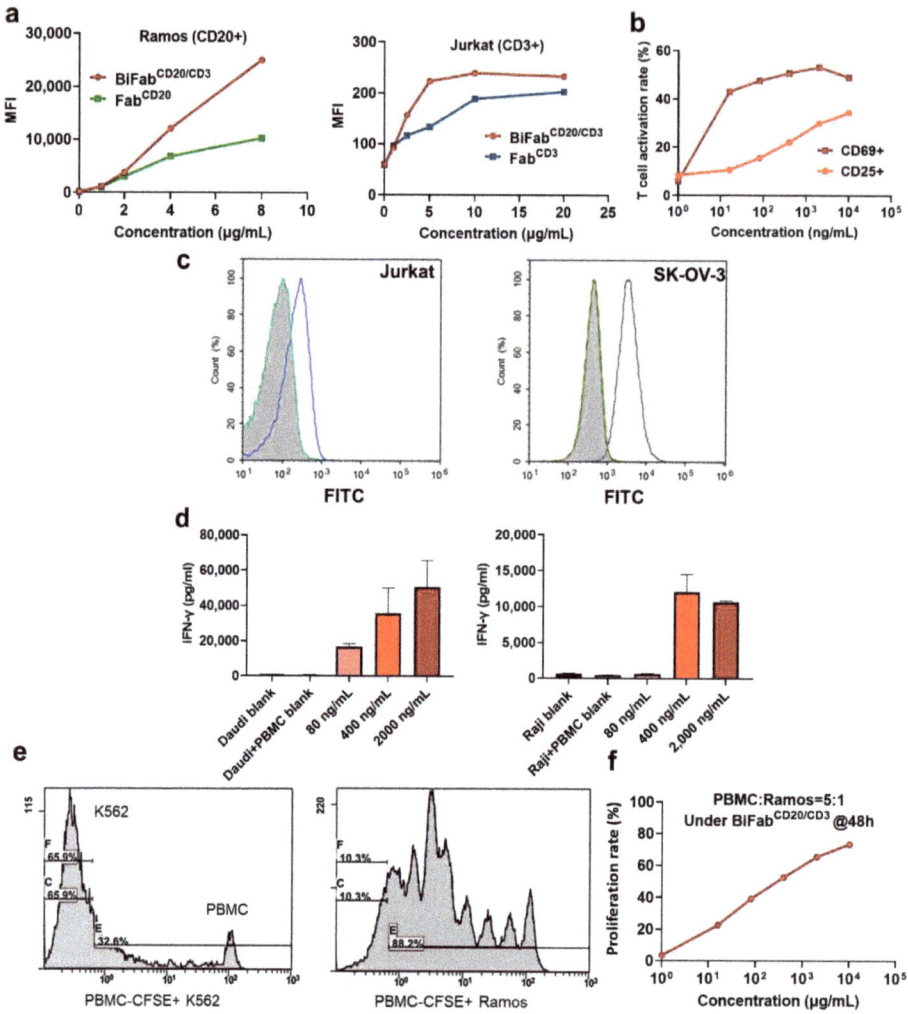

Figure 2. In vitro efficacy of BiFabs. (**a**) The binding abilities of Fabs and BiFab with CD20-positive Ramos and Jurkat cells. (**b**) The in vitro efficacy of the BiFab$^{CD20/CD3}$ on T cell activation. After CD20-positive B cell depletion, fresh PBMCs were treated with serial concentrations of BiFab$^{CD20/CD3}$ in the presence of target tumor cells at an E:T ratio of 5:1 for 48 h. The expression levels of CD69 and CD25 on T cells, two biomarkers for T cell activation, were evaluated after immuno-staining via flow cytometry. (**c**) Evaluation of the binding abilities of BiFab$^{Her2/CD3}$ with CD3-positive Jurkat cells and HER2-positive SK-OV-3 cells by flow cytometry. (**d**) The quantification of interferon-γ release from T cells activated by BiFab$^{CD20/CD3}$. Fresh PBMCs were incubated with Daudi or Raji cells at an E:T ratio of 5:1 for 48 h. The secreted interferon-γ from T cells was quantified by ELISA Kit. (**e**) BiFab$^{CD20/CD3}$ mediated T cell proliferation in the presence of CD20-negative K562 cells or CD20-positive Ramos cells at an E:T ratio of 5:1 for 48 h. (**f**) After treatment with various concentrations of BiFab$^{CD20/CD3}$ with an E:T ratio of 5:1 for 48 h, T cell proliferation was analyzed by flow cytometry.

3.5. BiFab$^{CD20/CD3}$ Eliminated B-Cell Lymphoma in Xenograft Mouse Model

We next evaluated the in vivo efficacy of BiFab$^{CD20/CD3}$ with mouse xenograft model of B-cell lymphoma. Mouse xenograft tumor model was successfully established by co-injection of Ramos and PBMC cells (E:T ratio = 4:1). The intravenous administration of

BiFabs was initiated 24 h after inoculation to facilitate T cell activation. The administration was repeated every two days for a total of four injections. Strikingly, the BiFab$^{CD20/CD3}$ completely suppressed the tumor growth at a dosage of 3 mg/kg, and there was only one mouse that underwent a recurrence in the 1 mg/kg group (Figure 3e). In contrast, the anti-CD3 Fab group did not show any significant efficacy in vehicle group, in which tumor grew rapidly. These results demonstrated that the BiFab$^{CD20/CD3}$ could efficiently mediate T cell killing in vivo.

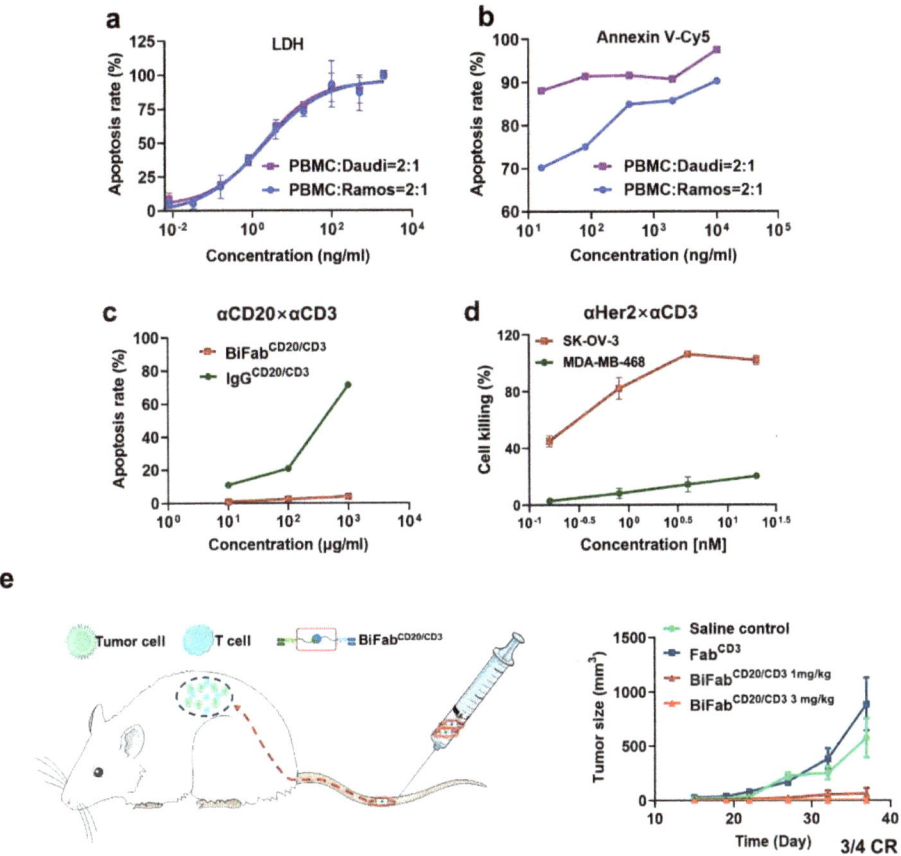

Figure 3. The in vitro and in vivo antitumor activities of BiFabs. (**a**) The in vitro efficacy of BiFab$^{CD20/CD3}$. Target cells (Ramos and Daudi) and active T cells (E:T = 2:1) were incubated with serial diluted BiFab$^{CD20/CD3}$ for 24 h (data shown as mean ± SD, n = 3). LDH release was determined by ELSIA kit and used to calculate cell viability. (**b**) The in vitro cytotoxicity of BiFab$^{CD20/CD3}$ was analyzed by Annexin V/PI apoptosis detection kit, by using the same condition as described in (**a**). (**c**) Study on potential Fc-related cytotoxicity of BiFab$^{CD20/CD3}$. The K562 cells and PBMCs were co-cultured with serial concentrations of non-binding IgG-based bispecific antibody or BiFab. The apoptosis rate was determined by Annexin V-Cy5 Apoptosis Detection Kit. (**d**) The in vitro cytotoxicity of BiFab$^{Her2/CD3}$. Target tumor cells (SK-OV-3 or MDA-MB-468) and PBMC (E:T = 4:1) were incubated with serial concentrations of BiFab$^{Her2/CD3}$ for 72 h, and the LDH release in the supernatant was determined by LDH detection kit. All data were shown as mean ± SD, n = 3. (**e**) The in vivo antitumor activities of BiFab$^{CD20/CD3}$ in mouse xenograft model. Mice were inoculated subcutaneously with 2.5 × 10^6 Ramos cells in the presence of 1 × 10^7 fresh human PBMC from healthy donors at an E:T ratio of 4:1. All samples were administered intravenously via the tail vein at following dosages, 1 mg/kg of Fab^{CD3} and 1 mg/kg or 3 mg/kg of BiFab$^{CD20/CD3}$ at every two days for four times.

4. Discussion

We have presented here a facile approach utilizing sortase A-mediated bio-click chemistry to generate BiFabs with potent antitumor activity. Paired click handles (e.g., N_3 and DBCO) was conjugated to the C-terminal LPETG tag of Fabs via sortase A mediated transpeptidation, followed by site-specific conjugation between two click handles-modified Fabs for BiFab generation. We have presented exemplary BiFabs against two different targets. First, the BiFab$^{CD20/CD3}$ exhibited superior efficacy in mediating T cells, from either PBMC or ATC, to kill multiple CD20-positive lymphoma cell lines while spared CD20-negative tumor cells in vitro (Supplemental Figure S2). The BiFab$^{CD20/CD3}$ also efficiently inhibited CD20-positive tumor growth in the mouse xenograft model (Figure 3e). Second, the BiFab$^{Her2/CD3}$ also showed potent in vitro antitumor activity against HER2-positive tumor cell lines (Figure 3d), demonstrating the robustness of sortase A-mediated bio-click chemistry in generating various potent BiFabs.

The first BiFab construct, termed as BsF(ab')2, was first generated by chemical conjugation of Fab'-SH with the thionitrobenzoate derivative of another Fab (Fab'-TNB), which was described by Paul Carter et al. from Genentech Inc. [19]. The BiFab$^{CD20/CD3}$ was also generated without Fc region. Fc region of IgG-based bispecific antibodies could potentially induce nonspecific T cell activation [33], causing off-target T cell engaging-related side effects. However, cytokine-related adverse effects, such as cytokine release syndrome (CRS), are probably inevitable for T cell engaging and activation related immunotherapy, e.g., CAR-T, BiTE [34,35]. At present, T-cell engagers targeting CD20 are mostly based on classical IgG-based antibody. Sun et al. [36] reported a T-cell recruiting bispecific antibody CD20-TDB with EC_{50} of 0.22–11 ng/mL at the E: T ratio of 10:1. Smith et al. [37] reported another anti-CD20/CD3 T cell engagers REGE2280 and REGN 1979, which showed favorable EC_{50} of 2.25–12.6 ng/L at the E:T ratio of 10:1 (ATC as effector cells). FBTA05 is a trifunctional chimeric rat/mouse CD3 × CD20 targeting bispecific antibody, and therefore it has higher immunogenicity [38]. In comparison with above anti-CD20/CD3 bispecific antibodies, our BiFab$^{CD20/CD3}$ showed a potent apoptosis-inducing ability at an E:T ratio of 2:1 using ATC as effector cells (EC_{50} = 0.26 ng/mL for Daudi cell lines and 0.275 ng/mL for Ramos) (Figure 3a). The in vivo antitumor efficacy of BiFab$^{CD20/CD3}$ was consistent with its in vitro efficacy, since four intravenous injections (3 mg/kg) of BiFab$^{CD20/CD3}$ completely suppressed tumor growth in the mouse tumor xenograft models (Figure 3e).

We previously reported a nucleic acid (i.e., left-handed DNA, L-DNA) mediated protein-protein assembly (NAPPA) approach to offer a general approach for preparing antibodies with higher-order specificity [39]. Similar to the NAPPA approach, our two-step conjugation strategy allows the preparation of modular Fab derivatives and the generation of customized Fab library thereof, which is the major difference comparing with the conventional BiFab construction methods (Figure 4). In addition, both BiFab$^{CD20/CD3}$ and BiFab$^{Her2/CD3}$ showed potent antitumor efficacies, regardless of different tumor target, suggesting the effectiveness and robustness of sortase A mediated chemo-enzymatic approach. Lawrence G Lum et al. [40] explored the application of anti-CD20/CD3 bispecific anti-body-armed activated T cells (aATC). The anti-CD20/CD3 aATC was a cell-based therapy that activated T cells from patients were armed with chemically conjugated anti-CD3 × anti-CD20 bispecific antibody, and then expanded and re-infused into patients. The aATC therapy was demonstrated to be safe and effective in a phase I clinical trial [40]. Inspired by this study, the sortase A-mediated bio-click chemistry could be further applied to personalized immunotherapy through ATC armed with combination-optimized BiFab. Since the efficacy of BiFab varies when using Fabs with different affinities or paratopes, the sortase A mediated transpeptidation reaction during BiFab generation facilitates the construction of Fab library for rapid efficacy evaluation of different BiFabs.

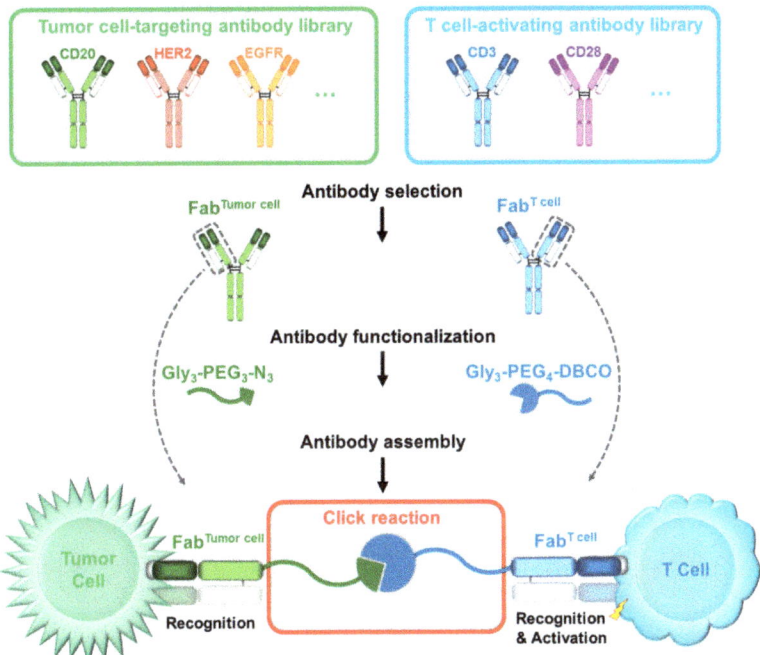

Figure 4. Schematic diagram of modular BiFab generation. Fabs could be adapted from full-length IgGs targeting tumor antigens or T cell/NK cell activating receptors. Fabs are genetically modified to have a C-terminal sortase A recognition motif (e.g., LPETG). Then, the paired click chemistry could be installed to the Fabs via sortase A mediated transpeptidation, followed by click reaction between two Fabs to generate BiFab.

5. Conclusions

We constructed BiFab$^{CD20/CD3}$ and BiFab$^{Her2/CD33}$ via sortase A-mediated bio-click chemistry and demonstrated their anti-tumor activity through engaging human immune cells. Our results shown here indicates that Sortase A-mediated click handle installation holds promise for facile generation of potent bispecific Fabs and further efficient construction of Fab derivative library for personalized tumor immunotherapy in the future.

Supplementary Materials: The following are available online at https://www.mdpi.com/article/10.3390/cancers13184540/s1, Figure S1: (**a**) Size exclusion chromatography (SEC) purification of BiFab$^{CD20/CD3}$. (**b**) SDS-PAGE analysis of peaks from (**a**); Line 1, high molecular weight protein marker; Line 2, reduced protein product from peak 1; Line 3, reduced protein product from peak 2. Figure S2: BiFab$^{CD20/CD3}$ activated T cells in the presence of target cell lines with different antigen expression level and mediated target cells killing in a T cell-dependent manner. (**a**) Target cell lines with different antigen expression level are measured by flow cytometry; (**b**) Target cell lines of different CD20 expression level and PBMC isolated from a healthy donor (1:5 cell ratio) were incubated with serial concentrations of BiFab$^{CD20/CD3}$ for 48 h.

Author Contributions: Conceptualization, L.P., S.C., W.L. and W.Z.; methodology, X.B. and W.Z.; validation, X.B.; formal analysis, W.L., X.B. and S.J.; resources, Y.X.; writing—original draft preparation, X.B.; writing—review and editing, L.P. and X.B.; supervision, L.P. and S.C.; funding acquisition, L.P., S.C. and Z.Z. All authors have read and agreed to the published version of the manuscript.

Funding: This research was funded by Joint Funds of the National Natural Science Foundation of China (Grant No. U20A20409), National Natural Science Foundation of China (Grant No. 82073750), key research and development project of Zhejiang province (No.2018C03022), the Fundamental

Research Funds for the Central Universities (No.2020QNA7005) and Zhejiang Province "Qianjiang Talent Plan".

Institutional Review Board Statement: For PBMC extraction assay, the experiment protocol was reviewed and approved by medical ethic committee in College of Pharmaceutical Sciences, Zhejiang University, China (2018-003). The animal experiments were carried out in compliance with the Public Health Service Policy on Human Care and Use of Laboratory Animals. The protocol was approved by the committee on the Ethics of Animal Experiments of Zhejiang University, China (19699).

Informed Consent Statement: Informed consent was obtained from all subjects involved in the study. Blood donors were recruited and informed of risks and discomforts of the donation process, and a signed informed consent document was obtained. The blood collection was done by standard phlebotomy.

Acknowledgments: We thank Ning Hu from Sun Yat-Sen University for helpful discussion.

Conflicts of Interest: The authors have filed a patent for the BiFab generation. Wenhui Liu is an employee of Hangzhou Biosun Pharmaceutical Co., Ltd.

References

1. Goebeler, M.-E.; Bargou, R.C. T cell-engaging therapies—BiTEs and beyond. *Nat. Rev. Clin. Oncol.* **2020**, *17*, 418–434. [CrossRef]
2. Hegde, P.S.; Chen, D.S. Top 10 Challenges in Cancer Immunotherapy. *Immunity* **2020**, *52*, 17–35. [CrossRef]
3. Singh, A.K.; McGuirk, J.P. CAR T cells: Continuation in a revolution of immunotherapy. *Lancet Oncol.* **2020**, *21*, e168–e178. [CrossRef]
4. Viardot, A.; Bargou, R. Bispecific antibodies in haematological malignancies. *Cancer Treat. Rev.* **2018**, *65*, 87–95. [CrossRef]
5. Slaney, C.Y.; Wang, P.; Darcy, P.K.; Kershaw, M.H. CARs versus BiTEs: A Comparison between T Cell–Redirection Strategies for Cancer Treatment. *Cancer Discov.* **2018**, *8*, 924–934. [CrossRef]
6. Blanco, B.; Domínguez-Alonso, C.; Álvarez-Vallina, L. Bispecific immunomodulatory antibodies for cancer immunotherapy. *Clin. Cancer Res. Off. J. Am. Assoc. Cancer Res.* **2021**, *6*, 3–17. [CrossRef]
7. Li, J.; Stagg, N.J.; Johnston, J.; Harris, M.J.; Menzies, S.A.; DiCara, D.; Clark, V.; Hristopoulos, M.; Cook, R.; Slaga, D.; et al. Membrane-Proximal Epitope Facilitates Efficient T Cell Synapse Formation by Anti-FcRH5/CD3 and Is a Requirement for Myeloma Cell Killing. *Cancer Cell* **2017**, *31*, 383–395. [CrossRef]
8. Wolf, E.; Hofmeister, R.; Kufer, P.; Schlereth, B.; Baeuerle, P.A. BiTEs: Bispecific antibody constructs with unique anti-tumor activity. *Drug Discov. Today* **2005**, *10*, 1237–1244. [CrossRef]
9. Nagorsen, D.; Kufer, P.; Baeuerle, P.A.; Bargou, R. Blinatumomab: A historical perspective. *Pharmacol. Ther.* **2012**, *136*, 334–342. [CrossRef]
10. Moore, P.A.; Zhang, W.; Rainey, G.J.; Burke, S.; Li, H.; Huang, L.; Gorlatov, S.; Veri, M.C.; Aggarwal, S.; Yang, Y.; et al. Application of dual affinity retargeting molecules to achieve optimal redirected T-cell killing of B-cell lymphoma. *Blood* **2011**, *117*, 4542–4551. [CrossRef]
11. Liu, L.; Lam, C.-Y.K.; Long, V.; Widjaja, L.; Yang, Y.; Li, H.; Jin, L.; Burke, S.; Gorlatov, S.; Brown, J.; et al. MGD011, A CD19 x CD3 Dual-Affinity Retargeting Bi-specific Molecule Incorporating Extended Circulating Half-life for the Treatment of B-Cell Malignancies. *Clin. Cancer Res.* **2017**, *23*, 1506–1518. [CrossRef]
12. Kipriyanov, S.M.; Moldenhauer, G.; Schuhmacher, J.; Cochlovius, B.; Von der Lieth, C.-W.; Matys, E.R.; Little, M. Bispecific tandem diabody for tumor therapy with improved antigen binding and pharmacokinetics. *J. Mol. Biol.* **1999**, *293*, 41–56. [CrossRef]
13. Reusch, U.; Duell, J.; Ellwanger, K.; Herbrecht, C.; Knackmuss, S.H.; Fucek, I.; Eser, M.; McAleese, F.; Molkenthin, V.; Le Gall, F.; et al. A tetravalent bispecific TandAb (CD19/CD3), AFM11, efficiently recruits T cells for the potent lysis of CD19 $^+$ tumor cells. *MAbs* **2015**, *7*, 584–604. [CrossRef]
14. Quintero-Hernández, V.; Juárez-González, V.R.; Ortíz-León, M.; Sánchez, R.; Possani, L.D.; Becerril, B. The change of the scFv into the Fab format improves the stability and in vivo toxin neutralization capacity of recombinant antibodies. *Mol. Immunol.* **2007**, *44*, 1307–1315. [CrossRef]
15. Cheng, M.; Ahmed, M.; Xu, H.; Cheung, N.-K.V. Structural design of disialoganglioside GD2 and CD3-bispecific antibodies to redirect T cells for tumor therapy. *Int. J. Cancer* **2015**, *136*, 476–486. [CrossRef]
16. Ha, J.-H.; Kim, J.-E.; Kim, Y.-S. Immunoglobulin Fc Heterodimer Platform Technology: From Design to Applications in Therapeutic Antibodies and Proteins. *Front. Immunol.* **2016**, *7*, 394. [CrossRef]
17. Kontermann, R.E. Dual targeting strategies with bispecific antibodies. *MAbs* **2012**, *4*, 182–197. [CrossRef] [PubMed]
18. Nitta, T.; Yagita, H.; Azuma, T.; Sato, K.; Okumura, K. Bispecific F(ab')2 monomer prepared with anti-CD3 and anti-tumor monoclonal antibodies is most potent in induction of cytolysis of human T cells. *Eur. J. Immunol.* **1989**, *19*, 1437–1441. [CrossRef]
19. Zhu, Z.; Lewis, G.D.; Carter, P. Engineering high affinity humanized anti-p185HER2/anti-CD3 bispecific F(ab')2 for efficient lysis of p185HER2 overexpressing tumor cells. *Int. J. Cancer* **1995**, *62*, 319–324. [CrossRef]
20. Graziano, R.F.; Guptill, P. Chemical production of bispecific antibodies. *Methods Mol. Biol. (Clifton N. J.)* **2004**, *283*, 71–85. [CrossRef]

21. Kim, C.H.; Axup, J.Y.; Dubrovska, A.; Kazane, S.A.; Hutchins, B.A.; Wold, E.D.; Smider, V.V.; Schultz, P.G. Synthesis of Bispecific Antibodies using Genetically Encoded Unnatural Amino Acids. *J. Am. Chem. Soc.* **2012**, *134*, 9918–9921. [CrossRef]
22. Lu, H.; Zhou, Q.; Deshmukh, V.; Phull, H.; Ma, J.; Tardif, V.; Naik, R.R.; Bouvard, C.; Zhang, Y.; Choi, S.; et al. Targeting Human C-Type Lectin-like Molecule-1 (CLL1) with a Bispecific Antibody for Immunotherapy of Acute Myeloid Leukemia. *Angew. Chem. Int. Ed.* **2014**, *53*, 9841–9845. [CrossRef]
23. Ramadoss, N.S.; Schulman, A.D.; Choi, S.; Rodgers, D.T.; Kazane, S.A.; Kim, C.H.; Lawson, B.R.; Young, T.S. An Anti-B Cell Maturation Antigen Bispecific Antibody for Multiple Myeloma. *J. Am. Chem. Soc.* **2015**, *137*, 5288–5291. [CrossRef]
24. Pan, L.; Zhao, W.; Lai, J.; Ding, D.; Zhang, Q.; Yang, X.; Huang, M.; Jin, S.; Xu, Y.; Zeng, S.; et al. Sortase A-Generated Highly Potent Anti-CD20-MMAE Conjugates for Efficient Elimination of B-Lineage Lymphomas. *Small* **2017**, *13*, 1602267. [CrossRef]
25. Pishesha, N.; Ingram, J.R.; Ploegh, H.L. Sortase A: A Model for Transpeptidation and Its Biological Applications. *Annu. Rev. Cell Dev. Biol.* **2018**, *34*, 163–188. [CrossRef]
26. Wagner, K.; Kwakkenbos, M.J.; Claassen, Y.B.; Maijoor, K.; Böhne, M.; van der Sluijs, K.F.; Witte, M.D.; van Zoelen, D.J.; Cornelissen, L.A.; Beaumont, T.; et al. Bispecific antibody generated with sortase and click chemistry has broad antiinfluenza virus activity. *Proc. Natl. Acad. Sci. USA* **2014**, *111*, 16820–16825. [CrossRef]
27. Becer, C.R.; Hoogenboom, R.; Schubert, U.S. Click chemistry beyond metal-catalyzed cycloaddition. *Angew. Chem. Int. Ed Engl.* **2009**, *48*, 4900–4908. [CrossRef]
28. Liu, W.; Zhao, W.; Bai, X.; Jin, S.; Li, Y.; Qiu, C.; Pan, L.; Ding, D.; Xu, Y.; Zhou, Z.; et al. High antitumor activity of Sortase A-generated anti-CD20 antibody fragment drug conjugates. *Eur. J. Pharm. Sci.* **2019**, *134*, 81–92. [CrossRef]
29. Woodle, E.S.; Thistlethwaite, J.R.; Jolliffe, L.K.; Zivin, R.A.; Collins, A.; Adair, J.R.; Bodmer, M.; Athwal, D.; Alegre, M.L.; Bluestone, J.A. Humanized OKT3 antibodies: Successful transfer of immune modulating properties and idiotype expression. *J. Immunol.* **1992**, *148*, 2756–2763.
30. Wishart, D.S.; Feunang, Y.D.; Guo, A.C.; Lo, E.J.; Marcu, A.; Grant, J.R.; Sajed, T.; Johnson, D.; Li, C.; Sayeeda, Z.; et al. DrugBank 5.0: A major update to the DrugBank database for 2018. *Nucleic Acids Res.* **2018**, *46*, D1074–D1082. [CrossRef]
31. Xu, Y.; Jin, S.; Zhao, W.; Liu, W.; Ding, D.; Zhou, J.; Chen, S. A Versatile Chemo-Enzymatic Conjugation Approach Yields Homogeneous and Highly Potent Antibody-Drug Conjugates. *Int. J. Mol. Sci.* **2017**, *18*, 2284. [CrossRef] [PubMed]
32. Labrijn, A.F.; Janmaat, M.L.; Reichert, J.M.; Parren, P.W.H.I. Bispecific antibodies: A mechanistic review of the pipeline. *Nat. Rev. Drug Discov.* **2019**, *18*, 585–608. [CrossRef] [PubMed]
33. Borlak, J.; Länger, F.; Spanel, R.; Schöndorfer, G.; Dittrich, C. Immune-mediated liver injury of the cancer therapeutic antibody catumaxomab targeting EpCAM, CD3 and Fcγ receptors. *Oncotarget* **2016**, *7*, 28059–28074. [CrossRef]
34. Wilke, A.C.; Gökbuget, N. Clinical applications and safety evaluation of the new CD19 specific T-cell engager antibody construct blinatumomab. *Expert Opin. Drug Saf.* **2017**, *16*, 1191–1202. [CrossRef] [PubMed]
35. Shimabukuro-Vornhagen, A.; Gödel, P.; Subklewe, M.; Stemmler, H.J.; Schlößer, H.A.; Schlaak, M.; Kochanek, M.; Böll, B.; von Bergwelt-Baildon, M.S. Cytokine release syndrome. *J. Immunother. Cancer* **2018**, *6*, 56. [CrossRef]
36. Sun, L.L.; Ellerman, D.; Mathieu, M.; Hristopoulos, M.; Chen, X.; Li, Y.; Yan, X.; Clark, R.; Reyes, A.; Stefanich, E.; et al. Anti-CD20/CD3 T cell–dependent bispecific antibody for the treatment of B cell malignancies. *Sci. Transl. Med.* **2015**, *7*, 287ra70. [CrossRef] [PubMed]
37. Smith, E.J.; Olson, K.; Haber, L.J.; Varghese, B.; Duramad, P.; Tustian, A.D.; Oyejide, A.; Kirshner, J.R.; Canova, L.; Menon, J.; et al. A novel, native-format bispecific antibody triggering T-cell killing of B-cells is robustly active in mouse tumor models and cynomolgus monkeys. *Sci. Rep.* **2016**, *5*, 17943. [CrossRef] [PubMed]
38. Stanglmaier, M.; Faltin, M.; Ruf, P.; Bodenhausen, A.; Schröder, P.; Lindhofer, H. Bi20 (fBTA05), a novel trifunctional bispecific antibody (anti-CD20 × anti-CD3), mediates efficient killing of B-cell lymphoma cells even with very low CD20 expression levels. *Int. J. Cancer* **2008**, *123*, 1181–1189. [CrossRef]
39. Pan, L.; Cao, C.; Run, C.; Zhou, L.; Chou, J.J. DNA-Mediated Assembly of Multispecific Antibodies for T Cell Engaging and Tumor Killing. *Adv. Sci. Weinh. Baden-Wurtt. Ger.* **2020**, *7*, 1900973. [CrossRef]
40. Lum, L.G.; Thakur, A.; Liu, Q.; Deol, A.; Al-Kadhimi, Z.; Ayash, L.; Abidi, M.H.; Pray, C.; Tomaszewski, E.N.; Steele, P.A.; et al. CD20-targeted T cells after stem cell transplantation for high risk and refractory non-Hodgkin's lymphoma. *Biol. Blood Marrow Transplant. J. Am. Soc. Blood Marrow Transplant.* **2013**, *19*, 925–933. [CrossRef]

Article

Systemic Treatment Initiation in Classical and Endemic Kaposi's Sarcoma: Risk Factors and Global Multi-State Modelling in a Monocentric Cohort Study

Lina Benajiba [1,*], Jérôme Lambert [2], Roberta La Selva [3], Delphine Cochereau [4], Barouyr Baroudjian [4], Jennifer Roux [4], Jérôme Le Goff [5], Cécile Pages [4], Maxime Battistella [6], Julie Delyon [4] and Céleste Lebbé [4,*]

1. Université de Paris, AP-HP, Clinical Investigations Center, INSERM U944, Saint Louis Hospital, 75010 Paris, France
2. Université de Paris, AP-HP, Biostatistics Department, Saint Louis Hospital, 75010 Paris, France; jerome.lambert@univ-paris-diderot.fr
3. A.O.U. Città della Salute e della Scienza di Torino, 10126 Turin, Italy; robertals@libero.it
4. Université de Paris, AP-HP, Dermatology Department, INSERM U976, Saint Louis Hospital, 75010 Paris, France; delphine_cochereau@hotmail.fr (D.C.); barouyr.baroudjian@aphp.fr (B.B.); Jennifer.roux@aphp.fr (J.R.); pageslaurent.cecile@iuct-oncopole.fr (C.P.); julie.delyon@aphp.fr (J.D.)
5. Université de Paris, AP-HP, Microbiology Department, Saint Louis Hospital, 75010 Paris, France; jerome.le-goff@aphp.fr
6. Université de Paris, AP-HP, Pathology Department, INSERM U976, Saint Louis Hospital, 75010 Paris, France; maxime.battistella@aphp.fr
* Correspondence: lina.benajiba@aphp.fr (L.B.); celeste.lebbe@aphp.fr (C.L.)

Citation: Benajiba, L.; Lambert, J.; La Selva, R.; Cochereau, D.; Baroudjian, B.; Roux, J.; Le Goff, J.; Pages, C.; Battistella, M.; Delyon, J.; et al. Systemic Treatment Initiation in Classical and Endemic Kaposi's Sarcoma: Risk Factors and Global Multi-State Modelling in a Monocentric Cohort Study. *Cancers* **2021**, *13*, 2519. https://doi.org/10.3390/cancers13112519

Academic Editors: Xianda Zhao and Subree Subramanian

Received: 22 April 2021
Accepted: 18 May 2021
Published: 21 May 2021

Publisher's Note: MDPI stays neutral with regard to jurisdictional claims in published maps and institutional affiliations.

Copyright: © 2021 by the authors. Licensee MDPI, Basel, Switzerland. This article is an open access article distributed under the terms and conditions of the Creative Commons Attribution (CC BY) license (https://creativecommons.org/licenses/by/4.0/).

Simple Summary: Over the past decades, clinical features and patients' outcome of iatrogenic and HIV-related KS epidemiological subtypes have been widely described in large cohort series. Due to their lower incidence and the limited resources available in endemic KS countries, classical and endemic KS epidemiological studies remain scarce, thus increasing the challenge of such clinically heterogeneous chronic diseases' management. In this large retrospective cohort study, six risk factors for treatment initiation were identified: time between first symptoms and diagnosis ≥1 year, endemic KS, total number of lesions ≥10, visceral or head/neck localization and edema. No response or treatment-free time difference was observed between the most frequently used therapeutic options: chemotherapy and interferon-alpha. Assessment for systemic treatment risk factors provides guidance for adequate follow-up and patients' information on disease outcome. Absence of efficacy difference between systemic regimens allows treatment choice based on fitness.

Abstract: Background: Although several studies described the clinical course of epidemic and post-transplant Kaposi's Sarcoma (KS), the lack of large cohorts of classic/endemic KS, precluded such characterization. Methods: We used multi-state modelling in a retrospective monocentric study to evaluate global disease evolution and identify risk factors for systemic treatment (ST) initiation. 160 classic/endemic KS patients consecutively diagnosed between 1990 and 2013 were included. Results: 41.2% of classic/endemic KS patients required ST. Cumulative incidence of ST after 2 years of follow-up was 28.4% [95% CI: 20.5; 35.5]. Multivariate analysis identified six risk factors for ST initiation: time between first symptoms and diagnosis ≥1 year, endemic KS, total number of lesions ≥10, visceral, head or neck localization and presence of edema. Type of ST, type of KS, age and time between diagnosis and ST were not associated with response. Mean treatment-free time during the first 5 years following ST was 44 months for interferon and 44.6 months for chemotherapy treated patients (Mean difference: −0.5 months [95% CI: −9.5; 4.9]). Conclusions: Our study reveals ST risk factors in classic/endemic KS and highlights the clinical aggressiveness of the endemic KS subtype. No efficacy difference was observed between standard of care treatments, enabling treatment choice based on patient's fitness.

Keywords: classical and endemic Kaposi Sarcoma; systemic treatment; multi-state modelling; treatment free interval; chemotherapy; interferon

1. Introduction

KS is an HHV-8 related lympho-angioproliferative disease with 4 clinical settings: iatrogenic (immunosuppressive therapy related), epidemic (HIV immune-deficiency related), endemic, and classic. Endemic KS develops in Sub-Saharian Africans whereas classic KS typically affects middle to elderly Mediterranean men with a male to female ratio ranging from 2:1 to 5:1 and an estimated incidence of 1.58 per 100,000 inhabitants per year in Sardinia [1,2].

HHV-8, also called KS-associated herpes virus (KSHV) is a herpes virus mainly transmitted through prolonged or repeated saliva contact during mother to child or sexual interactions. After an initial replicative phase, HHV-8 enables immune system evasion and establishes latency in the KS tumors [3]. T cell immune suppression is a well-recognized risk factor for HIV and transplant associated KS. Similarly, defects in NK cell and HHV-8-specific CD8 cells activity have been reported in classic and endemic KS [4].

The four KS epidemiological subtypes account for a wide clinico-pathological disease spectrum with some patients experiencing an indolent form of the disease while others present an aggressive disseminated pattern. Clinically, KS manifests mainly as purple-blue pigmented macules, plaques or nodules in the skin. More rarely, KS can also involve mucosa, lymph nodes or visceral organs such as gastro-intestinal tractus, lungs, bones and liver. Classic and endemic KS are typically indolent and mainly presents as limb lesions, with less than 10% mucosal, visceral or lymph node involvement [5]. HIV and iatrogenic KS are usually more disseminated in the skin and frequently involve mucosa, lymph nodes and visceral organs [6,7].

KS treatment remained mainly unchanged over the past 30 years. Management is based on: number and localization of lesions, presence of symptomatic lesions, disease progression and patient's fitness [8]. Patients with asymptomatic lesions are usually offered careful observation. Those presenting with symptomatic superficial or isolated skin lesions are treated locally, while more extensive, disseminated or visceral locations are treated systemically. Although such systemic strategies, based on interferon or chemotherapy, usually result in a 50 to 80% overall response rate, they are not curative and their efficacy is only transient [8].

The low incidence of classic/endemic KS, combined with the limited resources available in the endemic KS countries has precluded conducting large therapeutic studies in these KS subtypes. Non-HIV related KS subtypes management is thus mainly based on small retrospective clinical series, physicians experience and consensus based multidisciplinary guidelines [8,9]. No large real-life data are available to our knowledge to help clinicians' choice on systemic therapy decisions.

Aiming to improve endemic/classic KS management, we conducted a large monocentric retrospective study, in order to describe the disease's clinical course, identify risk factors for systemic treatment initiation and evaluate the rate and duration of response to systemic therapy, in a real-life cohort of classic/endemic KS patients. To our knowledge, this is the first study to report risk factors for systemic treatment initiation and to use multi-state modelling to evaluate global disease evolution in this rare KS subtypes. Our study should inform clinicians on the clinical course of classic/endemic KS and provide guidance in therapeutic management. Overall our findings should improve patients' quality of life and experience with this rare and chronic KS subtypes.

2. Materials and Methods

2.1. Study Design and Study Population

We performed a retrospective monocentric study, including all patients consecutively diagnosed with classic or endemic KS ($n = 160$) within one French dermato-oncology center between January 1990 and December 2013. Patients diagnosed prior to January 1990 have been excluded from the study to avoid discrepancies in the treatment modality due to absence of use of interferon-alpha as a first-line treatment in KS at the time. All histologically proven KS without any context of HIV or other causes of iatrogenic immunosuppression were enrolled in this study and analyzed. KS was then sub-classified into endemic or classical subtypes according to Lebbé et al. [8]. Patients work up at diagnosis included an exhaustive clinical examination, chest radiography, abdomen ultrasound as well as white blood cell count, protein electrophoresis and check for the negativity of HIV. Additional work up was performed on an individual basis depending on patient symptoms.

The study was performed in accordance with the ethical guidelines of the Declaration of Helsinki. Patients provided informed consent. A unique anonymized database was established and homed in a secured system, meeting the security standards required by the protection of personal data law promulgated on 20 June 2018 in France. A diagnosis of Compliance and Security Research was carried out and approved by the data protection reference department of Saint-Louis hospital.

2.2. Endpoints

Clinical and biological features at time of KS diagnosis, local and systemic therapies with respective responses and toxicity were collected using a specific case report form (CRF), based on patient's clinical records including both medical charts and electronic medical records. Clinical response was evaluated clinically and/or radiologically by the physician in charge of the patient. Complete Response (CR) was defined as complete clearance of all KS lesions, Partial Response (PR) was defined as a decrease in lesions area >50%, while Progressive Disease corresponded to patients with >25% lesions increase. All remaining patients were considered in Stable Disease (SD). The best overall response corresponds to the best response at any time during the assessment period. Local treatments were grouped as follows: surgery, local chemotherapy (imiquimod, fluorouracil or alitretinoin), radiotherapy or others (photodynamic therapy, cryotherapy, laser therapy, compression). Systemic treatments were grouped as follows: low dose interferon, chemotherapy (taxanes or anthracyclines-based regimens) or others (everolimus, thalidomide, lenalidomide, sunitinib, imatinib, ribavirine, ganciclovir or lopinavir).

2.3. Statistical Analysis

Quantitative variables are reported as median and interquartile range, while qualitative variables are reported as number and percentage.

Time to systemic treatment initiation was estimated using initial KS diagnostic date as origin. Since only 2 patients died without receiving any systemic treatment, this competing risk was not taken into account and these patients were censored at their time of death when estimating the cumulative incidence of systemic treatment initiation. We used a parametric modelling of the cumulative incidence of treatment initiation, using a Weibull distribution, to examine whether the risk of treatment initiation was constant across time. Association between baseline characteristics and systemic treatment initiation was estimated using a Cox proportional hazard model. To identify independent predictors of systemic treatment initiation, we then constructed a multivariate model, including all variables significantly associated in the univariate analysis and with less than 20% of missing data, and using a stepwise AIC-based variable selection.

We compared characteristics at time of systemic treatment initiation between responding patients (CR or PR as best overall response) and non-responding patients (SD or PD as best overall response) to first line of systemic treatment using Wilcoxon test or Fisher's exact test.

To assess the global evolution of the disease, we modelled the whole course of treatment using multi-state modelling. In this setting, a patient can go from the state "treated" to "non-treated" and vice versa several times during his follow-up, and ultimately go to the absorbing state of death. Transition probabilities between treated and non-treated were calculated using the Nelson Aalen estimator, and we calculated the mean time spent treatment free, truncated to some fixed limit tau. This mean time spent treatment-free was calculated starting from the beginning of first systemic treatment until either 1 year or 5 years following this treatment initiation. This treatment-free time was compared between groups according to KS subtype or first line of systemic treatment. The mean difference truncated at 1 or 5 years was calculated along with its bootstrapped 95% confidence interval.

All statistical analyses were performed using R software (version 3.6.1).

3. Results

This section may be divided by subheadings. It should provide a concise and precise description of the experimental results, their interpretation, as well as the experimental conclusions that can be drawn.

3.1. Demographic Characteristics

A total of 160 patients with histologically proven KS were included in the study. Patient's characteristics are described in Table 1; 131 patients (81.9%) had classic KS and 29 patients had endemic KS. Median age was 62.6 years (IQR: 54.5; 72.4) and 87.5% of patients were males. Lower limbs were the most commonly involved region (91.2%). The majority of patients (55.0%) had less than 10 lesions, and 10% had an extensive skin disease with over 100 KS lesions. Almost half of the patients (46.3%) presented with lymphedema, and 21.2% patients had symptomatic painful lesions.

Table 1. Patients' characteristics at KS diagnosis. UNL: Upper Normal Limit. IQR: Inter-Quartile Range. LDH: Lactate Dehydrogenase.

Patients Characteristics (Total)	n = 160
Age at diagnosis (years), median (IQR)	62.6 (54.5; 72.4)
Gender	
female	20 (12.5%)
male	140 (87.5%)
Subtype of Kaposi Sarcoma	
classic	131 (81.9%)
endemic	29 (18.1%)
Number of lesions	
0–10	88 (55%)
10–100	56 (35%)
>100	16 (10%)
Disease localisation	
lower limbs	146 (91.2%)
upper limbs	61 (38.1%)
trunk	32 (20.1%)
head or neck	21 (13.3%)
mucosa	12 (7.5%)
visceral	18 (11.2%)
Painful lesions	
yes	33 (21.2%)
no	123 (78.8%)
Lymphedema	
yes	74 (46.3%)
no	86 (53.7%)
Serum LDH (n = 97)	
<UNL	79 (81.4%)
>UNL	18 (18.6%)

Table 1. Cont.

Patients Characteristics (Total)	n = 160
HHV8 PCR (n = 118)	
positive	27 (22.9%)
negative	91 (77.1%)
viral load (log), median (range)	3.22 (2.50; 3.59)
Lymphocytes count (n = 117)	
>1500/mm^3	60 (51.3%)
<1500/mm^3	57 (48.7%)
CD4 count (nb/mm^3), median (IQR) (n = 76)	701 (507; 913)
Treatment	
observation	22 (13.8%)
local	72 (45%)
systemic	66 (41.2%)

HHV-8 viral quantification was available for 118 patients at diagnosis. Among them, 22.9% patients had a positive peripheral blood viral load with a median HHV-8 viral load of 3.22 log (IQR: 2.50; 3.59). Lymphopenia (total lymphocytes count <1500/mm^3) was present in 48.7% patients, and CD4 median count was 701/mm^3 (IQR: 507; 913). Serum LDH level was above upper normal limit in 18.6% of patients.

3.2. Risk Factors for Systemic Treatment Initiation

With a median follow-up of 4.8 years, 13.8% of patients did not require any treatment while 45.0% and 41.2% of them required local or systemic treatments respectively. Local treatment consisted in surgery (36.1%), local chemotherapy (29.5%), radiotherapy (26.2%) or other (8.2%). Systemic treatments included low dose interferon (50.0%), chemotherapy (taxanes or anthracyclines-based regimens) (45.5%) or other therapies (4.5%). Among the 66 patients who required systemic treatment, 53% received more than one line of treatment.

Cumulative incidence of systemic treatment initiation after 2 years of follow-up was 28.4% (95% CI: 20.5; 35.5), and median time from KS diagnosis to systemic treatment initiation was 8.8 years (95% CI: 4.7; 12.7) (Figure 1). Parametric modelling of the cumulative incidence showed that instantaneous risk of systemic treatment initiation decreases over time.

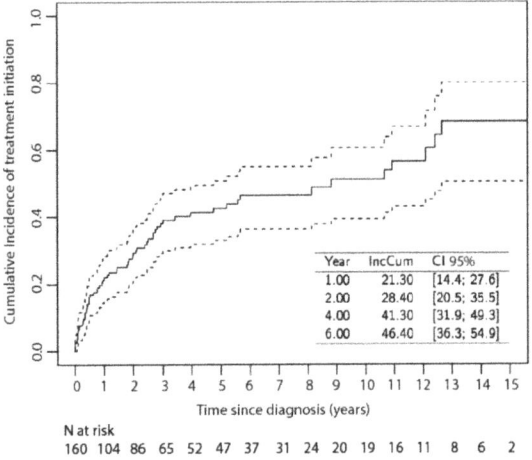

Figure 1. Cumulative Incidence of systemic treatment initiation in classic/endemic KS patients. Dashed lines correspond to the 95% Confidence Interval (CI). The inset table reports the cumulative incidences (IncCum) at 1, 2, 4 and 6 years after KS diagnosis, with their 95% CI.

Among baseline variables, endemic KS subtype, total number of cutaneous lesions, disease localization, painful lesions, lymphedema, and elevated serum LDH levels were significantly associated with systemic treatment initiation in univariate analysis (Table 2).

Table 2. Univariate and Multivariate COX model identifying risk factors for systemic treatment initiation in patients with classic/endemic KS. HR: Hazard Ratio. CI: Confidence Interval.

		Univariate Analysis			Multivariate Analysis		
		HR	CI 95%	p-Value	HR	CI 95%	p-Value
Age at diagnosis (years)		1	(0.98–1.02)	8.62×10^{-1}			
Time from first symptoms to diagnosis (months)				4.98×10^{-4}			6.82×10^{-3}
<1 m		1			1		
1–12 m		0.98	(0.47–2.05)		2.19	(0.98; 4.91)	
>12 m		2.61	(1.44–4.72)		2.7	(1.41; 5.17)	
Subtype of Kaposi Sarcoma				4.56×10^{-5}			7.59×10^{-4}
classic		1			1		
endemic		3.61	(2.07–6.3)		3.29	(1.71; 6.36)	
Number of lesions				2.26×10^{-9}			1.27×10^{-4}
0–10		1			1		
10–100		5.2	(2.82–9.6)		3.64	(1.81; 7.35)	
>100		6.17	(2.89–13.18)		4.56	(1.98; 10.54)	
Disease localization							
lower limbs	no	1		4.76×10^{-3}			
	yes	7.25	(1.01–52.24)				
upper limbs	no	1		2.82×10^{-3}			
	yes	2.1	(1.29–3.41)				
trunk	no	1		1.99×10^{-2}			
	yes	1.93	(1.14–3.29)				
head or neck	no	1		1.12×10^{-2}	1		2.34×10^{-2}
	yes	2.26	(1.26–4.06)		2.21	(1.15; 4.25)	
mucosa	no	1		8.50×10^{-2}			
	yes	1.96	(0.97–3.97)				
visceral	no	1		2.32×10^{-5}	1		3.98×10^{-2}
	yes	4.1	(2.3–7.31)		2.11	(1.06; 4.18)	
Painful lesions				1.17×10^{-3}			
no		1					
yes		2.41	(1.45–4.02)				
Lymphedema				2.52×10^{-7}			5.62×10^{-3}
no		1			1		
yes		3.89	(2.23–6.79)		2.3	(1.25; 4.26)	
Serum LDH				3.42×10^{-3}			
<ULN		1					
>ULN		2.76	(1.47–5.17)				
Lymphocytes count				2.99×10^{-1}			
>1500/mm^3		1					
<1500/mm^3		0.75	(0.43–1.3)				

Six risk factors for systemic treatment initiation were identified with multivariate analysis: endemic versus classic KS (HR: 3.29 [95% CI: 1.71; 6.36]), total number of lesions higher than 10 {HR: 3.64 (95% CI: 1.81; 7.35)}, visceral localization (HR: 2.11 [95% CI: 1.06; 4.18]), head or neck localization (HR: 2.21 [95% CI: 1.15; 4.25]), presence of edema (HR: 2.18 [95% CI: 1.18;4.04]) and a time between first symptoms and diagnosis longer than 1 year (HR: 2.70 [95% CI: 1.41; 5.17] for more than 1 year) (Figure 2a–f and Table 2).

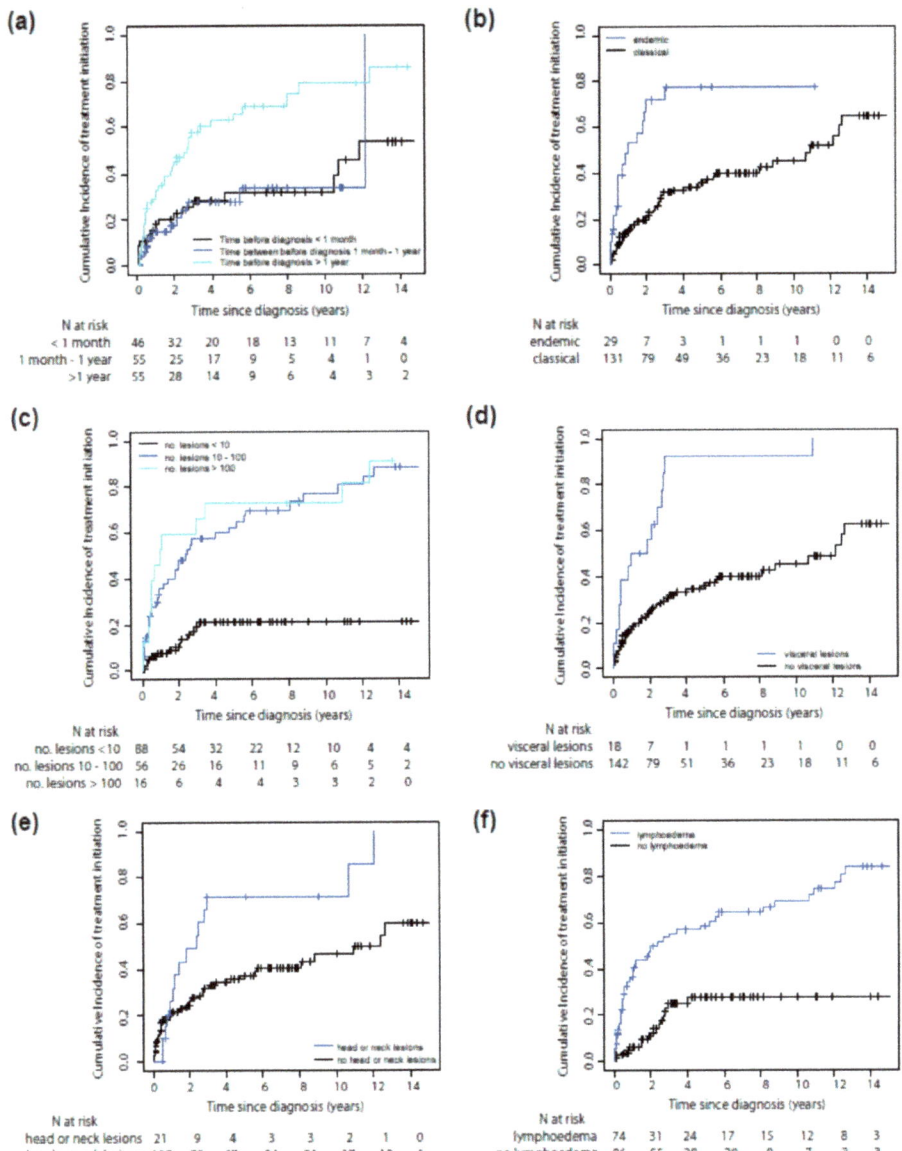

Figure 2. Cumulative Incidence of systemic treatment initiation in classic/endemic KS patients according to: time between first symptoms and diagnosis (a), KS subtype (endemic vs. classic) (b), total number of lesions (c), visceral localization (d), head/neck localization (e), or presence of lymphedema (f).

3.3. Therapeutic Response to KS Systemic Treatment

Among the 66 patients who received at least one line of systemic treatment for KS, best overall response (BOR) after the first line of systemic treatment was available for 64 patients. 14% of patients had a complete response (CR) while partial response (PR) was observed in 69%, stable disease (SD) in 8% and progressive disease (PD) in 9% of patients. BOR after first line of systemic treatment, according to type of therapy (interferon, chemotherapy or other regimens) is presented in Figure 3. 93.1% and 75.0% of patients treated respectively with chemotherapy or low-dose interferon achieved an objective response (CR or PR).

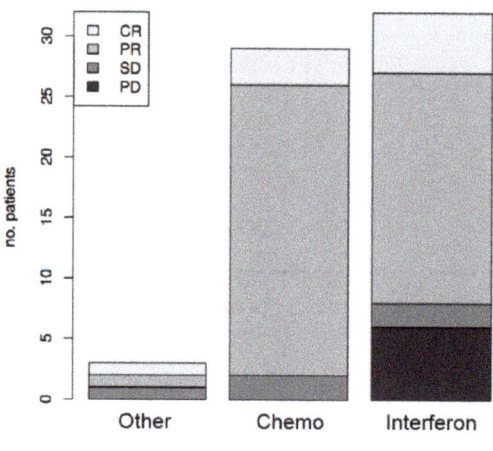

Figure 3. Best overall response after first line of systemic treatment in classic/endemic KS patients, according to type of therapy. Chemo: chemotherapy. CR: Complete Response. PR: Partial Response. SD: Stable Disease. PD: Progressive Disease.

Type of first line therapy (low dose interferon, chemotherapy or other), type of KS (endemic or classic), age at therapy initiation and time between diagnosis and systemic treatment initiation were not associated with BOR (Table A1).

3.4. Treatment Free Time after KS Systemic Treatment

Given the chronic evolution of KS and the impact of systemic treatment on the quality of life, we explored the treatment course of the 66 classic/endemic KS patients who received systemic treatment (Figure A1).

Multi-state modelling was used to study the whole course of treatment. The mean time spent treatment free was calculated starting from the beginning of first systemic treatment until either 1 year or 5 years following treatment initiation. The mean cumulative treatment-free time during the first year and the first 5 years following systemic treatment initiation was 5.4 months [95% CI: 4.4; 6.3] and 44.9 months [95% CI: 41.3; 48.1] respectively (Figure 4).

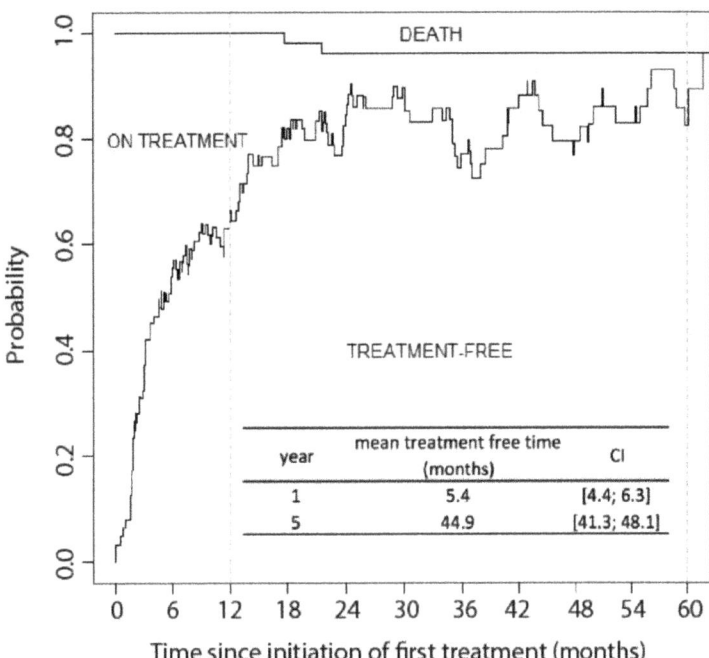

Figure 4. Treatment free time after systemic treatment initiation in classic/endemic KS. This figure was generated using state occupation probabilities: the area under the curve represents the mean time spent alive, on treatment and alive treatment-free during the first 5 years following the initiation of first systemic treatment. The inset table reports the mean times spent treatment free during the first year and the first 5 years following the initiation of first systemic treatment in the whole cohort, with their 95% CI. Dotted lines correspond to 1 and 5 years after first treatment.

The mean treatment-free time during the first and the 5 first years post-treatment initiation was not significantly different between classic (5.6 and 45.3 months at 1 and 5 years respectively) and endemic KS (4.8 and 43.5 months at 1 and 5 years respectively) (mean difference at 1 year: 0.8 months [95% CI: −4.9; 9.1], mean difference at 5 years: 1.7 months [95% CI: −1.4; 3.0]) (Figure 5a,b).

During the first-year post-treatment initiation, the mean treatment-free time was higher in chemotherapy-treated patients (7.3 months) compared to interferon-treated patients (3.5 months) (mean difference: −3.8 months [95% CI: −6.0; −2.7]). However, this difference was no longer observed at 5 years post-treatment initiation (chemotherapy: 44.6 months, interferon: 44.0 months, mean difference: −0.5 months [95% CI: −9.5;4.9]), suggesting that the early difference observed is mainly related to a difference in treatment regimens length, rather than a real impact of either regimen on the disease course (Figure 5c,d).

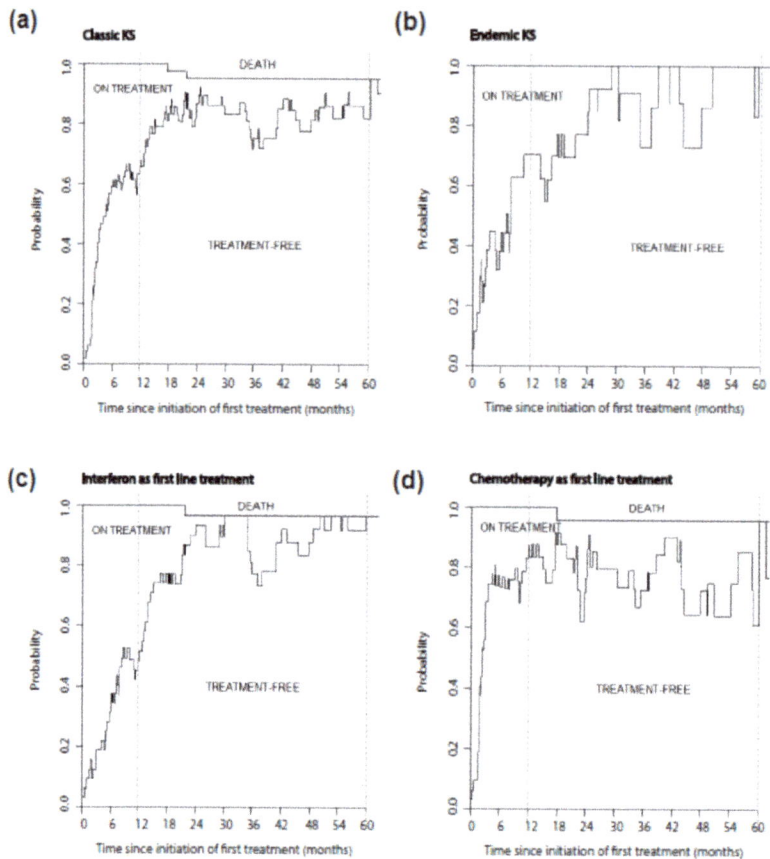

Figure 5. Treatment free time after systemic treatment initiation in classic/endemic KS patients stratified according to KS subtype: classic (**a**) or endemic (**b**), or according to type of first line treatment: interferon (**c**) or chemotherapy (**d**). This figure was generated using state occupation probabilities: the area under the curve represents the mean time spent alive, on treatment and alive treatment-free during the first 5 years following the initiation of first systemic treatment. Dotted lines correspond to 1 and 5 years after first treatment initiation.

4. Discussion

Our study provides an overview of clinical characteristics and therapeutic outcome in a large real-life endemic/classic KS cohort. Although classic/endemic KS are thought to have an indolent disease course, they can become symptomatic and require the use of systemic therapy [8]. In our study, 41.2% patients required the use of systemic therapy. This proportion reflects the recruitment of our specialized outpatients' clinic and may overestimate the incidence of systemic treatment requirement in a non-hospital dermatology consult.

Defining an adequate follow-up frequency adapted to the clinical course of endemic/classic KS, remain challenging for physicians taking care of KS patients [8].

There is no universally accepted staging classification for endemic/classic KS. Three staging systems have been proposed but deserve further validation and are not commonly used in the real-life. They take into account KS lesions localization and skin lesions number, extension and evolution [10–12]. Our results pinpoint a subpopulation of KS patients with high risk of systemic treatment requirement. Endemic KS, presence of more than 10 lesions, visceral or head/neck localization, or presence of lymphedema define objective criteria for systemic treatment initiation. These results should allow clinicians to better adjust follow-up schedules depending on susceptibility of systemic treatment need at diagnosis. Additionally, this will help physicians better inform patients on the clinical course of their disease.

Interestingly, our study shows that time from diagnosis to systemic treatment initiation does not influence response to treatment. This result allows physicians to adapt treatment initiation to patient's quality of life requirements, as KS is a chronic disease with no curative therapeutic options available to date, the main objective of systemic therapy remaining an improved quality of life [8]. Physical or psychological repercussions of KS lesions can indeed be responsible of a considerably reduced quality of life due to esthetic considerations, pain, edema or visceral symptoms, mainly gastro-intestinal and pulmonary [8]. Moreover, our data suggest that response to first-line systemic treatment does not depend on patients' age at treatment initiation. Thus, age should not preclude clinicians from treatment initiation if patients are considered fit to receive the proposed treatment. Systemic treatment initiation should thus be adapted to each patient, balancing its initial risk of systemic treatment requirement, patient's fitness and presence of quality of life impacting symptoms.

KS systemic therapy mainly relies on chemotherapeutic agents such as liposomal doxorubicin [13–16] or taxanes [17,18], and immune-modulating therapies such as low-dose interferon alpha and its pegylated derivatives, mainly used for younger patients with classic KS [19–21]. Our study did not reveal any significant difference in overall response rates, nor in treatment free time, between the two more frequently used regimens to treat endemic/classic KS: low dose interferon and taxanes or anthracyclines-based chemotherapy. Response rates in our study were in line with previous smaller scale studies [9]. Although both treatments are efficient, they are associated with significant relapse rates and are usually not curative. Aiming to highlight the therapeutic option offering the best quality of life, we further explored response to these two systemic treatment options in terms of response duration, through treatment free time evaluation. This is to our knowledge the first study comparing treatment-free time between both therapeutic options. Although treatment-free time was higher in the chemotherapy treated population during the first year, interferon- and chemotherapy-treated patients had similar treatment-free intervals on a long-term perspective. These results only reflect the longer duration of interferon regimens compared to short chemotherapeutic regimens and suggest that neither interferon nor chemotherapy-based regimens offer a longer duration of response. Further studies focusing on systemic treatment toxicities in the endemic/classic KS population should further inform on treatment choice depending on patient's fitness. Treatment choice should thus be adapted to patient's fitness to avoid specific toxicities and improve patient's quality of life.

Endemic KS develops in younger patients and tends to be difficult to treat and have poor clinical outcome with high lymphedema rates [22,23]. To our knowledge, this is the first report of a large series of non-HIV related and non-iatrogenic KS, highlighting the endemic subtype clinical aggressiveness. Although patients harboring the endemic subtype had a higher systemic treatment requirement rate in our study, they respond similarly to first-line therapy once initiated, both in terms of response rates and duration.

Finally, as KS therapy is now entering an exciting immunotherapy avenue, our data offers a baseline for clinical characteristics and outcome of endemic/classic KS in the pre-immunotherapy era. Indeed, KS tumors strongly express the T cell inhibitory molecules PD1 and PDL1 [24] and several promising results have recently emerged from pilot reports testing immune checkpoint inhibitors (ipilimumab and/or nivolumab), mainly in HIV-associated Kaposi Sarcoma (NCT02408861) [25–27]. Further anti-PD1/PDL1 clinical studies should inform on the efficacy of these agents compared to the standard chemotherapy and low-dose interferon regimens, especially in the long term.

Limitations of our study include its monocentric setting within a specialized outpatient's clinic. Proportion of patients requiring a systemic treatment may therefore be overestimated, and KS management heterogeneity across different centers needs to be taken into account in our study results interpretation. The wide period of patient's inclusion ranging from 1990 to 2013 also represents a limitation as this might be responsible for some degree of heterogeneity in terms of first-line treatment modalities.

5. Conclusions

In this study, we report the clinical characteristics and outcome of a large monocentric cohort of endemic and classic KS. The endemic subtype, total number of lesions, visceral or head/neck localization, presence of edema at diagnosis and a higher time between first symptoms and diagnosis were independent risk factors for systemic treatment initiation. No response difference was observed between the standard of care treatments, chemotherapy and interferon, thus promoting the guidance of therapeutic choice following patient's comorbidities.

Author Contributions: Conceptualization, L.B., J.D., M.B. and C.L.; methodology, J.L.; validation, L.B., J.L., R.L.S., D.C., B.B., J.R., C.P., J.L.G. and M.B.; formal analysis, J.L.; investigation, L.B., J.L., R.L.S., D.C., B.B., J.R. and C.P.; resources, L.B., R.L.S., D.C., B.B., J.R., C.P., J.L.G. and M.B.; data curation, L.B., J.L., R.L.S. and D.C.; writing—original draft preparation, L.B. and J.L.; writing—review and editing, L.B., J.L., M.B., J.D. and C.L.; visualization, L.B. and J.L.; supervision, C.L.; project administration, L.B. and C.L. All authors have read and agreed to the published version of the manuscript.

Funding: This research received no external funding.

Institutional Review Board Statement: The study was conducted according to the guidelines of the Declaration of Helsinki. A diagnosis of Compliance and Security Research was carried out and approved by the data protection reference department of Saint-Louis hospital.

Informed Consent Statement: Informed consent was obtained from all subjects involved in the study.

Data Availability Statement: All data supporting reported results is included in the main manuscript.

Acknowledgments: The authors would like to acknowledge all attending physicians, residents and fellows who participated to KS patients care in our specialized outpatient clinic.

Conflicts of Interest: The authors declare no conflict of interest.

Appendix A

Table A1. Factors associated to first systemic treatment line best overall response. CR: Complete Response. PR: Partial Response. SD: Stable Disease. PD: Progressive Disease.

	CR or PR (n = 53)	SD or PD (n = 11)	p-Value
Age at treatment (years), median (IQR)	64.5 (57.7; 73.4)	58.3 (51.8; 74.1)	0.48
Time from diagnosis to systemic treatment (days), median (IQR)	428 (142; 1026)	306 (86; 830)	0.54
Subtype of Kaposi Sarcoma			1
classic	39 (74%)	8 (73%)	
endemic	14 (26%)	3 (27%)	
Type of systemic treatment			0.11
chemotherapy	27 (51%)	2 (18%)	
interferon	24 (45%)	8 (73%)	
other	2 (4%)	1 (9%)	

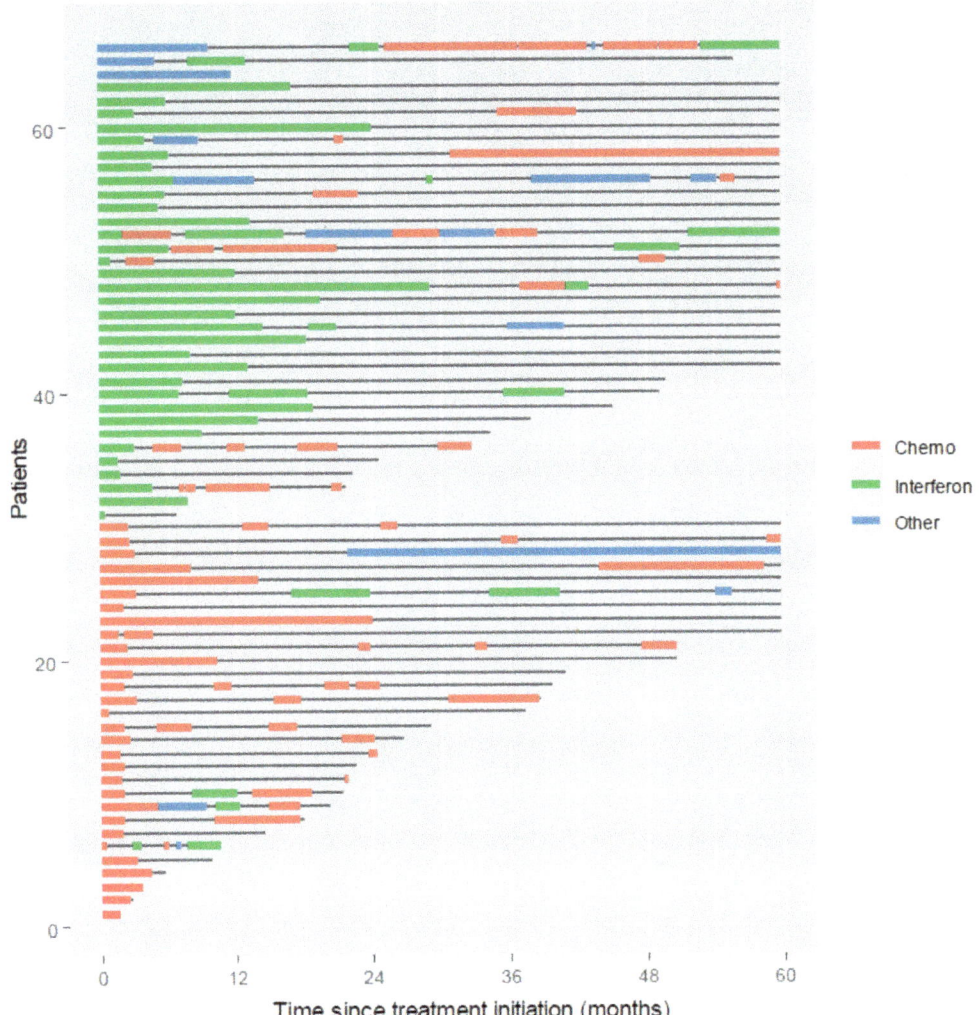

Figure A1. Lollipop plot representing treatment course of the 66 classic/endemic KS patients who received systemic treatment. Each line represents a single patient. Treatment course is represented during the first five years of follow-up. Treatment free periods are represented by a grey line while treatment intervals correspond to green, red or blue boxes when patients are treated with interferon, chemotherapy or other regimens respectively.

References

1. Cottoni, F.; De Marco, R.; Montesu, M.A. Classical Kaposi's sarcoma in north-east Sardinia: An overview from 1977 to 1991. *Br. J. Cancer* **1996**, *73*, 1132–1133. [CrossRef] [PubMed]
2. Dal Maso, L.; Polesel, J.; Ascoli, V.; Zambon, P.; Budroni, M.; Ferretti, S.; Tumino, R.; Tagliabue, G.; Patriarca, S.; Federico, M.; et al. Classic Kaposi's sarcoma in Italy, 1985–1998. *Br. J. Cancer* **2005**, *92*, 188–193. [CrossRef]
3. Chang, Y.; Cesarman, E.; Pessin, M.S.; Lee, F.; Culpepper, J.; Knowles, D.M.; Moore, P.S. Identification of herpesvirus-like DNA sequences in AIDS-associated Kaposi's sarcoma. *Science* **1994**, *266*, 1865–1869. [CrossRef]
4. Lambert, M.; Gannage, M.; Karras, A.; Abel, M.; Legendre, C.; Kerob, D.; Agbalika, F.; Girard, P.-M.; Lebbe, C.; Caillat-Zucman, S. Differences in the frequency and function of HHV8-specific CD8 T cells between asymptomatic HHV8 infection and Kaposi sarcoma. *Blood* **2006**, *108*, 3871–3880. [CrossRef]

5. Stratigos, J.D.; Potouridou, I.; Katoulis, A.C.; Hatziolou, E.; Christofidou, E.; Stratigos, A.; Hatzakis, A.; Stavrianeas, N.G. Classic Kaposi's sarcoma in Greece: A clinico-epidemiological profile. *Int. J. Dermatol.* **1997**, *36*, 735–740. [CrossRef] [PubMed]
6. Nasti, G.; Martellotta, F.; Berretta, M.; Mena, M.; Fasan, M.; Perri, G.D.; Talamini, R.; Pagano, G.; Montroni, M.; Cinelli, R.; et al. Impact of highly active antiretroviral therapy on the presenting features and outcome of patients with acquired immunodeficiency syndrome-related Kaposi sarcoma. *Cancer* **2003**, *98*, 2440–2446. [CrossRef]
7. Frances, C. Kaposi's sarcoma after renal transplantation. *Nephrol. Dial. Transpl.* **1998**, *13*, 2768–2773. [CrossRef]
8. Lebbe, C.; Garbe, C.; Stratigos, A.J.; Harwood, C.; Peris, K.; Del Marmol, V.; Malvehy, J.; Zalaudek, I.; Hoeller, C.; Dummer, R.; et al. Diagnosis and treatment of Kaposi's sarcoma: European consensus-based interdisciplinary guideline (EDF/EADO/EORTC). *Eur. J. Cancer* **2019**, *114*, 117–127. [CrossRef]
9. Regnier-Rosencher, E.; Guillot, B.; Dupin, N. Treatments for classic Kaposi sarcoma: A systematic review of the literature. *J. Am. Acad. Dermatol.* **2013**, *68*, 313–331. [CrossRef]
10. Mitsuyasu, R.T.; Groopman, J.E. Biology and therapy of Kaposi's sarcoma. *Semin. Oncol.* **1984**, *11*, 53–59. [PubMed]
11. Krigel, R.L.; Laubenstein, L.J.; Muggia, F.M. Kaposi's sarcoma: A new staging classification. *Cancer Treat Rep.* **1983**, *67*, 531–534.
12. Brambilla, L.; Boneschi, V.; Taglioni, M.; Ferrucci, S. Staging of classic Kaposi's sarcoma: A useful tool for therapeutic choices. *Eur. J. Dermatol.* **2003**, *13*, 83–86.
13. Gill, P.S.; Wernz, J.; Scadden, D.T.; Cohen, P.; Mukwaya, G.M.; von Roenn, J.H.; Jacobs, M.; Kempin, S.; Silverberg, I.; Gonzales, G.; et al. Randomized phase III trial of liposomal daunorubicin versus doxorubicin, bleomycin, and vincristine in AIDS-related Kaposi's sarcoma. *J. Clin. Oncol.* **1996**, *14*, 2353–2364. [CrossRef] [PubMed]
14. Northfelt, D.W.; Dezube, B.J.; Thommes, J.A.; Miller, B.J.; Fischl, M.A.; Friedman-Kien, A.; Kaplan, L.D.; Du Mond, C.; Mamelok, R.D.; Henry, D.H. Pegylated-liposomal doxorubicin versus doxorubicin, bleomycin, and vincristine in the treatment of AIDS-related Kaposi's sarcoma: Results of a randomized phase III clinical trial. *J. Clin. Oncol.* **1998**, *16*, 2445–2451. [CrossRef]
15. Stewart, S.; Jablonowski, H.; Goebel, F.D.; Arasteh, K.; Spittle, M.; Rios, A.; Aboulafia, D.; Galleshaw, J.; Dezube, B.J. Randomized comparative trial of pegylated liposomal doxorubicin versus bleomycin and vincristine in the treatment of AIDS-related Kaposi's sarcoma. International Pegylated Liposomal Doxorubicin Study Group. *J. Clin. Oncol.* **1998**, *16*, 683–691. [CrossRef] [PubMed]
16. Lichterfeld, M.; Qurishi, N.; Hoffmann, C.; Hochdorfer, B.; Brockmeyer, N.H.; Arasteh, K.; Mauss, S.; Rockstroh, J.K. German Clinical AIDS Working Group (KAAD) Treatment of HIV-1-associated Kaposi's sarcoma with pegylated liposomal doxorubicin and HAART simultaneously induces effective tumor remission and CD4+ T cell recovery. *Infection* **2005**, *33*, 140–147. [CrossRef]
17. Cianfrocca, M.; Lee, S.; Von Roenn, J.; Tulpule, A.; Dezube, B.J.; Aboulafia, D.M.; Ambinder, R.F.; Lee, J.Y.; Krown, S.E.; Sparano, J.A. Randomized trial of paclitaxel versus pegylated liposomal doxorubicin for advanced human immunodeficiency virus-associated Kaposi sarcoma: Evidence of symptom palliation from chemotherapy. *Cancer* **2010**, *116*, 3969–3977. [CrossRef]
18. Gill, P.S.; Tulpule, A.; Espina, B.M.; Cabriales, S.; Bresnahan, J.; Ilaw, M.; Louie, S.; Gustafson, N.F.; Brown, M.A.; Orcutt, C.; et al. Paclitaxel is safe and effective in the treatment of advanced AIDS-related Kaposi's sarcoma. *J. Clin. Oncol.* **1999**, *17*, 1876–1883. [CrossRef]
19. Northfelt, D.W.; Dezube, B.J.; Thommes, J.A.; Levine, R.; Von Roenn, J.H.; Dosik, G.M.; Rios, A.; Krown, S.E.; DuMond, C.; Mamelok, R.D. Efficacy of pegylated-liposomal doxorubicin in the treatment of AIDS-related Kaposi's sarcoma after failure of standard chemotherapy. *J. Clin. Oncol.* **1997**, *15*, 653–659. [CrossRef]
20. Costa da Cunha, C.S.; Lebbe, C.; Rybojad, M.; Ferchal, F.; Rabian, C.; Vignon-Pennamen, M.D.; Calvo, F.; Morel, P. Long-term follow-up of non-HIV Kaposi's sarcoma treated with low-dose recombinant interferon alfa-2b. *Arch. Dermatol.* **1996**, *132*, 285–290. [CrossRef] [PubMed]
21. Tur, E.; Brenner, S. Classic Kaposi's sarcoma: Low-dose interferon alfa treatment. *Dermatology* **1998**, *197*, 37–42. [CrossRef]
22. Pantanowitz, L.; Duke, W.H. Lymphoedematous variants of Kaposi's sarcoma. *J. Eur. Acad. Dermatol. Venereol.* **2008**, *22*, 118–120. [CrossRef]
23. El-Mallawany, N.K.; Villiera, J.; Kamiyango, W.; Peckham-Gregory, E.C.; Scheurer, M.E.; Allen, C.E.; McAtee, C.L.; Legarreta, A.; Dittmer, D.P.; Kovarik, C.L.; et al. Endemic Kaposi sarcoma in HIV-negative children and adolescents: An evaluation of overlapping and distinct clinical features in comparison with HIV-related disease. *Infect Agent Cancer* **2018**, *13*, 33. [CrossRef]
24. Chen, J.; Del Valle, L.; Lin, H.Y.; Plaisance-Bonstaff, K.; CraigForrest, J.; Post, S.R.; Qin, Z. Expression of PD-1 and PD-Ls in Kaposi's sarcoma and regulation by oncogenic herpesvirus lytic reactivation. *Virology* **2019**, *536*, 16–19. [CrossRef]
25. Uldrick, T.S.; Goncalves, P.H.; Abdul-Hay, M.; Claeys, A.J.; Emu, B.; Ernstoff, M.S.; Fling, S.P.; Fong, L.; Kaiser, J.C.; Lacroix, A.M.; et al. Assessment of the Safety of Pembrolizumab in Patients with HIV and Advanced Cancer-A Phase 1 Study. *JAMA Oncol.* **2019**, *5*, 1332–1339. [CrossRef]
26. Delyon, J.; Bizot, A.; Battistella, M.; Madelaine, I.; Vercellino, L.; Lebbe, C. PD-1 blockade with nivolumab in endemic Kaposi sarcoma. *Ann. Oncol.* **2018**, *29*, 1067–1069. [CrossRef]
27. Galanina, N.; Goodman, A.M.; Cohen, P.R.; Frampton, G.M.; Kurzrock, R. Successful Treatment of HIV-Associated Kaposi Sarcoma with Immune Checkpoint Blockade. *Cancer Immunol. Res.* **2018**, *6*, 1129–1135. [CrossRef]

Review

De Novo Carcinoma after Solid Organ Transplantation to Give Insight into Carcinogenesis in General—A Systematic Review and Meta-Analysis

Eline S. Zwart [1,2,†], Esen Yüksel [1,†], Anne Pannekoek [1], Ralph de Vries [3], Reina E. Mebius [2] and Geert Kazemier [1,*]

1. Amsterdam Universities Medical Centers, Cancer Center Amsterdam, Department of Surgery, VU University, 1081 HV Amsterdam, The Netherlands; e.zwart@amsterdamumc.nl (E.S.Z.); e.yuksel@amsterdamumc.nl (E.Y.); a.pannekoek@student.vu.nl (A.P.)
2. Amsterdam Universities Medical Centers, Department of Molecular Cell Biology and Immunology, VU University, 1081 HV Amsterdam, The Netherlands; r.mebius@amsterdamumc.nl
3. Medical Library, Vrije Universiteit, 1081 HV Amsterdam, The Netherlands; r2.de.vries@vu.nl
* Correspondence: g.kazemier@amsterdamumc.nl
† These authors contributed equally to this work.

Citation: Zwart, E.S.; Yüksel, E.; Pannekoek, A.; de Vries, R.; Mebius, R.E.; Kazemier, G. De Novo Carcinoma after Solid Organ Transplantation to Give Insight into Carcinogenesis in General—A Systematic Review and Meta-Analysis. *Cancers* 2021, *13*, 1122. https://doi.org/10.3390/cancers13051122

Academic Editors: Subree Subramanian and Xianda Zhao

Received: 26 January 2021
Accepted: 2 March 2021
Published: 5 March 2021

Publisher's Note: MDPI stays neutral with regard to jurisdictional claims in published maps and institutional affiliations.

Copyright: © 2021 by the authors. Licensee MDPI, Basel, Switzerland. This article is an open access article distributed under the terms and conditions of the Creative Commons Attribution (CC BY) license (https:// creativecommons.org/licenses/by/ 4.0/).

Simple Summary: Patients receiving a solid organ transplantation, such as a kidney, liver, or lung transplantation, inevitably have to take drugs to suppress the immune system in order to prevent rejection of the transplanted organ. However, these drugs are known to cause malignancies in the long term. This study focuses specifically on newly developed carcinomas in patients who use those drugs after a solid organ transplantation. This systematic review and meta-analysis of published data show a 20-fold risk to develop a carcinoma after solid organ transplantation compared to the general population, with specifically increased risks in patients who receive cyclosporine or azathioprine. By comparing the different pathways involved in immunosuppression and the occurrence of carcinoma development, new insights can be discovered for future research and understanding of carcinoma development in transplantation patients and the general population as well.

Abstract: Immunosuppressive therapy after solid organ transplantation leads to the development of cancer in many recipients. Analysis of the occurrence of different types of de novo carcinomas in relation to specific immunosuppressive drugs may give insight into their carcinogenic process and carcinogenesis in general. Therefore, a systematic search was performed in Embase and PubMed. Studies describing over five de novo carcinomas in patients using immunosuppressive drugs after solid organ transplantation were included. Incidence per 1000 person-years was calculated with DerSimonian–Laird random effects model and odds ratio for developing carcinomas with the Mantel–Haenszel test. Following review of 5606 papers by title and abstract, a meta-analysis was conducted of 82 studies. The incidence rate of de novo carcinomas was 8.41. Patients receiving cyclosporine developed more de novo carcinomas compared to tacrolimus (OR1.56, 95%CI 1.00–2.44) and mycophenolate (OR1.26, 95%CI 1.03–1.56). Patients receiving azathioprine had higher odds to develop de novo carcinomas compared to mycophenolate (OR3.34, 95%CI 1.29–8.65) and head and neck carcinoma compared to tacrolimus (OR3.78, 95%CI 1.11–12.83). To conclude, patients receiving immunosuppressive drugs after solid organ transplantation have almost a 20-fold increased likelihood of developing carcinomas, with the highest likelihood for patients receiving cyclosporine A and azathioprine. Looking into altered immune pathways affected by immunosuppressive drugs might lead to better understanding of carcinogenesis in general.

Keywords: organ transplantation; carcinoma; epidemiologic studies; immunosuppression

1. Introduction

Solid organ transplantation patients receive different immunosuppressive drugs to prevent graft rejection. Each of these drugs inhibits the immune system in a specific manner. Calcineurin inhibitors, such as cyclosporin A (CsA) and tacrolimus (TAC), inhibit the proliferation of T cells which is important to prevent graft rejection [1,2]. Another group of immunosuppressive drugs, such as azathioprine (AZA) and mycophenolate (MMF), are called antimetabolites and inhibit DNA synthesis, thereby preventing proliferation of T and B cells [1]. Studies have demonstrated that MMF has a superior ability to prevent allograft rejection compared to AZA, which caused AZA to be mostly replaced by MMF [3–6]. Newer, more potent suppressors of lymphocyte proliferation, such as sirolimus (SIR) and everolimus (EVER), are inhibitors of the mammalian target of rapamycin (mTOR), which is an intracellular kinase involved in cell metabolism, growth, and proliferation (Table 1).

Table 1. Class of inhibitors and main working mechanisms.

Class of Inhibitor	Main Mechanism of Action	Immunosuppressive Drug
Calcineurin inhibitor	Inhibition of T cell proliferation	Cyclosporine A Tacrolimus
Antimetabolites	Inhibition DNA synthesis	Azathioprine Mofetil mycophenolate
mTOR inhibitors	Inhibition of mTOR kinase, involved in metabolism, growth, and proliferation	Sirolimus Everolimus

Even though outcomes of solid organ transplantation have improved dramatically since the discovery of immunosuppressive drugs, their use comes with a drawback. Overall, a two to seven times higher risk for development of de novo malignancies can be found in transplant recipients compared to the general population [7,8]. Long-term use of immunosuppressive agents is considered to be the major contributing factor [8]. Post-transplant lymphoproliferative disorders, (non-)melanoma skin cancer, and Kaposi's sarcoma are among the most frequently occurring neoplasms after solid organ transplantation and they have been broadly investigated [9]. However, a large overview of occurrence of de novo carcinomas after solid organ transplantation is lacking. By analyzing the occurrence of different types of de novo carcinomas in relation to specific immunosuppressive drugs, insight can also be gained into the carcinogenesis process, providing new perspectives for translational cancer research. Therefore, the aim of this systematic review is to examine the overall and tumor-specific incidence of de novo carcinomas in varied solid organ transplant recipients using specific immunosuppressive drugs in order to gain insight into the pathways contributing to carcinogenesis in those patients, but also in the general population.

2. Materials and Methods

2.1. Search Strategy and Study Selection

This systematic review followed the Preferred Reporting Items for Systematic Reviews and Meta-Analyses (PRISMA) [10]. A literature search was conducted in the bibliographic databases of PubMed and Embase.com from inception up to September 10, 2020, in collaboration with a medical librarian. The following terms were used, including synonyms and closely related words, as index terms or free-text words: "Immunosuppression", "Organ Transplantation", and "Carcinoma". The full search strategies can be found in Table S1. Title and abstracts were independently reviewed by E.Z. and A.P. After contemplation about conflicts, full texts were screened by A.P. and E.Y., and in case of conflict, E.Z. was consulted. All screening was conducted with the use of Rayyan, a systematic web app [11]. Studies that included solid organ transplant recipients of 18 years and older, who received chronic immunosuppression and developed a de novo carcinoma, were considered eligible. Studies written in languages other than English, literature reviews, studies describing less than five de novo carcinomas, studies describing recurrent hepatocellular carcinomas, studies describing premalignant lesions, and studies that did not describe the specific immunosuppressive treatment regimen were excluded. In case of overlapping databases, the study with the largest and most complete dataset was included.

2.2. Data Collection and Interpretation

Data of the included articles were extracted using a standardized data extraction form, including study design, patient demographics, duration of follow-up, number of transplant recipients, transplantation period, and number of patients with de novo carcinomas. Corresponding authors were contacted by email regarding missing follow-up data by E.Z. De novo head and neck carcinomas were defined as ear, nose, pharynx, larynx, lip, oral (gland), buccal, tongue, or tonsil carcinomas. Likewise, de novo colorectal carcinoma was defined as colon and rectal carcinomas. De novo uterine carcinoma included uterus and cervix carcinomas. The types of immunosuppressive drugs recorded for the included articles were AZA, CsA, MMF, TAC, SIR, and EVER.

2.3. Quality Assessment

To assess the quality of the included articles, the Newcastle–Ottawa Scale was consulted as risk of bias tool [12]. The coding manual for cohort studies was used by allocating stars for included articles to assess bias in selection, comparability of the study groups, and outcome of interest. The assessment was performed by A.P. and E.Y. independently. Conflicts were solved through discussion.

2.4. Statistical Analysis

The incidence of de novo carcinomas per 1000 person-years was calculated and pooled with DerSimonian–Laird random effects model in RevMan 5 [13]. The odds ratio (OR) for developing a de novo carcinoma between different immunosuppressive drugs was calculated with the Mantel–Haenszel random effects test in RevMan 5. Forest plots display the included studies for each comparison, with the OR per solid organ transplantation type and the overall effect presented with a 95% confidence interval. Events were defined as the occurrence of a de novo carcinoma. A p value less than or equal to 0.05 was considered statistically significant. All possible comparisons of immunosuppressive drugs present in the included cohorts were tested. Outcomes of comparisons with two or more study cohorts were considered eligible.

3. Results

3.1. Study Selection

After duplicate removal, the search identified 6318 records. Based on title and abstract, 5569 records were excluded. Consequently, 749 full-text articles were assessed for eligibility. After exclusion of 667 articles, a total of 82 were included for qualitative and quantitative synthesis (Figure 1).

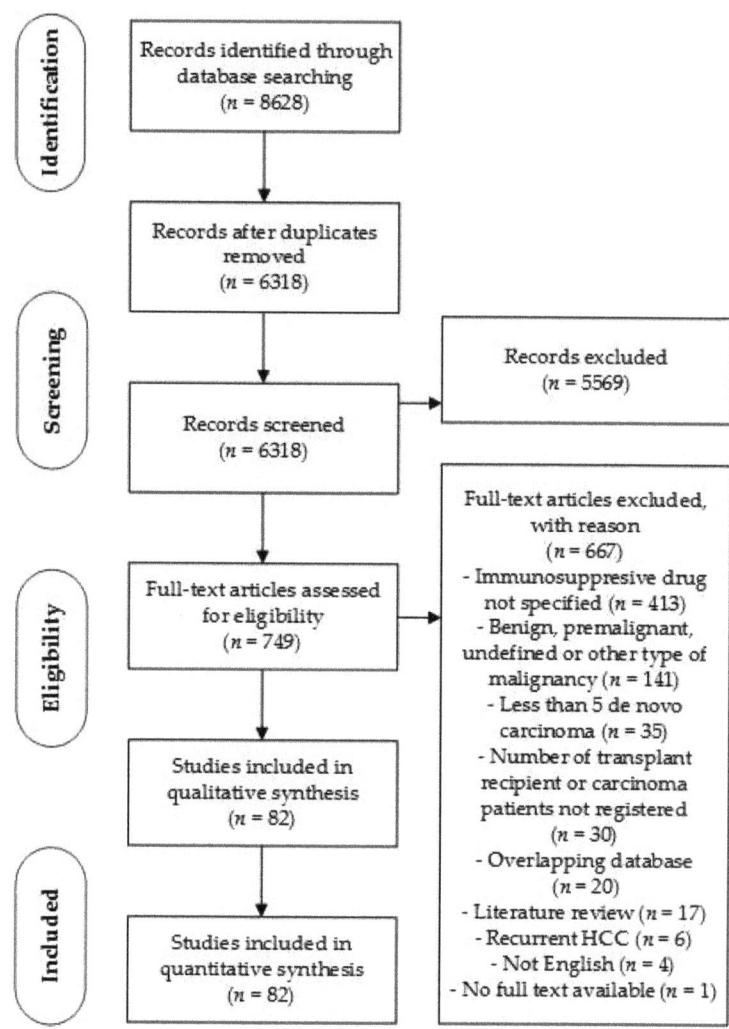

Figure 1. PRISMA Flowchart.

3.2. Study Characteristics

Overall, these 82 studies comprised a total of 237,540 recipients, who received 207,304 kidney, 21,404 liver, 5865 heart, and 2235 lung transplants. Transplant recipients were followed up for a mean period of 84.8 months after transplantation. Patients were diagnosed with a de novo carcinoma at a mean age of 52.3 years and after 66.8 months of follow-up. The baseline characteristics of the included cohorts are presented in Table 2. Most of the 82 studies were conducted in the United States ($n = 11$), followed by France ($n = 9$), Italy ($n = 7$), and Korea ($n = 6$) (Table S2). The vast majority of the studies ($n = 64$) were hospital-based, while others were database-guided or multicenter studies. Thirty-two authors were contacted regarding missing follow-up data, of which only 3 replied and 10 had invalid contact information.

Table 2. Baseline characteristics of transplant recipients.

Variables	Transplant Recipients
Total solid organ transplant recipients, n	237,540
Kidney transplant	207,304
Liver transplant	21,404
Heart transplant	5865
Lung transplant	2235
Other transplant	732
Follow-up (in months), mean	84.8
	Patients with PTC
Sex (M/F), n	1782/698
Time until diagnosis (in months), mean	66.8 (73/82)
Age at diagnosis, mean	52.3 (42/82)
Living/cadaveric donor, n	545/1642 (28/82)
Smokers, n	250
Induction therapy, n	172
Baseline immunosuppressive therapy, n	
AZA	723
MMF	741
CsA	1055
TAC	627
SIR	201
EVER	6
Combined triple therapies, n	
CsA + AZA + steroids	296
CsA + MMF + steroids	90
TAC + AZA + steroids	10
TAC + MMF + steroids	203
CsA + SIR + steroids	90
Survival	
1-year (%)	81.3 (12/82)
3-year (%)	75.5 (6/82)
5-year (%)	62.4 (16/82)

Numbers are calculated for studies including these variables, (n/82) shows number of studies used for calculation. Abbreviations: PTC: post-transplant de novo carcinoma | M/F: male/female | MMF: Mycophenolate mofetil | CsA: Cyclosporine A | TAC: Tacrolimus | SIR: Sirolimus | EVER: Everolimus.

3.3. Quality Assessment

Overall, 33 of the 82 studies scored 5 or 6 out of a maximum of 8 points, which represents a fair quality. A score of 5 or 6 was mainly due to missing information regarding follow-up in the outcome category and the ascertainment of exposure in the selection category. The remaining 49 articles were considered high-quality studies with a score equal to or over 7 (Table S3).

3.4. De Novo Carcinoma Occurrence

The incidence rate per 1000 person-years of solid organ transplant recipients developing de novo carcinomas was 8.41 (95% CI 7.40–9.43, $p < 0.00001$). De novo carcinoma occurrence in the included studies varied from 7.81 to 115.4 cases per 1000 person-years. Patients who underwent a heart transplantation developed more de novo carcinomas compared to kidney and liver transplantations, particularly de novo bladder and upper gastrointestinal tract carcinomas (Figure 2).

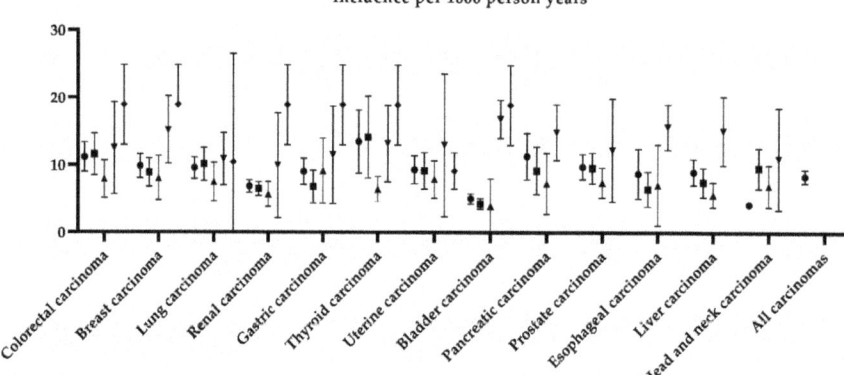

Figure 2. Pooled mean incidence of de novo carcinomas per 1000 person-years. Bars denote the 95% confidence interval.

3.5. CsA Versus TAC, AZA, MMF, and SIR

Patients who received CsA had a significantly higher likelihood of developing a de novo carcinoma compared to patients who received TAC, both calcineurin inhibitors (OR 1.56, 95% CI 1.00–2.44, $p = 0.05$) (Figure S1). No difference was found in the subgroup analysis for different types of de novo carcinomas. The odds for development of de novo carcinoma were not significantly different for patients who received CsA compared to AZA, one of the antimetabolites (OR 1.04, 95% CI: 0.90–1.21, $p = 0.59$), yet there appeared to be a trend towards higher occurrence of de novo esophageal and duodenal carcinoma in patients who received CsA (OR 2.47 95% CI 0.62–9.77, $p = 0.20$ and OR 4.05 95% CI 0.42–39.23, respectively) (Figure S2). Significantly more de novo carcinomas developed in patients who received CsA compared to patients who received MMF, another antimetabolite (OR 1.26, 95% CI 1.03–1.56, $p = 0.03$) (Figure S3). There was no difference observed in the subgroup analysis, but there appeared to be a trend of a higher likelihood of developing de novo head and neck carcinomas in patients who received CsA (OR 2.68, 95% CI 0.83–8.65, $p = 0.10$). There was no significant difference between patients using CsA and SIR, an mTOR inhibitor (OR 1.29, 95% CI 0.70–2.36, $p = 0.87$) (Figure S4).

3.6. AZA Versus TAC and MMF

Patients who received AZA had a higher likelihood of developing de novo head and neck carcinomas compared to patients who received TAC (OR 3.78, 95% CI 1.11–12.83, $p = 0.03$). Furthermore, there appeared to be a trend for a higher overall likelihood of developing a de novo carcinoma and for developing de novo lung carcinoma (OR 2.00, 95% CI 0.78–5.14, $p = 0.15$) (OR 7.28, 95% CI 0.93–56.73, $p = 0.06$) (Figure S5). Patients who received AZA had significantly higher odds of developing de novo carcinomas compared to patients who received MMF (OR 3.34, 95% CI 1.29–8.65, $p = 0.01$) (Figure S6).

3.7. MMF Versus TAC

No difference was observed in development of de novo carcinomas between patients who received MMF and patients who received TAC (OR 0.88, 95% CI 0.69–1.14, $p = 0.33$) (Figure S7).

4. Discussion

This systematic review shows that patients who receive immunosuppressive drugs after solid organ transplantation have a high incidence of de novo carcinomas, with an almost 20-fold increase compared to the age-corrected general population, as indicated by the WHO Global incidence of cancer between ages 30 and 69 (0.43 cases per 1000 person-years) [14]. This age range is comparable with the range described in the included studies. The incidence found in the analyzed cohorts for each specific type of de novo carcinoma resulted in a particularly high likelihood of de novo bladder and upper gastrointestinal tract carcinomas after heart transplantation. The incidence of bladder carcinomas in the general population between ages 30 and 69 is 0.075 per 1000 person-years, which is over 200 times lower than after a heart transplantation [14]. Heart transplantation recipients typically receive a higher dose of immunosuppressive drugs compared to kidney and liver transplantation recipients, causing a larger impairing effect on the immune system [7]. Furthermore, heart transplant recipients are described to be on average older and often tend to have a history of smoking, which are independent risk factors for bladder and upper gastrointestinal tract carcinomas [15–17]. However, the significant influence of the immunosuppressive therapy after heart transplantations cannot be ignored.

The current meta-analysis did not show a significant correlation between specific carcinomas and different immunosuppressive drugs, except for significantly less head and neck carcinomas in patients using the calcineurin inhibitor TAC or the antimetabolite MMF compared to the antimetabolite AZA. However, the comparison between MMF and AZA has to be interpreted carefully, as it was only described in two of the included studies, which consisted of unequal cohorts of patients. Many of the comparisons were rarely described in the included articles, which might cause those comparisons to be underpowered. Therefore, the lack of statistically significant results for those evaluations should not be considered irrelevant, but should warrant future epidemiological studies.

Furthermore, in agreement with previously published studies, the current results show that overall de novo carcinomas occur more often in solid organ transplant recipients using the calcineurin inhibitor CsA compared to MMF and TAC [18,19]. For instance, Tjon et al. showed that CsA treatment in comparison to TAC is the most important risk factor for de novo carcinoma in liver transplant recipients, supporting the results that this review provides for the whole transplant population [18]. Pathogenesis of specific types of cancer may be clarified by looking in depth into which immunosuppressive agents induce carcinogenesis.

Calcineurin inhibitors CsA and TAC are considered to have a similar working mechanism on the immune system via the calcineurin pathway. The main working mechanism is inhibition of the calcineurin activity in immune cells, thereby preventing the activation and nuclear translocation of nuclear factor of activated T-cells (NFAT), leading to inhibition of Interleukin-2 (IL-2) production in T cells [20]. IL-2 is an important factor for maintenance of CD4+ regulatory T cells, but also plays a critical role in the proliferation and differentiation of CD4+ T cells, promotion of CD8+ T cell and NK cell cytotoxic activity, and modulation of T cell differentiation programs in response to tumor antigens [21]. Inhibition of IL-2 production therefore has a profound effect on the immune system. In vitro and in vivo, calcineurin inhibitors inhibited degranulation of NK cells and reduced IFNγ production by NK cells [22]. Furthermore, the capability of dendritic cells to stimulate T cells and produce IL-12 and CXC-chemokine ligand 10 is reduced [23,24]. If dendritic cells are incubated with tacrolimus, they develop a tolerogenic phenotype, which has a suppressive effect on CD4+ T cell proliferation [25].

Both CsA and TAC promote tumor formation by inducing tumor growth factor-β (TGF-β) and inhibiting apoptosis and DNA repair. This results in enhanced growth and diminished apoptosis of cancer cells [18,26–28]. The discrepancy between tumor-promoting effects of CsA and TAC might be due to the lower level of TGF-β that is induced by TAC compared to CsA [26,29]. In a healthy cell, TGF-β is a multifunctional cytokine that hampers proliferation, promotes apoptosis, and induces differentiation and fibroblast growth. However, in carcinoma cells, TGF-β loses its controlling function, leading to enhanced proliferation, diminished differentiation, and apoptosis of carcinoma cells [30].

Additionally, in vivo CsA has been shown to induce tumor progression and angiogenesis, independent of calcineurin, by releasing mitochondrial reactive oxygen species, leading to stimulation of mitogenic pathways in tumor and stromal cells [31]. Another mechanism via which CsA promotes angiogenesis is stimulation of prolyl hydroxylase activity, causing hypoxia-inducible factor 1a (HIF-1α) destabilization. HIF1α increases the expression of vascular endothelial growth factor, leading to angiogenesis [32].

Moreover, patients using CsA also have a higher incidence of Kaposi sarcoma and lymphoma [33,34]. Taking into account the effects of oncogenic viruses such as Epstein–Barr virus on non-Hodgkin lymphoma and human immunodeficiency virus on Kaposi sarcoma, this result insinuates a greater role of the suppressed immune system on the oncogenic effect of CsA than a direct effect of CsA on epithelial cells.

Both CsA and AZA give a higher odds to develop de novo carcinomas compared to MMF, even though MMF is more potent in preventing graft rejection than AZA [6,35]. Both AZA and MMF inhibit the purine pathway, but do so via different metabolites. MMF ultimately inhibits the formation of guanine nucleotides [36]. Guanine is one of the purine nucleobases necessary to generate DNA and is thus required for cell replication. MMF is metabolized to mycophenolic acid (MPA), which inhibits the enzyme inosine monophosphate dehydrogenase, thereby reducing the amount of guanine nucleotides formed [36]. Cells are able to generate guanine nucleotides through two distinct pathways: the de novo pathway and the salvage pathway. Whereas other cells are able to use both pathways, lymphocytes are only able to use the de novo pathway, and thus their proliferation is inhibited. Not only lymphocytes but also fibroblasts are affected by MMF, which are suspected to also rely partly on the de novo pathway. AZA also has an effect on the de novo purine synthesis pathway. AZA is a prodrug of 6-mercaptopurine (6-MP) and is nonenzymatically cleaved into 6-MP and imidazole derivatives. The main therapeutic effect of AZA relies on its metabolism to cytotoxic thioguanine nucleotides via the 6-MP pathway, which also inhibit the de novo purine synthesis, by inhibiting amidotransferase enzymes and purine ribonucleotide interconversion [37,38]. In addition, toxic thioguanine nucleotides are incorporated into DNA and RNA [37,38], which is thought to mediate the cytotoxic effects of AZA. Furthermore, the imidazole derivatives potentially also have effects on lymphocyte function. Although this metabolite has not been investigated in relation to the therapeutic effects of AZA, imidazole derivatives can reduce T cell proliferation and NFAT signaling following T cell receptor activation in mice [39]. Moreover, AZA can directly promote apoptosis and inhibit proliferation pathways through inhibition of Rac1 and Bcl-xL [40,41], and via inhibition of Rac1, it also blocks CD28 signaling [42]. In addition to its action on T cells, 6-MP can also inhibit Rac1 in activated macrophages, which leads to a reduction in the expression of inducible nitric oxide synthase [40]. The Rac1 pathway is also targeted by 6-MP in nonimmune cells [43], as 6-MP decreases Rac1 activation in endothelial cells and reduces activation of nuclear factor κ-light-chain-enhancer of activated B cells (NF-κB), leading to decreased transcription of proinflammatory cytokines. Furthermore, 6-MP selectively decreases VCAM-1 protein levels in TNF-α stimulated endothelial cells [44]. The elaborate effects of AZA, not only on the lymphocytes but also on macrophages and endothelial cells, might cause the higher odds of developing a de novo carcinoma compared to MMF.

The final group of drugs in this meta-analysis are the mTOR inhibitors. Aside from lymphocytes proliferation inhibition, mTOR inhibitors play a role in the intracellular signaling pathways in all cells of the immune system. For instance, the lifespan of and expression of costimulatory molecules by dendritic cells are increased, while on the other hand, metabolic NK cell function is reduced by mTOR inhibition [45].

As many malignancies upregulate the mTOR pathway, the mTOR inhibitors are currently used as anticancer therapeutics [46]. In this meta-analysis, there was no significant difference between mTOR inhibitors and other immunosuppressive drugs. However, there were only five studies which described the use of sirolimus in comparison with another drug and none which described everolimus. Therefore, perhaps there was insufficient power to detect any differences.

In the first months after transplantation, the risk of graft rejection is highest. Therefore, patients receive induction therapy in the first period after solid organ transplantation. In the included articles, the induction therapy regimen is scarcely described. Each center decides the optimal regimen based on the patient's characteristics, but induction therapy mostly consists of a triple therapy combination of corticosteroids, IL-2 receptor antagonists, polyclonal antilymphocyte and antithymocyte preparations, and monoclonal antibody targeting. The effect of solely the induction therapy on carcinogenesis is still unknown. A Cochrane review from 2017 regarding the polyclonal and monoclonal antibody therapies showed an uncertain effect on malignancies [47]. It has also been described that basiliximab, an IL-2 receptor antagonist, does not increase the risk of malignancies [48]. The effect on carcinoma formation rather than malignancies in general has not been investigated separately. However, the increased odds of developing de novo carcinomas found in this meta-analysis are therefore most likely caused by the maintenance therapy.

Within the immunosuppressive regimen, patients often switch to different drugs in case of chronic rejection, adverse events, or the availability of new drugs. Four years after the transplantation, less than half of the patients still used the first prescribed combination [49]. However, this study also included the patients who received induction therapy, which explains why the majority of switches were found in the first year. In this meta-analysis, correction for switches in the maintenance therapy was not possible. Therefore, only studies describing longer periods of baseline therapy were included in the drug-specific comparisons. Furthermore, patients use combination triple therapies. For most of the patients, the baseline therapy was described in the included articles, and only for 689 patients the prescribed triple therapy combination. Even though only the baseline therapies were included in the drug-specific comparisons, one can assume that over time, multiple other drugs were simultaneously given. The effect of these switches on the outcome cannot be assessed, but as switches would have occurred in each group, these might partially cancel each other out.

There are many other factors contributing to the carcinogenesis process, such as smoking, alcohol, diet, and genetics. For certain carcinomas, including HCC and cervix carcinomas, viruses can also play a pivotal role in the carcinogenesis process. For HCC, hepatitis B virus (HBV) causes a 100-fold increase of the relative risk to develop HCC. The oncogenic role of HBV is not completely understood as it might be caused by both direct and indirect mechanisms, including immune-mediated hepatic inflammation leading to genetic damage, the induction of oxidative stress, and integration of the HBV DNA into the host genome that induces chromosomal instability [48]. Due to the immunosuppressive drugs, HBV can be reactivated. For cervix carcinoma, human papilloma virus HPV plays an important role. Furthermore, also vulva, vagina, penis, and anus carcinomas are associated with HPV [50]. In a systematic review by Grulich et al. [51], they reported increased standardized incidence ratios for these HPV-associated carcinomas in patients who underwent a solid organ transplantation. Unfortunately, the occurrence of viruses was not well described in the included articles. One might assume that this plays an additional role in the carcinogenesis which could not be corrected for in the meta-analysis.

Most studies had a long inclusion period, which might lead to differences in treatment regimens. The longest study had an inclusion period of 47 years and in another study, the first included patient received their solid organ transplantation in 1963. Since then, a lot has changed in the knowledge and possibilities of immunosuppressive drugs. A major breakthrough was the discovery of cyclosporine A, which was first given to patients in 1978 [52]. Furthermore, the detection of de novo carcinomas has improved with better imaging techniques and standardized follow-up protocols. In this study, there was no trend towards a higher or lower incidence per 1000 person-years of de novo carcinomas based on year of publication. In the comparisons of different immunosuppressive drugs, the longer inclusion period probably has a minimal influence as the inclusion period is equal for both drugs within one article.

This systematic review has several limitations. Many of the cohort studies included were conducted retrospectively, based on small groups of transplant recipients, and were often hospital-based. Small cohorts might lead to overestimation of the effect. Even though articles with less than five de novo carcinomas were excluded, this might still have introduced some overestimation. Using the NOS score, most articles were deemed to be of fair quality. One of the major problems was lacking information regarding follow-up data. Even though all authors were contacted regarding missing follow-up data, this might still introduce some risk of bias as not all authors replied to supply the follow-up data. These missing follow-up data also limit direct comparability of calculated incidences to the general population. However, the clear trend towards a higher incidence of carcinoma after solid organ transplantation cannot be ignored. Furthermore, large (inter-)national registries and studies based on International Classification of Diseases (ICD) codes might have missed a few de novo carcinomas as diagnosis might not always be coded correctly in the patient records. This could lead to underrepresentation of de novo carcinoma occurrence and bias in the outcome category. Additionally, changes in the immunosuppressive drug regimens could have been missed. A thorough check through each individual patient record is the only way to prevent missing changes in therapy leading to ascertainment bias and missing de novo carcinomas leading to assessment of exposure bias. Many studies were not eligible due to the strict requirements that the immunosuppressive drug regimen and the total solid organ transplantation group described. Finally, the changes in maintenance therapies were rarely described, while this might also influence the cancer development in the long term.

To determine the specific correlation between immunosuppressive drugs and cancer development, a combination of a large prospective cohort with sufficient follow-up for carcinomas to develop and translational research is needed. Important confounders can be determined from the prospective cohorts and further examined in in vitro and in vivo models.

Looking in depth into pathways of calcineurin inhibitors, such as IL-2, TGF-B, and HIF1α, and antimetabolites pathways may lead to enhanced comprehension of carcinogenesis in transplant recipients. Additionally, as described, these pathways are also contributing to carcinogenesis in the general population. Exploring these pathways would thus be an interesting topic for translational research and could in the long term give rise to preventive and therapeutic options for specific types of cancer, both in patients who underwent a solid organ transplantation and in the general population.

5. Conclusions

This systematic review and meta-analysis show an almost 20-fold higher likelihood of de novo carcinoma development in patients using immunosuppressive drugs after solid organ transplantation. The likelihood is highest for patients receiving cyclosporine A and azathioprine. By looking in depth into the pathways affected by these immunosuppressive drugs, a deeper understanding of carcinogenesis can be achieved and new starting points for translational and clinical research might be found.

Supplementary Materials: The following are available online at https://www.mdpi.com/2072-6694/13/5/1122/s1, Table S1: Search strategy, Table S2: Study characteristics, Table S3: Newcastle–Ottawa Scale risk of bias score, Figure S1: Cyclosporine A versus tacrolimus, Figure S2: Cyclosporine A versus azathioprine, Figure S3: Cyclosporine A versus mycophenolate mofetil, Figure S4: Cyclosporine A versus sirolimus, Figure S5: Azathioprine versus tacrolimus, Figure S6: Azathioprine versus mycophenolate mofetil, Figure S7: Mycophenolate mofetil versus tacrolimus.

Author Contributions: Conceptualization, E.S.Z. and G.K.; methodology, E.S.Z., E.Y. and R.d.V.; formal analysis, E.S.Z. and E.Y.; data curation, E.S.Z., E.Y. and A.P.; writing—original draft preparation, E.S.Z., E.Y. and A.P.; writing—review and editing, E.S.Z., E.Y., A.P., R.d.V., R.E.M. and G.K.; visualization, E.S.Z. and E.Y.; supervision, R.E.M. and G.K. All authors have read and agreed to the published version of the manuscript.

Funding: This research was funded by Cancer Center Amsterdam Foundation, The Netherlands, grant number 2017-4-09.

Institutional Review Board Statement: Not applicable.

Informed Consent Statement: Not applicable.

Conflicts of Interest: The funders had no role in the design of the study; in the collection, analyses, or interpretation of data; in the writing of the manuscript, or in the decision to publish the results.

References

1. Wiseman, A.C. Immunosuppressive Medications. *Clin. J. Am. Soc. Nephrol.* **2016**, *11*, 332–343. [CrossRef]
2. Noble, S.; Markham, A. Cyclosporin. A review of the pharmacokinetic properties, clinical efficacy and tolerability of a microemulsion-based formulation (Neoral). *Drugs* **1995**, *50*, 924–941. [CrossRef] [PubMed]
3. Placebo-controlled study of mycophenolate mofetil combined with cyclosporin and corticosteroids for prevention of acute rejection. European Mycophenolate Mofetil Cooperative Study Group. *Lancet* **1995**, *345*, 1321–1325. [CrossRef]
4. Keown, P.; Landsberg, D.; Hardie, I.; Rigby, R.; Isoniemi, H.; Häyry, P.; Morris, P. A blinded, randomized clinical trial of mycophenolate mofetil for the prevention of acute rejection in cadaveric renal transplantation. The Tricontinental Mycophenolate Mofetil Renal Transplantation Study Group. *Transplantation* **1996**, *61*, 1029–1037.
5. Mycophenolate mofetil for the prevention of acute rejection of primary cadaveric kidney transplants: Status of the MYC 1866 study at 1 year. The U.S. Mycophenolate Mofetil Study Group. *Transplant. Proc.* **1997**, *29*, 348–349. [CrossRef]
6. Wagner, M.; Earley, A.K.; Webster, A.C.; Schmid, C.H.; Balk, E.M.; Uhlig, K. Mycophenolic acid versus azathioprine as primary immunosuppression for kidney transplant recipients. *Cochrane Database Syst. Rev.* **2015**, Cd007746. [CrossRef] [PubMed]
7. Chiu, B.; Sergi, C. Malignancy after Heart Transplantation: A Systematic Review of the Incidence and Risk Factors Compared with Other Solid Organ Transplants. *J. Clin. Exp. Cardiol.* **2013**, *S9*, 5. [CrossRef]
8. Liu, Z.N.; Wang, W.T.; Yan, L.N. De Novo Malignancies After Liver Transplantation With 14 Cases at a Single Center. *Transplant. Proc.* **2015**, *47*, 2483–2487. [CrossRef]
9. Herrero, J.I. De novo malignancies following liver transplantation: Impact and recommendations. *Liver Transplant.* **2009**, *15* (Suppl. S2), S90–S94. [CrossRef] [PubMed]
10. Moher, D.; Liberati, A.; Tetzlaff, J.; Altman, D.G. Preferred reporting items for systematic reviews and meta-analyses: The PRISMA statement. *PLoS Med.* **2009**, *6*, e1000097. [CrossRef]
11. Ouzzani, M.; Hammady, H.; Fedorowicz, Z.; Elmagarmid, A. Rayyan—A web and mobile app for systematic reviews. *Syst Rev.* **2016**, *5*, 210. [CrossRef] [PubMed]
12. Wells, G.; Shea, B.; O'Connell, D.; Peterson, J.; Welch, V.; Losos, M.; Tugwell, P. The Newcastle-Ottawa Scale (NOS) for Assessing the Quality of Nonrandomised Studies in Meta-Analyses. Available online: http://www.ohri.ca/programs/clinical_epidemiology/oxford.asp (accessed on 2 March 2021).
13. *Review Manager (RevMan) [Computer Program].* Version 5.3; The Nordic Cochrane Centre, The Cochrane Collaboration: Copenhagen, Denmark, 2014.
14. Ferlay, J.; Ervik, M.; Lam, F.; Colombet, M.; Mery, L.; Piñeros, M.; Znaor, A.; Soerjomataram, I.; Bray, F. *Global Cancer Observatory: Cancer Today*; International Agency for Research on Cancer: Lyon, France, 2020; Available online: https://gco.iarc.fr/today/home (accessed on 2 March 2021).
15. Freedman, N.D.; Silverman, D.T.; Hollenbeck, A.R.; Schatzkin, A.; Abnet, C.C. Association between smoking and risk of bladder cancer among men and women. *JAMA* **2011**, *306*, 737–745. [CrossRef] [PubMed]
16. Fan, Y.; Yuan, J.M.; Wang, R.; Gao, Y.T.; Yu, M.C. Alcohol, tobacco, and diet in relation to esophageal cancer: The Shanghai Cohort Study. *Nutr. Cancer* **2008**, *60*, 354–363. [CrossRef] [PubMed]
17. Botha, P.; Peaston, R.; White, K.; Forty, J.; Dark, J.H.; Parry, G. Smoking after cardiac transplantation. *Am. J. Transplant.* **2008**, *8*, 866–871. [CrossRef] [PubMed]
18. Tjon, A.S.; Sint Nicolaas, J.; Kwekkeboom, J.; De Man, R.A.; Kazemier, G.; Tilanus, H.W.; Hansen, B.E.; Van der Laan, L.J.; Tha-In, T.; Metselaar, H.J. Increased incidence of early de novo cancer in liver graft recipients treated with cyclosporine: An association with C2 monitoring and recipient age. *Liver Transplant.* **2010**, *16*, 837–846. [CrossRef]
19. Campistol, J.M.; Eris, J.; Oberbauer, R.; Friend, P.; Hutchison, B.; Morales, J.M.; Claesson, K.; Stallone, G.; Russ, G.; Rostaing, L.; et al. Sirolimus therapy after early cyclosporine withdrawal reduces the risk for cancer in adult renal transplantation. *J. Am. Soc. Nephrol.* **2006**, *17*, 581–589. [CrossRef] [PubMed]
20. Lee, J.U.; Kim, L.K.; Choi, J.M. Revisiting the Concept of Targeting NFAT to Control T Cell Immunity and Autoimmune Diseases. *Front. Immunol.* **2018**, *9*, 2747. [CrossRef]
21. Jiang, T.; Zhou, C.; Ren, S. Role of IL-2 in cancer immunotherapy. *Oncoimmunology* **2016**, *5*, e1163462. [CrossRef] [PubMed]
22. Morteau, O.; Blundell, S.; Chakera, A.; Bennett, S.; Christou, C.M.; Mason, P.D.; Cornall, R.J.; O'Callaghan, C.A. Renal transplant immunosuppression impairs natural killer cell function in vitro and in vivo. *PLoS ONE* **2010**, *5*, e13294. [CrossRef] [PubMed]

23. Sauma, D.; Fierro, A.; Mora, J.R.; Lennon-Duménil, A.M.; Bono, M.R.; Rosemblatt, M.; Morales, J. Cyclosporine preconditions dendritic cells during differentiation and reduces IL-2 and IL-12 production following activation: A potential tolerogenic effect. *Transplant. Proc.* **2003**, *35*, 2515–2517. [CrossRef] [PubMed]
24. Tiefenthaler, M.; Hofer, S.; Ebner, S.; Ivarsson, L.; Neyer, S.; Herold, M.; Mayer, G.; Fritsch, P.; Heufler, C. In vitro treatment of dendritic cells with tacrolimus: Impaired T-cell activation and IP-10 expression. *Nephrol. Dial. Transplant.* **2004**, *19*, 553–560. [CrossRef] [PubMed]
25. Ren, Y.; Yang, Y.; Yang, J.; Xie, R.; Fan, H. Tolerogenic dendritic cells modified by tacrolimus suppress CD4(+) T-cell proliferation and inhibit collagen-induced arthritis in mice. *Int. Immunopharmacol.* **2014**, *21*, 247–254. [CrossRef]
26. Weischer, M.; Rocken, M.; Berneburg, M. Calcineurin inhibitors and rapamycin: Cancer protection or promotion? *Exp. Dermatol.* **2007**, *16*, 385–393. [CrossRef]
27. Durnian, J.M.; Stewart, R.M.; Tatham, R.; Batterbury, M.; Kaye, S.B. Cyclosporin-A associated malignancy. *Clin. Ophthalmol.* **2007**, *1*, 421–430. [PubMed]
28. Andre, N.; Roquelaure, B.; Conrath, J. Molecular effects of cyclosporine and oncogenesis: A new model. *Med. Hypotheses* **2004**, *63*, 647–652. [CrossRef] [PubMed]
29. Euvrard, S.; Ulrich, C.; Lefrancois, N. Immunosuppressants and skin cancer in transplant patients: Focus on rapamycin. *Dermatol. Surg.* **2004**, *30*, 628–633. [CrossRef] [PubMed]
30. Jakowlew, S.B. Transforming growth factor-beta in cancer and metastasis. *Cancer Metastasis Rev.* **2006**, *25*, 435–457. [CrossRef] [PubMed]
31. Zhou, A.Y.; Ryeom, S. Cyclosporin A promotes tumor angiogenesis in a calcineurin-independent manner by increasing mitochondrial reactive oxygen species. *Mol. Cancer Res.* **2014**, *12*, 1663–1676. [CrossRef] [PubMed]
32. Denko, N.C. Hypoxia, HIF1 and glucose metabolism in the solid tumour. *Nat. Rev. Cancer* **2008**, *8*, 705–713. [CrossRef]
33. Bieber, C.P.; Reitz, B.A.; Jamieson, S.W.; Oyer, P.E.; Stinson, E.B. Malignant lymphoma in cyclosporin A treated allograft recipients. *Lancet* **1980**, *1*, 43. [CrossRef]
34. Cattaneo, D.; Gotti, E.; Perico, N.; Bertolini, G.; Kainer, G.; Remuzzi, G. Cyclosporine formulation and Kaposi's sarcoma after renal transplantation. *Transplantation* **2005**, *80*, 743–748. [CrossRef] [PubMed]
35. Rigotti, P.; Cadrobbi, R.; Baldan, N.; Sarzo, G.; Parise, P.; Furian, L.; Marchini, F.; Ancona, E. Mycophenolate mofetil (MMF) versus azathioprine (AZA) in pancreas transplantation: A single-center experience. *Clin. Nephrol.* **2000**, *53*, S52–S54.
36. Ransom, J.T. Mechanism of action of mycophenolate mofetil. *Ther. Drug Monit.* **1995**, *17*, 681–684. [CrossRef]
37. Van Scoik, K.G.; Johnson, C.A.; Porter, W.R. The pharmacology and metabolism of the thiopurine drugs 6-mercaptopurine and azathioprine. *Drug Metab. Rev.* **1985**, *16*, 157–174. [CrossRef]
38. Van Os, E.C.; Zins, B.J.; Sandborn, W.J.; Mays, D.C.; Tremaine, W.J.; Mahoney, D.W.; Zinsmeister, A.R.; Lipsky, J.J. Azathioprine pharmacokinetics after intravenous, oral, delayed release oral and rectal foam administration. *Gut* **1996**, *39*, 63–68. [CrossRef]
39. Jung, E.J.; Hur, M.; Kim, Y.L.; Lee, G.H.; Kim, J.; Kim, I.; Lee, M.; Han, H.K.; Kim, M.S.; Hwang, S.; et al. Oral administration of 1,4-aryl-2-mercaptoimidazole inhibits T-cell proliferation and reduces clinical severity in the murine experimental autoimmune encephalomyelitis model. *J. Pharmacol. Exp. Ther.* **2009**, *331*, 1005–1013. [CrossRef] [PubMed]
40. Maltzman, J.S.; Koretzky, G.A. Azathioprine: Old drug, new actions. *J. Clin. Invest.* **2003**, *111*, 1122–1124. [CrossRef] [PubMed]
41. Marinković, G.; Hamers, A.A.; De Vries, C.J.; De Waard, V. 6-Mercaptopurine reduces macrophage activation and gut epithelium proliferation through inhibition of GTPase Rac1. *Inflamm. Bowel Dis.* **2014**, *20*, 1487–1495. [CrossRef] [PubMed]
42. Tiede, I.; Fritz, G.; Strand, S.; Poppe, D.; Dvorsky, R.; Strand, D.; Lehr, H.A.; Wirtz, S.; Becker, C.; Atreya, R.; et al. CD28-dependent Rac1 activation is the molecular target of azathioprine in primary human CD4+ T lymphocytes. *J. Clin. Invest.* **2003**, *111*, 1133–1145. [CrossRef] [PubMed]
43. Marinković, G.; Kroon, J.; Hoogenboezem, M.; Hoeben, K.A.; Ruiter, M.S.; Kurakula, K.; Otermin Rubio, I.; Vos, M.; De Vries, C.J.M.; Van Buul, J.D.; et al. Inhibition of GTPase Rac1 in Endothelium by 6-Mercaptopurine Results in Immunosuppression in Nonimmune Cells: New Target for an Old Drug. *J. Immunol.* **2014**, *192*, 4370–4378. [CrossRef] [PubMed]
44. Hessels, A.C.; Rutgers, A.; Sanders, J.S.F.; Stegeman, C.A. Thiopurine methyltransferase genotype and activity cannot predict outcomes of azathioprine maintenance therapy for antineutrophil cytoplasmic antibody associated vasculitis: A retrospective cohort study. *PLoS ONE* **2018**, *13*, e0195524. [CrossRef] [PubMed]
45. Kahan, B.D. Sirolimus: A comprehensive review. *Expert Opin. Pharmacother.* **2001**, *2*, 1903–1917. [CrossRef] [PubMed]
46. Hua, H.; Kong, Q.; Zhang, H.; Wang, J.; Luo, T.; Jiang, Y. Targeting mTOR for cancer therapy. *J. Hematol. Oncol.* **2019**, *12*, 71. [CrossRef] [PubMed]
47. Hill, P.; Cross, N.B.; Barnett, A.N.R.; Palmer, S.C.; Webster, A.C. Polyclonal and monoclonal antibodies for induction therapy in kidney transplant recipients. *Cochrane Database Syst. Rev.* **2017**, *1*, CD004759. [CrossRef] [PubMed]
48. Chapman, T.M.; Keating, G.M. Basiliximab: A review of its use as induction therapy in renal transplantation. *Drugs* **2003**, *63*, 2803–2835. [CrossRef]
49. Meier-Kriesche, H.U.; Chu, A.H.; David, K.M.; Chi-Burris, K.; Steffen, B.J. Switching immunosuppression medications after renal transplantation–a common practice. *Nephrol. Dial. Transplant.* **2006**, *21*, 2256–2262. [CrossRef] [PubMed]
50. Saraiya, M.; Unger, E.R.; Thompson, T.D.; Lynch, C.F.; Hernandez, B.Y.; Lyu, C.W.; Steinau, M.; Watson, M.; Wilkinson, E.J.; Hopenhayn, C.; et al. US assessment of HPV types in cancers: Implications for current and 9-valent HPV vaccines. *J. Natl. Cancer Inst.* **2015**, *107*, djv086. [CrossRef] [PubMed]

51. Grulich, A.E.; Van Leeuwen, M.T.; Falster, M.O.; Vajdic, C.M. Incidence of cancers in people with HIV/AIDS compared with immunosuppressed transplant recipients: A meta-analysis. *Lancet* **2007**, *370*, 59–67. [CrossRef]
52. Calne, R.Y.; White, D.J.; Thiru, S.; Evans, D.B.; McMaster, P.; Dunn, D.C.; Craddock, G.N.; Pentlow, B.D.; Rolles, K. Cyclosporin A in patients receiving renal allografts from cadaver donors. *Lancet* **1978**, *2*, 1323–1327. [CrossRef]

MDPI
St. Alban-Anlage 66
4052 Basel
Switzerland
Tel. +41 61 683 77 34
Fax +41 61 302 89 18
www.mdpi.com

Cancers Editorial Office
E-mail: cancers@mdpi.com
www.mdpi.com/journal/cancers

www.ingramcontent.com/pod-product-compliance
Lightning Source LLC
LaVergne TN
LVHW070051120526
838202LV00102B/2043